Foundations of
ELECTRICAL
ENGINEERING

Foundations of
ELECTRICAL
ENGINEERING

J. R. Cogdell

Department of Electrical and Computer Engineering
University of Texas at Austin

PRENTICE HALL, Englewood Cliffs, New Jersey 07632

Library of Congress Cataloging-in-Publication Data

Cogdell, J. R.
   Foundations of electrical engineering / J.R. Cogdell.
      p.   cm.
   Includes index.
   ISBN 0-13-329525-7
   1. Electric engineering.   I. Title.
TK146.C59 1990
621.3--dc20   89-8563
                                                   CIP

Editorial/production supervision: Patrice Fraccio
Cover and interior design: Linda J. Den Heyer Rosa
Manufacturing buyer: Robert Anderson
Page layout: Gail Collis

Printed in the United States of America
10 9 8 7 6 5 4 3 2

ISBN 0-13-329525-7

Prentice-Hall International (UK) Limited, *London*
Prentice-Hall of Australia Pty. Limited, *Sydney*
Prentice-Hall Canada Inc., *Toronto*
Prentice-Hall Hispanoamericana, S.A., *Mexico*
Prentice-Hall of India Private Limited, *New Delhi*
Prentice-Hall of Japan, Inc., *Tokyo*
Simon & Schuster Asia Pte. Ltd., *Singapore*
Editora Prentice-Hall do Brasil, Ltda., *Rio de Janeiro*

Dedicated
to the Glory of God
to my wife
Ann
and to my children
Amy, Thomas, and Christina

Contents

2 The Analysis of DC Circuits   24

PART II   ELECTRONICS   212

Contents

Preface

This work surveys the foundations of electrical engineering. Applications are as up to date as possible, but emphasis is directed toward those principles that are enduring in their influence and usefulness. Specifically, we have emphasized energy, basic physical laws, analog and digital representation of information, and the time-domain/frequency-domain representation of signals.

This book expands our earlier work, *Introduction to Circuits and Electronics*, to include sections on systems and electromechanics. Since relatively few electrical engineers study electrical motors these days, we have more material on this subject than other introductory books. Our conviction is that all engineers need to understand this important topic.

The mathematics used in this book would be taught in a good freshman calculus course. We use some differential equations, complex arithmetic, and vector-field expressions, but these subjects are explained when introduced. A sound knowledge of mechanics and electrical physics is essential, but in Chapter 12 we review the electrical physics required for the section on electromechanics.

We are convinced that an engineering education should build strong problem-solving skills. The text and many of the problems at the end of the chapters are designed to exercise the reader's skills in problem solving.

I gratefully acknowledge the assistance of many colleagues for helpful suggestions, particularly Paul Wildi, W. C. Duesterhoeft, Bill Hamilton, David Brown, Lee Baker, Ben Streetman and Jon Valvano. Special thanks go to my son, Thomas, for producing many of the computer graphs and for reading the entire manuscript and suggesting many improvements. My editor, Elizabeth Kaster, has served me and Prentice Hall very well by insisting on a rework of the entire manuscript after the author thought he was finished.

This book is the product of many hours of work during the past six years. My hope is that its readers derive a portion of the benefit and pleasure that I had in writing the book.

1 Electric Charge and Electrical Energy

1.1 INTRODUCTION TO ELECTRICAL ENGINEERING

1.1.1 Place of Electrical Engineering in Modern Technology

What is electrical engineering? Electrical engineering is, in one sense, the opposite of lightning. Lightning unleashes electrical energy unpredictably, seldom to the benefit of human society. Electrical engineering harnesses electrical energy for human good—for transporting energy and information, for lifting the burdens of toil and tedium. When electrical energy is important for its own sake, to turn motors or illuminate department stores, we think of the electrical power industry. When energy is important for its symbolic (information) content, we think of the electronics industry. Both use electrical energy for beneficial purposes.

This book explains the fundamental ideas and techniques of electrical engineering. Our goal is to provide you with a strong foundation to solve basic and practical problems and to furnish you with a repertoire of words and concepts for clear thinking and clear communication. With this course for preparation, you may advance to further study of circuits, electronics, or electrical energy conversion.

Electrical engineering is considered one of the more difficult subjects on most college campuses. One reason is that electrical phenomena are invisible. You cannot see the current in a wire, the magnetic field causing a motor to turn, the radio waves passing through your body at this moment. You must, therefore, develop the ability to visualize the invisible in order to master electrical concepts.

Another reason for EE's reputation is that advanced mathematical methods are frequently used to solve EE problems. Although the mathematical scope of this book has been limited, we have found it necessary to use complex numbers and vector calculus now and then.

Despite these difficulties, electrical engineering is increasingly prevalent in modern technology. Just as electronic watches and calculators have replaced mechanical watches and calculators, electrical parts will in the future replace many mechanical parts in appliances, automobiles, manufacturing processes, instrumentation, and other areas. This technological revolution, still in its infancy, arises from social, historical, and physical causes.

Why electrical engineering increasingly dominates technology.
Electrical science has developed in recent times. Perhaps it was the invisibility of electrical phenomena that retarded scientific investigation; whatever the reason, knowledge of electrical phenomena was vague and fragmentary until about one hundred years ago when Maxwell unified electrical theory with his famous equations. Understanding of the electrical properties of matter, particularly matter in the solid state, is a twentieth-century accomplishment.

Significant technology usually precedes scientific understanding. Excellent swords were manufactured before scientific metallurgy developed; violins came before acoustics; thoroughbreds preceded genetics. Except for the magnetic compass, however, little significant technology predated electrical science. Again we might blame the invisibility and resulting abstraction of electrical things for retarding the intuitive invention of useful gadgets. It is historical fact that electrical science and technology have marched side by side. At the present time, continued scientific advances are producing steady and spectacular progress in electrical tools and toys.

Consider next the physical factors underlying this electrical dominance. Matter, the stuff making up the entire physical creation, is fundamentally electrical, being comprised of charged nuclei surrounded by shells of charged electrons. Most of the macroscopic properties of matter arise from electrical properties. When one bumps a toe against a chair in the dark, does matter touch matter? No: when matter touches matter, we get nuclear reactions, not sore toes. What we experience as toe-striking-chair is basically the electrons in our toe being repelled, without physical contact, by the electrons in the chair. Similarly, stress-strain relationships in materials are largely macroscopic manifestations of the stretching of the electrical bonds holding the matter in crystalline order. As mentioned above, recent gains in the understanding of matter in the solid state have opened the door to new electrical technologies: transistors, lasers, integrated circuits, liquid crystal displays.

Consider further that electric charge is bipolar, that is, that electrical engineers work with positive and negative charges. Think of the wonderful machines one could invent with balls that rolled uphill in addition to ordinary balls that roll downhill. That is in effect what exists in electricity—positive charges move one way, negative charges the other. Transistors depend on this type of action, and modern electronics exploits fully the bipolarity of charged matter.

Consider as a third physical factor the number 1.76×10^{11} Coulombs/kilogram. This large number is the charge-to-mass ratio for the electron in the MKS system, as if the electric part dominates the mechanical part of the electron by this enormous ratio. Because of this ratio, we can control electrons electrically and move them about without supplying much mechanical energy; hence electric charges respond rapidly to electrical forces. The enormous range of frequencies that electrical engineers employ in communication networks, and the amazing speed with which electronic computers operate, testify to the significance of the charge-to-mass ratio of the electron.

The other large number favoring electrical technology over alternatives is 2.998×10^8 meters/second. This you might recognize as the speed of light in the MKS system of units. This is as fast as things can go in this world, and this is exactly the speed at which electrical energy travels. When you turn on the light switch at the wall, your action is transmitted to the light bulb with the speed of light. More to the point, when the electrons in a radio antenna in Paris, France, are shaken by a radio transmitter, the electrons in the repeater satellite midway over the Atlantic respond after a delay equal to the distance (22,000 miles) divided by the velocity of light, a delay of about 0.118 second. No wonder, is it, that our communication systems depend on electrical means?

Finally, there are social factors pushing electrical technology toward dominance. There are more people all the time, and an increasing amount of information is compiled, processed, communicated, and stored per person in all modern societies. The fruits of electrical engineering, computers and communication systems, process and transport all this information.

Summary. Historical, physical, and social factors enliven electrical engineering at this time. The scientific understanding of the electrical nature of matter, the charge-to-mass ratio of the electron and the velocity of light, the bipolarity of electric charge, and the population and information explosions—all these interact to press electrical engineering to the fore.

1.1.2 Role of Circuit Theory

The first part of this book is about electric circuit theory. There are three reasons why we start with circuit theory. It is neither as abstract nor as mathematically sophisticated as other branches of electrical engineering, so it is an easy place to start. More important, practically every electrical device is a circuit. A radio is a circuit, as is the power distribution system that runs your air conditioner. A computer is also a vastly complicated circuit. Hence an understanding of the language and methods of circuits gives a foundation to the study of all areas of electrical engineering.

Finally, circuits provide a logical starting point because circuit theory has generated the language of electrical engineering. Even devices that are more sophisticated than an electrical circuit (an aircraft radar, for example), are described by electrical engineers in the language of circuits. Hence we study circuits as an introduction to the language and ideas underlying much of electrical engineering.

1.2 PHYSICAL BASIS OF CIRCUIT THEORY

1.2.1 Energy and Charge

Charge is a fundamental physical quantity.   When something electrical happens, a traffic light changes or an electric motor begins turning, it means that electric charges are in motion. Charge is a property of matter, like mass; indeed, charge joins mass, length, and time as one of the fundamental units from which all scientific units are derived. The unit of electric *charge* is the coulomb (abbreviated C), named in honor of Charles Augustin de Coulomb (1736–1806). There are two types of charge, *positive* and *negative*. The names derive from the fact that the two types of charge produce opposite effects. Thus equations describing the effects of charges encompass both types of charges if we associate a positive number with one type and a negative number with the other. Traditionally, the electron has been assigned a negative sign and the proton a positive. The magnitude of the charge of the electron is the smallest possible charge; in the MKS system of units this is†

$$e = -1.602 \times 10^{-19} \text{ C} \tag{1.1}$$

Since the mass of the electron is 9.11×10^{-31} kg, the charge-to-mass ratio of the electron is 1.76×10^{11}, in the MKS system. The charge on the proton is positive in sign and equal in magnitude to that of the electron, but the proton mass is 1846 times greater. Since the charge-to-mass ratio of the fundamental bits of matter is so great, electrical effects usually dominate mechanical inertia effects. Hence we usually talk about charge as if it were not tied to mass—as if it were massless.

Forces between charges.   We know about electric charges because charges exert forces on other charges. There are two types of forces between charges. The equation

$$\vec{F} = \vec{F}_e(Q_1, Q_2, R) + \vec{F}_m(Q_1, \vec{u}_1, Q_2, \vec{u}_2, R) \tag{1.2}$$

gives the force between two charges when they are moving with velocity $\vec{u}_1$ and $\vec{u}_2$, as shown in Fig. 1.1. The first type of force, $\vec{F}_e$, describes repulsion or attraction, depending on whether the charges are alike or opposite. This type of force is called an *electrostatic* force because it depends on only the position and not the motion of the charges. This type of force is responsible for lightning, because charges are separated in clouds by droplet separation; and electrostatic forces are used in photocopy machines to form images on glass drums with charged bits of dry ink.

The second type of force, $\vec{F}_m$, is motional, that is, it depends on the location and motion of the charges; these are called *magnetic* forces. Magnetic forces turn

† Physical constants are tabulated in Appendix A.

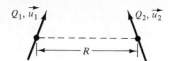

Figure 1.1 Two charges in motion.

motors, deflect electron beams in TV tubes, and effect energy conversion in generators. Electrical engineers thus have both electrostatic and magnetic effects to use in manipulating electrical energy.

The importance of energy. Energy is the medium of exchange in a physical system, like money in an economic system. Energy is exchanged whenever one physical thing affects another. In mechanics it takes force and movement to do work (exchange energy), and in electricity it takes electrical force and movement of charges to do work (exchange energy). The electrical force is represented by the voltage and the movement of charge by the current in an electrical circuit.

1.2.2 What Is Circuit Theory?

A circuit problem. To see how voltage and current describe energy exchanges, and to illustrate what circuit theory involves, imagine that we remove the battery and one headlight from a car and connect them together with wire, as shown in Fig. 1.2. From our experience with batteries, wire, and headlights, we expect that the light will glow and also get hot, which suggests that the battery is supplying energy to the headlight. Most people assume that the energy flows through the wire, but it is more accurate to say that the wire guides the energy from the battery to the headlight. In Fig. 1.3 we show an electric circuit representing the physical situation depicted in Fig. 1.2: a 12.6-volt (V) battery symbol and a 5.25-ohm (Ω) resistor (modeling the headlight), with lines representing the wire. From Ohm's law, we calculate the current to be 2.40 amperes (A). The power is voltage multiplying current, or 30.2 watts (W). This power indicates the rate at which energy is flowing from the battery into the headlight, where it is converted to heat and light energy.

The nature of electrical circuit theory. Soon we will carefully define current, voltage, and Ohm's law. Here we intend to show only what circuit theory

Figure 1.2 Physical circuit.   **Figure 1.3** Circuit theory model.

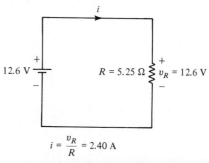

$$i = \frac{v_R}{R} = 2.40 \text{ A}$$

$$P = 12.6 \times 2.40 = 30.2 \text{ W}$$

encompasses. The solution of an engineering problem normally proceeds through four stages: first, the problem is identified; second, the problem is modeled; third, the model is analyzed; and fourth, the results are applied to the original physical problem. In the case of the battery and headlight, we skipped the first and last steps, but we did model a physical situation (battery and headlight) with common circuit symbols and we did analyze the circuit model (2.40 A and 30.2 W). Circuit theory consists only in the third step: taking a given circuit model and solving for certain results through the application of known circuit laws such as Ohm's law. This third step is what you will learn to do in the first part of this book. In the remainder you use circuit theory in learning the principles of electromechanics and electronic systems. We begin with the definitions of current and voltage.

1.3 CURRENT AND KIRCHHOFF'S CURRENT LAW

1.3.1 Definition of Current

Current is charge in motion. In the experiment with the battery and headlight described above, we recognize that the headlight glows because of charges moving through the wires. If we had connected our circuit with a non-conductor of electricity, say, a piece of string, the headlight would not glow. What makes an electrical conductor is the presence of conduction electrons in the material; a string has no such conduction electrons, whereas a copper wire does. A nonconductor has plenty of electrons but its electrons are bound to the nuclei and cannot move.

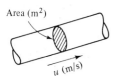

Area (m²)

u (m/s)

Figure 1.4 Wire with current.

The movement of charge in a wire comprises an electric current. Consider a wire with a cross section of A square meters, and an electron concentration n_e of conduction electrons per cubic meter, moving with a velocity u from left to right, as pictured in Fig. 1.4. Recalling that e is the charge on the electron, we define the current to be

$$i = n_e e A u \quad \text{C/s} \quad \text{or} \quad \text{ampere} \quad (1.3)$$

Note that the units of current are coulombs per second, but to honor André Ampère (1775–1836) we give this unit a special name, the ampere (A). The electron concentration for copper is about 1.13×10^{29} electrons/m³ and hence the charge density is $n_e \times e$ or -1.81×10^{10} C/m³. Traveling through No. 12 wire (0.081 in. in diameter) at a snail's pace of 0.1 mm/s, the charges constitute a current of $i = -5.8$ A (left to right). We could also express this result by saying that the current is $i = +5.8$ A (right to left). In this book we use a lowercase italic letter such as i to represent a quantity that may vary in time and uppercase italic letter such as I to indicate a quantity that must be constant.

Reference directions. That we can express the current two ways reveals the importance of reference directions. When we write "left to right," we are indicating the direction to which we are referring current flow. To specify a current, we require a reference direction plus a numerical value, which may be positive or negative. We will call current a *signed* quantity, since it can be numerically

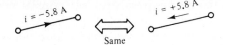

Figure 1.5 The same physical current.

positive or negative. In this sense charge is also a signed quantity. In this book the current reference directions are indicated by arrowheads drawn on the lines representing the conductors as in Fig. 1.5, or with arrows beside the wires. The relationship between the reference direction and the sign of the current is shown in Fig. 1.5, which expresses the same physical current in two ways and with both symbols for current reference directions. The problem solver is responsible for assigning current reference directions as a beginning step in analyzing a circuit. These reference directions may be assigned without regard for the direction of physical charge flow; they are assigned for bookkeeping purposes. This freedom in assigning reference directions will be clarified below when we state Kirchhoff's current law.

To summarize, current describes the movement of charges in conductors. To specify the current in a conductor, we need both a reference direction and a numerical value, which can be positive or negative.

A mechanical analogy.   For most of us, an understanding of mechanics comes easier than an understanding of electrical phenomena. We have all experienced forces and motion, springs and inertia. Also, mechanics is taught early in most curricula, so we have had more instruction and more time to ponder. Because our intuition for mechanics is relatively well developed, in this book we present mechanical analogs for many electrical quantities and phenomena.

Velocity is the simplest analog for electric current, and displacement would therefore be analogous to charge. By these analogies, we mean only that the equations relating these quantities are similar.

For example, if we know the velocity of an object, $u(t)$, the change in displacement, Δx, during the period $t_1 < t < t_2$ would be

$$\Delta x = \int_{t_1}^{t_2} u(t)\, dt \tag{1.4}$$

Similarly, if we know the current in a wire, $i(t)$, the charge Δq passing a cross section of the wire during the period $t_1 < t < t_2$ would be

$$\Delta q = \int_{t_1}^{t_2} i(t)\, dt \tag{1.5}$$

Velocity and displacement in general are vector quantities. Charge motion is also in general a vector quantity; but in circuits, the wires channel the current in established directions, and a plus or minus sign is all that remains of the "vectorness" of the current. In other words, charge flow in circuit theory corresponds to linear motion in mechanics.

Example.   A steady current of $+10^{-6}$ A flows toward the right in a copper wire of 0.001 in. diameter. Find the speed and direction of the electron flow. How many electrons pass a cross section of the wire in 1 μs?

Because the current to the right is positive, the electrons must be traveling to the left. Their velocity would be

$$u = \frac{I}{An_e |e|} = \frac{10^{-6}}{\pi(0.0005 \times 0.0254)^2 \times 1.81 \times 10^{10}} = 1.09 \times 10^{-7} \text{ m/s}$$

The total charge passing a cross section in 10^{-6} s would be

$$\Delta q = \int_0^{10^{-6}} i \, dt = 10^{-6} \text{ A} \times 10^{-6} \text{ s} = 10^{-12} \text{ C}$$

The number of electrons would be 10^{-12} C/1.60×10^{-19} C/electron $= 6.24 \times 10^6$ electrons.

1.3.2 Kirchhoff's Current Law

Conservation of charge and charge neutrality. All the evidence suggests that the universe was created charge-neutral; that is, there exists somewhere a positive charge for every negative charge. As was implied in our discussion on electrostatic forces [Eq. (1.2)], positive and negative charges can be separated by natural causes (leading to lightning) or man-made causes (TV tubes), but most matter does not contain surplus charge; that is, most matter is charge-neutral. This fact, combined with the fact that charge is conserved, leads to a constraint on the currents at a junction of wires.

KCL at a junction of wires. The junction of two or more wires is called a *node*. The constraint imposed by conservation of charge and charge neutrality is known as *Kirchhoff's current law* (KCL) and can be stated as follows: Because charge is conserved, the sum of the currents leaving a node is zero at all times. We could have written "currents entering a node" just as well. Another equivalent statement would be that the sum of the currents entering a node is equal to the sum of the currents leaving that node. Figure 1.6 illustrates KCL at node *a*. Note that a + sign is used for currents referenced departing from the node and a − sign is used for currents referenced toward the node. In general form, KCL is

$$\sum_{\text{node}} \pm i_n = 0 \qquad (1.6)$$

The signs come from the reference directions; the i's are counted + if referenced departing from the node and − if referenced entering the node.

KCL applied to groups of nodes. Kirchhoff's current law is one of the fundamental laws of electric circuits because it expresses the conservation of

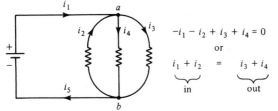

$$-i_1 - i_2 + i_3 + i_4 = 0$$

or

$$\underbrace{i_1 + i_2}_{\text{in}} = \underbrace{i_3 + i_4}_{\text{out}}$$

Figure 1.6 Kirchhoff's current law at node *a*.

charge. Not only does it apply to a simple node, but it applies also to groups of nodes. For example, if in Fig. 1.6 we add KCL for node a to KCL for node b, the i_2 term must appear with opposite signs in the two equations; thus the two i_2 terms will cancel. For the same reason, the i_3 and i_4 terms will also cancel. Clearly, this cancellation will always occur for the internal currents in a group of nodes. Thus in Fig. 1.6 we can easily show that i_1 and i_5 are equal by adding KCL for nodes a and b.

Example. Two sources are connected to the same load as shown in Fig. 1.7. The load current is 5 A, and the first source applies 8 A. How much current is supplied by the second source?

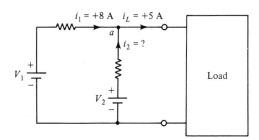

Figure 1.7

We may apply Kirchhoff's current law to node a. Summing currents leaving the node, we have

$$-i_1 - i_2 + (+5) = 0$$

Thus

$$i_2 = -i_1 + 5 = -3 \text{ A}$$

As we will see, this implies that the first source is supplying all the load current and also giving energy to the second source. If the second source were a battery, the battery would be charging, not discharging.

1.3.3 Check Your Understanding

1. The charge on the electron is $e = -1.60 \times 10^{-19}$ C. What is the current in a wire in which 10^8 electrons pass a cross section in 15 μs?
2. How many electrons pass a cross section in 15 μs in a wire carrying 55 μA?

   Answers.   1. 1.07 μA, sign not required;   2. $5.15 \times 10^{+9}$ electrons.

1.4 VOLTAGE AND KIRCHHOFF'S VOLTAGE LAW

1.4.1 Definition of Voltage

Voltage and energy exchanges.   Voltage expresses the potential of an electrical system for doing work. *Voltage* is defined to be the work done per unit charge by the electrical system in moving a charge from one point to another in

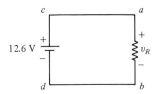

Figure 1.8 Battery circuit marked for definition of voltage.

a circuit. In Fig. 1.8 we repeat the circuit modeling the battery-headlight experiment described earlier. In the figure we have identified points in the circuit with the letters a, b, c, and d. The motion of charges around the circuit effects the transfer of energy from the battery to the headlight, which is represented by the resistor. The work done by the electrical system in moving a charge from a to b is measured by the voltage: the higher the voltage, the more the work that will be done. Specifically, the voltage from a to b is defined to be

$$v_{ab} = \frac{\text{work done by } q \text{ in moving } a \rightarrow b}{q} \tag{1.7}$$

The units of voltage are energy/charge, joules per coulomb in the MKS system, but to honor Count Alessandro Volta (1745–1827) we use the special name volt (V) for this unit. In our hypothetical experiment, the voltage between a and b will be 12.6 V because of the car battery.

Voltage as potential. Although our example implies that charges must move for the voltage to exist, such motion is not required. If the bulb were burned out and no current existed, the voltage would still be present between a and b. The voltage expresses the *potential* for doing work; that is, it measures how much work would be done if a charge were moved from a to b. It is similar to a gravitational potential, which expresses the work that would be done in a gravitational field if a mass were moved from one point to another. In other words, the voltage is defined to be the work on a hypothetical unit charge in moving between two points in a circuit. As for a gravitational potential, the path of the charge does not matter. The charge can be moved from a to b through the bulb or it can be moved outside the bulb, or it can be moved from a to the moon and then from the moon to b. In all cases the work, and hence the voltage, would be the same.

Voltage is a signed quantity. Both the work and the charge are signed quantities; hence voltage must also be a signed quantity. In the case where the battery turns on the headlight, energy is delivered by the electrical circuit to the thermodynamic system (heat and radiation); and in this case the work done by the electrical system (by q) would be considered positive. On the other hand, chemical processes in the battery are delivering energy to the electrical circuit; in this case we would consider the work done by the electrical system as negative in our definition of the voltage involving the battery. Voltage is thus the signed work done by the electrical system divided by the signed charge involved in doing that work.

There is yet another way in which the sign of the voltage can be affected, because the direction in which the charge is moved also must be considered. If

we were to move a hypothetical positive charge from a to b in Fig. 1.8, posit
work would be done and the voltage would be positive. If we were to move
hypothetical positive charge from b to a, negative work would be done and
voltage would be negative. To speak meaningfully of voltage as a signed quanti
we must have a clearly defined reference direction to indicate the assumed
rection of travel.

Example.   For our battery in Fig. 1.8, $V_{cd} = +12.6$ V. Hence if we mo
a charge of $+1$ C from c to d with our hand, the electrical system (the batte
would do $+12.6$ joules (J) of work (on us). This means that the charge would p
on us. But if we moved a charge of -1 C from c to d, the work done by t
electrical system would be -12.6 J, revealing that we would have to push on t
charge to move it from c to d.

Voltage reference directions.   Because voltage is a signed quantity,
need a way to assign reference directions in writing voltage equations. We w
use two conventions for defining the reference directions of voltages. The mo
explicit convention uses subscripts, as in Eq. (1.7), to define the beginning poi
(a), the ending point (b), and thus the direction traveled (from a to b). By th
convention, V_{gh} would be the work done by the electrical system per charge
moving a charge from g to h. Often in a circuit, however, we desire to expre
the voltage across a single element, as in the present case with the battery or t
headlight. In this case we can mark both ends of the circuit element with polari
symbols: a $+$ at one end and a $-$ at the other end. With this simpler notatio
the convention is that you move from the $+$ to the $-$; that is, the $+$ would l
the first subscript and the $-$ the second subscript. Figure 1.9 shows the circu
marked in this manner, and Fig. 1.10 shows an equally legitimate notation. The
retically, we could also reverse the polarity mark on the battery, but we sha
always consider batteries as having a numerically positive voltage.

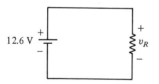
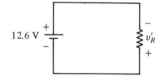

Figure 1.9 Resistor voltage defined.

Figure 1.10 Another definition of resistor voltage.

A mechanical analog.   Force is a mechanical analog for voltage, in two
senses. As external forces impart energy to mechanical systems and thus make
things happen, so voltage sources (batteries, for example) impart energy to electrical circuits. Beyond this subjective analogy, however, we will show later in
this chapter that equations involving voltage are similar in form to those involving
force in mechanical systems.

1.4.2 Kirchhoff's Voltage Law

Conservation of energy.   Kirchhoff's voltage law expresses conservation of energy in electrical circuits. The charges traveling around a circuit effect the transfer of energy from one circuit element to another but do not receive energy themselves on the average. This means that if you were to move a hypothetical test charge around a complete loop in a circuit, the total energy exchanged would add to zero. During part of the loop you would have to push on the charge to move it, but during other parts of the loop it would pull on you.

Forms of KVL.   Because the energy sum is zero, it follows from the definition of voltage that the voltage sum around a closed loop is zero also. This is *Kirchhoff's voltage law* (KVL):

$$\sum_{\text{loop}} \text{voltages} = 0 \tag{1.8}$$

We can apply KVL to the circuit in Fig. 1.8 with the result

$$v_{dc} + v_{ca} + v_{ab} + v_{bd} = 0 \tag{1.9}$$

The direct connections between c and a and between b and d imply ideal connections, and hence no voltage, between these points. Therefore, $v_{ca} = 0$ and $v_{bd} = 0$, and Eq. (1.9) reduces to

$$+v_{dc} + v_{ab} = 0 \tag{1.10}$$

When we apply KVL to voltages whose reference directions are indicated with the $+$ and $-$ polarity convention, we find that we cannot guarantee, as in Eq. (1.9), that all the signs in the resulting equation will be positive. Thus we must rewrite KVL in the form

$$\sum_{\text{loop}} \pm v\text{'s} = 0 \tag{1.11}$$

In writing KVL equations with $+$ and $-$ polarity symbols, we write the voltage with a positive sign if the $+$ is encountered before the $-$ and with a negative sign if the $-$ is encountered first. Applying this rule to Fig. 1.9, we express KVL as

$$-12.6 + v_R = 0 \tag{1.12}$$

Example.   We have a weak battery with a voltage of 11.5 V and a strong battery with a voltage of 13.6 V. (This includes the benefits of the alternator charging the battery during the jump-start.) We will consider jump-starting the car with the weak battery (a) when we make the correct connections with the jumper cables, and (b) when we make the incorrect connections.

Figure 1.11 shows the correct connections. According to the recommended procedure, we should first connect the positive terminals, as shown in the figure, and then complete the circuit by connecting between points on the auto chassis away from the batteries, to reduce the danger of explosion. Thus we are left with a voltage across the gap before the final connection is made, due to the differing voltages of the two electrical systems. Using KVL, we determine the gap voltage to be

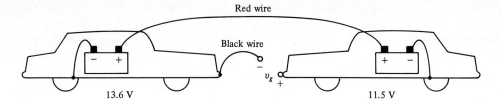

Figure 1.11

$$-v_g - 11.5 + 13.6 = 0 \Rightarrow v_g = -2.1 \text{ V} \qquad (1.13)$$

where v_g is the voltage across the gap. The same voltage would exist between the jumper cable and the negative terminal of the battery.

Figure 1.12 shows the incorrect connection, where we connect the negative of one battery to the positive of the other battery. From this bad start, we can continue in two ways. If we take the second jumper cable and proceed to connect between the two remaining battery terminals (A), we will have a gap voltage of

$$-v_g' + 11.5 + 13.6 = 0 \Rightarrow v_g' = +25.1 \text{ V} \qquad (1.14)$$

where v_g' is the gap voltage with the incorrect connection. This large voltage presents a danger to both batteries and to the person making the connection. On the other hand, if we follow the recommended procedure and connect between chassis (B), we have no problem because one positive battery terminal is isolated. After the chassis are connected, the two electrical systems are still unconnected, and no harm is done.

Example. As a second application of KVL, let us determine the unknown voltages, v and v_{dc}, in Fig. 1.13. Here we have deliberately mixed the reference-

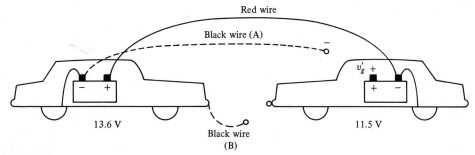

Figure 1.12

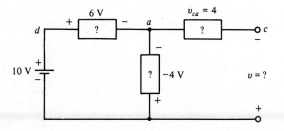

Figure 1.13 Example with mixed polarity conventions.

direction conventions. The boxes with the ?'s might be resistors or batteries or some other circuit element that we have not yet introduced. These boxes do not concern us, for we can write KVL from the given information without knowing what is in the boxes. We note that all the voltages around the leftmost loop are specified. Let us confirm that the given voltages satisfy KVL. We write a KVL loop equation going clockwise around the loop starting with the bottom of the battery:

$$-(10) + (+6) - (-4) = 0 \tag{1.15}$$

In this equation, the signs outside the parentheses come from the reference direction and the signs inside come from the numerical value of the voltages. Thus the minus sign before the first term is there because we encounter first the $-$ polarity symbol on the battery, and the next sign is positive because we then encounter the $+$ polarity marking on the topmost box. We note that KVL is satisfied. What would it mean if KVL were not satisfied around this loop? It would mean that your author, or the printer, had made an error; for a circuit in which KVL is not satisfied is not a circuit at all—it is nonsense, like $1 + 1 = 3$.

Let us proceed to determine the unknown v by writing KVL around the rightmost loop. What loop? There is indeed a loop even though no path exists for current, for we could still move our hypothetical test charge around this closed path. Thus KVL must be satisfied even when there is no closed path for charge to flow. Let us start at point a and go counterclockwise:

$$-(-4) + (v) + v_{ca} = 0$$
$$v = -v_{ca} - 4 = -4 - 4 = -8 \text{ V} \tag{1.16}$$

In Eq. (1.16) we have written the v_{ca} term with a $+$ sign because we are moving from the first subscript (c) to the second (a). Finally, we can determine v_{cd} by writing KVL from c to d to a to c:

$$v_{cd} + v_{da} + v_{ac} = 0$$
$$v_{cd} = -v_{da} - v_{ac} = -(+6) + (4) = -2 \text{ V} \tag{1.17}$$

Note that in the second form of Eq. (1.17) we wrote v_{da} as $+6$ because d and the $+$ symbol mark the same point in the circuit. That is, the convention for the $+$ and $-$ marking is that we move from the $+$ to the $-$, which in this case is the same as moving from d to a; hence v_{da} is $+6$ V. Note also that $v_{ac} = -v_{ca}$, as shown below.

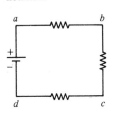

Figure 1.14 Circuit marked for subscript notation.

Some rules for subscripts. As a third application of KVL, we will demonstrate two properties of the subscript notation for voltages. For this purpose we refer to Fig. 1.14. One application of KVL is to go from a to b and then return to a. The resulting equation is

$$v_{ab} + v_{ba} = 0 \quad \text{or} \quad v_{ab} = -v_{ba} \tag{1.18}$$

Thus we see that reversing the subscripts changes the sign of a voltage. This reflects that the physical work done in moving a charge between two points is the negative of the work done in moving the same charge in the opposite direction.

Let us now determine v_{ac} by moving from a to c to b and back to a:

$$v_{ac} + v_{cb} + v_{ba} = 0$$

$$v_{ac} = -v_{ba} - v_{cb} \qquad (1.19)$$

$$= v_{ab} + v_{bc}$$

We eliminated the minus signs in the next-to-last form by reversing the subscripts. Note the pattern in the subscripts: the second subscript (b) of the first voltage (v_{ab}) is identical to the first subscript of the second term (v_{bc}). The final form of v_{ac} suggests that the middle point drops out, that is, does not matter. You can verify this for yourself by writing v_{ac} in terms of v_{ad} and v_{dc}. You will get the equation

$$v_{ac} = v_{ad} + v_{dc} \qquad (1.20)$$

Again the middle point drops out. This makes sense: it is like saying that the distance from New York to Los Angeles is the distance from New York to St. Louis plus the distance from St. Louis to Los Angeles. Clearly, "St. Louis" is a variable; we could have just as easily said Amarillo or Tucson. These two properties of the subscript notation will prove useful in later chapters.

1.4.3 Check Your Understanding

1. If $v_{ab} = -5$ V, how much energy is required by an *external agent* to move -2 C of charge from b to a?
2. If $v_{ab} = +2$ V and $v_{cb} = -1$ V, find v_{ba} and v_{ca}.
3. A tape player requires three 1.5-V batteries. What voltage is developed if one battery is inserted backward?
4. Is it always true that $v_{ab} + v_{bc} + v_{ca} = 0$?

   Answers. 1. $+10$ J; 2. -2 V, -3 V; 3. 1.5 V; 4. Yes.

ENERGY FLOW IN ELECTRICAL CIRCUITS

1.5.1 Voltage, Current, and Power

Power from voltage and current. At the beginning of this chapter we described electrical engineering as the useful control of electrical energy. Usually, energy is being exchanged between parts of a circuit, and the rate of energy flow, the power, must be calculated. We will now show that knowledge of voltage and current everywhere in an electrical circuit allows computation of energy flow (or power) throughout that circuit. Indeed, merely restating the definitions of voltage and current reveals their relationship to power flow. Voltage is the energy exchanged per charge and current is the rate of charge flow. Thus the rate of energy exchanged, the power, is by definition the product of voltage and current:

$$\frac{\text{work}}{\text{charge}} \times \frac{\text{charge}}{\text{time}} = \frac{\text{work}}{\text{time}} = \text{power} \qquad (1.21)$$

Thus if we know the voltage across a circuit component and the current through that element, we can determine the power into (or out of) that component by multiplying voltage and current (Volts × Amperes = Watts).

Load sets and source sets. Power is a signed quantity and we must again consider reference directions. The sign in the power formula depends on the combination of the voltage and current reference directions. We show both possibilities in Fig. 1.15. In a *load set* the current reference direction is directed *into* the + polarity marking (or the first subscript) of the voltage reference direction. This is the convention normally used with a resistor or some other element that receives power from the circuit. In a *source set* the current reference direction is directed *out of* the + polarity marking (or the first subscript) of the voltage. This combination is normally used for a source of energy to the circuit such as a battery. However, in either case in Fig. 1.15 we can speak of the power out of the component and the power into the component as meaningful quantities.

We must stress that the formulas given in Fig. 1.15 relate solely to reference directions. Any of the numerical values of the voltages, currents, or powers can be positive or negative. The direction of energy flow at a component will be established only after the sign depending on the reference directions is combined with the numerical signs of the v's and i's.

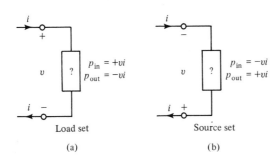

Load set   Source set

(a)   (b)

Figure 1.15 Sign conventions for power calculations: (a) load set; (b) source set.

Example. The use of load and source sets will be clarified in the example in Fig. 1.16. With the battery we have a source set since the current reference arrow is out of the + polarity marking. Hence the power out of the battery is $+vi$ or $+(+10)(+2) = +20$ W. In the preceding sentence, the + of the "$+vi$" is from the source set of reference directions and the + of the "$+20$" comes

Figure 1.16 The signs outside the parentheses come from the reference directions.

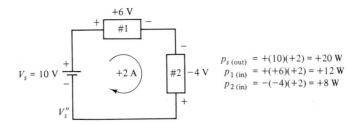

$$P_{s\,(out)} = +(10)(+2) = +20 \text{ W}$$
$$P_{1\,(in)} = +(+6)(+2) = +12 \text{ W}$$
$$P_{2\,(in)} = -(-4)(+2) = +8 \text{ W}$$

from the combination of the reference directions and the numerical values of the voltage and current. The interpretation of the positive power out of the source is that the battery is indeed delivering energy to the circuit. If the power out of the battery had been negative, we would learn that the battery is being recharged by some other source in the circuit. Box 1 has a load set and proves to be receiving energy from the circuit. Box 2 is also receiving energy from the circuit because a positive sign is produced by the combination of the minus sign from the reference directions and the minus sign of the voltage. Thus in calculating power we must continue to distinguish between signs arising from the reference directions and signs arising from the numerical values of the voltages and currents. Only the meaning of the power variable ("into" or "out of") and the numerical sign of the result allow the final conclusion as to which way energy is flowing at a given instant.

Mechanical analogies. We have presented mechanical force, velocity, and displacement as counterparts of voltage, current, and charge, respectively. Here we will review some of the relationships relating these mechanical variables to energy exchanged in a mechanical system. Then we will show the corresponding equations for the electrical variables.

The amount of energy exchanged, dW, by a force F operating through a distance, dx, would be

$$dW = F \, dx \quad \text{J} \tag{1.22}$$

When continuous motion is involved, the rate of energy exchanged, the power (p), would be

$$p = \frac{dW}{dt} = F\frac{dx}{dt} = Fu \quad \text{W} \tag{1.23}$$

where u is the velocity. Finally, the energy exchanged in a period of time $t_1 < t < t_2$ can be computed by integration:

$$W = \int_{t_1}^{t_2} p \, dt = \int_{t_1}^{t_2} Fu \, dt \quad \text{J} \tag{1.24}$$

These relationships have electrical counterparts. When a charge dq is moved from a to b, the energy given by the electrical circuit would be

$$dW = v_{ab} \, dq \quad \text{J} \tag{1.25}$$

When continuous charge flow, a current, is involved, the rate of energy exchanged, the power (p), would be

$$p = \frac{dW}{dt} = v_{ab}\frac{dq}{dt} = v_{ab}i \quad \text{W} \tag{1.26}$$

where i is the current referenced from a to b. Finally, the energy exchanged in a period of time $t_1 < t < t_2$ can be computed by integration:

$$W = \int_{t_1}^{t_2} p \, dt = \int_{t_1}^{t_2} v_{ab}i \, dt \quad \text{J} \tag{1.27}$$

Example. A battery has a voltage of 1.1 V when in a discharged state but after being recharged at 0.1 A for 2 hours (T) has a full-charge voltage of 1.3 V. We will compute the total charge given to the battery and the energy required to recharge the battery.

The charge received can be computed by integration:

$$i = \frac{dq}{dt} \Rightarrow q = \int_0^T i\, dt = iT = 0.1 \text{ A} \times 2 \text{ h} \times 3600 \text{ s/h} = 720 \text{ C}$$

We will assume that the voltage increases linearly with time:

$$v(t) = 1.1 + 0.2\left(\frac{t}{T}\right) \quad \text{V}$$

The energy required would thus be

$$W = \int_0^T \left[1.1 + 0.2\left(\frac{t}{T}\right)\right](0.1)\, dt = 0.1\left(1.1t + \frac{0.1}{T}t^2\right)_0^T$$

$$= 864 \text{ J}$$

1.5.2 Check Your Understanding

1. A battery charger puts 5 A into a 12.6-V auto battery. What is the power into the battery?
2. Find the power *out of* box 1 and box 2 in Fig. 1.16.
3. A 1.5-V flashlight battery puts out 300 mA for 10 minutes. How much energy does this represent?

Answers. 1. 63.0 W; 2. -12 W, -8 W; 3. 270 J.

Summary. Voltage and current allow us to monitor energy flow in an electrical circuit. Voltage and current prove to be easily measured in electrical circuits. Voltage and current also obey simple laws, KVL and KCL. For these reasons voltage and current are universally used by electrical engineers in describing the state of an electric network.

In the next chapter we will define resistors and state Ohm's law. This addition will allow us to develop many techniques for analyzing electrical circuits. Although the knowledge gained in that chapter will allow us to analyze many practical problems, our main interest lies in the techniques of solution; for these are general enough to permit assault of many circuit problems of great complexity.

PROBLEMS

Section 1.3: Current and Kirchhoff's Current Law

P1.1. In a copper wire 0.006 in. in diameter, electrons are moving 1 ft every minute. What is the magnitude of the current in the wire?

Electric Charge and Electrical Energy   Chap. 1

P1.2. The allowed safe current for a No. 10 copper wire, 0.01019 in. in diameter, is 30 A. At 30-A direct current, how fast are the electrons traveling in mm/s?

Answer. $u = 31.5$ mm/s

P1.3. For the circuit in Fig. P1.3:
(a) Write KCL for node a.
(b) Write KCL at node b and add to KCL at node a to show that i_1 and i_4 are equal.

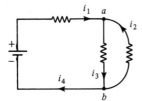

Figure P1.3

P1.4. Use KCL to determine i_1, i_2, and i_3 in the circuit shown in Fig. P1.4.

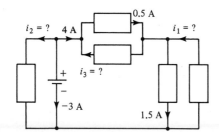

Figure P1.4

P1.5. For the circuit in Fig. P1.5, compute the unknown currents using KCL.

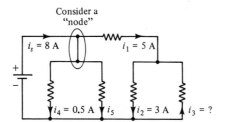

Figure P1.5

P1.6. In a simulated lightning bolt in a laboratory, shown in Fig. P1.6, the current increases linearly from 0 to 1000 A in 1 μs and then decreases linearly back to 0 in 4 μs. What is the total charge required?

Answer. 2.5×10^{-3} C

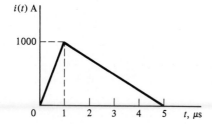

Figure P1.6

Section 1.4: Voltage and Kirchhoff's Voltage Law

P1.7. How much work is done by a 12.6-V battery if one electron is pulled off the negative terminal and placed on the positive terminal?

P1.8. In a cathode ray tube (CRT), a beam of electrons is accelerated through a certain voltage (V_c), then allowed to drift through an evacuated space until it strikes the tube surface, making a small spot of light. If $|V_c| = 10,000$ V, what velocity do the electrons acquire from the voltage source? *Hint:* Equate the final kinetic energy of an individual electron with the electrical energy delivered by the source to one electron.

Answer. $u = 5.93 \times 10^7$ m/s

P1.9. Write KVL to determine v in the circuit in Fig. P1.9.

Answer. $v = -15$ V

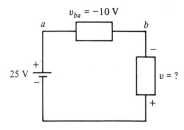

Figure P1.9

P1.10. For a circuit, $v_{ab} = 7$ V, $v_{cb} = -2$ V, $v_{dc} = 6$ V, and $v_{de} = -2$ V. Using the rules for adding subscripts, determine v_{ba}, v_{bd}, and v_{ae}. *Hint:* Draw a "circuit" showing the lettered points to aid in visualizing the patterns.

Answers. $v_{ba} = -7$ V, $v_{bd} = -4$ V, and $v_{ae} = +1$ V

P1.11. For the circuit shown in Fig. P1.11, find v_1, v_{ad}, v_{bc}, and v_{ac}.

Answers. $v_1 = +2$ V, $v_{ad} = -8$ V, $v_{bc} = +13$ V, and $v_{ac} = +6$ V

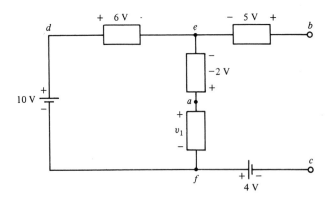

Figure P1.11

P1.12. For the circuit shown in Fig. P1.12, use KCL and KVL to determine i_1, i_2, v_{ad}, and v_x.

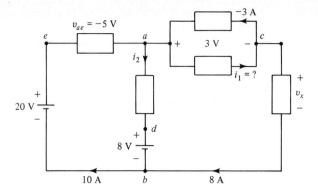

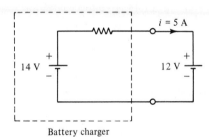

Figure P1.12

Section 1.5: Energy Flow in Electrical Circuits

P1.13. The MKS unit for energy is the joule, which is 1 watt-second, but the unit in common use by electrical utilities is the kilowatthour (kWh). How many joules are there in 1 kWh? At 6 cents/kWh, how many joules can one buy for a penny?

Answer. 6×10^5 J/cent

P1.14. The circuit shown in Fig. P1.14 represents a battery charger. Find the power into the battery being charged and the time it would take to impart 1 kilocoulomb (10^3 C) to the battery. What energy is given to the battery in this period of time?

Answer. 200 s, 12,000 J

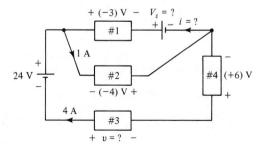

Battery charger   **Figure P1.14**

P1.15. For the circuit shown in Fig. P1.15:
  (a) Use KCL and KVL to find V_s, v, and i.
  (b) Calculate the power *into* every box and the power *out of* every source. Show that energy is conserved, that is, that $p_{out} = p_{in}$.

Figure P1.15

P1.16. In Fig. P1.16, the power into #2 is -10 W. Determine the power into #1 and the power out of the battery.

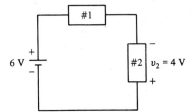

Figure P1.16

P1.17. In the circuit in Fig. P1.17 determine:
(a) v_{ab}
(b) i_x
(c) The sum of the powers into boxes 1, 2, and 3.

Answers. (a) $+ 10$ V, (b) 5 A, (c) 24 W

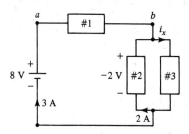

Figure P1.17

P1.18. In the circuit shown in Fig. P1.18, give the values for:
(a) v_{ab}
(b) v_x
(c) Power out of the 4-V battery.

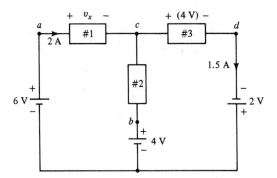

Figure P1.18

P1.19. For the circuit shown in Fig. P1.19, compute the unknown voltages and currents. (You supply the reference directions.) Compute the power *into* the two boxes and the power *out of* both sources. Show conservation of power.

Answers. $P_{\text{out of sources}} = 16$ W, $P_{\text{into resistors}} = 16$ W

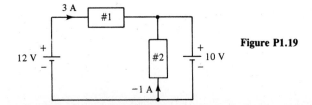

Figure P1.19

General Problems for Chapter 1

P1.20. In a semiconductor such as silicon, current is carried both by conduction electrons and by mobile "holes" which behave like positive charges having a charge equal in magnitude to that of the electron. Find the magnitude and direction of the current in a silicon diode of cross-sectional area 0.0001 in.^{2} with electrons (n_e of 10^{23} electrons/m^3) traveling with 0.3 mm/s to the right and holes (n_h of 1.5×10^{22} holes/m^3) traveling to the left at 0.1 mm/s.

P1.21. A lightning bolt might carry a current of 1000 A and last for about 1 ms. How much charge is exchanged between ground and cloud in such a lightning bolt? How many raindrops had to fall, each having a deficiency of 10 electronic charges, to create the initial charge separation neutralized by the lightning?

P1.22. A battery can be rated in "A-h," that is, ampere-hours. For example, a 60 A-h battery ought to put out 60 A for 1 h or 1 A for 60 h, or some other combination multiplying to 60. How many coulombs of charge may one hope to get out of a fully charged 80 A-h battery? How many electrons?

P1.23. A fully charged auto battery might have a voltage of 12.6 V. As the battery discharges, the voltage drops. The voltage will drop to about 10.0 V as total discharge is approached, at which point it drops to essentially zero if you try to draw current from it. Consider the case where we are discharging a 60-A-h battery over a 4-h period at a 15-A rate.
(a) Describe the voltage as a function of time during discharge.
(b) Compute the total energy output of the battery for this discharge rate.
(c) Show that the energy output is independent of the discharge rate. *Hint:* Let T be the time for discharge and $60/T$ be the current.

P1.24. Figure P1.24 shows the voltage and current for a semiconductor switch closure.
(a) How much energy is used up in each switching operation? (Assume source set of reference direction.)
(b) If the same energy were used in opening the switch, and the switch went through 5000 cycles each second, what is the average power lost in the switch? Each cycle involves opening and closing the switch.

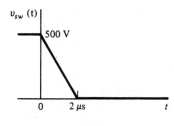

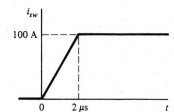

Figure P1.24

2 The Analysis of DC Circuits

2.1 DC CIRCUIT THEORY AND OHM'S LAW

2.1.1 Physics of Resistors

Charge, fields, and current. In Chapter 1 we described electrical engineering as the utilization of electrical energy for useful purposes. We presented definitions of voltage and current and showed that we can monitor energy flow in a circuit if we know the voltage and current throughout the circuit. We also stressed that conservation of charge and conservation of energy impose constraints on the currents and voltages in a circuit, leading to Kirchhoff's current law and Kirchhoff's voltage law.

In this chapter we show the principal techniques for analyzing circuits. Although we present these for direct current (dc) circuits, these techniques are widely applicable to other types of circuits, such as alternating current (ac) circuits and pulse circuits. For this reason, we will continue to represent voltage and current by lowercase symbols, to remind us that these would be time varying if the sources were time varying. Our sources, however, will be represented by uppercase symbols because these are constant in time.

First we must define resistance, voltage sources, and current sources. Let us once more consider our battery–headlight experiment. We know that chemical action produces charge separation within the battery. This charge separation appears at the battery terminals, and electrostatic forces are experienced by all the charges in the vicinity of the battery, significantly by charges in the wire and the headlight. Because of these forces, electrons move in the wire and headlight, and a current exists in the circuit.

It will not surprise you that, the greater the voltage, the greater will be the resulting current. For a large class of conductors, the current increases in direct proportion to the voltage. Physical experimentation leads to the following equation, known as *Ohm's law*:

$$i = \frac{v}{R} \quad \text{or} \quad v = Ri \tag{2.1}$$

In Eq. (2.1) R is the *resistance* and has the units volts per ampere, but to honor Georg Ohm (1787–1854) we use the unit ohm, abbreviated by the uppercase Greek letter omega, Ω. The resistance of a piece of wire is directly proportional to its length and inversely proportional to its cross-sectional area. The physical material from which the wire is drawn is also a factor: copper is a good conductor; iron not so good. The resistance of a wire also depends on its temperature, but the resistance may often be considered constant. In electronic circuits we use resistors made out of a carbon-impregnated binder, high-resistance wire, or metals deposited on nonconducting surfaces. The physical behavior of such resistors leads directly to the circuit theory definition of an ideal resistance.

Circuit theory of resistance. The circuit symbol for a resistance is given in Fig. 2.1. Note that we have arranged the reference directions of the voltage and current to be a load set. We could also write Ohm's law for a source set, which would introduce a minus sign into Ohm's law, but this is usually avoided.

Ohm's law is presented in graphical form in Fig. 2.2. We have shown voltage to be the independent variable (cause) and current to be the dependent variable (effect). The slope of the line is the reciprocal of resistance. Customarily, Ohm's law is also stated in terms of reciprocal resistance, which is called *conductance*:

$$i = Gv, \quad \text{where} \quad G = \frac{1}{R} \tag{2.2}$$

The unit of reciprocal resistance, or conductance, is amperes per volt, but we use siemens (the official unit) or mho (the unofficial but more common term). The symbol for siemens is S, and the common symbol for the mho, "ohm" spelled backward, is the upside-down omega ($\mho$).

Because we have a load set, the power into the resistor is $+vi$. We can express the power into a resistor in several ways, as follows:

Figure 2.1 Circuit symbol for a resistance.

Figure 2.2 Ohm's law in graphical form.

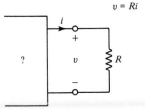

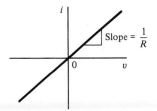

$$p = +vi = (Ri)i = i^2R = \frac{v^2}{R} \quad \text{watts} \qquad (2.3)$$

where v is the voltage across the resistor and i is the current through the resistor. Note that the power into a resistor is always positive. Resistance therefore always removes electrical energy from a circuit. As for our battery–headlight experiment, the energy lost to the electrical circuit usually appears as heat (also some light in that case).

2.1.2 Voltage and Current Sources

Voltage sources. Before we can analyze circuits, we must also define voltage and current sources. In Fig. 2.3 we show the circuit symbol, mathematical, and graphical definitions for an ideal dc voltage source. The ideal voltage source maintains constant voltage, independent of its output current. In general, a voltage source may have positive or negative voltage, and it may be constant or time varying. In this chapter we will use the battery symbol for a constant (dc) voltage source and always consider the battery voltage a positive number. Note that we have used a source set for the voltage source, although this is not absolutely necessary. Normally, a voltage source would produce a positive current out of the + terminal and thus act as a source of energy for the circuit; but it is possible that some other, more powerful source might force the current out of the + terminal to be negative, thus delivering energy to the source. When this happens for a battery, we say that the battery is being charged.

The dc voltage source of circuit theory models an ideal battery. We recognize that a real battery does not maintain constant voltage under heavy load—the car lights dim when the battery also has to start the motor. Real batteries store a finite amount of energy and need to be recharged or replaced frequently. By contrast, the ideal voltage source can deliver energy without limit. A physical battery can often be represented as an ideal voltage source in a circuit problem.

Current sources. Figure 2.4 shows the circuit theory symbol, mathematical definition, and graphical characteristic of a current source. The current source produces constant current, independent of its output voltage. Like the ideal voltage source, an ideal current source will deliver any required amount of energy. Unlike the voltage source, there is no physical device at your local hard-

Figure 2.3 (a) Symbol and mathematical definition for a voltage source; (b) graphical definition of a voltage source.

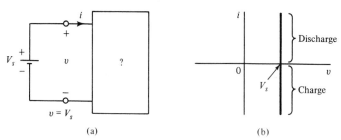

(a)   (b)

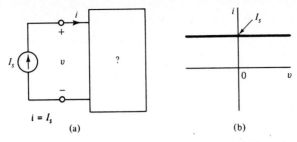

Figure 2.4 (a) Symbol for and mathematical definition of a current source; (b) graphical definition of a current source.

ward store whose electrical properties resemble a current source; however, we can build electronic devices that act like current sources.

2.1.3 Analysis of DC Circuits

Battery–headlight circuit solved carefully. We now have defined sufficient concepts and conventions to present and begin analyzing dc circuits. We will start with a careful solution of the battery–headlight problem. The circuit model is repeated in Fig. 2.5. We have used the + and − polarity notation and assigned reference directions such that the current reference direction is the direction in which current would be numerically positive. Shortly we will show that the choice of reference directions does not matter. Note that we have a load set for the resistor and a source set for the voltage source.

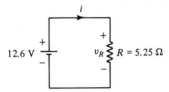

Figure 2.5 Battery–headlight circuit.

Logically, our next step is to count unknowns because we require as many equations as unknowns. We count two unknowns, i and v_R, and we therefore seek two equations. One equation comes from Ohm's law, the second from KVL:

$$v_R = 5.25i \quad \text{and} \quad -12.6 + v_R = 0$$

The solution is trivial:

$$v_R = +12.6 \text{ V} \quad \text{and} \quad i = +2.40 \text{ A}$$

The power into the resistor (headlight) is $+vi$ or 30.2 W, and the power out of the voltage source (battery) is $+30.2$ W.

Same circuit, changed reference directions. To show that the assumed reference directions do not matter, we will solve the problem with reversed reference directions (Fig. 2.6). The relevant equations are still Ohm's law and KVL:

$$v_R' = 5.25i' \quad \text{and} \quad -12.6 - v_R' = 0$$

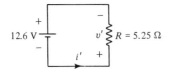

Figure 2.6 Battery–headlight circuit with different reference directions.

and the solution is again trivial:

$$v'_R = -12.6 \text{ V} \quad \text{and} \quad i' = -2.4 \text{ A}$$

Because the resistor still has a load set, the power into the resistor is still $+vi$, $+(-12.6)(-2.4)$ or 30.2 W, but this time the voltage source also has a load set. Hence the power out of the voltage source is $-vi'$ or $-(12.6)(-2.4)$ or 30.2 W. We must accept, therefore, that the assignment of reference directions is an arbitrary matter, a preliminary we must complete to begin writing equations.

A second example. In Fig. 2.7 we show a circuit with three resistors and a current source. The resistors are called *branches* in the circuit because they connect the nodes together. Branch currents and branch voltages are defined, including reference directions. We were careful to use load sets for the resistors and a source set for the source, but otherwise the reference directions were assigned arbitrarily. First let us count unknowns: There are four unknown v's and three unknown i's, so we require seven equations. Three of these equations come from Ohm's law:

$$v_1 = 2i_1 \quad \text{and} \quad v_2 = 4i_2 \quad \text{and} \quad v_3 = 6i_3$$

Two equations come from KVL around the two loops (*dabd* and *dbcd*):

$$-V_s + v_1 - v_2 = 0 \quad \text{and} \quad v_2 + v_3 = 0$$

We can write another KVL equation from the outer loop, but this equation can be derived from the other two and thus contains no new information. Similarly, we can write two KCL equations, but they are identical except for a minus sign; hence we get one equation from KCL at node b:

$$-i_1 - i_2 + i_3 = 0$$

The final equation comes from the definition of a current source:

$$i_1 = 2 \text{ A}$$

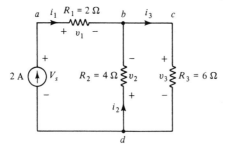

Figure 2.7 Seven equations are required to solve for the seven unknowns.

The solution is mathematically trivial, but we challenge you to go through it. The solution is

$$v_1 = 4 \text{ V}, \qquad v_2 = -4.8 \text{ V}, \qquad v_3 = 4.8 \text{ V}$$

$$i_2 = -1.2 \text{ A}, \qquad i_3 = 0.8 \text{ A}$$

If you did attempt the solution of the seven equations in seven unknowns, whether or not you got the correct answer, you probably said to yourself, somewhere along the way, "Surely there's a better way." That is a valid hope; indeed, this chapter develops many methods that are better than the one you just used. For of all the existing methods for analyzing that circuit, you just used a non-method. What was wrong with this "method?" Too many equations, for one thing. But we needed all those equations because we had all those unknowns, so we conclude that defining many unknowns led to trouble. And once we had the equations, there was no system for solving them—no logical procedure to follow.

Both of these defects are addressed by the methods we present in the remainder of this chapter. We will develop ways to reduce the number of unknowns and we will discover systematic procedures for solving the equations that result.

2.1.4 Check Your Understanding

1. Ohm's law is written with a + sign when the reference directions of the voltage across and current through the resistor are related by a load set or source set?
2. A 1000-Ω resistor is rated at 5 W. What is the maximum current that should go through this resistor? The maximum voltage across the resistor?
3. A 5-A current source is connected in series with a 10-V battery, with the current source going into the minus on the battery. What is the power into the battery?

Answers. 1. Load set; 2. 70.7 mA, 70.7 V; 3. −50 W.

2.2 SERIES AND PARALLEL RESISTORS; VOLTAGE AND CURRENT DIVIDERS

2.2.1 Series Resistors and Voltage Dividers

Resistors in series. Two resistors are said to be connected *in series* when the same current flows through them. Figure 2.8 shows a series connection of three resistors and a battery. It is important in this method to anticipate which

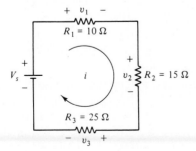

Figure 2.8 Three resistors connected in series.

way the current is positive, so we have defined the current reference direction such that the current will be positive out of the + terminal of the voltage source; and we put the + and − polarity symbols on the resistors according to load-set conventions. Because i will be numerically positive, the v's across the resistors will also be positive. We can write KVL around the loop, going clockwise.

$$-V_s + v_1 + v_2 + v_3 = 0 \Rightarrow V_s = v_1 + v_2 + v_3 \qquad (2.4)$$

Because the v's are positive and their sum is V_s, V_s divides between v_1, v_2, and v_3. To show how this division depends on the values of the resistors, we introduce Ohm's law:

$$V_s = R_1 i + R_2 i + R_3 i$$
$$= (R_1 + R_2 + R_3)i \qquad (2.5)$$
$$= R_{eq} i$$

where R_{eq} is an equivalent resistance. That the resistors carry the same current, and thus are in series, is clearly important to the reduction of Eq. (2.5) from the first to the second form. The difference between the second and the third forms of Eq. (2.5) is that we replaced the sum of the resistors by one equivalent resistor, R_{eq}; for the current will be the same with a single 50-Ω resistor as with three series resistors totaling 50 Ω. Figure 2.9 shows a circuit that is equivalent to that in Fig. 2.8, except that the three resistors have been replaced by R_{eq}.

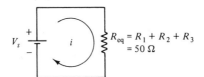

Figure 2.9 The equivalent resistor replaces three resistors in series.

Voltage dividers. We set out to learn how the voltage, V_s, divides between the three resistors. We have determined the current

$$i = \frac{V_s}{R_{eq}} = \frac{V_s}{50}$$

We can now obtain the voltage across the individual resistors from Ohm's law applied to the original circuit:

$$v_1 = R_1 i = R_1 \frac{V_s}{R_{eq}} = \frac{R_1}{R_{eq}} \times V_s = \frac{10}{50} V_s \qquad (2.6)$$

We can write formulas for v_2 and v_3 similarly. For example, if V_s were 60 V, v_1 would be 12 V, v_2 would be 18 V, and v_3 would be 30 V. Notice that the 60 V divides between the series resistors. Our results are easily generalized to include an arbitrary number of series resistors; indeed, the voltage across the ith resistor would be

$$v_i = \frac{R_i}{R_{eq}} V_s, \quad \text{where} \quad R_{eq} = R_1 + R_2 + \cdots + R_i + \cdots \text{ (all series resistors)}$$

As an intermediate result we learned that series resistors can be replaced by a single resistor whose value is the sum of the values of the resistors in series.

2.2.2 Parallel Resistors and Current Dividers

Parallel resistors. Resistors are said to be connected *in parallel* when they have the same voltage across them. Figure 2.10 shows a parallel combination of three resistors and a current source. We could have defined voltages for each resistor, but because the lines represent ideal connections, wires of zero resistance, the tops of the resistors are connected together and bottoms of the resistors are connected together, as we pictured them in Fig. 1.6. We were careful, however, to define the currents through the resistors with reference directions such as the currents would be positive. We write KCL for node a:

$$-I_s + i_1 + i_2 + i_3 = 0 \Rightarrow I_s = i_1 + i_2 + i_3 \tag{2.7}$$

Because the current reference symbols are directed into the $+$ end of v, we have load sets; hence we introduce Ohm's law into Eq. (2.7) in the usual form:

$$I_s = \frac{v}{R_1} + \frac{v}{R_2} + \frac{v}{R_3}$$

$$= v\left(\frac{1}{R_1} + \frac{1}{R_2} + \frac{1}{R_3}\right) \tag{2.8}$$

$$= v\,\frac{1}{R_{\text{eq}}}$$

where R_{eq} is an equivalent resistance. That the resistors have the same voltage and thus are connected in parallel is important in reducing Eq. (2.8) to the second form. The third form of Eq. (2.8) introduces a resistance, R_{eq}, that is equivalent to the three parallel resistors:

$$\frac{1}{R_{\text{eq}}} = \frac{1}{R_1} + \frac{1}{R_2} + \frac{1}{R_3} \Rightarrow R_{\text{eq}} = \frac{1}{(1/R_1) + (1/R_2) + (1/R_3)} \tag{2.9}$$

Many students have memorized the special case of Eq. (2.9) that applies to two resistors in parallel—the product over the sum. We would encourage use of Eq. (2.9) only for two reasons: (1) Eq. (2.9) is the correct form regardless of the number

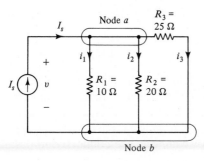

Figure 2.10 Three resistors connected in parallel.

of resistors; and (2) Eq. (2.9) is easier to implement on a calculator than the "product over the sum" for two parallel resistors.

Current dividers. Equation (2.9) is important, and we will discuss it later; but now we wish to use it in determining how the current I_s divides between the three resistors. We can find the voltage from the last form of Eq. (2.8) and solve for the current in, say, R_1 from Ohm's law:

$$v = R_{eq}I_s \Rightarrow i_1 = \frac{v}{R_1} = \frac{R_{eq}I_s}{R_1} = \frac{(1/R_1)}{(1/R_1) + (1/R_2) + (1/R_3)} I_s \quad (2.10)$$

The last form of Eq. (2.10) looks awkward but is the easiest form to implement on a calculator. Substituting the numbers, we learn that R_{eq} is about 5.3 Ω and that the current through R_1 is thus $0.53I_s$, or 53% of the total current through the three parallel resistors. You can confirm for yourself that 26% passes through R_2 and 21% through R_3. These results are easy to generalize: For any number of resistors that are connected in parallel, the current through the ith resistor, R_i, is

$$i_i = \frac{R_{eq}}{R_i} I_T \quad (2.11)$$

where I_T is the total current entering the parallel combination and

$$\frac{1}{R_{eq}} = \frac{1}{R_1} + \frac{1}{R_2} + \cdots + \frac{1}{R_i} + \cdots \text{ (all parallel resistors)} \quad (2.12)$$

Equation (2.12) shows how to combine resistors in parallel. The addition of reciprocals is awkward to write but easy to accomplish on a calculator. Indeed, adding reciprocals with a calculator is just as simple as adding numbers except that you must hit the 1/x key on the calculator after entering the resistor values. After summing the reciprocals, you then hit the 1/x key to display the equivalent resistance of the parallel combination. If we desire a neat equation, we must use the conductances

$$G_{eq} = G_1 + G_2 + \cdots + G_i + \cdots \text{ (all parallel resistors)} \quad (2.13)$$

For practice, add up the parallel combination of the three resistors in Fig. 2.10 using Eq. (2.9) to see if you get the right answer, 5.26 Ω.

Parallel resistors are common in electrical circuits, so it is important to have a simple way to indicate a parallel combination in an equation. We will use the notation $R_1 \parallel R_2$ to indicate that R_1 and R_2 are connected in parallel: thus

$$R_1 \parallel R_2 = \frac{1}{(1/R_1) + (1/R_2)}$$

Generalizing this notation to include more than two resistors, we calculate 12 $\Omega \parallel 16 \Omega \parallel 5 \Omega$ to be 2.89 Ω (check it). We note that resistors connected in parallel combine to produce a smaller equivalent resistance than any of the original resistors, whereas series resistors combine to make a larger equivalent resistance. All the lights in your house, for example, are connected in parallel since they all

The Analysis of DC Circuits   Chap. 2

require the same voltage, and their combined effect is to draw more current (smaller resistance) than any one of them would individually draw.

Example. We will calculate the voltage across the four resistors in Fig. 2.11. When we have two branches of multiple resistors in a circuit in parallel, the voltage across the two branches divides independently in the two branches. In Fig. 2.11, the equivalent resistance seen by the current source is $R_{eq} = (4 + 4) \| (8 + 4) = 4.8 \, \Omega$, and hence the total voltage across the parallel combination is $5 \times 4.8 = 24$ V with $+$ at the top. This voltage divides independently in the two branches by a routine application of Eq. (2.6), with the results shown.

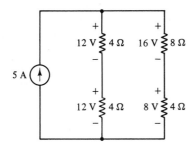

Figure 2.11 The voltage divides independently in the two parallel branches.

Reworking earlier examples with voltage dividers. In the preceding sections we have learned about series and parallel combinations and how voltage and current divide in them, respectively. Let us now use these concepts to solve the problem we worked in Section 2.1.3, repeated in Fig. 2.12. We will combine resistors in order to simplify the circuit. Notice that R_2 and R_3 are connected in parallel. We can replace them with an equivalent resistance of $4 \| 6$ or $2.4 \, \Omega$. This reduces the network to that shown in Fig. 2.13a. The 2-Ω resistor is connected

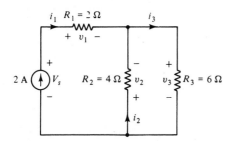

Figure 2.12 Same circuit as Fig. 2.7, to be solved by equivalent resistance method.

Figure 2.13 Combining resistors simplifies the network.

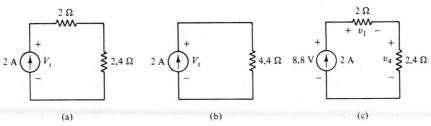

(a)   (b)   (c)

in series with the 2.4-Ω resistor and they can be combined into an equivalent resistance of 4.4 Ω (Fig. 2.13b). Clearly, the voltage across the current source, V_s, is 8.8 V.

We will now restore the original circuit. Figure 2.13c is the same as Fig. 2.13a, but because we are planning to divide the 8.8 V between the two resistors we have defined voltages with the proper polarity markings for that purpose. The original v_1 already has the desired polarity marking, and we have introduced a v_4 for the voltage across the 2.4-Ω resistor. "Wait" (you might object), "that is a current source, not a voltage source you are dividing." True, but the voltage created by the current source divides in the series circuit. It does not matter whether we have a 2-A current source producing 8.8 V or an 8.8-V voltage source producing 2 A—the circuit will respond in the same way. (The type of source would matter if we changed one of the resistors in the circuit; for in one case the source current would remain constant and in the other case the source voltage would remain constant.) So we can divide the 8.8 V whether it is produced by a voltage or a current source. The results are

$$v_1 = 8.8 \times \frac{2}{4.4} = 4.00 \text{ V} \quad \text{and} \quad v_4 = 8.8 \times \frac{2.4}{4.4} = 4.80 \text{ V}$$

Because v_4 is the voltage across the parallel combination of the 4-Ω and 6-Ω resistors in the original circuit, v_4 is the same as v_3 in the original problem, and i_3 follows from Ohm's law to be 0.80 A. The voltage across the 4-Ω resistor, v_2, is also 4.8 V, but the polarity marking in Fig. 2.12 is opposite to that of v_4. Thus v_2 is -4.8 V and i_2 is -1.2 A.

Reworking with current dividers. We could have used current dividers instead of voltage dividers. The 2 A passes through R_1, producing 4 V across it, and then divides between the R_2 and R_3. The current through R_3 can be determined directly by current division, as also can be the current through R_2. Notice, however, that the defined reference direction for i_2 is opposite to that of the resulting current and we must account for this with a minus sign. From the currents, we can calculate the voltages and complete the problem. On more complicated circuits, both voltage and current dividers might be used to learn how voltage and current distribute throughout the circuit.

Summary. In this method, we use series and parallel combinations of resistors to simplify a circuit, beginning far away from and coming toward the one source in the circuit. Ultimately, the entire circuit can be reduced to a single resistor across the source, although often the process does not have to be carried that far. The circuit is then restored, voltages and currents being divided progressively to yield the branch currents and voltages.

Advantages of this method. The genius of this method is that you are guided in the solution by the geometry of the circuit. This important information is neglected in our "nonmethod"—you were faced with all those equations and had no such guide. This method also avoids defining unnecessary variables. You merely introduce those you need to perform the voltage and current division. As a final advantage we might suggest that this method, after you use it a while, will develop your intuition about what is happening in circuits. The method directs your attention toward first one part of the circuit, then another; and it never forces you to consider the entire circuit at once.

Weaknesses of this method. Alongside these advantages we must place certain limitations. As we have presented it, the method works for circuits having only one source, although we will soon overcome this limitation. More significantly, this method does not handle the sign for you; you are expected to bring to the problem sufficient insight to know which way the current is positive. For only one source, this is frequently no problem; but you do have to remember to supply the sign and not look for the mathematics to produce it. Also, this method often requires that you analyze the entire circuit, solve for virtually every branch voltage and current, to determine any single unknown. Finally, there are circuits where this method fails.

Figure 2.14 shows such a circuit. Simple though this circuit is, no two resistors are in series or parallel. There is no place to start combining resistors. The next method we present represents a significant extension of the present method, but the revised method cannot handle the circuit in Fig. 2.14 either. However, the rest of the methods in this chapter easily solve the circuit in Fig. 2.14.

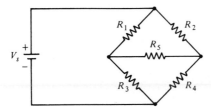

Figure 2.14 Circuit containing no parallel or series connections.

2.2.3 Check Your Understanding

1. Two circuit elements are said to be in parallel if they share the same voltage or current?
2. In a current divider consisting of three resistors, the resistors are in the ratio $1:2:3$. What percent of the total current goes through the largest resistor?
3. What are $2 + 5 \parallel 7$ and $10 \parallel 12 + 7 \parallel (1 + 2)$?
4. Three resistors having the same value, R, are connected together to have an equivalent resistance of $1.5R$. How are they connected?
5. Estimate the current to one 5-W automobile dome light.

Answers. 1. same voltage; 2. 18.2%; 3. 4.92, 7.55; 4. one in series with the other two in parallel; 5. about 0.40 A.

.3 SUPERPOSITION

2.3.1 Superposition Illustrated

Example. The principle of superposition extends the method taught in Section 2.2. Before giving a formal definition, we will illustrate the principle of superposition. The circuit in Fig. 2.15 will serve; we are to solve for v_2. Because

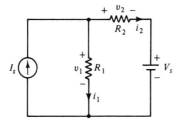

Figure 2.15 Solve for v_2 using KVL and KCL.

the circuit has two sources, we cannot use voltage and current dividers. Thus we will use the nonmethod of Kirchhoff's voltage and current laws, plus Ohm's law. These are

$$\text{Ohm's law:} \quad v_1 = i_1 R_1 \quad \text{and} \quad v_2 = i_2 R_2$$

$$\text{KVL:} \quad -v_1 + v_2 + V_s = 0 \tag{2.14}$$

$$\text{KCL:} \quad -I_s + i_1 + i_2 = 0 \tag{2.15}$$

First we eliminate i_1 and i_2 with Ohm's law, so Eq. (2.15) becomes

$$\frac{v_1}{R_1} + \frac{v_2}{R_2} = I_s \tag{2.16}$$

When we eliminate v_1 between Eqs. (2.14) and (2.16), we obtain the following result:

$$v_2 = \frac{I_s - (V_s/R_1)}{(1/R_1) + (1/R_2)} = I_s(R_1 \parallel R_2) - V_s \times \frac{R_2}{R_1 + R_2} \tag{2.17}$$

Some observations based on the result. The first form of Eq. (2.17) emerges from the algebra; the second form is more easily interpreted. Examination of the result suggests the following:

1. There are two components of v_2, one for each source. One part of v_2 is caused by the current source and the other part is caused by the voltage source.

2. The part of the voltage due to the current source appears to be what would be caused by the current source acting alone, provided the voltage source is replaced by a short circuit. This is true because the two resistors are in parallel if the voltage source is considered a short circuit. By *short circuit* we mean an ideal connection having no voltage across it.

3. The part of the voltage due to the voltage source appears to be what would be caused by the voltage source acting alone, provided the current source is replaced by an open circuit. By *open circuit* we mean that no path exists for current flow. Note that if there is no current flow through the current source, the two resistors are connected in series and the voltage-divider form in Eq. (2.17) is easily identified.

4. The total voltage is the sum of the separate effects of the two sources, provided the polarity of the effects of the sources is considered. In this case

the current source produces a voltage with a polarity the same as the reference direction of v_2 and it appears in the summation with a positive sign. The voltage source produces a voltage with a polarity opposite to that of v_2, and this component appears in the summation with a negative sign.

These observations suggest the general principle of superposition, which we state below.

2.3.2 Principle of Superposition

Principle stated.   The response of a circuit due to multiple sources can be calculated by summing the effects of each source considered separately, all others being turned OFF. By OFF we mean that current sources are replaced by open circuits and voltage sources are replaced by short circuits.

Turned-OFF sources.   The concept of turning OFF a source is important and bears elaboration. We will start with the graphical presentation of the i–v characteristics of a resistor in Fig. 2.16. Because the slope of the resistor characteristic is $1/R$, a zero-resistance line would have an infinite slope. The interpretation is that any amount of current can pass without producing a voltage. On the other hand, the line for an infinite resistance would have zero slope, implying zero current no matter how large the voltage. Notice that in Fig. 2.16 we have called zero resistance a short circuit and infinite resistance an open circuit. The circuit symbol for a short circuit is merely a line, and the circuit symbol for an open circuit is a break in the lines, indicating that there is no path for current. From our earlier definition of a voltage source, we know that a voltage source establishes a certain voltage at its terminals, independent of the current. In Fig. 2.17 this is represented by a vertical line at the voltage of that source. It follows that a voltage source of zero volts would be represented by a vertical line of infinite slope through the origin, which is the same as the characteristic of a short circuit. Thus a turned-OFF voltage source is a short circuit.

Figure 2.18 shows the definition of a current source. This is a horizontal line at the current of the source. A current source of zero value produces a horizontal line through the origin, the same characteristic as an open circuit. Thus a turned-OFF current source is identical to an open circuit.

Figure 2.16   The characteristic of a short circuit is vertical and the characteristic of an open circuit is horizontal.

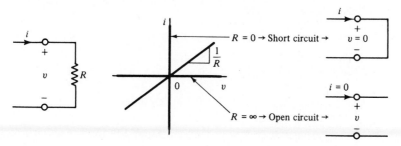

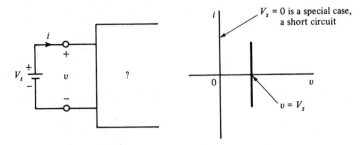

Figure 2.17 The voltage source has a vertical characteristic. Zero voltage ($V_s = 0$) is equivalent to a short circuit.

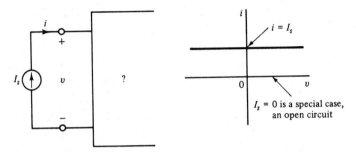

Figure 2.18 The current source has a horizontal characteristic. Zero source current ($I_s = 0$) is equivalent to an open circuit.

Superposition restated. The response of a circuit due to multiple sources can be calculated by summing the effects of each source considered separately, all others being turned OFF. A turned-OFF voltage source is equivalent to a short circuit, and a turned-OFF current source is equivalent to an open circuit.

Another example of superposition. We will solve for the voltage across the 2-Ω resistor in the circuit shown in Fig. 2.19. There are three sources, so we will have to solve three circuits that have a single source. In Fig. 2.20 we calculate the voltage due to the 4-A current source. We have turned OFF the 5-A current

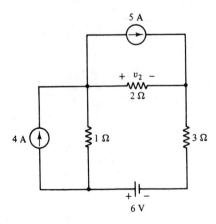

Figure 2.19 Each source will contribute to v_2.

The Analysis of DC Circuits   Chap. 2

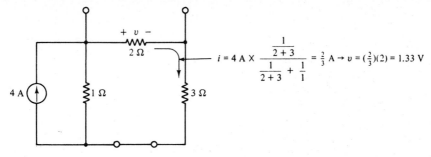

$$i = 4\,A \times \frac{\dfrac{1}{2+3}}{\dfrac{1}{2+3}+\dfrac{1}{1}} = \tfrac{2}{3}\,A \to v = (\tfrac{2}{3})(2) = 1.33\ V$$

Figure 2.20 Contribution of the 4-A source with the other sources OFF.

source, replacing it with an open circuit; and we have turned OFF the 6-V source, replacing it with a short circuit. The resulting circuit shows the 2-Ω and 3-Ω resistors connected in series with each other and together in parallel with the 1-Ω resistor. We use a current divider, Eq. (2.11), and then calculate the voltage across the 2-Ω resistor from Ohm's law. The polarity resulting from the 4-A source is the same as that of the original polarity markings for v_2. Hence this contribution will appear with a + sign in the final summation.

Figure 2.21 shows the circuit with the 4-A and 6-V sources turned OFF. From the perspective of the 5-A source, the 1-Ω and the 3-Ω resistors are in series and together they are in parallel with the 2-Ω resistor. The voltage resulting from the 5-A source is thus calculated from the total equivalent resistance of the circuit as seen by the 5-A source. Here the polarity due to the source is opposite to the reference direction of v_2, so we must use a minus sign in the final summation.

Finally, we calculate in Fig. 2.22 the contribution of the voltage source. With both current sources turned OFF, the three resistors are connected in series and the resulting voltage due to the 6-V source is readily calculated by voltage division. Note that the sign will be positive in the final summation because the polarity of the voltage due to the voltage source is the same as that of v_2. The principle of superposition states that we can now obtain the total voltage across the 2-Ω resistor by adding up the contributions of the three sources:

$$v_2 = +1.33 - 6.67 + 2 = -3.33\ V \tag{2.18}$$

Figure 2.21 Contribution of the 5-A source with the other sources OFF.

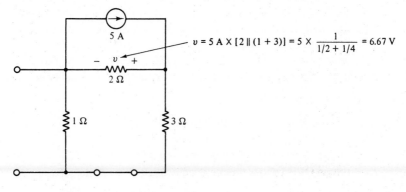

$$v = 5\,A \times [2 \,\|\, (1+3)] = 5 \times \frac{1}{1/2 + 1/4} = 6.67\ V$$

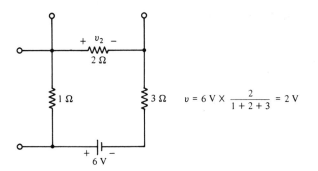

$$v = 6\,\text{V} \times \frac{2}{1 + 2 + 3} = 2\,\text{V}$$

Figure 2.22 Contribution of the voltage source with both current sources OFF.

2.3.3 Limitations of Superposition

Superposition works because Kirchhoff's laws and Ohm's law are linear equations. In all KVL and KCL equations, voltage and current appear in the first power; there are no squares or square roots or functions as e^{kv}. Superposition does not work for direct power calculations, however, because power calculations involve multiplication of voltage by current. Thus you will not get the correct total power by adding the power due to each source considered separately. Superposition can be used indirectly for power calculations, however, because the total current through a resistor can be computed using superposition and this total current can be then used to calculate correctly the power in that resistor. You can confirm for yourself the failure of superposition to compute power correctly by calculating the power in the 2-Ω resistor due to each source acting alone and testing if these powers add up to the true power calculated from the total voltage given in Eq. (2.18). Neither will superposition yield correct answers for circuits containing nonlinear electronic devices such as diodes or transistors.

Superposition is valid for computing the effects of individual sources on the voltages or currents throughout a circuit. As such, it is a useful principle and allows us to extend the method of voltage and current dividers to circuits having multiple sources. Because this method develops intuition and thus aids in design, superposition is often used by electrical engineers.

On the other hand, this method has weaknesses. For one you often have to solve for every voltage and current, perhaps several times, to calculate a single unknown current or voltage. With the simple example we have given, this is fairly easy but in a large circuit this method becomes an ordeal. It would be much better to solve for the unknown without having to calculate every voltage and current along the way. Another weakness is the multiple solutions—much better to solve for the effects of all sources at once. Yet another weakness is that this method is still somewhat unsystematic—many decisions have to be made along the way. The methods to be presented in the remainder of this chapter eliminate these weaknesses.

2.3.4 Check Your Understanding

1. An infinite resistance is equivalent to a turned-OFF voltage or current source?
2. A turned-OFF voltage source is equivalent to a resistor of what value?

3. Does superposition give the correct value for computing the power out of a dc voltage source?

   Answers. 1. Current source; 2. zero ohms; 3. yes, because the current is correct and the voltage is constant.

4 NODE-VOLTAGE ANALYSIS

2.4.1 Basic Idea

Node voltages. The weaknesses of the preceding method are eliminated by the method of node-voltage analysis, or nodal analysis. This method is both efficient and systematic. The basic idea is to write KCL at all nodes except one, but we write these current-law equations in such a way that branch current variables are never formally defined. We avoid defining branch current variables by expressing the currents in terms of the "node voltages" in a special way.

   Figure 2.23 shows a simple circuit with two voltage sources and a resistor. The point at the bottom we have denoted r for *reference node* and the points at the ends of the resistor we have called a and b. Here for clarity we use the double-subscript notation for the current: i_{ab} means the branch current flowing from node a to node b. Using KVL and Ohm's law, we may express i_{ab} thus:

$$-v_{ar} + v_{ab} + v_{br} = 0 \Rightarrow v_{ab} = v_{ar} - v_{br}$$

$$i_{ab} = \frac{v_{ab}}{R_{ab}} = \frac{v_{ar} - v_{br}}{R_{ab}} \tag{2.19}$$

We will change the appearance of Eq. (2.19) by simplifying our notation. First we will drop the r in the voltage variables. The reference node, labeled r, is like an elevation datum in surveying: We measure all elevations (voltages) relative to it. Thus when we speak of the elevation of St. Louis as being 413 ft or at the Salton Sea as being -287 ft, we are talking about the elevation relative to mean sea level, to which we assign an elevation of zero. Similarly, when we talk about the voltage at node a (v_a), we are to understand that we are talking about v_{ar}, the voltage between node a and the reference node, which is thus assigned a voltage of zero. Later we will discuss how to identify the reference node; here we are interested in the pattern of subscripts in Eq. (2.19). With the change in notation, the form becomes

$$i_{ab} = \frac{v_a - v_b}{R_{ab}} \tag{2.20}$$

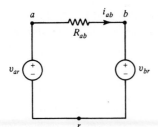

Figure 2.23 The reference node is marked r.

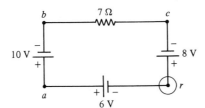

Figure 2.24 Determine the current in the resistor, referenced to the left.

Pattern of subscripts. Equation (2.20) can be stated in words as follows: The branch current from a to b is the voltage at a, minus the voltage at b, divided by the resistance between a and b. (This applies only when there is a resistor between a and b.) The pattern established in Eq. (2.20) is so simple and intuitive that with it we can express branch currents without defining current variables. That is, we can keep the "current from a to b" part in our heads and write on the paper the voltage at a, minus the voltage at b, divided by the resistance between a and b.

We can practice using this pattern by determining the current through the 7-Ω resistor, referenced to the left, in Fig. 2.24. This would be the voltage at c, minus the voltage at b, divided by 7Ω. We may use KVL to determine v_c and v_b. We begin at b, go to r, then to a and back to b: $v_{br} - (6) + (10) = 0$; hence $v_b = -4$ V. To find v_c, we go from c to r (around the battery) and then return through the battery: $v_{cr} + (8) = 0$; or $v_c = -8$ V. So the current toward the left through the 7-Ω resistor would be: $[-8 - (-4)]/7 = -0.57$ A.

2.4.2 Node-Voltage Technique

We will find the voltage across the 3-Ω resistor in Fig. 2.25. We will do this through the following method.

- *Define a reference node.* The circuit has three nodes. We may choose any of them as the reference node. Normally, however, the node with the most wires coming into it is chosen as the reference node. In this case the node at the bottom will be chosen. We mark the reference node with r, as shown in Fig. 2.26.

Figure 2.25 Find the voltage across the 4-Ω resistor.

Figure 2.26 Circuit with labeled nodes. The unknowns are v_a and v_b.

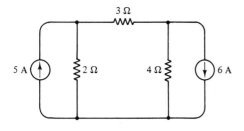

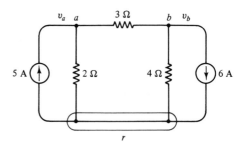

- *Step 2.* Count the independent nodes. Here we have three nodes, the reference node plus two independent nodes. Thus two equations will be written. A full discussion of this step follows on page 45.
- *Label the independent nodes.* We have already labeled the reference node; now we label the other nodes, as shown in Fig. 2.26. The node voltages, v_a and v_b, are our secondary unknowns from which we will calculate the primary unknown, the voltage specified when the problem was defined. In nodal analysis, we first calculate the node voltages; then we calculate from these the specified unknown in the problem, a current or voltage or power of interest.
- *Write KCL in a special form.* The special form is

$$\sum \text{currents leaving the node in } R\text{'s}$$
$$= \sum \text{currents entering the node from sources}$$

Note that we avoid defining current variables by expressing currents with the node voltages:

$$\frac{v_a - (0)}{2} + \frac{v_a - v_b}{3} = +(+5) \tag{2.21}$$

In Eq. (2.21) the left side represents the currents leaving node a in the two resistors connected directly to node a and the right side represents the current entering from the 5-A source. The two terms on the left side come from the two resistors. Note that the current flowing through the 2-Ω resistor to the reference node is merely the node voltage, v_a, divided by the resistance between node a and the reference node. We wrote the (0) for the voltage of the reference node as a reminder. Similarly, we can write KCL for node b in terms of the node voltages:

$$\frac{v_b - v_a}{3} + \frac{v_b - (0)}{4} = -(+6) \tag{2.22}$$

where the first term on the left side represents the current flowing from node b to node a. Note that this is the negative of the second term in Eq. (2.21). This change in sign is appropriate because we are now expressing the current referenced in the opposite direction from before. Note also that the current source term on the right side has a negative sign because we are summing the currents *entering* the node from the sources.

Equations (2.21) and (2.22) are rewritten below.

$$(\tfrac{1}{2} + \tfrac{1}{3})v_a - (\tfrac{1}{3})v_b = 5$$

$$-(\tfrac{1}{3})v_a + (\tfrac{1}{3} + \tfrac{1}{4})v_b = -6$$

We may solve these equations by any method, such as Cramer's rule for determinants, with the result

$$v_a = 2.44 \text{ V} \quad \text{and} \quad v_b = -8.89 \text{ V}$$

- *Compute the primary unknown.* We can now calculate the original unknown from the resulting node voltages. The voltage across the 3-Ω resistor was

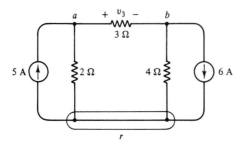

Figure 2.27 Using v_a and v_b to determine v_3.

sought, but we cannot calculate it without a polarity marking. If we wish the + polarity symbol at node a, we have the marking shown in Fig. 2.27. We can determine v_3 by writing KVL around the loop *rabr*:

$$v_{ra} + v_3 + v_{br} = 0$$
$$v_3 = -v_{ra} - v_{br} = v_{ar} - v_{br} = v_a - v_b \tag{2.23}$$

The second line of Eq. (2.23) was converted to a form where r was the second subscript and then r was dropped. Thus v_3 with the polarity marking of Fig. 2.27 is 2.44 − (−8.89) or 11.3 V. Had we marked the + at node b, we would have followed a similar procedure. You can confirm for yourself that this would have reversed the right sides of Eq. (2.23) and resulted in v_3' being −11.3 V, where the primed v_3 has the + at node b.

2.4.3 Some Refinements

How to handle voltage sources.   Nodal analysis deals routinely with current sources. The circuit of Fig. 2.28, however, is currently beyond the reach of the method. In Fig. 2.28 we have identified a reference node and labeled the other two nodes a and b. Look at the circuit carefully and you will likely see what to do with node a. If the reference node is considered to be zero volts, it constrains v_a to be +10 V. To assure you, we will write KVL from r to a and back to r through the voltage source:

$$v_{ra} + 10 = 0 \Rightarrow v_{ar} = v_a = +10 \text{ V}$$

Thus when we write KCL for node b, we treat v_a as known rather than unknown:

$$\frac{v_b - (+10)}{6} + \frac{v_b - (0)}{5} = +(+1)$$

The solution for v_b is 7.27 V.

In Fig. 2.29 we show another circuit with a constrained node. Here we have one node constrained to another independent node rather than to the reference node. We recognize that although the voltages at nodes a and b are unknown, they are not independent. If we knew the voltage at either of them, we could determine the voltage at the other; thus it would be inappropriate to treat them as independent unknowns. We can determine the relationship between them by writing KVL from r to a to b and back to r:

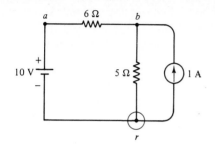

Figure 2.28 Node-voltage problem with a voltage source. The unknown is v_b.

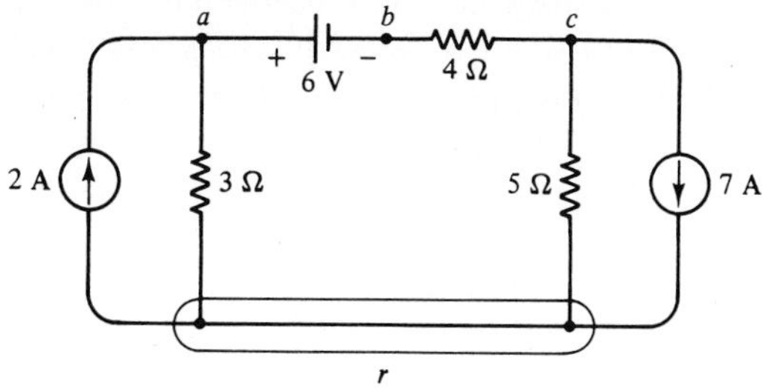

Figure 2.29 Nodes a and b are constrained by the 6-V source.

$$v_{ra} + 6 + v_{br} = 0 \Rightarrow -v_a + 6 + v_b = 0$$
$$v_b = v_a - 6 \tag{2.24}$$

In the second line of Eq. (2.24) we have written the equation with v_a as our unknown and v_b expressed from v_a. Now let us write KCL for node c:

$$\frac{v_c - (0)}{5} + \frac{v_c - (v_a - 6)}{4} = -7$$

In the second term on the left side, we write $(v_a - 6)$ for v_b in expressing the current referenced out of node c in the 4-Ω resistor. When we write KCL for node a, we find that our familiar pattern breaks down when we try to write the current departing node a for b because we would have zero in numerator and denominator. This does not mean that the current is indefinite; it means that the method does not apply. To express the current departing node a for b, we must recognize that this current is identical to the current departing node b for c, which we can express

$$\frac{v_a - (0)}{3} + \frac{(v_a - 6) - v_c}{4} = +2$$

We now have two equations in two unknowns, v_a and v_c, which yield -2.75 V and -20.4 V, respectively. From these we could calculate any primary unknown that was required.

Summary. A voltage source will establish a constraint between two nodes. We can express the node voltage at one end of a voltage source in terms of the node voltage at the other end plus or minus the source value, depending on the polarity of the source. To determine the sign, we may have to write KVL around a loop containing the reference node and the two constrained nodes.

Independent nodes. The preceding paragraph implies that a first count of the nodes of a circuit could overestimate the number of unknown node voltages to be determined. If there are voltage sources, some nodes will be constrained and the number of unknowns (and equations to be solved) will be reduced. We will designate an *independent node* as a node whose voltage cannot be derived from the voltage of another node. When we analyze a circuit using the method of node voltages, we will have as many unknowns (and equations to solve) as we have independent nodes. Here is a rule for counting independent nodes: Turn OFF all sources and count the nodes that remain. The number of independent nodes is one less than the number of remaining nodes. What we mean by "turning OFF" all sources is that voltage sources are replaced by short circuits and current sources are replaced by open circuits.

This concept of an independent node leads us back to the second step in our method for node-voltage analysis. Step 2 is to turn OFF all sources and determine the number of independent nodes. The third step is to label only the independent nodes, because these are the only secondary unknowns.

Practice for step 2.   Let us practice by counting the independent nodes for the circuit in Fig. 2.30. When we turn OFF the sources, all the nodes are tied to the reference node with short circuits except for the node where the 5-Ω and 6-Ω resistors connect. Thus we have only two nodes, one of which is the reference node. We therefore have only one unknown node voltage and one equation to write, KCL at the independent node. See if you can correctly solve for the unknown node voltage (14.6 V).

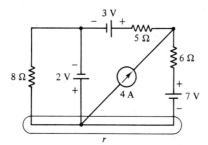

Figure 2.30 Counting the independent nodes is an important step.

But what happended to the 8-Ω resistor? It got shorted out when we turned OFF the 2-V source. This means that the voltage across the 8-Ω resistor is fixed by the 2-V source, independent of what goes on in the rest of the circuit. We can thus compute its voltage, current, and power independent of the remainder of the circuit. The 8-Ω resistor merely *looks* like part of the same circuit. The 2-V source, being an ideal voltage source, acts on the 8-Ω resistor independent of the rest of the circuit.

2.4.4 Critique

A good method.   Nodal analysis is our first really systematic method for analyzing circuits. It can be implemented somewhat routinely and always works. It is probably the favorite method of the electrical engineer for analyzing a circuit, all other factors being equal. Loop currents, the next method we present, also qualifies as a popular and powerful method.

What about KVL?   We might pause to ask: How can we solve a circuit without consideration of KVL? This is certainly what we appear to do when we use the method of node voltages. *Answer:* Kirchhoff's voltage law refers to branch voltages around a loop. Node voltages, on the other hand, are not branch voltages but are all referred to the same point. Hence KVL does not apply directly. If we were to use the node voltages (secondary unknowns) to compute the branch voltages (primary unknowns) around a loop and if we then add up the derived voltages, we would find that they added to zero as required by KVL.

The reference node versus "ground."   When the node voltage method is presented in books and used in practice, the reference node is often called the *ground node* and given the symbol "⏚." Strictly speaking, the ground in an electrical circuit usually identifies the point that is physically connected via a thick wire to the moist earth, usually for safety purposes. We will discuss grounding in Chapter 5; here we only comment on the relationship between the reference node of nodal analysis and the physical ground of an electrical system. The grounded portion of an electrical circuit usually has many wires connecting to it. Because many wires connect to the ground, the electrical ground is often designated the reference node in a nodal analysis. But this is mere coincidence: the reference node and the ground are totally different concepts. We have avoided referring to the reference node as the ground to establish the concept of the reference node without confusion with the concept of electrical grounding. You should be aware, however, that many people use the terms interchangeably when discussing nodal analysis.

Node voltages and electrical potential.   The analogy we made earlier between node voltages and elevations above mean sea level is in fact more than an analogy. Elevation is a measure of gravitational potential, and the node voltages are a measure of the electrical potential of the various nodes in an electrical circuit, relative to the reference node. What we have defined as the voltage between two points can also be called the *potential difference* between the points. Likewise, we can define *potential rises* and *potential drops* in a circuit; for example, the potential rises across a battery (going from − to +) and drops across a resistor (going from + to −). However, we will not speak of potential rises and drops in our development of circuit theory. Our definition of voltage is that of a potential drop.

2.4.5 Check Your Understanding

1. What is the name of the point in a circuit defined to have zero volts?
2. If $v_{rb} = +5$ V, where r is the reference node, what is the voltage at node b?
3. If node a is connected to the reference node by a 40-V voltage source with the − at node a, what is v_a?
4. Two nodes are said to be constrained together when connected by a voltage source or a current source. (Which?)

   Answers.   1. The reference node (not "ground"); 2. −5 V; 3. −40 V; 4. voltage source.

2.5.1 Simple Method of Loop-Current Analysis

Consider the circuit shown in Fig. 2.31; we are to solve for the voltage across the 2-Ω resistor using loop currents. We give step-by-step instructions.

- Label the independent loops 1, 2, 3, . . . and define loop currents i_1, i_2, i_3, . . . going clockwise (or counterclockwise, just be consistent) around the loops. An independent loop is a loop that does not pass through a current source. In this circuit both loops are independent loops and we get the picture of Fig. 2.32.
- Write KVL going with the currents around all loops in a special form:

$$\sum \text{voltages across } R\text{'s} = \sum (+ \text{ or } -) \text{ voltage sources} \qquad (2.25)$$

On the right side of Eq. (2.25) we use $+$ if the voltage aids the loop current for that loop and $-$ if the voltage source opposes the loop current for that loop. That is, if the voltage source were very large, would it force the loop current in that loop to be large and positive or to be large and negative relative to the reference direction? If the source forces its loop current to be positive, use a plus for that source term; if negative, use a minus.
- Solve for the loop currents and from them compute the branch currents and voltages as required. Note that because we write one KVL equation for each loop and we have one loop current for each loop, we always get the same number of equations and unknowns.

We will now follow this plan for the circuit in Fig. 2.31. The loop currents have already been defined in Fig. 2.32. The KVL equation for loop 1 is

$$i_1(1 \ \Omega) + (i_1 - i_2)(2 \ \Omega) = + 7 - (+6) \qquad (2.26)$$

The first term in Eq. (2.26) is the voltage across the 1-Ω resistor. We are going with the current and we are using a load set for the voltage across the resistor; thus we automatically get a $+$ sign for that voltage. The second term is the voltage across the 2-Ω resistor. The current in that resistor is $i_1 - i_2$, the difference between the two loop currents. Because i_1 is referenced down and i_2 is referenced

Figure 2.31 The voltage across the 2-Ω resistor is to be determined.

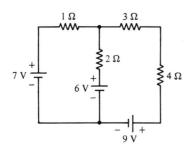

Figure 2.32 Same circuit with loop currents i_1 and i_2 defined.

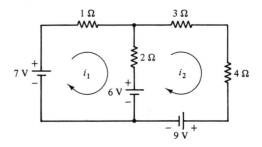

The Analysis of DC Circuits   Chap. 2

up, the current in the resistor is their difference. We are going with i_1, so we write the voltage across that resistor as $+i \times R$, where i is the current referenced in the direction we are going, $i_1 - i_2$. Thus the + sign in the second term is automatic, just like the + sign on the first term. The +7 on the right side is due to the 7-V source. It has a + sign because that source tends to force the loop current in that loop (i_1) in its positive direction. The $-(+6)$ is due to the 6-V source. It has the minus sign outside the parentheses because it opposes the positive flow of i_1.

Of course, there are not two physical currents in the 2-Ω resistor, one going up and the other going down. The loop "currents" are mathematical variables that may or may not be identical to a physical current somewhere in the circuit. In this case i_1 is the physical current in the 1-Ω resistor, i_2 is the physical current in the 3-Ω and 4-Ω resistors, and $i_1 - i_2$ is the physical current referenced downward in the 2-Ω resistor.

The KVL equation for the second loop is

$$i_2(3 \ \Omega) + i_2(4 \ \Omega) + (i_2 - i_1)(2 \ \Omega) = +6 - (+9) \tag{2.27}$$

We are now going with i_2 around loop 2. The first two terms in Eq. (2.27) are due to i_2 going through the 3-Ω and 4-Ω resistors. We automatically get + signs because we are going the same direction as the loop current. When we get to the 2-Ω resistor, we write the current as $i_2 - i_1$ because now we are going up, in the same direction as i_2. This term in Eq. (2.27) is the negative of the corresponding term in Eq. (2.26). The sign is changed because in both cases we are writing voltages across resistors going clockwise: in loop 1 this requires going downward through the 2-Ω resistor and in loop 2 this requires going up.

We now have two equations in two unknowns:

$$(1 + 2)i_1 - (2)i_2 = +1$$

$$-(2)i_1 + (3 + 4 + 2)i_2 = -3$$

Therefore, i_1 is 0.130 A and i_2 is -0.304 A. This result is not the end of the problem, however, for we set out to calculate the voltage across the 2-Ω resistor. We were silent about the reference direction on this voltage because we wanted to consider both possibilities. If we put the + polarity symbol at the top of the 2-Ω resistor, the voltage would be $+(i_1 - i_2)(2 \ \Omega)$ or $[0.130 - (-0.304)](2) = +0.870$ V. If we put the + polarity symbol at the bottom of the 2-Ω resistor, the voltage would be $(i_2 - i_1)(2 \ \Omega)$ or $[(-0.304) - (+0.130)](2) = -0.870$ V.

Example. As a second example, we invite you to write the equations for the three-loop circuit shown in Fig. 2.33. We have already labeled the loop currents for you so that you can check your results with those below. Turn back to the step-by-step instructions on page 48 and follow the directions. The KVL loop equations are

$$2i_1 + 8(i_1 - i_2) + 3(i_1 - i_3) = +10 - 9$$

$$4i_2 + 5(i_2 - i_3) + 8(i_2 - i_1) = -11 \tag{2.28}$$

$$(6 + 7)i_3 + 3(i_3 - i_1) + 5(i_3 - i_2) = +12 + 9$$

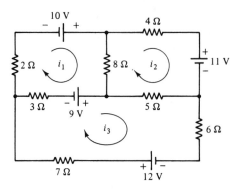

Figure 2.33 Circuit with three loops.

Note the form of the equations. On the left side all the signs outside the current differences are positive because you are writing KVL *going with the loop current* in that loop and because we use load sets for resistors. The only minus signs appear in front of the loop currents in the loops adjacent to the one you are going around. These minus signs occur because all the loop currents are referenced clockwise and thus go opposite directions in the resistors that are common to two loops. Consequently, the signs on the left side exhibit a simple pattern. On the right side of the equations the signs result from the direction of the sources: Do they aid or oppose the loop current of that loop?

Solving three equations in three unknowns is no fun, but you may have a fancy calculator that solves the problem automatically. If so, it ought to give these results: $i_1 = 0.0892$ A, $i_2 = -0.330$ A, and i_3 is 0.934 A.

2.5.2 Some Extensions and Fine Points

How to handle current sources. Perhaps you have noticed that, as it now stands, our loop-current method cannot handle current sources. Current sources require a modest extension of the standard procedure. Consider the circuit in Fig. 2.34. The circuit has a current source; but, not knowing any better, we have defined and labeled our loop currents nevertheless. The current source would constrain the two loop currents passing through it; for by definition of a current source we must require that

$$i_2 - i_1 = 2 \text{ A} \tag{2.29}$$

Having two unknowns, we require another equation. The second equation comes from KVL around the outer loop:

$$5i_1 + 8i_2 - 10 = 0 \tag{2.30}$$

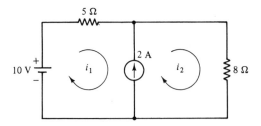

Figure 2.34 The current source constrains i_1 and i_2.

In writing Eq. (2.30) we have departed from the standard procedure in two ways: we went around the outer loop, which does not follow any single loop current; and we wrote KVL with all terms on the left side of the equation, which is the way we originally learned to write KVL equations. We went around the outer loop because we do not know how to handle the current source in a KVL equation. Furthermore, we have already accounted for its effect in Eq. (2.29). Equations (2.29) and (2.30) yield $i_1 = -0.462$ A and $i_2 = 1.538$ A. We note that i_1 is the current in the voltage source and the 5-Ω resistor and i_2 is the current in the 8-Ω resistor. Note that $i_2 - i_1$ is 2 A, as required by the current source.

The solution presented above gives the correct answer but is not the recommended way to handle current sources. We now present two additional solutions that use the concept of a "constrained loop current." Figure 2.35 shows the same circuit with one unknown loop current defined, i_1', and one known loop current of 2 A defined to flow around the right loop. "Wait," you should say, "that current will divide at the top and go both ways." Yes, that is true, but bear with us until we finish and then you will understand this approach. Because we now have only one unknown, we need only one equation, which comes from KVL around the loop of i_1':

$$5i_1' + 8(i_1' + 2) - 10 = 0 \qquad (2.31)$$

In Eq. (2.31) the term for the 8-Ω resistor includes the effect of both the unknown and the constrained loop currents. They add because both are referenced in the same direction through that resistor. The solution of Eq. (2.31) is routine: $i_1' = -0.462$ A. Using this and the constrained loop current, we calculate the current in the 8-Ω resistor to be $i_1' + 2 = 1.538$ A. Thus by using the constrained loop concept to handle the current source, we get the same results for the branch currents as we obtained before with the first, more straightforward method.

The third solution is similar to the second, except now we constrain the 2-A loop current from the source to flow through the 5-Ω resistor and voltage source, as shown in Fig. 2.36. (We realize that we have not yet answered your objection to our authority over the 2-A current. Please bear with us a little longer.) The KVL equation around the loop of i_1'' is

$$5(i_1'' - 2) + 8i_1'' - 10 = 0 \qquad (2.32)$$

Figure 2.35 The loop current i_1' is routed to avoid the current source.

Figure 2.36 Again the loop current misses the current source. The constrained loop current is re-routed.

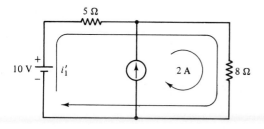

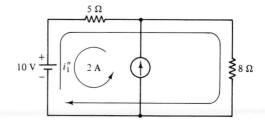

Solution of Eq. (2.32) is routine: $i_1'' = 1.538$ A, and the current referenced left to right in the 5-Ω resistor is $i_1'' - 2 = -0.462$ A. Again we have the same answers for the branch currents.

We have shown how the constrained loop current works, but why does it work? Who are we to make the current from the current source go wherever we wish? The answer is: We are the ones defining variables in the problem. We are defining variables by numbering loops and by drawing currents going clockwise around these loops; and when there are current sources we are defining variables by assigning the paths by which those currents are defined to flow. Look at Figs. 2.35 and 2.36. The unknown loop current appears to be the same in both, but by changing the path of the 2-A current, we changed the meaning of the unknown loop current. Thus we can handle current sources by defining their currents to flow in certain paths, modifying our interpretation of the unknown loop currents accordingly.

Counting independent loops. The first step in our standard procedure was to number the independent loops. We identified independent loops by turning OFF all sources, as we did when we wished to identify independent nodes in the method of node voltages. For example, when we turn OFF both sources in Fig. 2.34, the current source becomes an open circuit and the voltage source becomes a short circuit. We are left with one loop containing two resistors. Thus we have one independent loop, requiring one unknown and one KVL equation.

Loop currents and mesh currents. What we have been using up to now are mesh currents, a special class of loop currents. In circuit terminology, a loop is any closed path. A *mesh* is a special loop, namely, the smallest loop one can have. A mesh is thus a loop that contains no other loops. In the fuller sense of the loop-current method, we can define the loop currents with great freedom, allowing them to go wherever we wish within certain guidelines. For our relatively simple circuits, the guidelines are that we must define the correct number of currents and that we must go through each resistor with at least one loop current.

We will rework the circuit in Fig. 2.31 as an example of this more general loop-current method. The circuit is redrawn in Fig. 2.37 with true loop† currents drawn as shown. Notice that we define one current clockwise and the other counterclockwise. We write KVL around the loops of the unknown currents with the following results:

$$-7 + 1(i_1' - i_2') + 2i_1' + 6 = 0$$

$$+7 - 9 + (3 + 4)i_2' + 1(i_2' - i_1') = 0$$

Therefore, $i_1' = 0.435$ A and $i_2' = 0.304$ A. Because i_1' is the only current through the 2-Ω resistor, we could have solved for it only in order to compute the voltage across that resistor, the original unknown. The result is $0.435 \times 2 = 0.870$, + at top.

† Actually i_1' is also a mesh current but is called a loop current here because it is used in the more general method.

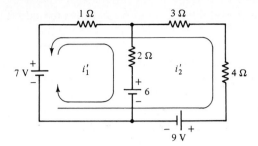

Figure 2.37 These are true loop currents, not mesh currents.

The more general loop-current method is useful when we wish to determine only one branch current or voltage, because we can define all the unknown loop currents to avoid that branch except one loop current. The resulting equations can then be solved for that one loop current, which will be the desired branch current. The trouble with the generalized loop-current method is that all the nice symmetries and automatic sign patterns of the mesh-current method vanish, and once again we are required to pay careful attention to the signs.

How we can ignore KCL. With the loop-current method we solve for the currents of a circuit by writing only KVL equations. How can we ignore KCL? *Answer:* We can ignore KCL because we defined the currents to flow in complete loops rather than from one point to another in the circuit. In Fig. 2.38 we show two loop currents flowing through a node. If we wrote KCL for such a node, each loop current would contribute two equal and opposite terms: each loop current must add to zero at every node. Thus KCL is satisfied automatically when we use loop currents to describe the circuit; we have only to satisfy KVL to find the solution.

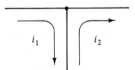

Figure 2.38 Node with two loop currents.

2.5.3 Summary of the Methods Presented So Far

Which method to use? We now have three methods for solving circuit problems: the method of current and voltage dividers, the method of node voltages, and the method of loop currents. How can we select the best one to use on a given circuit? Here are some of the factors to consider:

• One important consideration is: Are you trying to analyze or design the circuit? If the circuit is completely specified and we are trying to determine some aspect of its response, say, the power out of a source or the voltage across some resistor, the nodal or loop methods are favored. These are general methods of analysis, which solve for the entire circuit response all at once. Special attention is not given toward the effect of a single resistor or source; rather, everything is incorporated into the equations at the beginning.

On the other hand, if you are designing the circuit in some regard, determining a resistor here or there to yield a desired result, the method of voltage and current dividers is favored. This method focuses on the effects of individual resistors and sources at specific points in the circuit. Design must, of course, deal in such details and this method is well suited for allowing the designer to control the way voltage and current distribute throughout a circuit.

- Another important consideration is: How many equations must be solved? It is possible for a circuit to have more independent nodes than loops, or vice versa. The circuit shown in Fig. 2.39, for example, has two independent loops but only one independent node. Thus you would favor nodal analysis for simpler mathematics. The solution, by the way, is 1.73 A.

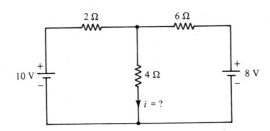

Figure 2.39 The circuit has two independent loops but only one independent node.

- Finally, there are numerous minor considerations that would suggest a method if those discussed above fail to dictate a choice. If there are many current sources, nodal analysis is favored, but if many voltage sources, loops might be easier. If the unknown is a voltage, nodes might be best, but if the unknown is a current, loops might be more efficient. If there is only one source and the circuit is not too complicated, the method of voltage and current dividers is favored. These decisions come easily as a result of much experience in circuit analysis. As a beginner, you will have to practice the various methods on a number of circuits before you develop the ability to choose the most efficient method.

2.5.4 Check Your Understanding

1. How many independent loops are there in the circuit in Fig. 2.14?
2. What is the name of a loop current passing through a current source?
3. If a circuit has five independent nodes and two independent loops, what method of analysis does this suggest?
4. In the standard method for mesh (loop) current analysis, what is the coefficient of the mesh current in the equation for that mesh?

Answers. 1. Three independent loops; 2. a constrained loop current; 3. loop currents (two equations in two unknowns); 4. the sum of the resistors around its mesh.

2.6.1 Example to Justify the Concept

We will analyze the circuit shown in Fig. 2.40 by a method that you will surely think strange. We are to determine the current in the variable resistor R as a function of that resistance. Our method uses an equivalent circuit first proposed by a French telegrapher named Thévenin. Thévenin's equivalent circuit leads to one of the most useful ideas of electrical engineering, namely, the idea of the "output resistance" of a circuit. This concept influences the thinking of electrical engineers in much of the work they do. We will now justify the method as we analyze the circuit in Fig. 2.40. Later we will distill the approach into a simple procedure.

- Remove the resistor. Yes, the first step is to dismantle the circuit you are trying to analyze. In the lab, you could literally cut out the resistor; on paper, you merely erase it or redraw the circuit without it. The resulting circuit is shown in Fig. 2.41.
- Measure (in the lab) or calculate (on paper) the open circuit voltage between a and b, v_{ab}. We call v_{ab} the *open-circuit voltage* because this voltage appears between a and b with the original resistor gone, that is, with the circuit "open" between a and b. This is a good opportunity for you to calculate v_{ab} by a method of your choosing. If you get 16 V, you are correct.
- Connect to b, but not to a, a voltage source equal to the open-circuit voltage (a 16-V battery in this case), as shown in Fig. 2.42. The voltage across the gap, $v_{aa'}$, is now zero. If you do not believe that, write KVL around a to a' to b back to a and you ought to get zero for $v_{aa'}$.

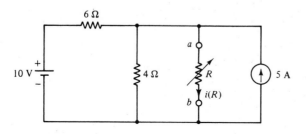

Figure 2.40 Solve for the current in R as a function of R, $i(R)$.

Figure 2.41 The open-circuit voltage is $v_{ab} = 16$ V. **Figure 2.42** With the 16-V source inserted, $v_{aa'} = 0$ V.

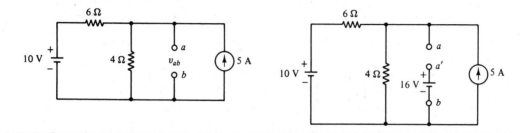

- Now connect the original resistor between a and a'. Because there is no voltage across the gap, replacing the resistor will disturb nothing and the voltage across the resistor will remain at zero. Hence *no current will exist in the resistor*. This is an important conclusion and deserves careful attention. By replacing the resistor we have restored our original circuit, except that now we have inserted a voltage source in series with the resistor. That voltage source has a polarity to oppose the flow of current through the resistor and has the exact magnitude to prevent any current from flowing. We might say that it "bucks out" the current in the resistor. In hydraulics, it would be like inserting a certrifugal pump to stop fluid flow.

Now consider superposition. Normally, we would calculate the total current in the modified circuit as the combined results of all three sources, but this time we will distinguish between the original sources and the inserted source, as indicated in the equation

$$i_{total} = i_{original} - i_{inserted} = 0 \qquad (2.33)$$

In Eq. (2.33) $i_{original}$ stands for the current through R from the original sources in the circuit, the 10-V battery and the 5-A current source; and $i_{inserted}$ stands for the current due to the open-circuit voltage source that we inserted, a 16-V battery in this case. The minus sign comes from the polarity with which we inserted the battery, namely, so as to reduce the current to zero.

The two components must be equal. The component due to the original sources is what we set out to calculate in the first place. Because this is equal to the current due to the inserted source, we reason that we can calculate the current due to one source instead of calculating the current due to multiple sources (two in this case).

When we use superposition, we turn all the sources OFF except the source we are considering at the moment. Hence we calculate the effect of the inserted source by turning OFF the original sources. This produces the circuit shown in Fig. 2.43. Now R is seen to be connected in series with an equivalent resistance of $R_{eq} = 4 \parallel 6 = 2.4 \ \Omega$ and the resulting current is easily seen to be

$$i(R) = \frac{v_{open\ circuit}}{R + R_{eq}} = \frac{16}{R + 2.4} \qquad (2.34)$$

Equation (2.34) gives the current in R as a function of R, which was what we set out to find. Because this was an easy problem to begin with, you may wonder why we solved it by this roundabout method. The point is this: We would have derived the same simple equation as Eq. (2.34) no matter how complicated the

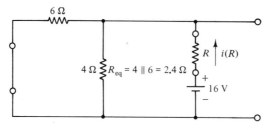

Figure 2.43 With the original sources OFF, the circuit is reduced to an equivalent resistance.

The Analysis of DC Circuits   Chap. 2

original circuit. There could have been hundreds of sources and thousands of resistors in the original circuit and we still would have reduced the circuit to two parameters, an open-circuit voltage and an equivalent resistance.

2.6.2 Thévenin's Equivalent Circuit

Basic concept. Equation (2.34) suggests the simple equivalent circuit shown in Fig. 2.44. The Thévenin equivalent circuit consists of a voltage source, V_T, in series with an equivalent resistance. The magnitude and polarity of V_T are identical to the open-circuit voltage at a-b, the terminals of the resistor of interest. The equivalent resistor (R_{eq}) is computed at the load terminals with all sources in the circuit turned OFF.

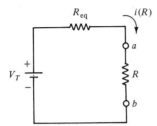

Figure 2.44 Equivalent circuit suggested by Eq. (2.34).

Example. Let us solve another problem using Thévenin's equivalent circuit, this time skipping the justifying steps. The circuit in Fig. 2.45 is drawn with everything in a box except the resistor of special interest, which is usually called the "load." We wish to calculate the voltage across that load resistor. We will replace the circuit in the box with the simpler circuit shown in Fig. 2.46. First we calculate the Thévenin voltage source (V_T), the open-circuit voltage between terminals x and y with R removed. Because the circuit has only one loop with R gone, we can employ the method of loop currents to find the current in the 20-Ω resistor and hence the required voltage. This analysis shows V_T to be 15 V with the + polarity symbol at terminal x. The polarity is important because we require that the two circuits in the boxes in Figs. 2.45 and 2.46 be fully equivalent to the load. This requires that the polarity in the Thévenin equivalent circuit be identical to that in the original circuit. If, for example, the open-circuit voltage had come

Figure 2.45 Replace the circuit in the box by a Thévenin equivalent circuit.

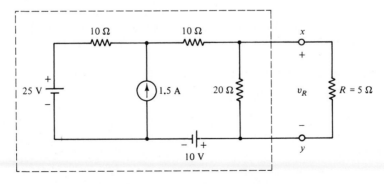

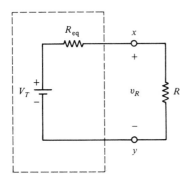

Figure 2.46 The Thévenin equivalent circuit consists of the open-circuit voltage in series with the equivalent resistance.

out -15 V with the $+$ polarity symbol at terminal x, we would have turned V_T around in Fig. 2.46 or else put x at the bottom and y at the top.

The other step is to compute R_{eq}, the equivalent resistance. To make this computation, we turn OFF all three sources within the box, as shown in Fig. 2.47. Thus with $R_{eq} = 10\ \Omega$, we now have the Thévenin equivalent circuit shown in Fig. 2.48. The solution to our original problem is easy: v_R is $+5$ V. Of course, with modest effort we could have computed this directly from the original circuit in Fig. 2.45. With this method, however, we gain the freedom to ask many additional questions such as: What value of R makes the voltage 20 V? or what value of R withdraws the most power from the circuit? These questions lead to mathematical complexities with the original circuit but can be answered simply with the equivalent circuit. The answer to the first question is that no value of R will give 20 V. The investigation of the second question leads to an interesting and important result, to which we will soon turn.

First, let us consider further what we mean by an equivalent circuit. Thévenin's equivalent circuit replaces the circuit within the box only for effects *external* to the box. We can no longer ask questions about what goes on in the box after we have replaced it by an equivalent circuit. For example, if we are interested in the current in the 20-Ω resistor or the total power consumed by the resistors in the box, the equivalent circuit is useless.

Maximum power transfer. Let us now investigate the question of maximizing the power in R. This is a straightforward problem in differential calculus. The power in R as a function of R, $P(R)$, is

Figure 2.47 Turning OFF the three sources leaves a series–parallel combination.

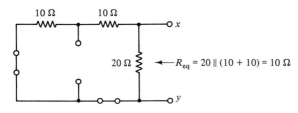

$$R_{eq} = 20 \parallel (10 + 10) = 10\ \Omega$$

Figure 2.48 Thévenin equivalent circuit for the circuit in Fig. 2.45.

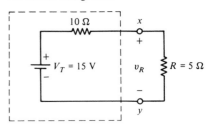

The Analysis of DC Circuits   Chap. 2

$$P(R) = i^2R = \frac{V_T^2 R}{(R + R_{eq})^2} = \frac{(15)^2 R}{(R + 10)^2} \qquad (2.35)$$

To maximize $P(R)$ as given in Eq. (2.35), we take the derivative with respect to R and set it equal to zero. You can confirm that the maximum (or minimum) occurs at R equals 10 Ω. To confirm that we have a maximum and not a minimum, a second derivative can be taken, but a better way in this case is to make a simple sketch of the function to show that we have found a maximum. Figure 2.49 shows such a sketch reflecting the result of our computation. Note that the maximum power is $V_T^2/4R_{eq}$, which is the value of $P(R_{eq})$ in Eq. (2.35).

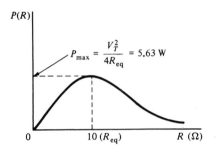

Figure 2.49 The power in R is maximum when $R = R_{eq}$.

Importance of maximum power transfer. The maximum power out of a circuit is important because often we deal in electronics with small amounts of power and wish to make full use of the power that is available. On a TV set, for example, we pull out the "rabbit ears" antenna to receive power from radio waves originating at a transmitter many miles away. The antenna does not collect much power, so the TV receiver is designed to make maximum use of the power provided by the antenna. Although our results were derived for a simple battery and resistor, they can be applied to a TV antenna. Our results show that we should design the receiver input circuit, represented here by a load resistor, to have a special value for withdrawing the maximum power from the antenna. In general, the equivalent resistance of a source of power is called its *output resistance* because it is the resistance the source presents to a load.

We might point out that the maximum power that can be extracted from a circuit is given by $V_T^2/4R_{eq}$, as shown in Fig. 2.49. This is known as the available power from the source, the "source" being in this case the entire circuit represented by the equivalent circuit. If the source has a low output resistance, it can supply much power to an external load.

2.6.3 Norton's Equivalent Circuit

An American engineer named Norton came up with a similar equivalent circuit. Norton's equivalent circuit consists of a current source connected in parallel with an equivalent resistance, as shown in Fig. 2.50. The derivation of the values for the resistance and the current source, I_N, is similar to that for Thévenin's circuit and will not be repeated here. Indeed, the equivalent resistance is the same as before, namely, the resistance of the circuit presented to the load after all internal

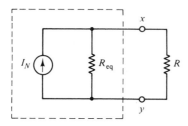

Figure 2.50 The Norton equivalent circuit appears in the box.

sources are turned OFF. The Norton current source has a magnitude identical to the current that would flow in a short circuit of the output terminals.

As an example, we will find the Norton equivalent circuit for the circuit in Fig. 2.45. The value of R_{eq} is the same as before, 10 Ω. The value of the Norton current source, I_N, can be determined by replacing the load with a short circuit. This gives the circuit in Fig. 2.51. The short circuit effectively removes the 20-Ω resistor from the circuit because it forces its voltage, and hence its current, to be zero. The current flowing through the short circuit is easily calculated. The result is 1.5 A in the short circuit; hence the Norton equivalent circuit is as shown in Fig. 2.52. From this simple circuit, v_R is seen to be 1.5 × 10 ∥ 5 or +5 V, as before. Notice that the polarity of the equivalent source must produce the current in the proper direction through the load.

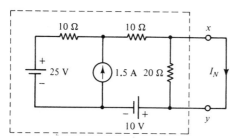

Figure 2.51 Short the output to find I_N. The 20-Ω resistor is effectively removed.

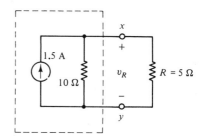

Figure 2.52 Norton equivalent circuit.

2.6.4 Relationship between Thévenin's and Norton's Equivalent Circuits

If two circuits are equivalent to the same circuit, they must be equivalent to each other. If the Norton's circuit in Fig. 2.52 is open-circuited, the voltage must be the same as the voltage source in the Thévenin circuit of Fig. 2.48. You can see that this proves true in this case, as it must be in all. In general, it must be true that

$$V_T = I_N R_{eq} \tag{2.36}$$

Equation (2.36) is sometimes useful in theoretical work, but it also can be applied in the laboratory to find the output resistance of a source. That is, we may be unable to get inside and turn OFF the internal sources, particularly in an electronic circuit, but we can measure the open-circuit output voltage and the current that

The Analysis of DC Circuits   Chap. 2

flows when the output terminals are shorted. From the measured voltage and current, we can compute the output resistance from the equation

$$R_{eq} = \frac{V_T}{I_N} \tag{2.37}$$

Loading of a circuit. The output resistance of a circuit indicates the power-producing capabilities of the circuit. We also are concerned about the *loading* of circuits, as illustrated by Fig. 2.53a. Here we show a voltage divider creating a voltage v and a voltmeter to measure the voltage. We assume that the voltmeter has an input resistance of 10 MΩ, and hence it will change the voltage to be measured. We may determine the loaded voltage, v', from the Thévenin equivalent circuit in Fig. 2.53b. The loaded voltage proves to be 3% lower than the unloaded voltage, which is the Thévenin voltage of 3.85 V. In practice, we can correct for loading error if we know the output resistance of the circuit and the input resistance of the meter.

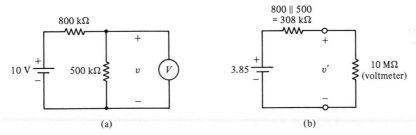

(a)   (b)

Figure 2.53 (a) The amount of loading from the meter depends on the output resistance of the circuit at the point of measurement and the resistance of the meter; (b) Thévenin equivalent circuit showing the effect of meter loading.

2.6.5 Source Transformations

Transformations between Thévenin's and Norton's equivalent circuits can be used to handle "wrong" kinds of sources in nodal and loop analysis. In the nodal analysis of the circuit in Fig. 2.54a, for example, the branch from a to c can be

Figure 2.54 (a) The voltage source is inconvenient for nodal analysis (Fig. 2.29 repeated); (b) after the branch is transformed to a Norton equivalent circuit, the nodal analysis is routine.

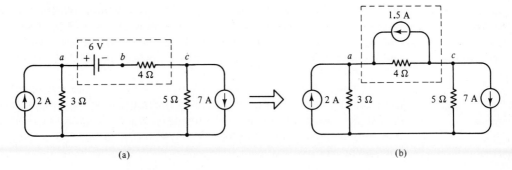

(a)   (b)

converted to a Norton circuit to convert the 6-V source to an equivalent current source, as shown in Fig. 2.54b. Notice that the circuits in the boxes are equivalent. With the transformed circuit, the node voltage equations can be written without using constrained nodes.

In closing this section, we wish to compare the method of equivalent circuits with our first method of voltage and current dividers. With voltage and current dividers, the strategy was to combine resistors until we have represented the entire circuit to the *source* as a single equivalent resistance. We then restore the original circuit, dividing voltage and current as we go. Eventually, we can calculate the current or voltage across a particular resistor of interest. If there are multiple sources, we must repeat this process for each source.

The method of equivalent circuits works in the opposite direction. Here we represent the entire circuit, including multiple sources, to the *load* of interest. We bring everything, as it were, to a particular point in the circuit that has special importance. For this reason, the method of equivalent circuits plays an important role in both design and analysis of electrical circuits.

2.6.6 Check Your Understanding

1. A 1.5-V dry cell has a maximum current of 300 mA. What is the internal resistance of the battery?
2. A circuit has an output voltage of 20 V and a short-circuit current of 0.5 A. What is the maximum power that can come out of this circuit?
3. Normally we adjust a load for maximum power in an electronics or a power circuit?
4. A circuit has a variable load, R. Power measurements show that the power into R is maximum at 10 W with $R = 2 \Omega$. What would be the open-circuit voltage, V_T?

Answers. 1. 5Ω; 2. 2.5 W; 3. in electronics—power circuits want constant voltage; 4. 8.94 V.

2.7 TIME-VARYING AND DEPENDENT SOURCES

2.7.1 Time-Varying Sources

Many of the sources in electrical circuits vary with time. Typically, constant sources (dc) find application in electronic circuits and automotive electrical systems; but most power circuits use alternating current (ac), and information signals must be time varying to carry information. Because Kirchhoff's laws and Ohm's law are independent of time, all equations and techniques presented in this chapter can be used for resistive circuits with time-varying sources. This statement includes the power relationships, provided that these are understood as instantaneous power as a function of time, not time-average power.

In Chapter 3 we introduce capacitors and inductors, whose definitions, unlike resistors, contain time derivatives. These call for new concepts and techniques of analysis; nevertheless, many of the concepts of this chapter remain valid and relevant, for example, superposition, nodal and loop analysis, and Thévenin equivalent circuits.

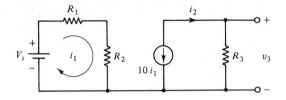

Figure 2.55 Circuit containing a dependent source.

2.7.2 Dependent Sources

Frequently, in modeling a physical device with a circuit, we must use a voltage or current source whose strength is controlled by a voltage or current somewhere in the circuit distant from the source. Figure 2.55 shows such a circuit (actually, two connected circuits, for the single wire can carry no current and permits no voltage difference). The current source in the right part of the circuit is shown to be controlled by the current in the left part. For example, if $i_1 = 1$ mA, then $i_2 = -10$ mA. The current source is a *dependent* source, specifically, a current-controlled current source.

Controlled sources appear mainly in electronics. When we listen to a telephone, for example, we rightly suppose that the current in the handset is controlled by a distant voice. "Well," you may object, "the telepone wires coming to the telephone are part of a circuit—there is no mystery here." True, but this "circuit" may include a telephone in London, a microwave repeater satellite midway over the Atlantic, and an interstate transmission system. For purposes of circuit analysis, such a "circuit" is best modeled with controlled sources.

The four possible types of controlled sources are summarized in Table 2.1. For the voltage source we use the general symbol for a voltage source rather than the battery symbol. The units of the K's are useful in checking dimensional consistency of results. In Fig. 2.55, for example, $K_1 = 10$ and is dimensionless. The

TABLE 2.1 DEFINING EQUATIONS AND SYMBOLS FOR THE FOUR POSSIBLE TYPES OF DEPENDENT SOURCES

Controlling Quantity	Controlled Source	Equation	Units of K	Symbol
Current, i_c	Current, i_s	$i_s = K_1 i_c$	None	
Current, i_c	Voltage, v_s	$v_s = K_2 i_c$	Ohms	
Voltage, v_c	Current, i_s	$i_s = K_3 v_c$	Siemens (mhos)	
Voltage, v_c	Voltage, v_s	$v_s = K_4 v_c$	None	

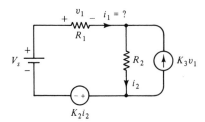

Figure 2.56 Circuit with two dependent sources. The interaction between sources and variables makes this circuit difficult.

circuit in Fig. 2.55 is typical of many in electronics because it can be analyzed in stages: First the left part is solved and then the right part. The analysis is easy because the right part has no influence on the left part. More difficult are problems like that shown in Fig. 2.56, where interaction occurs between the dependent sources and their controlling variables. This circuit has one independent source (the battery), and two dependent sources. The current source depends on the voltage across R_1, and the voltage source is controlled by the current in R_2. We are required to solve for i_1. In Section 2.7.3 we apply the various methods presented in this chapter. The discussion focuses on the modifications required by the presence of the dependent sources.

2.7.3 Analyzing Circuits with Dependent Sources

The nonmethod.   Straightforward application of KVL, KCL, and Ohm's law must be successful on such circuits. There are four unknowns, and one can write two equations for the resistors, one KVL equation for the loop, and one KCL equation for either of the nodes. Solution of the resulting equations would be straightforward.

Voltage and current dividers; series and parallel combinations.   This method is of limited value in problems involving controlled sources. Certainly, we can combine series or parallel resistors, provided that *none of the controlling variables disappear* in the process. But probably the circuit cannot in this way be much simplified. Furthermore, superposition can be applied to the *independent* sources but cannot be applied to the controlled sources. The controlled sources have in part the character of resistors, as suggested by K_2 in Table 2.1 having the dimensions of ohms. In short, this method offers many disadvantages and no advantages when the circuit contains controlled sources.

Node voltages and loop currents.   Used with care, nodal and loop analysis can be applied to circuits with dependent sources. As an example, we will analyze the circuit in Fig. 2.56 by using node voltages.

The first step is to determine the number of independent nodes. For this purpose, *all* sources are turned OFF. For our circuit, turning OFF all three sources yields a circuit with one loop and two nodes, one of which is an independent node. Thus we have one unknown. In Fig. 2.57 we choose a reference node and define v_a as our secondary unknown. (Recall that we are to solve for i_1.) We write KCL at node a. To write the current departing node a through the current-con-

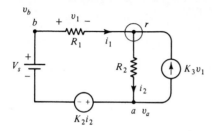

Figure 2.57 Analysis by the nodal technique.

trolled voltage source, we must use Ohm's law on R_1. For this purpose we have identified node b, which is constrained to node a by two sources. The constraint is given by the equation

$$v_b = v_a - K_2 i_2 + V_s \qquad (2.38)$$

To eliminate i_2, we note that $i_2 = -v_a/R_2$. Thus the current departing node a and passing through R_1 is

$$i_1 = \frac{v_a - K_2(-v_a/R_2) + V_s - (0)}{R_1}$$

$$= \frac{v_a(1 + K_2/R_2) + V_s}{R_1} \qquad (2.39)$$

We now can write KCL at node a:

$$\frac{v_a - (0)}{R_2} + \frac{v_a(1 + K_2/R_2) + V_s}{R_1} = -K_3\left[v_a\left(1 + \frac{K_2}{R_2}\right) + V_s\right] \qquad (2.40)$$

Note that we had to substitute the simplified version of Eq. (2.38) for circuit $v_1(= v_b)$ as the controlling variable for the current source. The solution of Eq. (2.40) follows from

$$v_a\left[\frac{1}{R_2} + \frac{1 + K_2/R_2}{R_1} + K_3\left(1 + \frac{K_2}{R_2}\right)\right] = -\frac{V_s}{R_1} - K_3 V_s \qquad (2.41)$$

Now that we have solved for our secondary unknown, we must further manipulate the equations to solve for i_1, our primary unknown. We leave this for a homework problem. At the risk of understatement, we note that one must in such problems proceed with care because the dependent sources introduce unexpected inter-actions between circuit variables.

For this circuit, loop current analysis works out better than nodal analysis because the unknown is a current. We leave this for a homework problem.

Thévenin and Norton equivalent circuits.   Equivalent circuits can be used to analyze circuits with dependent sources, again with new cautions and complexities. In Fig. 2.58 we show the circuit of Fig. 2.56 with R_1 removed for the open-circuit calculation. We can determine $V_T(= v_1)$ by writing KVL around the opened loop

$$-V_s + V_T + R_2(K_3 V_T) + K_2(K_3 V_T) = 0 \qquad (2.42)$$

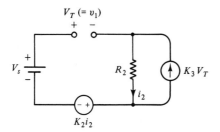

$V_T (= v_1)$

V_s

R_2

$K_3 V_T$

i_2

$K_2 i_2$

Figure 2.58 Open-circuit voltage calculation.

Note that we can write i_2 as $K_3 V_T$ because, with R_1 removed, the current from the dependent current source must pass totally through R_2. Equation (2.42) can be solved for V_T.

$$V_T = \frac{V_s}{1 + R_2 K_3 + K_2 K_3} \qquad (2.43)$$

We cannot combine resistors in series or parallel to determine the output resistance, R_{eq}. This approach fails because we must leave the dependent sources turned ON, since they affect the output resistance. Here we have two possible approaches. One is to leave all sources ON, short the output to calculate I_N, and compute the output resistance from Eq. (2.37). The other approach is to turn OFF the independent source(s) in the circuit (V_s in this case) while leaving the dependent sources operating, excite the circuit at its output with an external source (say, connect a 1-V voltage source or a 1-A current source), and compute the circuit response at the output. If, for example, we excite the circuit with a 1-A current source and the resulting voltage is 12 V, the output resistance of the circuit would be 12 Ω.

We will use the first approach and leave the second for a homework problem. Figure 2.59 shows the circuit with the output shorted. Note that this eliminates the dependent current source. Thus I_N is identical to i_2 and can be determined by the application of KVL around the loop:

$$-V_s + R_2 I_N + K_2 I_N = 0$$

Therefore, the Norton current is

$$I_N = \frac{V_s}{R_2 + K_2} \qquad (2.44)$$

Finally, we may compute the equivalent output resistance from the ratio, $R_{eq} = V_T / I_N$:

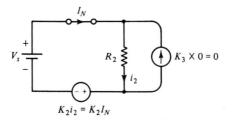

I_N

V_s

R_2

$K_3 \times 0 = 0$

i_2

$K_2 i_2 = K_2 I_N$

Figure 2.59 Short-circuit current calculation. Note that the shorting of the output effectively removes one of the dependent sources.

The Analysis of DC Circuits   Chap. 2

$$R_{eq} = \frac{R_2 + K_2}{1 + R_2 K_3 + K_2 K_3} \qquad (2.45)$$

We leave for a homework problem the verification of the dimensional correctness of this result.

From the values of V_T and R_{eq}, we derive the Thévenin equivalent circuit in Fig. 2.60, from which the calculation of i_1 is straightforward.

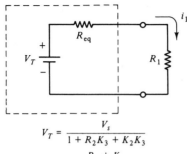

$$V_T = \frac{V_s}{1 + R_2 K_3 + K_2 K_3}$$

$$R_{eq} = \frac{R_2 + K_2}{1 + R_2 K_3 + K_2 K_3}$$

Figure 2.60   Thévenin equivalent circuit. We find R_{eq} from $R_{eq} = V_T/I_N$.

Summary.   We may use Thévenin and Norton equivalent circuit concepts for circuits with dependent sources. However, we must leave ON the dependent sources for the calculation of the equivalent-circuit parameters. Such calculations may require careful consideration.

PROBLEMS

Section 2.1: DC Circuit Theory and Ohm's Law

P2.1. A resistor capable of handling safely 2 W is to be placed across the terminals of a 12.6-V battery. What range of resistances will not exceed the 2-W limit?

Answer. R must be greater than or equal to 79.4 Ω

P2.2. A car radio designed to operate from a 6.3-V system uses 2 A of current. What resistance should be placed in series with this radio if it is to be used in a 12.6-V system? What should be the power rating of this resistor?

P2.3. For the circuit shown in Fig. P2.3, use the nonmethod shown in Section 2.1.3 to solve for v_5, i_7, and i_{20}. Compute the power out of each source and into each resistor, and show that power is conserved.

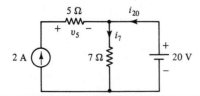

Figure P2.3

P2.4. Using the nonmethod described in Section 2.1.3, solve for all voltages and currents in the circuit in Fig. P2.4. The polarity directions have been assigned arbitrarily. Bear in mind as you proceed that this method is being displayed as a standard to make attractive the more efficient methods of the remainder of this chapter.

Answers. (currents only) $i_1 = -4$ A, $i_2 = 0.8$ A, $i_3 = 0.8$ A, and $i_4 = 4.8$ A

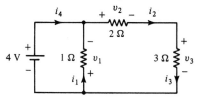

Figure P2.4

Section 2.2: Series and Parallel Resistors; Voltage and Current Dividers

P2.5. **(a)** Three resistors connected in series have resistance values in the ratio 1:2:3 and combine to an equivalent resistance of 120 Ω. What is the smallest resistor?
(b) The three resistors, still in series, are placed across a 300-V dc voltage source. What is the voltage that will appear across the largest resistor?
(c) Two series resistors are to work as a voltage divider, with the smaller getting 30% of the total voltage. What are the resistors, given that their combined resistance is to be 200 Ω?

Answers. (a) 20 Ω; (b) 150 V; (c) 60 Ω, 140 Ω

P2.6. For the circuit shown in Fig. P2.6, the resistor R is variable; hence the voltage across R, $v(R)$, will depend on the value of R, as the notation indicates. Determine $v(R)$ and, from that, the value of R to make the voltage 5.5 V.

Answers. $R = 8.46$ Ω

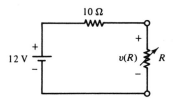

Figure P2.6 The arrow means that R can vary.

P2.7. **(a)** Find the equivalent resistance of the parallel combination shown in Fig. P2.7.
(b) If 10 A enters the parallel combination, referenced in at a and out at b, what is the current referenced downward in the 5-Ω resistor? What is the current referenced upward in the 15-Ω resistor? What is the voltage, v_{ab}?

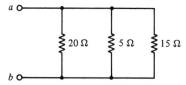

Figure P2.7

P2.8. For the circuit shown in Fig. P2.8:
(a) Find R such that $i_R = 0.5$ A.
(b) Find R (a different R) such that the power in R is 50 W.

Answers. (a) 300 Ω; (b) 17.2 Ω or 583 Ω

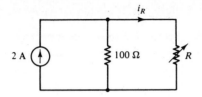

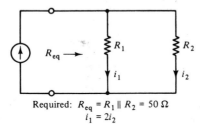

2 A · 100 Ω · i_R · R

Figure P2.8 The arrow means that R can vary.

P2.9. Design a current divider that has an equivalent resistance of 50 Ω and divides the current in the ratio of 2:1. This problem is summarized in Fig. P2.9.

R_{eq} → · R_1 · i_1 · R_2 · i_2

Required: $R_{eq} = R_1 \| R_2 = 50$ Ω
$i_1 = 2i_2$

Figure P2.9

P2.10. (a) Evaluate $R_{eq} = 7 \| (5 + 6 \| 8 + 1 \| 2)$.
   (b) Draw the circuit corresponding to this expression.
   (c) What resistance in parallel with 90 Ω reduces the equivalent parallel resistance to 70 Ω?

Answers. (a) 3.96;   (c) 315 Ω

P2.11. For the circuits shown in Fig. P2.11, solve for the indicated unknown using voltage- and current-divider techniques.

Answers. (a) $i = 5.38$ A, $v = 46.2$ V;   (b) $i = 0.182$ A, $v = 3.15$ V;
(c) $i = -2.76$ A;   (d) $P_{out} = -20$ W, $v_{10} = -2.22$ V

Figure P2.11

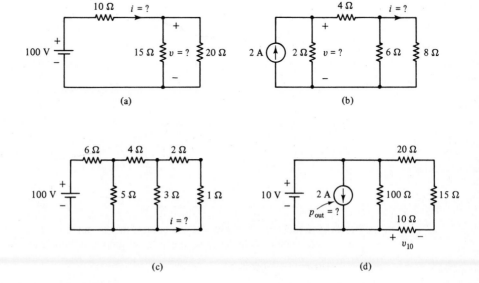

(a)

(b)

(c)

(d)

P2.12. For the circuits in Fig. P2.12, find the indicated unknowns using voltage and current dividers.

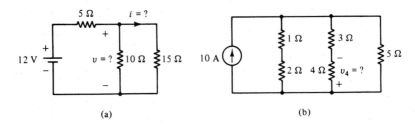

(a)   (b)

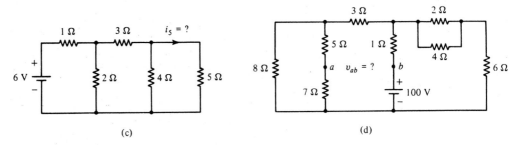

(c)   (d)

Figure P2.12

Section 2.3: Superposition

P2.13. Find the current in the 100-Ω resistor in Fig. P2.13 using superposition.

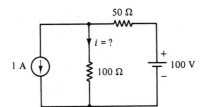

Figure P2.13

P2.14. In Fig. P2.14, find i using the principle of superposition.

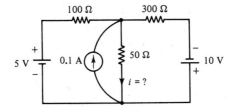

Figure P2.14

P2.15. For the circuit in Fig. P2.15, determine the current in the 10-Ω resistor with the reference direction shown using the principle of superposition.

Answer.   +0.667 A

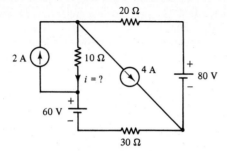

Figure P2.15

Section 2.4: Node-Voltage Analysis

P2.16. For the circuit shown in Fig. P2.16, solve for v_a and v_b using node-voltage techniques. Current dividers would provide an easy check on your results.

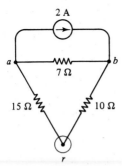

Figure P2.16

P2.17. For the circuit shown in Fig. P2.17, write the KCL equation for node b using the node-voltage patterns and solve for $v_{br} = v_b$. Check your solution using the voltage-divider method.

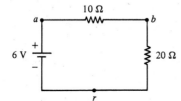

Figure P2.17

P2.18. For the circuit in Fig. P2.18:
(a) Write KVL to show $v_{br} = v_b = +5$ V and $v_{cr} = v_c = -10$ V.
(b) Find the current downward in the 50-Ω resistor by first solving for v_a using nodal analysis.
Hint: Nodes b and c are constrained to the reference node.

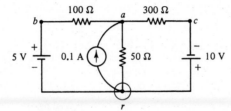

Figure P2.18

P2.19. Using node-voltage analysis, solve for the indicated unknowns in Fig. P2.19. *Hints:* In parts (a) and (b), note that kΩ, mA, and volts make a consistent set of units; in part (c), find v_a, then v_{10} from the voltage divider.

Answers. (a) 3.33 V; (b) $v_{20} = +22.5$ V; (c) $i = 1.40$ A, $v_{10} = 14.0$ V

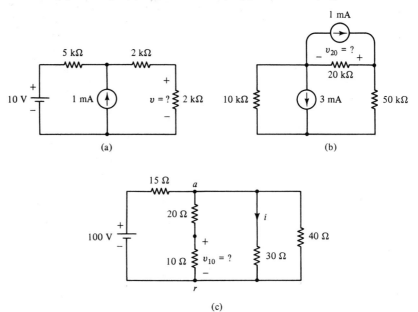

(a)

(b)

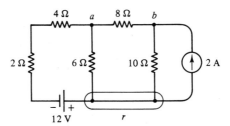

(c)

Figure P2.19

P2.20. For Fig. P2.20, write the nodal equations, using the notation given. Do not solve the equations.

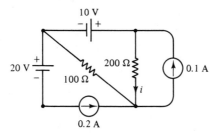

Figure P2.20

P2.21. For Fig. P2.21, determine the current in the 200-Ω resistor with the reference direction given. Use nodal analysis.

Figure P2.21

Section 2.5: Loop-Current Analysis

P2.22. Solve for all branch currents in the circuit in Fig. P2.22, using loop current analysis.

Answer. $i_4 = +1$ A upward

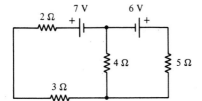

Figure P2.22

P2.23. Find the power out of the 10-V source in Fig. P2.23 using loop currents to analyze the circuit.

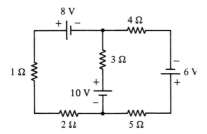

Figure P2.23

P2.24. Solve for the current referenced downward in the 100-Ω resistor in the circuit in Fig. P2.13 using the constrained loop concept to handle the current source.

P2.25. Solve for i in Fig. 2.25 using the loop-current method.

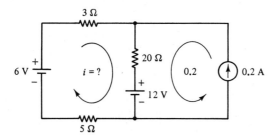

Figure P2.25

P2.26. For the circuit in Fig. P2.26, solve for i_4 using mesh-current variables to analyze the circuit. Note that you may direct the constrained loops to miss the 4-Ω resistor. Be sure to count independent loops before defining variables.

Answer. $i_4 = -1.33$ A

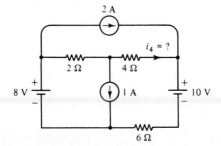

Figure P2.26

P2.27. Solve for the power out of the 80-V source in the circuit in Fig. P2.27, using loop-current variables. Remember to count independent loops first.

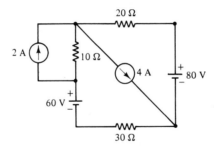

Figure P2.27

P2.28. True loop (not mesh) currents are defined in Fig. P2.28 satisfying the rules given in Section 2.5.2. Write the KVL equations for the circuit using these loop variables. You are not required to solve the resulting equations.

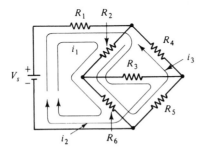

Figure P2.28

Section 2.6: Thévenin's and Norton's Equivalent Circuits

P2.29. Place a connection (a short circuit, no resistor) between a and a' in Fig. 2.42. Solve for the current flowing down from a to a' by the most efficient method. The answer should be zero current, as argued in Section 2.6.1.

P2.30. Develop a Thévenin equivalent circuit for the part of the circuit shown in the box in Fig. P2.30. Use this equivalent circuit to solve for i, as shown. *Hint:* After you have removed the load, use nodal analysis with v_{ar} your only unknown and the 3-Ω and 4-Ω resistors in series, then determine v_{br} (which is V_T) with a voltage divider.

Answers. $V_T = 0.889$ V (+ at y), $R_{eq} = 2.22\ \Omega$, $i = -0.123$ A

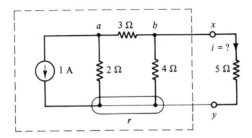

Figure P2.30

P2.31. For the circuit in Fig. P2.31:

(a) Replace the circuit in the box by a Thévenin equivalent circuit. Solve for V_T by the most efficient method.

(b) Find v_{ab} for $R = 3\ \Omega$.

(c) What value of R receives maximum power from the circuit?

(d) What values of R will receive 15 W from the circuit?

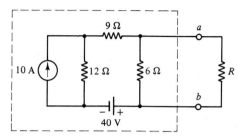

Figure P2.31

P2.32. A black box with a circuit in it was connected to a variable resistor and the power in the resistor was measured as the resistance was varied. The results are shown in Fig. P2.32. From this graph, determine the Thévenin equivalent circuit for the circuit in the black box.

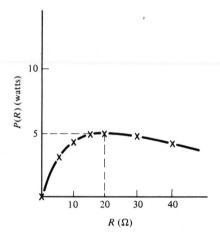

Figure P2.32

P2.33. Rework Problem P2.30, this time using a Norton equivalent circuit for the portion of the circuit in the box.

P2.34. A mysterious black box is found in the electrical engineering lab. A curious student measured the output voltage to be 12.6 V. Then he shorted the output through an ammeter (consider the ammeter to have zero resistance) and read a current of 56 A. Give the Norton equivalent circuit for the box. How much power can be gotten out of the box if a variable resistor is connected to its terminals and adjusted for maximum power?

Answers. $R_{eq} = 0.225\ \Omega$, $P_{available} = 176$ W

P2.35. For the circuit shown in Fig. P2.35:

    (a) Find the value of R to receive maximum power.

    (b) For the value of R in part (a), find the power out of the 12-V source.

Answer. 46.9 W

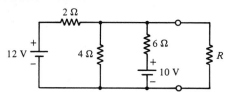

Figure P2.35

P2.36. The starter motor on a car draws 80-A starting current, which lowers the battery voltage from 12.6 to 8.8 V. What would be the battery voltage if it were being charged at 30 A?

P2.37. For the circuit shown in Fig. P2.37:

    (a) Find the current in the 1-Ω resistor.

    (b) What resistance, replacing the 1-Ω resistor, would draw one-half the current in part (a)?

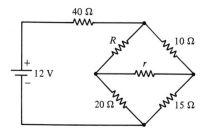

Figure P2.37

P2.38. For the circuit shown in Fig. P2.38:

    (a) What value of resistance, R, will reduce the current through r to zero?

    (b) For the value of R in part (a), what is the current through the voltage source?

Answers. (a) 13.33 Ω, (b) 0.221 A.

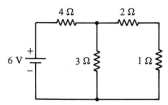

Figure P2.38

P2.39. A circuit has a variable load, R. Power measurements show that the power into R is maximum at 100 W with $R = 6 \ \Omega$. What current would flow if R were replaced by a short circuit?

P2.40. What value I_s reduces the circuit in Fig. P2.40 to a simple resistor from the viewpoint of the output terminals?

Answer. 2 A

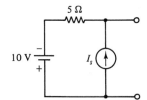

Figure P2.40

Section 2.7: Time-varying and Dependent Sources

P2.41. (a) Using the results in Eq. (2.41) solve for i_1 in Fig. 2.57.
(b) Show that the result in part (a) is dimensionally correct.

P2.42. Using loop-current analysis, solve for i_1 in Fig. 2.56.

P2.43. Connect a 1-A current source to the open circuit in Fig. 2.58 (V_s OFF) and compute the resulting voltage. This would be the value of R_{eq} and hence should agree with Eq. (2.45).

P2.44. Show that Eq. (2.45) is dimensionally correct.

General Problems for Chapter 2

P2.45. For Fig. P2.45, find R such that the power into the 200-Ω resistor is 12.5 W.

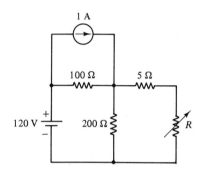

Figure P2.45

P2.46. A circuit, as shown in Fig. P2.46a, has a variable load and meters to monitor load voltage and current. The table in Fig. P2.46b shows partial results of a series of tests. Fill in the blank spaces in the table with the missing data.

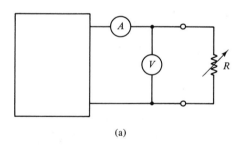

(a)

V	I	R
	0.1 A	0
12 V		300 Ω
		∞

(b)

Figure P2.46

P2.47. A 16-Ω loudspeaker draws maximum power from the output of its amplifier, which is capable of producing 25 W in the speaker. What would be the power produced in an 8-Ω speaker? Represent the loudspeakers as resistors of 16 Ω and 8 Ω, respectively.

P2.48. An electric stove (dc or ac, it does not matter, because the same power formulas apply) requires 230 V for the line voltage. The stove uses two heater elements that can be switched in one at a time or placed in series or parallel, making four heating temperatures. If the highest setting requires 2000 W and the lowest 444 W, what are the powers for the two intermediate settings?

P2.49. A voltmeter must draw some current from the circuit it is measuring in order to operate (see Fig. P2.49). The amount of current it draws depends on the meter scale and is specified in terms of the "ohms per volt" of the meter. For example, a 10-kΩ/V meter would have a resistance of 10 kΩ on the 1-V scale, 100 kΩ on the 10-V scale, and so on. For this problem, assume that we are measuring with a 10-kΩ/V meter.

   (a) If we measure 5 V on the 10-V scale, what current does the meter draw from the circuit?

   (b) If the output resistance of the circuit is 600 Ω and we measure 5 V on the 10-V scale, what is the true open-circuit voltage of the source, that is, what would be measured by an ideal meter that drew no current from the circuit?

   (c) With our 10-kΩ/V meter, on an unknown circuit we measure 24 V on the 30-V scale but 30 V on the 100-V scale. Explain the reason for this apparent discrepancy and determine from these measurements the Thévenin equivalent circuit for the source we are measuring.

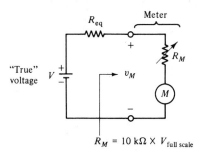

$R_M = 10\ \text{k}\Omega \times V_{\text{full scale}}$

v_M = voltage indicated   **Figure P2.49**

P2.50. A potentiometer or "pot" is a variable resistor connected so as to produce a voltage that is adjustable. This is the device commonly used, for example, as a volume control in a radio. The model of a pot is given in Fig. P2.50. Determine the Thévenin equivalent circuit and make a graph of $V_T(x)$ and $R_{\text{eq}}(x)$ versus x for $R = 10\ \text{k}\Omega$ and $V_s = 100$ V.

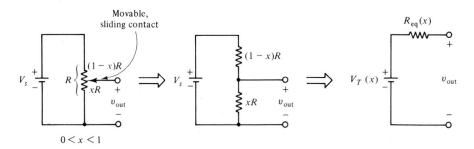

Figure P2.50 Thévenin equivalent circuit for a potentiometer. Note that the output voltage and output resistance depend upon the setting of the potentiometer, x.

P2.51. The ladder network shown in Fig. P2.51, if terminated with the proper value of R_t, has the property that the input current is divided by 2 at each node, as shown. What should be the value of R_t for this property? What would be the input resistance to the ladder if the R–$2R$ pattern were continued infinitely?

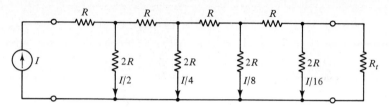

Figure P2.51

P2.52. After Norton died and went to heaven, he chanced to encounter Thévenin one day. They got into a discussion about whose equivalent circuit was better. To maintain peace, they proposed the "Thevenort" circuit shown in Fig. P2.52.

  (a) Give values of V_h, I_h, and R_h that correspond to an open-circuit voltage (V_{oc}) of 5 V and a short-circuit current (I_{sc}) of 2 A. (The answer may not be unique.)

  (b) Give general relationships that relate V_h, I_h, and R_h to the open-circuit voltage (V_{oc}) and the short-circuit current (I_{sc}).

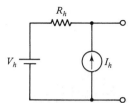

Figure P2.52 "Thevenort" circuit.

3 The Dynamics of Circuits

3.1 THEORY OF INDUCTORS AND CAPACITORS

3.1.1 Time and Energy

Statics and dynamics. In mechanics, dynamics is usually taught after statics. Statics deals with the distribution of forces in a structure; time is not a factor. Dynamics deals with the exchange of energy between components in a system, and time is an important factor because energy cannot be exchanged except as a time process.

In our study of electrical circuits, we have not yet considered time as an important quantity. Our dc circuits involve only KVL, KCL, and Ohm's law, and none of these equations have time as a factor. Indeed, even if we had allowed one of our voltage or current sources to have an output that varied with time, the solution would not become more complicated. It would be like letting the force in a statics problem vary slowly with time: the method of solution would be valid provided energy exchanges between components of the system remain small. In a true dynamics problem, the rate of energy transfer between components must be considered.

With this chapter we begin the study of electrical circuits in which rates of energy exchanged between circuit components are an important factor. We begin by introducing the two circuit components that store energy in electric circuits. The presence of inductors or capacitors in an electric circuit suggests a true dynamics problem. We first will identify the two types of energy that may be stored in a circuit.

Electric energy and magnetic energy storage. To understand what we mean by magnetic energy and electric energy, let us reconsider Eq. (1.2), which describes the form of the vector force between two charges:

$$\vec{F} = \vec{F}_e(Q_1, Q_2, R) + \vec{F}_m(Q_1, \vec{u}_1, Q_2, \vec{u}_2, R)$$

Equation (1.2) describes the force Q_2 experiences in the presence of Q_1, the charges being a distance R apart and having velocities $\vec{u}_1$ and $\vec{u}_2$. When we first presented Eq. (1.2) we called attention to the two types of force, one positional and the other motional. Here we wish to emphasize energy. From mechanics, you recall that when a displacement is made against a force, work is done (mechanical energy is exchanged). Similarly, if we displace a charge in the presence of a motional (magnetic) force ($\vec{F}_m$), magnetic energy is exchanged. To store magnetic energy, we must bring many moving charges close together. This is what an inductor does. On the other hand, when we move charges in the presence of positional (electrostatic) forces ($\vec{F}_e$), electric energy is exchanged. To store electric energy, we must separate charges, yet keep them close together. This is what a capacitor does.

Analogy between mechanical and electrical energy. Magnetic and electric energy are the two forms of electrical energy, arising from the two types of electrical force. In a mechanical system we also have two types of force and two types of energy. Forces that depend on position, as in a spring or a gravitational field, store potential energy. Forces that depend on changes in velocity lead to kinetic energy. Potential energy and kinetic energy in a mechanical system are analogous to electric energy and magnetic energy in an electrical system. We must emphasize, however, that this is only an analogy. Magnetic energy is not kinetic energy; it is merely analogous to kinetic energy. We will continue to identify analogies between mechanical and electrical systems because your understanding of mechanics can help you understand the behavior of electrical systems.

3.1.2 Inductor Basics

Physical inductor. In Fig. 3.1 we show a coil of wire, which will act as an inductor. When current flows in the wire, moving charges are close together and hence magnetic energy is stored.

Figure **3.1** A coil of wire gets moving charges close together and acts as an inductor.

Circuit-theory definition of an inductor. Figure 3.2 shows the circuit symbol for an inductor. The equations describing the voltage–current characteristics of an inductor are based on Faraday's law of induction and are given in

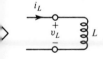

Figure **3.2** Circuit model for an inductor.

$$v_L(t) = +\frac{d}{dt}[Li_L(t)] \quad \text{and} \quad i_L(t) = i_L(0) + \frac{1}{L}\int_0^t v_L(t')\,dt' \qquad (3.1)$$

The circuit symbol and the accompanying equations together define a circuit-theory model for an inductor. Normally, the inductance is considered a constant and brought outside the derivative. Chapter 12 discusses the calculation of the inductance of a coil. Note that v_L and i_L form a load set.

Analogy with Newton's law.   The differential form of Eq. (3.1) is analogous to Newton's second law. Just as changes in the velocity of a mass require a force, or vice versa, changes in the current through an inductor produce a voltage, or vice versa. If the current is increased (positive di/dt), the inductor voltage opposes the change by being numerically positive with the polarity shown. If the current is decreased (negative di/dt), the inductor acts momentarily as a source trying to keep the current going. Thus the physical polarity of the inductor voltage will tend to keep the current constant, just as a mass will tend to maintain constant velocity.

The integral form of Eq. (3.1) is analogous to determining the velocity by integrating the acceleration, which is proportional to an exciting force. We note that only changes in current can be determined from the voltage across the inductor, just as only changes in velocity can be computed from the force acting on a mass. To determine the current fully, we need to know the inductance, the voltage as a function of time, and the initial current, $i_L(0)$.

Stored magnetic energy in an inductor.   We pointed out that a load set of v–i variables is used to define the equation of an inductor. In Eq. (3.2) we compute the energy stored in the inductor by integrating the input power, $+v_L i_L$.

$$p_L = \frac{dW_m}{dt} = +v_L i_L = \left(L \frac{di_L}{dt} \right) i_L = \frac{d}{dt}\left(\tfrac{1}{2} L i_L^2 \right) \tag{3.2}$$

$$W_m = \int p_L\, dt = \int \frac{d}{dt}\left(\tfrac{1}{2} L i_L^2 \right) dt = \tfrac{1}{2} L i_L^2 \tag{3.3}$$

where W_m is the stored magnetic energy in the inductor and p_L is the power into the inductor. The constant of integration is set to zero because there is no stored energy when the current is zero. The result in Eq. (3.3) is analogous to the kinetic energy in a moving mass, with inductance playing the role of mass and current playing the role of velocity.

Units of inductance.   The inductance, L, depends on coil size and the number of turns of wire in the coil. The inductance also depends on the material (if any) located near the coil. In particular, the magnetic properties of iron increase greatly the inductance of a coil wound on an iron core. The units of inductance are volt-seconds per ampere, but to honor Joseph Henry (1797–1878) we use the name "henry" (H) for this unit; millihenries (10^{-3} H or mH) and microhenries (10^{-6} H or μH) are also in common use.

Example.   We will use the definition of an inductor to compute the voltage across, the power into, and the energy stored in a 2-H inductor with current shown in Fig. 3.3. Because the current function is piecewise continuous, we cannot represent it by a single mathematical formula but rather must represent it sepa-

The Dynamics of Circuits   Chap. 3

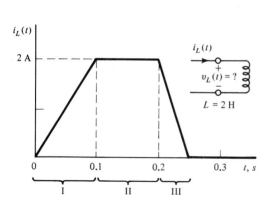

Figure 3.3 A voltage will result from the changing current.

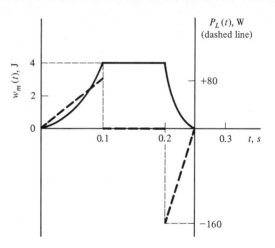

Figure 3.4 The power curve (dashed line) is the slope of the stored energy.

rately in its several regions. During interval I, $0 < t < 0.1$, the slope is constant at 20 A/s; thus during this period the inductor voltage will be 2 H $\times$ $(+20$ A/s) or $+40$ V. During interval II, $0.1 < t < 0.2$, the slope is zero and hence the voltage will be zero also. In interval III, $0.2 < t < 0.25$, the slope is -40 A/s, and the inductor voltage will be -80 V.

When we compute the power into the inductor as the product of v_L and i_L, we note that the power is positive during interval I, zero during II, and negative during III. The magnetic energy stored in the inductor can be computed by integrating the input power, but the easier way uses Eq. (3.3), with the results shown in Fig. 3.4. During interval I, when the current is increasing in magnitude, we must supply power to the inductor to increase the stored magnetic energy. During interval II, the current is constant and hence no energy is exchanged between the inductor and source: the system is "coasting" like a mass in constant motion. During interval III, the inductor acts as a source, giving energy back to the circuit. The net energy exchanged is zero, for the ideal inductor is lossless; that is, it does not convert electrical energy to nonelectrical form. The wire in a physical inductor would have resistance and would not give back all the energy delivered to it. Some of the energy would heat up the wire and be lost to the electrical circuit. This energy is not lost to the entire physical system because it appears as thermal energy. The circuit model for a real coil of wire, shown in Fig. 3.5, is more complicated than a pure inductance because the model must account for loss as well as the storage of magnetic energy.

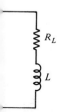

gure 3.5 The circuit odel for a coil of wire ust contain a resistance to account for ss.

3.1.3 Capacitor Basics

Physical capacitor. Electric (electrostatic) forces arise from interactions between separated charges, and the associated energy is called electric energy. To store electric energy, we must separate charges, as in Fig. 3.6.

Sec. 3.1   Theory of Inductors and Capacitors

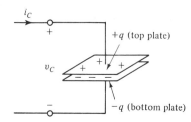

Figure 3.6 Structure having capacitance.

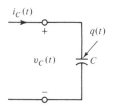

Figure 3.7 Circuit symbol for capacitance.

Circuit–theory definition of a capacitor. Capacitance is defined as the constant relating charge and voltage in a structure that supports a charge separation. If q is the charge on the $+$ side of the capacitor and the voltage is v_C, as shown in Fig. 3.7, the capacitance C is defined as

$$q = Cv_C \qquad (3.4)$$

Chapter 12 discusses the physics of capacitance. For every charge arriving at the $+$ side of the capacitor, a charge of like sign will depart from the $-$ side and the structure as a whole will remain charge-neutral. Thus KCL will be obeyed because the charge flowing into the $+$ terminal side must flow out of the $-$ terminal, as for a resistor. If current is positive into the $+$ terminal, positive charge will accumulate there and will be increasing in proportion to the current. From the definition of current we can relate the charge in Eq. (3.4) and current as follows:

$$i_C = \frac{dq(t)}{dt} \qquad (3.5)$$

where $q(t)$ is the charge in the $+$ side of the capacitor. We may thus define the relationship between current and voltage as

$$i_C = \frac{d}{dt}(Cv_C) \qquad (3.6)$$

Note that we have again used a load set for the voltage and current variables.

Units of capacitance. The capacitance, C, appears in Eq. (3.6) as a constant relating the current to the derivative of the voltage across the capacitor. The units of capacitance can be derived from fundamentals, but to honor Michael Faraday (1791–1867) we use the name "farad" (F). Realistic capacitor values come small and usually are specified in microfarads (10^{-6} F or μF), nanofarads (10^{-9} F or nF), or picofarads (10^{-12} F or pF). When a capacitor is constructed from parallel plates, as in Fig. 3.6, the capacitance depends on the area, separation, and material (if any) lying between the plates.

Figure 3.8 Mechancial analog for capacitance.

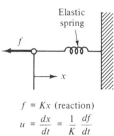

$f = Kx$ (reaction)

$$u = \frac{dx}{dt} = \frac{1}{K}\frac{df}{dt}$$

Mechanical analog of capacitance. The mechanical analog of a capacitor is a spring, as shown in Fig. 3.8. The analog of velocity is current and thus displacement of the spring is analogous to charge. The voltage across a capacitor is analogous to the force produced by the spring. Comparison of the equation in Fig. 3.8 with Eq. (3.6) shows that capacitance corresponds to the inverse of the stiffness constant, K, and is thus analogous to the compliance of a spring.

The Dynamics of Circuits   Chap. 3

Integral *i–v* equation. Equation (3.6) is useful in determining the current through a capacitor, given the voltage as a function of time. If we know the current and wish to determine the voltage, we must integrate. Let us consider that we know the capacitor voltage at some time, say, $t = 0$, and wish to determine the voltage at a later time, t. We can integrate Eq. (3.6) from 0 to t, with the result

$$v_C(t) = v_C(0) + \frac{1}{C} \int_0^t i_C(t')dt' \tag{3.7}$$

Note that we have used t' as the dummy variable of the integration process because t is one limit of the integral.

Example. As an example of the use of Eq. (3.7), consider a capacitor with current as shown in Fig. 3.9. Here we know the voltage at the beginning time, $v_C(0) = +5$ V, and the current through the capacitor. We wish to compute the voltage for all $t > 0$. We will integrate, noting that we must divide the integration into several regions due to the piecewise nature of the current. During interval I, $0 < t < 1$ ms, the current is constant at -75 mA. Because the initial voltage is $+5$ V, the voltage is

$$v_C(t) = +5 + \frac{1}{10 \ \mu\text{F}} \int_0^t -75 \times 10^{-3} \ dt' = +5 - 7500t, \qquad \text{I: } 0 < t < 1 \text{ ms}$$

During interval I, the voltage starts at $+5$ V and decreases linearly, reaching -2.5 V at $t = 1$ ms, the end of interval I. At this time we must change formulas for i_C and start integrating again. Our initial value is now -2.5 V and the current is

$$i_C(t) = -0.150 + 75t \text{ A}, \qquad \text{II: } 1 < t < 3 \text{ ms}$$

Thus, during interval II, the voltage will be

$$v_C(t) = -2.5 + \frac{1}{10 \ \mu\text{F}} \int_{1 \text{ ms}}^t (-0.150 + 75t') \ dt', \qquad \text{II: } 1 < t < 3 \text{ ms}$$

Figure 3.9 (a) The capacitor has an initial voltage; (b) current into the capacitor for positive time.

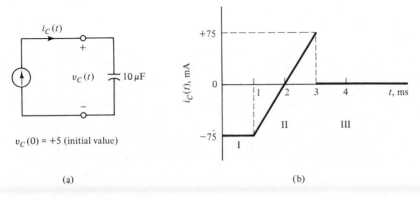

(a)

(b)

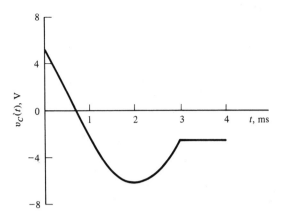

Figure 3.10 Capacitor voltage for positive time.

As shown in Fig. 3.10, the voltage continues in the negative direction until it reaches -6.25 V at $t = 2$ ms, and then increases until it reaches -2.5 V again at the end of interval II.

During interval III, which starts at $t = 3$ ms and continues indefinitely, the current is zero, indicating that the charge on the capacitor is constant. Hence the voltage will remain at -2.5 V. This constant voltage applies for an ideal capacitor; a physical capacitor would discharge eventually due to leakage current.

The voltage curve in Fig. 3.10 can be related to the physical processes in the capacitor. The negative current represents a discharging of the capacitor and the voltage decreases accordingly. At the moment when the voltage is zero, the capacitor is totally discharged, but the negative current continues and an excess of negative charge begins to accumulate on the + side (excess positive on the − side). This continues into interval II, although at a slower rate, until the time in the middle of interval II when the current is zero. At this time the + side has its maximum negative charge and the charge is temporarily constant because the current is zero. During the second half of interval II, the current is positive and the negative charge on the + side begins to diminish, resulting in a decrease of the negative voltage between the + side and the − side. At the end of interval II, the current stops and the voltage remains unchanged thereafter.

Mechanical analogy. The process of integrating current through a capacitor to determine voltage is analogous to integrating velocity to determine displacement in a mechanics problem. The only difference is that the value of the capacitance scales the integral of the current, whereas the scale factor is unity in the mechanical analog.

Stored energy in a capacitor. The charge separation in a capacitor stores electric energy. This energy is analogous to potential energy stored in a stressed spring. We may derive the stored electric energy in a capacitor by integrating the power into the capacitor, as shown in

$$W_e = \int p_C \, dt = \int v_C \times C \frac{dv_C}{dt} \, dt = \int \frac{d}{dt} \left(\tfrac{1}{2} C v_C^2\right) dt = \tfrac{1}{2} C v_C^2 \qquad (3.8)$$

where p_C is the power into the capacitor and W_e is the stored electric energy in the capacitor. The constant of integration must be zero because the uncharged capacitor stores no energy. We see from Eq. (3.8) that the stored energy depends uniquely on the voltage (or the charge) and the capacitance. For example, if we take a 10-μF capacitor and connect it briefly to a 12.6-V battery, the capacitor will receive $\frac{1}{2} \times 10^{-5}(12.6)^2 = 7.94 \times 10^{-4}$ J from the battery, which it will store until it is discharged or perhaps until the charge neutralizes over a period of time. Although this is not much energy, it would take only about 1 μs to charge the capacitor. Hence the rate of energy flow would be rather high, about 800 W.

Mechanical analog for resistance. A mechanical analog for a resistance is frictional loss of a special type. Voltage is analogous to force, and current analogous to velocity. Voltage/current, or resistance, would thus imply a force that is proportional to velocity. We experience such a force when we try to move underwater. A common mechanical component having this property would be an automotive shock absorber. Table 3.1 summarizes the analogies between mechanical and electrical quantities.

TABLE 3.1 SUMMARY OF MECHANICAL AND ELECTRICAL ANALOGS

Mechanical	Electrical
Force	Voltage
Velocity	Current
Displacement	Charge
Mass	Inductance
Spring compliance	Capacitance
Shock absorber	Resistance

3.1.4 Check Your Understanding

1. An inductor has a stored energy of 5 J and an inductance of 0.1 H. What is the current through the inductor?
2. If an ideal 10-V battery were connected for 1 s to an ideal 1-H inductor, how much energy would be given to the inductor?
3. A capacitor has a stored energy of 500 μJ and a capacitance of 0.15 μF. What is the voltage across this capacitor?
4. What is the mechanical analog of an inductor?

Answers. 1. 10 A; 2. 50 J; 3. 81.6 V; 4. mass.

2 FIRST-ORDER TRANSIENT RESPONSE OF *RL* AND *RC* CIRCUITS

3.2.1 Type of Problem We Will Solve

In this chapter we will show how to solve an important class of problems. Typical problems of this class are shown in Fig. 3.11. These circuits are characterized by having a single energy storage element, one capacitor or inductor, and having

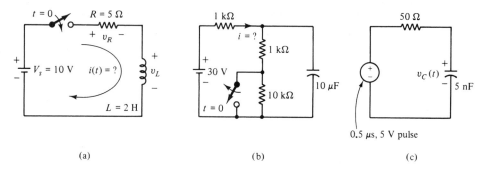

Figure 3.11 Each circuit has one energy storage element, a constant source, and a sudden change.

loss, represented by one or more resistors. They have dc sources and a switch†
that either opens or closes at a known time, normally defined to be $t = 0$. The
circuit will have one dc state before the switch action and another dc state long
after the switch action. Consequently, the state of the circuit goes through a
transition. This transition lasts for a brief period of time; thus these problems are
often called *transient problems*.

These problems are important because they represent switching something
on or off, which is often a critical period in the operation of a device. Also included
in this class of problems are many digital signals, such as a computer uses in
processing information. Such digital signals can often be represented as a dc
source being turned on and off.

3.2.2 Classical Differential Equation Solution

Deriving the differential equation.   Our approach will be first to analyze
a representative problem using the techniques of differential equations (DEs).
Then we will develop a much simpler method of solution. Our simple method can
also be applied to mechanical, thermal, or chemical problems that fall into the
class described in Section 3.2.1.

We will analyze the circuit in Fig. 3.11a. With the switch open, the full 10
V appears across the switch because there can be no current to cause a voltage
across the resistor or the inductor. After the instant at which the switch is closed,
KVL for the circuit is

$$-V_s + v_R(t) + v_L(t) = 0 \qquad (3.9)$$

Equation (3.9) becomes a DE when we express v_R and v_L in terms of the unknown
current.

$$L \frac{di(t)}{dt} + Ri(t) = V_s, \qquad t > 0 \qquad (3.10)$$

Form of the solution.   We now must solve this DE. Equation (3.10) is a
linear DE with constant coefficients and a constant forcing term on the right side.
The solution of a DE of this type usually proceeds by separating the unknown

† The pulse in Fig. 3.11c can be considered a voltage switched on, then off.

solution into two parts: the homogeneous part and the particular integral, also called the *forced response* or the *steady-state response*. These names are appropriate because this portion of the solution is caused by the forcing function on the right side. In this problem, and all problems we will solve in this chapter, this response must be constant because the forcing function is constant in time.

The form of the homogeneous response is determined by the left side of the equation. This part of the solution, also called the *natural response*, satisfies the equation with the forcing function set to zero. A linear DE with constant coefficients is always satisfied by a function of the form $e^{-t/\tau}$, where τ (Greek lowercase tau) is an unknown constant with the dimensions of time. The solution must therefore be of the form

$$i(t) = A + Be^{-t/\tau} \tag{3.11}$$

where A, B, and τ are unknown constants.

Determining the unknown constants. We can determine A and τ by substituting Eq. (3.11) back into Eq. (3.10). The coefficient of the exponential term must vanish if the equation is valid for all times. Hence we determine τ to be

$$LB\left(\frac{-1}{\tau}\right)e^{-t/\tau} + R(A + Be^{-t/\tau}) = V_s$$

$$B\left(\frac{-L}{\tau} + R\right)e^{-t/\tau} + AR = V_s \Rightarrow \tau = \frac{L}{R} = 0.4 \text{ s}$$

Setting the coefficient of the exponential term to zero leads also to the value of A, the steady-state response.

$$AR = V_s \Rightarrow A = \frac{V_s}{R} = 2 \text{ A}$$

To determine B, we must consider the initial condition. The initial condition for this, and for all such systems, arises from consideration of energy. Time is required for energy to be exchanged and hence processes involving energy carry the system from one state to another, particuarly when, as here, a sudden change occurs. The closing of the switch will eventually allow the inductor to store energy, but at the instant after the switch is closed, the stored energy must still be zero. Zero energy implies zero current because the stored energy in an inductor [Eq. (3.3)] is $\frac{1}{2}Li^2$; hence, $i(0^+)$ is zero, where 0^+ denotes the instant after the switch is closed. This condition leads to the value for B:

$$0 = A + Be^{-0/\tau} \Rightarrow B = -A = -2 \text{ A}$$

The final solution is given in Eq. (3.12) and plotted in Fig. 3.12.

$$i(t) = \frac{V_s}{R} - \frac{V_s}{R}e^{-t/\tau} = 2 - 2e^{-t/0.4} \text{ A} \tag{3.12}$$

Physical interpretation of the solution. The mathematical solution is complete, but we wish to examine carefully the physical interpretation of the

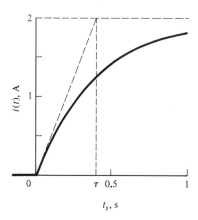

Figure 3.12   Inductor current.

response, for we will develop our simpler method from a physical understanding of this type of problem. The current is zero before the switch is closed and approaches V_s/R as a final value. This final value does not depend on the value of the inductor, but would be the current if no inductor were in the circuit. The inductor cannot matter in the end because we have a dc excitation in the circuit and eventually the current must reach a constant value. The constant current renders the inductor to be invisible since an inductor will exert itself only when its current is changing.

Without the inductor, the current would change instantaneously from zero to V_s/R when the switch is closed. The inductor effects a smooth transition between the initial and final values of the current. The smooth change caused by the inductor takes place over a *characteristic time* of τ, which is also called the *time constant*. Note from Fig. 3.12 that the current initially increases with a rate as if to arrive at the final value in one time constant. This time constant for the *RL* circuit, L/R, can be interpreted from energy considerations. The inductor will require little energy if L is small or R is large; hence the transition will be rapid for small τ. But if the inductor is large or R is small, much energy will eventually be stored in the inductor; hence the transition period will be much longer for large τ.

Summary.   We see that the response consists of a transition between two constant states, an initial state and a final state. The energy storage element effects a smooth transition between these states. Because of the form of the DE for this class of problems, the transition between states is exponential and is characterized by a time constant. The stored energy in the inductor carries the state of the circuit across the instant of sudden change and thus leads directly or indirectly to the initial condition.

3.2.3 A Simpler Method

We will now rework this problem using a more efficient method. The goal is to write the circuit response directly based on physical understanding. There are four steps.

Find the time constant. We already know the time constant for this type of problem, L/R, but in general we can determine the time constant for a first-order system by putting the DE into the special form of Eq. (3.13). We divide by the coefficient of the derivative term, and place the coefficient of the other term on the left side entirely in the denominator.

$$\frac{dx}{dt} + \frac{x}{\tau} = \text{constant} \tag{3.13}$$

In Eq. (3.13), x refers to the physical quantity being determined. It would represent a voltage or current in a circuit problem, but in other physical systems x might represent a temperature, a velocity, or something else. When we put the DE in this form, τ will always be the characteristic time, or time constant. To put Eq. (3.10) into this form, we have to divide by L and flip R into the denominator, but you can see τ works out in this case to L/R, as it must. Since the purpose of this method is to avoid DEs, we hesitate to suggest that you must write the DE to get started. Usually, you will know the equation of the time constant for the circuit or system from previous experience. For an RL circuit, for example, the time constant is always L/R. Because we solve circuits having only one energy storage element with this method, we have only one L. In Section 3.2.5 we will show you how to handle circuits with more than one resistor. Returning to the problem we are solving, we now know our time constant: L/R is 0.4 s for this circuit.

Find the initial condition. The initial condition (x_0 in general, i_0 in this case) always follows from consideration of the energy condition of the system at the beginning of the transient. In this case, we argued earlier that because the inductor had no energy before the switch was closed, it must have no energy the instant after the switch is closed. Hence the initial current is zero (i.e., $i_0 = 0$). Later we will discuss generally how to determine initial values.

Find the final value. The final value (x_∞ in general, i_∞ in this case) follows from the steady-state solution of the system. Earlier we argued that the inductor eventually becomes invisible. This is true because the final state of the circuit is a dc state, and the inductor has no voltage across it for a constant current. Thus application of Ohm's law to the circuit in Fig. 3.11a shows that the final current must be 2 A.

Substitute the time constant and the initial and final values into a standard form

$$x(t) = x_\infty + (x_0 - x_\infty)e^{-t/\tau} \tag{3.14}$$

Equation (3.14) is the solution for all first-order systems having dc (constant) excitation. In our case, the x's are currents of known numerical value, so our solution is

$$i(t) = \frac{V_s}{R} + \left(0 - \frac{V_s}{R}\right)e^{-t/\tau}$$

$$= 2 + (0 - 2)e^{-t/0.4} \text{ A}$$

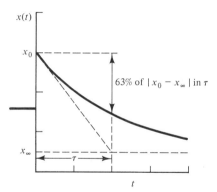

Figure 3.13 Generalized response in Eq. (3.14).

You can confirm that this is the same solution we obtained directly by solving the DE. This response is shown in Fig. 3.12.

Figure 3.13 shows a generalized plot of Eq. (3.14). The curve starts at x_0 and asymptotes to x_∞. Its initial rate is such as to move from x_0 to x_∞ in one time constant, but it only reaches $(1 - e^{-1})$ or 63% of the way during the first time constant. You might note that we have shown a discontinuity at the origin. This can happen in general, although it does not happen in our present example.

Summary. Our efficient method consists of determining three constants and substituting these into a standard form. The time constant can be determined from the DE, but usually the formula for τ is known from prior experience. The other two constants are the initial value of the unknown, which is determined from energy considerations, and the final value, which is determined from the dc solution. The energy storage element effects a smooth transition between the initial and final values and influences the initial value through energy considerations.

Example. As a second example of our simple method, we will solve the same circuit again, except that this time we will determine the voltage across the inductor. We already know the time constant. The initial condition follows indirectly from energy considerations. The initial energy in the inductor is zero, which implies zero current. Turn back to Fig. 3.11a and consider the state of the circuit at the instant after the switch is closed. Because of the inductor there can be no current at this initial instant; hence there can be no voltage across the resistor. If there is no voltage across the resistor and no voltage across the switch, the full 10 V of the battery must appear across the inductor. Hence the initial value of the inductor voltage is 10 V. The final value of the inductor voltage must be zero because we have a dc source and eventually the current becomes constant. Consequently, di/dt approaches zero, and the inductor voltage must also approach zero. We now know the time constant, the initial value, and the final value, and we can therefore write the full solution with our standard form,

$$v_L(t) = 0 + (10 - 0)e^{-t/0.4} = 10e^{-t/0.4} \text{ V} \tag{3.15}$$

The plot of Eq. (3.15) is shown in Fig. 3.14. Notice that we have shown a discontinuity in the inductor voltage at the origin. Before the switch closes, the

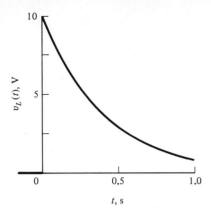

Figure 3.14 Inductor voltage.

inductor has no voltage. When the switch closes, the current begins to increase, but the inductor voltage instantly jumps to 10 V to oppose that increase. The initial rate of increase of the current is limited to

$$10 = 2 \left. \frac{di}{dt} \right|_{t=0} \Rightarrow \frac{di}{dt} = \frac{10}{2} = 5 \text{ A/s}$$

As the current increases, the voltage across the resistor increases accordingly, and the voltage across the inductor must decrease. Finally, the current increases to a level where it is limited by the resistor. Thus the inductor dominates the beginning and the resistor dominates the end of the transient.

3.2.4 An *RC* Circuit

We will now apply our method to the *RC* circuit shown in Fig. 3.15. We will determine the voltage across the resistor, $v_R(t)$, assuming that the energy in the capacitor initially is zero.

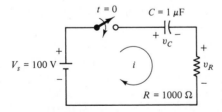

Figure 3.15 The capacitor is uncharged when the switch closes.

Write the DE. We must write the DE because we have never analyzed an *RC* circuit before. After the switch is closed, KVL is

$$-V_s + v_C + v_R = 0 \tag{3.16}$$

To derive the DE for v_R, we must express v_C in terms of v_R. One approach is to differentiate Eq. (3.16) and then substitute v_R/RC for dv_C/dt. This substitution is a consequence of Ohm's law and the definition of a capacitor:

$$\frac{dv_C}{dt} = \frac{i}{C} = \frac{v_R/R}{C} = \frac{v_R}{RC} \tag{3.17}$$

The differentiation eliminates the constant, V_s, and after the substitution from Eq. (3.17) we have

$$\frac{v_R}{RC} + \frac{dv_R}{dt} = 0 \Rightarrow \frac{dv_R}{dt} + \frac{v_R}{RC} = 0 \tag{3.18}$$

Comparison of Eq. (3.18) with Eq. (3.13) shows that the time constant for an RC circuit is RC. So τ is $1000 \times 10^{-6} = 1$ ms in this circuit. This is the only time we will write the DE for an RC circuit; henceforth we know that the time constant is RC.

Find the initial condition. The initial condition emerges from consideration of energy. The capacitor is unenergized at $t = 0^-$, the instant before the switch is closed. Hence it remains unenergized at $t = 0^+$, the instant after the switch is closed, because delay is required for energy (and hence charge) to be stored in the capacitor. The electric energy stored in a capacitor is given in Eq. (3.8) as $\frac{1}{2}Cv_C^2$; thus the voltage across the capacitor will be zero before and after the closing of the switch. At $t = 0^+$, therefore, the full voltage of the battery appears across the resistor, and the initial value for v_R must be 100 V.

Find the final value. The final value of the resistor voltage follows from the requirement that the voltage and current eventually become constant. This means that dv_C/dt is zero, i is zero, and v_R must therefore be zero.

Substitute the time constant and the initial and final values into the standard form. Now that we know the time constant, the initial value, and the final value for v_R, we substitute into the standard form of Eq. (3.14).

$$v_R(t) = 0 + (V_s - 0)e^{-t/\tau}$$
$$= 100e^{-t/1 \text{ ms}} \tag{3.19}$$

Equation (3.19) is plotted in Fig. 3.16. As the current flows, as evidenced by the voltage across the resistor, charge accumulates and builds up voltage across the

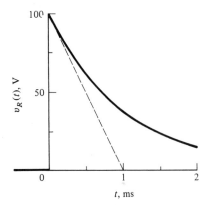

Figure 3.16 Resistor voltage.

capacitor. The voltage across the resistor must diminish with time and eventually vanish. When the full voltage of the battery appears across the capacitor, the current stops.

3.2.5 Circuits with Multiple Resistors

Thévenin equivalent circuit. Circuits with more than one resistor can be analyzed with the concept of equivalent circuits. The circuit in Fig. 3.17a, for example, can be reduced to the equivalent circuit in Fig. 3.17b. From this equivalent circuit, it is clear that the time constant is $R_{eq}C$, where R_{eq} is the equivalent resistance of the circuit with all sources turned OFF and the switch closed. Thus R_{eq} is the output resistance of the circuit to the capacitor as a load. In this case, R_{eq} is $30 + 20 \parallel 10 = 36.7 \ \Omega$, and thus τ is $36.7 \times 0.03 \times 10^{-6}$ or $1.1 \ \mu s$.

We may or may not be interested in the Thévenin equivalent circuit as a means for calculating the initial or final values of the unknown. That is a separate problem to be approached by the most efficient method. It is the *concept* of the Thévenin circuit, specifically the concept of the output resistance, that leads to the time constant for a transient circuit with multiple resistors. In general, the relevant resistance for the time constant is the output resistance presented to the energy storage element as a load.

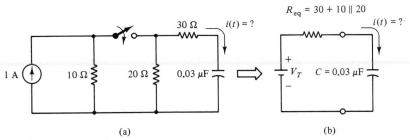

Figure 3.17 (a) The three resistors can be combined using a Thévenin equivalent circuit; (b) the time constant is easily identified in this equivalent circuit.

Example. The circuit in Fig. 3.11b can be approached with this technique. Here we clearly should not replace the circuit external to the capacitor with a Thévenin equivalent circuit, for in that case we would eliminate the unknown of interest, $i(t)$. We evoke the concept of the equivalent circuit only to calculate the equivalent resistance seen by the capacitor with the switch open, $(1 \ k\Omega + 10 \ k\Omega) \parallel 1 \ k\Omega$ or $917 \ \Omega$. We continue this example below.

3.2.6 Initial and Final Conditions

What we mean by "initial" and "final." We now look generally at how to calculate the initial and final values. By "initial" we mean the instant after a change occurs in the circuit, usually a switch closing or opening. This does not have to be the time origin, although often we define $t = 0$ as the time of switch action. By "final" we mean the steady-state condition of the circuit, its state after a large amount of time. The final state may be hypothetical because the circuit may never reach that state due to subsequent switch action. That is, we may first close a switch and then open it before the circuit reaches the final state. The circuit cannot anticipate the second switch action and hence reacts to the first switch action as if it would reach steady state. In this section we develop guidelines for determining the final and initial states of the circuit.

Determining final values. Final values arise out of the eventual steady state of the circuit. This implies that all time derivatives must eventually vanish. Consequently, the current through capacitors and the voltage across inductors must approach zero, as suggested in

$$v_C \rightarrow \text{constant} \Rightarrow i_C = C \frac{dv_C}{dt} \rightarrow 0$$

$$i_L \rightarrow \text{constant} \Rightarrow v_L = L \frac{di_L}{dt} \rightarrow 0$$

(3.20)

This means that capacitors act as open circuits and inductors act as short circuits in establishing final values, as shown in Figs. 3.18 and 3.19, respectively.

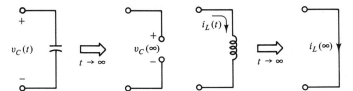

Figure 3.18 The capacitor acts as an open circuit as time becomes large.

Figure 3.19 The inductor acts as a short circuit as time becomes large.

Determining initial values. Initial values always follow from energy considerations. The fundamental principle is that the stored electric energy in a capacitor and the stored magnetic energy in an inductor must be continuous functions of time. From continuity of energy we conclude that the voltage across a capacitor and the current through an inductor must be continuous functions of time. Thus we always calculate capacitor voltage or inductor current before the switch is thrown and then carry this value over to the moment after the switch is thrown. From these known values of the capacitor voltage or inductor current, other circuit variables can be calculated.

A model for an energized capacitor at the instant after the switch action is shown in Fig. 3.20. For that first instant, the charged capacitor acts like a voltage source because the voltage across the capacitor cannot change instantaneously. As current flows through the capacitor, its voltage will change, and hence the capacitor acts as a voltage source only at that first instant. Similarly, a current source models an energized inductor at the instant after switch action, as shown in Fig. 3.21. Although these models are valid only for the first instant of the new regime, they suffice for the calculation of the initial values of the circuit variables of interest.

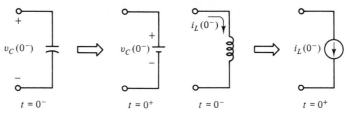

Figure 3.20 A battery models the initial voltage of the capacitor.

Figure 3.21 A current source models the initial current of the inductor.

Example. We will illustrate this principle in the circuit of Fig. 3.22. The switch is opened after having been closed for a time long enough to establish steady state throughout the circuit. We are interested in the initial voltage across the inductor after the switch is opened. Following the suggestions made above, we first calculate the inductor current at the instant before the switch is opened. We note that with the switch closed and with the circuit in steady state, the 50-Ω resistor is short-circuited by the switch and the 150-Ω resistor is shorted by the switch and the dormant inductor. Thus the current from the battery is 10 V/ 100 Ω or 0.1 A and this current flows through the switch and inductor. This information does not give us the initial value of $v_L(t)$ directly, but it leads indirectly to the initial value through the analysis of the circuit in Fig. 3.23. We have replaced the inductor by a current source and eliminated the switch because it is now an open circuit. As you can see, we have notated the circuit to suggest solution by nodal analysis. You may confirm that v_a is zero at $t = 0^+$. From this information we can calculate v_L from KVL around the right-hand loop, with the result that $v_L(0^+) = -5$ V. The inductor acts initially as a source to keep the current going at the same rate, even though it must now flow through the 50-Ω resistor.

Having made all these preparations, let us finish the problem. The time constant is L/R_{eq}, where R_{eq} is the equivalent resistance seen by the inductor with the switch open. Thus R_{eq} is 50 + 150 $\|$ 100 or 110 Ω, and the time constant is 0.1/110 or 909 μs. Since the final value of the inductor voltage is zero, the full solution is

$$v_L(t) = 0 + (-5 - 0)e^{-t/909\mu s} = -5e^{-t/909\mu s} \quad V$$

Figure. 3.22 After being closed a long time, the switch is opened.

Figure 3.23 Equivalent circuit at $t = 0^+$.

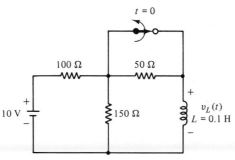

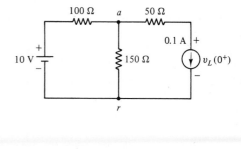

Summary. An energized capacitor acts as a voltage source in the calculation of initial values. A special case is an unenergized capacitor, which acts as a short circuit. In the calculation of final values, a capacitor acts as an open circuit. An energized inductor acts as a current source in the calculation of initial values. A special case is an unenergized inductor, which acts as an open circuit. For final values, an inductor acts as a short circuit. We can solve for initial and final values by replacing capacitors and inductors by these models. The initial and final values thus are derived from the analysis of a circuit containing only resistors and sources.

These principles are valid for circuits containing multiple inductors and capacitors, which lead to higher-order DEs that cannot be solved by our four-step procedure. We discuss the transient response of such circuits in Chapter 11, where we use the above principles to determine initial and final values.

Example. Consider the circuit in Fig. 3.11b. We have already determined the equivalent resistance to be 917 Ω, so the time constant is $\tau = RC = 9.17$ ms. To find the initial value, we must determine the voltage across the capacitor at $t = 0^+$. At the instant before the opening of the switch, the circuit will be in steady state; indeed, this is the final state from previous actions that established the circuit. Thus the capacitor will act as an open circuit, as shown in Fig. 3.24. The voltage across C is 15 V because the 30 V of the source divides equally between the two 1-kΩ resistors (the 10-kΩ resistor being shorted by the switch).

Consequently, the circuit that must be analyzed for the initial value of $i(t)$ is shown in Fig. 3.25. Notice that $i(0^+)$ can be determined by writing KVL around the outer loop containing the two sources, with the result $i(0^+) = 15$ mA.

The final value of $i(t)$ can be established from the equivalent circuit in Fig.

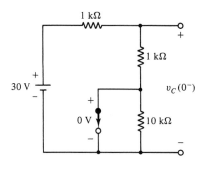

Figure 3.24 The capacitor acts as an open circuit in the steady state before the switch is opened.

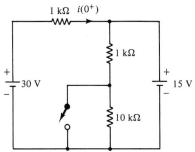

Figure 3.25 The capacitor now acts as a battery in the initial-value calculation.

The Dynamics of Circuits   Chap. 3

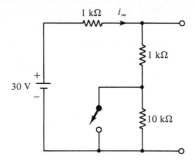

Figure 3.26 Equivalent circuit for final-value calculation.

3.26. Note that the capacitor is treated as an open circuit in this calculation of the final value. The value of i_∞ is easily derived from this series circuit: $i_\infty = 30$ V/12 kΩ = 2.5 mA.

We now have determined the time constant, the initial value, and the final value; hence the solution follows from Eq. (3.14).

3.2.7 Pulse Problem

Pulse circuits. Finally, we consider the circuit in Fig. 3.11c as detailed in Fig. 3.27. Although this problem appears to differ significantly from the others, we can analyze it with the same techniques.

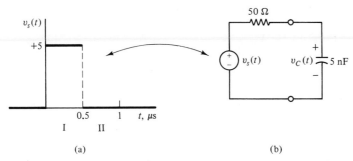

Figure 3.27 (a) Voltage pulse; (b) the capacitor is uncharged when the pulse begins.

Charging transient. We assume that the capacitor is initially unenergized. When the source voltage jumps to +5 V, current will flow and the capacitor will begin charging toward +5 V. Bear in mind that the circuit cannot anticipate the end of the pulse, but will respond as if a 5-V battery has been permanently attached. Thus the transient proceeds toward a final value, even though that value will never be attained. During interval I, the time constant will be $RC = 0.25$ μs, the initial value will be zero, and the final value will be +5 V. Thus the capacitor voltage during interval I will be

$$v_C(t) = 5 + (0 - 5)e^{-t/0.25\ \mu s} = 5(1 - e^{-t/0.25\ \mu s})\quad \text{V}$$

Discharging transient. When the trailing edge (the end) of the pulse comes along 0.5 μs after the leading edge, the capacitor now has a voltage across

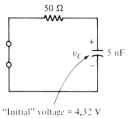

50 Ω

v_C 5 nF

"Initial" voltage = 4.32 V

Figure 3.28 During interval II the "initial" voltage is 4.32 V.

it. Since the time constant is half the pulse width, the value of the capacitor voltage at the beginning of the second transient would be $5(1 - e^{-2}) = 4.32$ V. This becomes the initial value for the transient during interval II. During this interval, the voltage source is OFF and thus is equivalent to a short circuit. The situation in interval II is therefore that shown in Fig. 3.28. We know the initial voltage, the final value is clearly zero, and the time constant is unchanged. Hence the capacitor voltage during interval II is

$$v_C(t) = 0 + (4.32 - 0)e^{-t'/0.25 \; \mu s}$$

$$= 4.32 \exp\left(-\frac{t - 0.5 \; \mu s}{0.25 \; \mu s}\right) \quad V, \qquad t > 0.5 \; \mu s$$

where t' is time counted from the beginning of interval II. For example, at $t = 1$ μs we calculate the voltage to be

$$v_C(1 \; \mu s) = 4.32e^{-(1.0-0.5)/0.25} = 0.585 \; V$$

Putting the two parts of the transient together, we plot the results in Fig. 3.29.

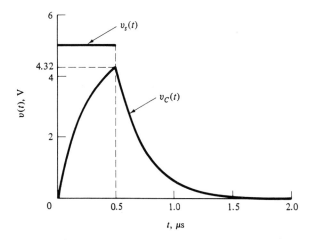

Figure 3.29 The capacitor voltage is a distorted version of the input pulse.

Comparing a pulse with a battery-switch. It is interesting to contrast the pulse problem with that shown in Fig. 3.30. Here we simulate the pulse with a battery and a switch that closes for 0.5 μs. The charging part of the transient will be the same as for the pulse problem, but when the switch is opened, there is no discharge path for the capacitor. Hence the capacitor will charge up to 4.32 V in both cases, but with the battery-switch no discharge will occur.

We can understand the difference between the pulse source and the battery-

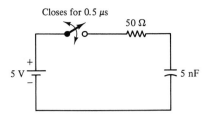

Closes for 0.5 μs

50 Ω

5 V

5 nF

Figure 3.30 The battery switch has the same Thévenin voltage but different output resistance from the pulse source.

The Dynamics of Circuits   Chap. 3

switch source by considering the Thévenin equivalent circuits for each. The Thévenin voltage source is the same for both, but the output resistance differs. With the switch closed, the output resistance is 50 Ω for both cases, but with the switch open the output resistance becomes infinite (an open circuit) for the battery-switch source.

A puzzle. While we are considering battery-switch sources, let us consider what would happen in the circuit in Fig. 3.11a if we opened the switch after the current is established. Energy considerations require that the current in the inductor continue after the switch is opened. But no current can flow through an open switch. What will happen?

This question can be answered on two levels. At the theoretical level, we must outlaw this situation. We have created our dilemma by violating the definitions of the circuit elements we are using. Strictly speaking, we can no more open the switch on an energized inductor than we can short circuit an ideal voltage source—the definitions of these elements are contradictory. On the theoretical level, opening the switch on an inductor is like setting $1 = 0$ in mathematics—it is nonsense.

But this answer does not fully satisfy us, does it? There are, after all, real inductors and real switches. What happens when we perform the experiment? If you try it, you will witness a spark when you open a switch connected in series with an energized inductor. The voltage across the inductor, and hence the voltage across the switch, rises instantaneously to a high value, such that the air between the switch contacts becomes ionized. The ionized air provided a resistive path for deenergizing the inductor. Indeed, this is the principle behind the conventional automotive ignition system: the coil is the inductor, the points are the switch, and you know where the spark occurs.

3.2.8 Check Your Understanding

1. In the differential equation, $5dx/dt + 2x = 3$, what is the time constant? What is the final value for x?

2. If the initial current in a capacitor is 5 mA, what is the current one time constant later in time?

3. In using a Thévenin equivalent circuit to determine the equivalent resistance in an RL transient problem, what must be considered the load?

4. Which, during a switch action, is constant: voltage, current, stored energy, or power?

5. In a simple circuit containing a resistor and an energy storage element (L or C), when the value of the resistor was doubled, the transient lasted twice as long. What was the other element?

6. At $t \to \infty$, what circuit element approaches a short circuit?

7. A capacitor will allow an instantaneous change in its current or voltage?

8. An RC circuit has a time constant of 10 ms. The capacitor is replaced by a 50-mH inductor and the time constant changes to 8 ms. What was C?

9. In a simple transient the initial energy in an inductor is 10 mJ and the final energy is 0. What is the energy after two time constants?

Answers. 1. 2.5 s, 1.5; 2. 1.84 mA; 3. the inductor; 4. stored energy; 5. a capacitor; 6. an inductor; 7. its current; 8. 1600 μF; 9. 0.183 J.

Section 3.1: The Theory of Inductors and Capacitors

P3.1. As shown in Fig. P3.1, a 50-mH inductor has an ac current of

$$i_{ac} = 10 \cos (500t) \quad A$$

Calculate the voltage across the inductor with the polarity shown in Fig. P3.1.

Answer. $v_{ac}(t) = -250 \sin (500t) = 250 \cos (500t + 90°)$ V

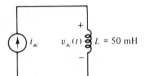

Figure P3.1

P3.2. A 6-V battery is connected to an ideal 0.2-H inductor, as shown in Fig. P3.2.
(a) Calculate the current as a function of time after the switch is closed.
(b) At what time does the stored energy in the inductor reach 10 J? Verify the stored energy by integrating the input power (product of v and i) from $t = 0$ to the time you calculate.

Answer. $t = 0.333$ s

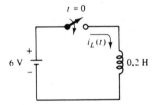

Figure P3.2

P3.3. Prove from the defining equation of inductance that two inductors (L_1 and L_2) in series combine like resistors in series, that is, $L_{eq} = L_1 + L_2$. Show also that two inductors in parallel combine like resistors in parallel, that is,

$$L_{eq} = L_1 \parallel L_2 = \frac{1}{(1/L_1) + (1/L_2)}$$

P3.4. If an ideal 5-V battery were connected to an ideal 1-H inductor, how much energy would be given to the inductor in one second?

P3.5. How much charge and how much energy are stored in a 1000-μF capacitor that has been charged to 200 V? If this energy were converted to kinetic energy and distributed to the excess electrons on one side of the capacitor, what would be their velocity?

Answer. $v = 5.93 \times 10^6$ m/s

P3.6. A 0.1-μF capacitor is charged with a 1-μs pulse of current, as shown in Fig. P3.6. Find the voltage across the capacitor with the polarity shown, as a function of time.

Answer. Starts at 0, charges to 1 V in 1 μs, and stays at 1 V thereafter because with $i_s = 0$ charge is trapped on the ideal capacitor.

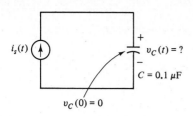
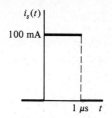

$v_C(0) = 0$

Figure P3.6

P3.7. To move the spot of a CRT across the screen, the voltage across a pair of deflection plates must be increased in a linear fashion, as shown in Fig. P3.7. If the capacitance of the plates is 0.001 μF, sketch the resulting current through the capacitor.

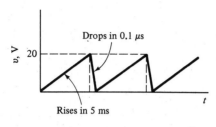

Figure P3.7

P3.8. Prove from the defining equation of capacitance that two capacitors (C_1 and C_2) connected in parallel are equivalent to a single capacitor of value $C_{eq} = C_1 + C_2$, but in series combine like resistors in parallel,

$$C_{eq} = C_1 \| C_2 = \frac{1}{(1/C_1) + (1/C_2)}$$

P3.9. An inductor and a capacitor are placed in series with a current source whose current increases with time as shown in Fig. P3.9.
(a) Find the voltage across the inductor, $v_L(t)$.
(b) Assuming no initial charge on the capacitor, find its voltage, $v_C(t)$.
(c) Calculate the instant when the stored energy in the capacitor first exceeds that in the inductor.

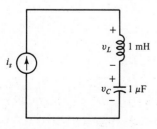

Figure P3.9

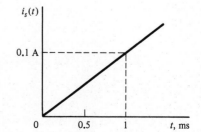

P3.10. For Fig. P3.10, in the series *RLC* circuit, the voltage across the inductor is shown.
  (a) Determine and sketch the voltages across the resistor and the capacitor.
  (b) What is the value of v_s at $t = 3$ s?

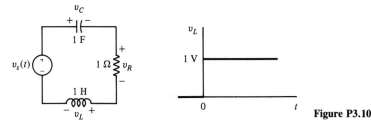

Figure P3.10

P3.11. For the circuit shown in Fig. P3.11, the voltage across the resistor is 0 V for negative time and $v_R = t$ V for positive time. The capacitor is initially unenergized.
  (a) What is the stored energy in the inductor at $t = 2$ s?
  (b) What is the voltage across the capacitor at the same time?
  (c) What is the voltage across the source at the same time?

  Answers.   (a) 2 J,   (b) 2 V,   (c) 5 V

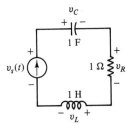

Figure P3.11

Section 3.2: First-order Transient Response of *RL* and *RC* Circuits

P3.12. The solution of a first-order DE with constant coefficients and a constant term on the right side is

$$x(t) = 5 - 5e^{-t/0.1 \text{ s}}$$

  (a) Plot this solution. What is the value at the origin and as time becomes large?
  (b) What is the DE that this solution satisfies? Verify by substitution that your DE is correct.

P3.13. Verify that the standard solution [Eq. (3.14)] satisfies Eq. (3.13) and has the required values at $t = 0$ and t very large.

P3.14. A system is described by the DE

$$10 \frac{dx}{dt} + 5x = 50$$

  (a) What is the time constant for this system?
  (b) What would be the final value, x_∞?
  (c) If $x_0 = 12$, find and sketch $x(t)$ for $t > 0$.

P3.15. Figure P3.15 shows a switch that is changed from a to b at $t = 0$. Assume that the switch has been in position a for a long time before the switch action. Find and sketch $v_L(t)$, the voltage across the inductor with the polarity shown.

Answer. $v_L(t) = -10e^{-t/15 \text{ ms}}$ V

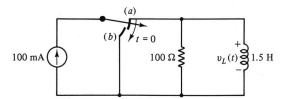

Figure P3.15

P3.16. A current source is placed in series with a resistor and inductor, as shown in Fig. P3.16. During this period the switch is open. Then the switch is closed, and the circuit is separated into two independent loops that share a common short circuit but do not interact.
(a) Calculate the current in the right-hand loop after the switch closes at $t = 0$.
(b) Plot the current in the switch, referenced downward.

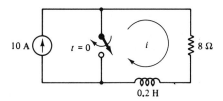

Figure P3.16

P3.17. For the circuit shown in Fig. P3.17, assume that the switch has been in the position a for a long time and then is changed to position b.
(a) Find and sketch $i(t)$.
(b) Calculate by integration the energy lost in the 10-Ω resistor during positive time. Confirm that all the energy stored initially in the capacitor is accounted for by the loss in the resistor.

Answer. (a) $i(t) = 1e^{-t/10 \ \mu\text{s}}$ A

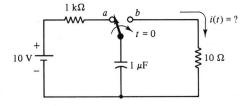

Figure P3.17

P3.18. The switch in Fig. P3.18 is in position a for negative time, moved to b at $t = 0$, and to c at $t = 1$ s. Sketch the voltage across the capacitor for $0 < t < 3$ s.

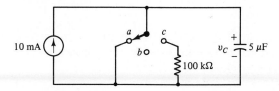

Figure P3.18

P3.19. A capacitor is charged to 100 V and disconnected. Leakage reduces its voltage to 25 V in 10 min. Estimate the additional time required for the voltage to drop to 8 V. *Hint*: The leakage is represented by a resistor in parallel with the capacitor.

P3.20. After being closed for a long time, the switch in Fig. P3.20 is opened at $t = 0$. Find:

(a) The voltage across the capacitor with + at top as a function of time.

(b) The integral of the current during the period $0 < t < \infty$.

(c) The total energy given to the circuit by the battery during the period $0 < t < \infty$.

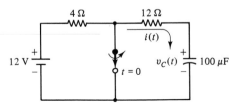

Figure P3.20

P3.21. For the circuit shown in Fig. P3.21, the switch has been open a long time and is closed at $t = 0$.

(a) What is the time constant of the circuit with the switch closed?

(b) Determine the capacitor voltage with the polarity shown and sketch.

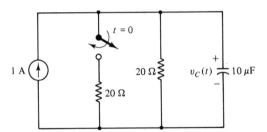

Figure P3.21

P3.22. For the circuit shown in Fig. P3.22, the switch is closed at $t = 0$. Find:

(a) The voltage across the 10-Ω resistor for $t > 0$.

(b) The stored energy in the capacitor in the steady-state condition.

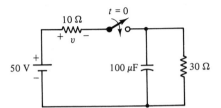

Figure P3.22

P3.23. For the circuit shown in Fig. P3.23, the switch has been closed for a long time and then is opened at $t = 0$. Determine the voltage across the capacitor with the polarity shown.

The Dynamics of Circuits   Chap. 3

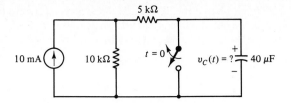

Figure P3.23

P3.24. After being closed a long time, the switch in Fig. P3.24 is opened at $t = 0$. Find:
  (a) The time constant.
  (b) The initial value of $i(t)$.
  (c) The final value of $i(t)$.
  (d) The time function, $i(t)$.

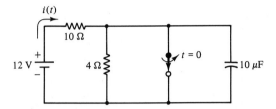

Figure P3.24

P3.25. For the circuit shown in Fig. P3.25, find:
  (a) τ
  (b) i_0
  (c) i_∞
  (d) $i(t)$

Answers. (a) 2 ms; (b) -0.125 A; (c) 0; (d) $-0.125\, e^{-t/2 \text{ ms}}$ A

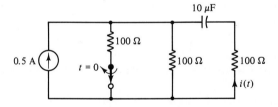

Figure P3.25

P3.26. After being open a long time, the switch in Fig. P3.26 is closed at $t = 0$. Determine the voltage across the 120-Ω resistor.

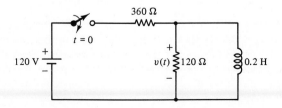

Figure P3.26

P3.27. For the circuit shown in Fig. P3.27, the switch is open a long time, closed for 5 ms, and then opened again. Find and sketch $i(t)$.

Answers. $i(t) = 2.5(1 - e^{-t/10 \text{ ms}})$ mA for $0 < t < 5$ ms, and $i(t) = 0.984 \exp[-(t - 5 \text{ ms})/20 \text{ ms}]$ mA for $t > 5$ ms

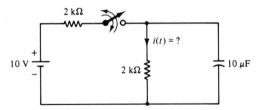

Figure P3.27

P3.28. For the circuit shown in Fig. P3.28, the switch has been closed for a long time and then is opened at $t = 0$. Determine the voltage across the inductor with the polarity shown.

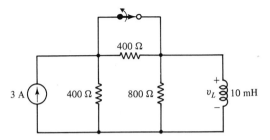

Figure P3.28

P3.29. The parallel RL circuit shown in Fig. P3.29 is excited by the current pulse as shown. Calculate and sketch the resulting voltage, v_L. Assume that the inductor is initially unenergized.

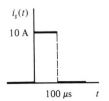

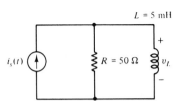

Figure P3.29

P3.30. A pulse can be modeled as two sources, one switching the voltage on and the other switching it off, as shown in Fig. P3.30. Note that the two sources add to zero except during the period $0 < t < t_1$. The voltage across the capacitor can be calculated by superposition. Use this model to rework the problem in Fig. 3.27. Specifically, solve for the voltage across the capacitor at $t = 1.0$ μs and verify the result given on page 100.

The Dynamics of Circuits   Chap. 3

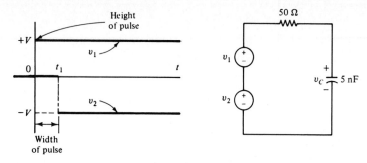

Figure P3.30 The pulse can be made by superposing two sources.

General Problem for Chapter 3

P3.31. When the switch in the *RC* circuit of Fig. P3.31 is closed, energy will for a period of time flow from the battery into the circuit. Once the current stops, the energy flow will cease.

(a) Calculate by integration the total energy given to the circuit by the battery for $t > 0$. Show that this is Vq, where q is the charge on the capacitor.

(b) Show by direct calculation that one-half this total energy is stored in the capacitor and the other half is lost to the circuit as losses in the resistor.

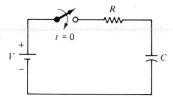

Figure P3.31 The capacitor initially is unenergized.

4 The Analysis of AC Circuits

4.1 INTRODUCTION TO ALTERNATING CURRENT

4.1.1 Importance of AC

Thomas A. Edison was a clever, determined inventor whose activities excited the public imagination toward the practical uses of electricity. He pioneered in, among other things, the generation and distribution of electrical power for lighting. But Edison was committed to direct current. The power plants built by the Edison Electric Lighting Company produced dc.

Edison had a lot of young inventors and scientists working for him. His was, in fact, the first industrial research laboratory. One of these underlings was Nikola Tesla, a young engineer from Central Europe. Tesla appears to have been the first person to recognize the possibilities of ac and he is credited with inventing the ac induction motor. But his efforts to convince Boss Edison of the benefits of ac were in vain, and Tesla eventually quit.

Tesla went to a rival company and battle was pitched: Was it to be dc or ac? Nasty ads were placed in the newspapers by Edison's group, claiming that ac was unsafe. From our perspective, these warnings of the dangers of ac seem ridiculous, but at the time they created a serious debate.

Edison was wrong and Tesla was right. Today the vast majority of all electrical power is generated, distributed, and consumed in the form of ac power. Before World War II, to major in electrical engineering meant to study ac generators, motors, transformers, transmission lines, and the like. The ac power industry currently employs a mature technology and is a vital part of modern civilization. Count, for example, the number of electric motors in your dwelling. Of course you will think first of the air conditioner or heater blower and the refrigerator and other large appliances, but do not overlook the hair dryer, the electric clocks, the timer in the dishwasher, the phonograph turntable.

In this chapter you will learn how to analyze ac circuits and, more broadly, how electrical engineers think about ac waveforms. This chapter may tax your patience and your intellect, for the methods by which electrical engineers attack ac problems initially appear abstract and roundabout. We begin by deriving a simple DE, only to lead you into a review of complex numbers. But these methods, once mastered, become powerful, both in solving problems and in stimulating insight. We begin by acquainting you with the central figure in our development, the sinusoidal waveform.

4.1.2 Sinusoids

Physical model for a sinusoid.   Most of us were introduced to sines and cosines through the study of triangles. Later we learned that circular motion leads to sine and cosine functions. Figure 4.1 shows a crank. The horizontal projection of the crank is the length times the cosine of the angle ϕ. If the rotation speed is uniform, the horizontal projection becomes a sinusoidal function of time. We may describe this waveform mathematically as a sine function or a cosine function, but we will simply call it a sinusoid, or sinusoidal waveform.

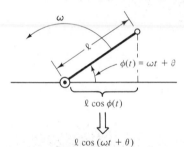

Figure 4.1   The projection of the rotating lever is a sinusoid.

Mathematical form for a sinusoid.   Electrical engineers have adopted the cosine function as the standard mathematical form for sinusoidal waveforms. In Fig. 4.2 we show the peak value, period, and phase of a sinusoidal waveform. The corresponding mathematical form of such a waveform is

$$v(t) = V_p \cos(\omega t + \theta) \qquad (4.1)$$

The peak value of the voltage is V_p. The period, T, is related to the frequency of the sinusoid. The "event" frequency, namely, the number of cycles during a period of time, is the reciprocal of the period:

$$f = \frac{1}{T} \qquad (4.2)$$

For example, if the period were 0.02 s, the event frequency would be 1/0.02 or 50 cycles per second. Because the unit "cycles per second" is awkward to say,

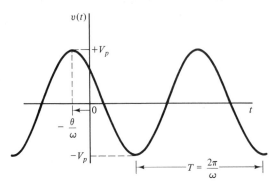

Figure 4.2 The sinusoidal function is defined by its peak value, its phase, and its period (or frequency).

most people tend to shorten it to "cycles," which is misleading to the novice and offensive to the purist. To honor Heinrich Hertz (1857–1894), the unit hertz, abbreviated Hz, has replaced cycles per second as the common unit for event frequency. So 50 hertz means 50 cycles/second.

When most people talk about a frequency, they mean the event frequency. When we write mathematical expressions for sinusoidal waveforms, however, we more often deal with the proper mathematical measure, the angular frequency, ω (the Greek lowercase omega), in radians per second. Because there are 2π radians in a full circle (a cycle), the relationship between event frequency, f, and radian frequency is

$$\omega = 2\pi f = \frac{2\pi}{T}$$

The scientific dimensions of frequency are reciprocal seconds. The numerator (radians or cycles) represents an angle and is therefore dimensionless. Just as we retain radians or degrees to remind us which measure of an angle we are using, so we need to state the units for frequency to make explicit which frequency we mean, f or ω.

Units of phase. The phase, θ (the Greek lowercase theta), is what permits the waveform in Fig. 4.2 to represent a general sinusoid. We have drawn the curve for a phase of $+50°$, but we can shift the position of the sinusoid by varying the phase. The phase is related to the time origin when we use a mathematical description, but in an ac problem significance belongs only to the relative phases of the various sinusoidal voltages and currents. Here you must tolerate one of the traditional inconsistencies of electrical engineers. The proper mathematical unit for ωt is radians and hence the correct unit for phase should also be radians. For example, we may wish to compute the time when the voltage reaches its positive peak. The cosine is maximum for zero angle, so the peak occurs when the total angle is zero:

$$\omega t_{\text{peak}} + \theta = 0 \Rightarrow t_{\text{peak}} = -\frac{\theta}{\omega} = -\frac{\theta}{2\pi} T \tag{4.3}$$

When we solve an equation like Eq. (4.3), we must use radian measure for the phase, as the last form of Eq. (4.3) suggests. But electrical engineers usually speak of phase in degrees, as we did above ($+50°$). This inconsistency is tolerable because only relative phase is significant in most situations. Electrical engineers, like most people, still are more comfortable thinking about and sketching angles in degree measure. Probably you also think best in degree measure, so we will continue to express phase in degrees unless the mathematics demands radians.

Some familiar frequencies.   Frequency is a familiar concept. When we speak of a motor speed as 4000 rpm, for example, we are indicating a frequency of 66.7 Hz, and each spark plug would be firing with a frequency of 33.3 Hz. The power system frequency is 60 Hz in this country, although 50 Hz is used in some other parts of the world, and 400 Hz is used in airborne and some naval applications.† When the radio announcer tells you that you are "tuned to 1200 on your radio dial," he is giving his station frequency, 1200 kHz (kilohertz) or 1.2×10^6 Hz. The FM stations broadcast at frequencies of about 100 MHz (megahertz) or about 10^8 Hz. The UHF TV band extends to about 800 MHz, and communication satellites relay signals at about 5 GHz (gigahertz) or 5×10^9 Hz. The highest frequencies currently used for radio signals are about 300 GHz, 3×10^{11} Hz. We find infrared, optical, and x-ray radiation at even higher frequencies.

Later in this book we will explore more fully the importance of frequency in electrical engineering, particularly in communication systems. For now, we wish merely to introduce the concept of frequency and to show how ac waveforms are represented for ac circuit analysis.

4.1.3 AC Circuit Problem

Figure 4.3 shows the ac circuit we will analyze. The ac source has a frequency of 60 Hz or 120π, about 377 rad/s, and is connected at $t = 0$. We wish to solve for the current, $i(t)$. After the switch is closed, we write KVL as

$$-v_s(t) + v_R(t) + v_L(t) = 0$$

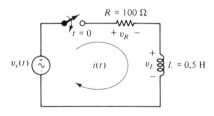

$v_s(t) = V_p \cos(\omega t + \theta)$
$V_p = 100$ V, $f = 60$ Hz, $\theta = 30°$

Figure 4.3   Solve for $i(t)$ for positive time. The switch closes at $t = 0$.

We may introduce $i(t)$ through Ohm's law and the definition of inductance, with the result

$$L \frac{di}{dt} + Ri = V_p \cos(\omega t + \theta) \tag{4.4}$$

† Because motors and transformers are smaller and lighter at the higher frequency.

Equation (4.4) is a linear first-order DE, but the technique we used in Chapter 3 does not apply because we do not have a dc term on the right side. The character of the response, however, is similar to our earlier results. Here also there will be a transition from one state of the circuit, before the switch was closed, to another state corresponding to the closed switch. Here also the transition period is expressed in terms of the L/R time constant of the circuit. In fact, the form of the solution is

$$i(t) = Ae^{-t/\tau} + i_p(t), \qquad \tau = \frac{L}{R} \tag{4.5}$$

where A is an unknown constant, τ the time constant, and $i_p(t)$ the particular integral, or steady-state current. The final state, $i_p(t)$, is no longer a constant but results from a dynamic equilibrium between source, resistor, and inductor. Methods for determining the steady-state response are the focus of this chapter. Equation (4.5) can be solved with several methods. The equation can be integrated directly with the aid of an integrating factor. This approach fails, however, when we try it on more complicated circuits and, besides, our goal is to learn how electrical engineers analyze ac problems. No electrical engineer would integrate this equation directly to find the steady-state solution. To learn the best procedure, you have to proceed through a long chain of reasoning. We will lead you down the traditional path—and please be patient. Once we arrive, you will be amazed how easily we can solve ac circuit problems.

4.1.4 Check Your Understanding

1. What is the frequency in hertz of the ticking of a pendulum clock if it ticks twice every second?
2. If eight cycles of a sinusoidal waveform take 2 ms, find the angular frequency, ω.
3. What is the value of $\cos(100t + 30°)$ at $t = 10$ ms?
4. What is the time between positive peaks for the sinusoidal waveform in the previous question?

   Answers. 1. 2 Hz; 2. 25,133 rad/s; 3. 0.0472; 4. 6.28×10^{-2} s.

4.2 REPRESENTING SINUSOIDS WITH PHASORS

4.2.1 Sinusoids and Linear Systems

Sinusoid in, sinusoid out. An important idea is suggested in Fig. 4.4. Think of the circuit as a linear system, linear because the equations of R and L are linear equations and a "system" because the circuit is an interconnection of such elements. (We could also have capacitors in our circuit, but that would complicate this first effort.) Think of the voltage source as an input to this system and the current $i(t)$ as an output of the system. The important idea is the following: If the input is a sinusoid, the output is also a sinusoid at the same frequency. This assertion can be justified through examination of Eq. (4.4) and reflection on the properties of sinusoids. The sinusoidal steady-state solution of Eq. (4.4) must be

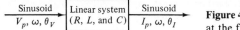

Figure 4.4 A linear system responds at the frequency of excitation.

a function which, when differentiated and added to itself, will result in a sinusoid of frequency ω. The only mathematical function that qualifies is a sinusoid of the same frequency because the "shape" of the sinusoidal function is invariant to linear operations such as addition, differentiation, and integration. To get a feeling for this, think about a system of springs and masses. If you shake such a system with a certain frequency (the ac input is an electrical shaking), the entire system will shake at the same frequency. In other words, no new frequencies are generated in the system; the only frequency that exists is the one you applied externally. So if we apply a voltage at 60 Hz, the current will respond at 60 Hz.

New unknowns. The input, being a sinusoid, is completely described by three numbers: the amplitude (V_p), the frequency (ω), and the phase (θ_V). The output must also be a sinusoid, and hence can also be described by an amplitude (I_p), frequency (ω), and phase (θ_I). Because the output frequency is known, only the amplitude and phase need to be derived to determine the steady-state current. Our object, therefore, is to develop an efficient method for finding the amplitude and phase of the output, which thus become our new unknowns. We will now develop a mathematical model suited to finding the unknown amplitude and phase. The first step is to represent the amplitude and phase of a sinusoid by a complex number.

A mathematical model. We will model a sinusoid as a rotating point in the complex plane. An initial difficulty is that many students are not up to speed in the complex number system without review. So now we must follow a detour in this development to refresh your knowledge of the complex plane. If you do not need the detour, jump to Section 4.2.3.

4.2.2 Mathematics of the Complex Plane

Most of us were introduced to complex numbers through the study of quadratic equations. We discovered that the solution to certain equations such as $x^2 = -4$ required the introduction of a new type of number:

$$x = \sqrt{-4} = i2$$

The new numbers are called *imaginary* and the symbol i is chosen (by mathematicians) to identify these new numbers, like "$-$" is used to identify negative numbers. The connotation of the word "imaginary" is unfortunate, because these new numbers are no less the product of mathematical imagination than "real" numbers. The solutions of other quadratic equations are combinations of the real and imaginary numbers, such as $2 \pm i\sqrt{2}$. These numbers are called *complex*. Complex numbers are perhaps well named because the rules for manipulating them are more complicated than those for real numbers, but on the whole the complex numbers represent a reasonable and useful extension of our number system.

We expect that you already have some skill in dealing with complex numbers. However, we will review the properties of complex numbers, because we will need many of these properties to understand fully why a rotating point in the complex plane is an ingenious tool for analyzing ac circuits. We now list and illustrate some of the important properties of complex numbers.

Complex plane.   All numbers can be represented as points in a complex plane, such as we show in Fig. 4.5. The horizontal axis represents real numbers, and the vertical axis represents imaginary numbers. You will note that we have begun using j instead of i to indicate imaginary numbers. This is customary among electrical engineers to avoid confusion between imaginary numbers and currents, which have traditionally been symbolized by i. In Fig. 4.5 we show two complex numbers, z_1 and z_2, each having a real and imaginary component. In general we can state that a complex number has a real and imaginary part,

$$z = x + jy$$

Below, we give the principal algebraic and geometric properties of complex numbers, for we must understand these properties to accomplish our goal of representing sinusoids as rotating points in the complex plane.

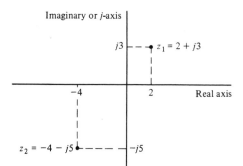

Figure 4.5   A complex number can be represented by a point in the complex plane.

Addition, subtraction, multiplication, and equality of complex numbers.   A good first approximation to the algebra of complex numbers is to use ordinary algebra plus two rules: (1) keep real and imaginary parts separate and (2) treat j^2 as -1. For example, when we add complex numbers, we add the real parts and the imaginary parts separately:

$$z_1 + z_2 = (2 + j3) + (-4 - j5) = (2 - 4) + j(3 - 5) = -2 - j2$$

and similarly for subtraction. Thus complex numbers add and subtract like vectors in a plane. This is one of the few properties common between complex numbers and vectors.

As a second example, we will find the square root of $-1 + j2$. We have two unknowns, the real and the imaginary parts of the root, so we can proceed as follows:

The Analysis of AC Circuits   Chap. 4

$$(x + jy)^2 = -1 + j2$$

$$x^2 + 2x(jy) + (jy)^2 = -1 + j2$$

$$x^2 - y^2 + j(2xy) = -1 + j2$$

$$x^2 - y^2 = -1 \quad \text{and} \quad 2xy = 2$$

Notice that we treated j^2 as -1 and we kept real and imaginary parts separate. Thus an equation involving complex numbers is equivalent to two ordinary equations, one for the real part and one for the imaginary part. We can continue the problem by solving simultaneouslly for x and y, but we will stop here. As we will soon see, there is a better way to find the roots of complex numbers.

Division, conjugation, and absolute values. Division requires a trick to get the results into a standard form:

$$\frac{z_2}{z_1} = \frac{-4 - j5}{2 + j3} = \frac{-4 - j5}{2 + j3} \times \frac{2 - j3}{2 - j3}$$

$$= \frac{(-4)(2) + (-j5)(-j3) + (-4)(-j3) + (-j5)(2)}{(2)^2 - (j3)^2}$$

$$= \frac{-8 - 15 + j(+12 - 10)}{4 + 9} = -\frac{23}{13} + j\frac{2}{13}$$

The first form we wrote to the right of the first equal sign is considered nonstandard because it contains a complex number in the denominator. To force the demoninator to be real, we multiply top and bottom by the *complex conjugate* of z, which is the same complex number except that the sign of the imaginary part is changed. This causes the cross term in the product in the denominator to drop out and thus forces the denominator to be real and positive. Meanwhile the numerator requires lots of careful work, but the rules are simple: Keep real and imaginary parts separate and let $j^2 = -1$. The final form is now considered standard because we can at a glance identify the real and imaginary parts of the quotient.

If these tedious manipulations of complex numbers discourage you, take heart—there is a better way to multiply and divide complex numbers. Before we present this better way, however, we must look more closely at the *complex conjugate*. In Fig. 4.6 we show the complex conjugate of z_1, denoted by z_1^*. We

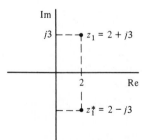

Figure 4.6 The complex conjugate of z_1 is z_1^*.

note that z_1^* is the mirror image of z_1. We have already shown that the product of a complex number with its conjugate is a real and positive number:

$$zz^* = (x + jy)(x - jy) = x^2 - (jy)^2 = x^2 + y^2$$

This suggests the definition of the absolute value (or the magnitude) of a complex number

$$|z| = \sqrt{zz^*} = \sqrt{x^2 + y^2} \tag{4.6}$$

The absolute value of a complex number thus is the Pythagorean sum of the real and imaginary parts. Geometrically, we identify this with the distance from the origin to the point representing the complex number, as shown in Fig. 4.7.

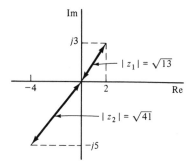

Figure 4.7 The magnitude of a complex number is its distance from the origin.

Trigonometric form. Figure 4.8 shows a general complex number. We have shown that z, a point in the complex plane, can be located either by its x, y location or by a distance and an angle. We have used r for the distance; clearly, r is $|z|$, the absolute value of z. We call the x, y form of z the *rectangular* form and the r, θ form the *polar* form. We can symbolize the two forms as

$$z = x + jy = r\underline{/\theta} = |z|\,\underline{/\theta} \tag{4.7}$$

where the symbol $\underline{/}$ is read "at an angle of." The right triangle yields simple transformations between rectangular and polar form:

$$x, y \quad \Leftrightarrow \quad r, \theta$$

$$
\begin{array}{ccc}
x = r \cos \theta & & r = \sqrt{x^2 + y^2} \\
& \Leftrightarrow & \\
y = r \sin \theta & & \theta = \tan^{-1} \dfrac{y}{x}
\end{array}
\tag{4.8}
$$

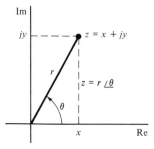

Figure 4.8 Complex number in rectangular and polar forms.

The Analysis of AC Circuits   Chap. 4

Most engineering calculators have built-in programs to accomplish this transformation both ways. Presumably, you have used this feature in working with vectors.

At this stage of our review, the polar form represents only a notation. If we wanted to multiply or take the square root of a complex number, we would have to do it with the rectangular form. But the polar form is closely related to the *exponential* form for a complex number and because of this relationship, the polar form proves to be extremely useful, not merely as a notation but also as a computational aid. All this arises out of Euler's theorem, an amazing relationship between the exponential function and angles in the complex plane.

Exponential form.   The Swiss mathematician Euler (pronounced "oiler," as in Houston Oilers) discovered an important property of complex numbers. He began with the series expansion for the function e^x,

$$e^x = 1 + x + \frac{x^2}{2!} + \frac{x^3}{3!} + \cdots$$

and made a series expansion for e^x when x is imaginary, $x = j\theta$.

$$e^{j\theta} = 1 + (j\theta) + \frac{(j\theta)^2}{2!} + \frac{(j\theta)^3}{3!} + \frac{(j\theta)^4}{4!} + \cdots$$

Euler simplified with $(j\theta)^2 = -\theta^2$, $(j\theta)^3 = -j\theta^3$, . . . , and followed the rule of grouping together real and imaginary parts. The results were

$$e^{j\theta} = \left(1 - \frac{\theta^2}{2!} + \frac{\theta^4}{4!} - \cdots \right) + j \left(\theta - \frac{\theta^3}{3!} + \frac{\theta^5}{5!} + \cdots \right)$$

The series in the parentheses Euler identified as the expansions for cosine and sine. We would speculate that he wrote something like

$$e^{j\theta} = \cos\theta + j\sin\theta \quad (?) \tag{4.9}$$

The question mark is not there because of some suspicion about the mathematics—the algebra is impeccable—but it is there because Euler must have wondered what this could mean. Look, for example, at the strange combination of mathematical symbols on the left side of Eq. (4.9). A special number from differential calculus (e), the square root of -1, and the ratio of the arc to the radius of a circle (θ)—what can these have to do with each other? And on the right side we find the ratios of the legs to the hypotenuse of a right triangle. What can these have to do with e and j? The right side of Eq. (4.9) is interpreted in Fig. 4.9.

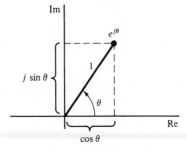

Figure 4.9 Implication of Eq. (4.9).

Since the Pythagorean sum of cosine and sine is unity, the right side of Eq. (4.9) must be a point in the complex plane one unit from the origin located at an angle θ counterclockwise from the positive real axis, as shown. Euler's theorem requires that this point also be expressed by the function $e^{j\theta}$, as indicated by Eq. (4.9).

The best way to get acquainted with a new mathematical relationship often is to try it on some familiar specific cases to see how it works. Let us try Euler's formula on the identity $\sqrt{-1} = j1$. We will express the left side of this equation with Euler's formula and follow the rules of exponents:

$$\sqrt{-1} = \sqrt{e^{j\pi}} = (e^{j\pi})^{1/2} = e^{j\pi/2} = \cos\frac{\pi}{2} + j\sin\frac{\pi}{2}$$

$$= 0 + j1$$

The first subsitution follows from the fact that -1 in the complex plane is one unit from the origin at an angle π from the positive real axis. The second change replaces the square root symbol by the one-half power, and the third change follows an ordinary rule of exponents. The last form on the first line is a direct application of Euler's formula, and the second line gives the cosine and sine of $\pi/2$ radians. We conclude that Euler's formula leads to the correct answer for this special case. If we tried it on other powers and roots, it would work out there also.

Let us try Euler's formula on another, more demanding calculation. In Fig. 4.10 we show two complex numbers to be multiplied. The numbers are chosen to give common right triangles so that we can use familiar angles for the Euler formula. Note that these numbers are not on the unit circle, so we must multiply the complex exponential by a constant to move it out to the correct distance from the origin. In other words, because the exponential function represents only the angle of the complex number, we may multiply it by a magnitude to represent any complex number.

$$\sqrt{2}\, e^{j\pi/4} = 1 + j1$$

$$2e^{j2\pi/3} = -1 + j\sqrt{3}$$

$$(\sqrt{2}\, e^{j\pi/4})(2e^{j2\pi/3}) \stackrel{?}{=} (1 + j1)(-1 + j\sqrt{3})$$

$$2\sqrt{2}\, e^{j11\pi/12} = (1)(-1) + (j1)(j\sqrt{3}) + (j1)(-1) + (1)(j\sqrt{3})$$

$$2\sqrt{2}\left(\cos\frac{11\pi}{12} + j\sin\frac{11\pi}{12}\right) = (-1 - 1.732) + j(1.732 - 1)$$

$$-2.732 + j0.732 = -2.732 + j0.732$$

On the left side we have combined the exponential forms according to the laws of exponents, and on the right side we have followed the usual rules for complex numbers in rectangular form. We again conclude that Euler's formula works. Indeed, the strange formula in Eq. (4.9) is valid and provides the foundation for many important techniques in dealing with complex numbers. As you will see, our model for an ac waveform arises directly out of Euler's formula. Before returning to the main road from this detour, we wish first to show how the ex-

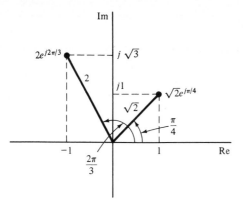

Figure 4.10 Two complex numbers in exponential form.

ponential form of complex numbers, based on Euler's formula, provides a simple way to multiply, divide, and take roots of complex numbers.

The relationship between our polar form and the exponential form follows from a comparison between Eqs. (4.8) and (4.9).

$$z = r\underline{/\theta} = re^{j\theta} \qquad (4.10)$$

In Eq. (4.10) we normally express the angle of the polar form in degrees, whereas in the exponential form the angle *must* be expressed in radians for the Euler formula to be valid. This inconsistency arises from our traditions but does not diminish the fact that both forms give the same information, namely, that the complex number falls in the complex plane at a radius r and an angle θ counterclockwise from the positive real axis. We stress that the polar form is merely a compact notation, whereas the exponential form is a respectable mathematical function. The laws of exponents reveal that the magnitude of the product of two complex numbers is the product of the magnitudes of the numbers and the angle of the product is the sum of their angles.

$$(z_1)(z_2) = (r_1 e^{j\theta_1})(r_2 e^{j\theta_2}) = r_1 r_2 e^{j(\theta_1 + \theta_2)}$$

Although the proof rests on the exponential form, the results are more easily expressed in polar form:

$$(r_1\underline{/\theta_1})(r_2\underline{/\theta_2}) = r_1 r_2 \underline{/\theta_1 + \theta_2}$$

Hence to multiply two complex numbers, multiply their magnitudes and add their angles. In a similar way, we can show that dividing two complex numbers requires dividing the magnitudes and subtracting the angles:

$$\frac{r_1\underline{/\theta_1}}{r_2\underline{/\theta_2}} = \frac{r_1}{r_2}\underline{/\theta_1 - \theta_2}$$

Summary of how to calculate with complex numbers.   When we add and subtract complex numbers, the rectangular form must be used. This is true because we add and subtract real and imaginary parts separately. When we multiply (or divide) complex numbers, the polar form should be used because the magnitudes multiply (or divide) and the angles add (or subtract).

For the analysis of ac circuits, we must master the manipulation of complex numbers. Our techniques require frequent changes between rectangular and polar form. In the days of slide rules, these conversions were a bother, but with modern calculators they usually require a simple push of a special button on the calculator. If you currently do not know about those buttons,† locate the manual on your calculator and look it up. This review of complex numbers is vital to your gaining skill in solving ac circuit problems. These operations, like the vector manipulations required in statics, are simple in principle, but care is required to produce the correct answer consistently.

Some examples.   Here are two additional examples of complex number manipulation, which give practice and show some new principles.

$$(2 - j6)(5\underline{/+30°})^* = (6.32\underline{/-71.6°})(5\underline{/-30°})$$

$$= 31.6\underline{/-101.6°}$$

$$= -6.34 - j31.0$$

Note that complex conjugation merely changes the sign of the angle.
  Find z, where $z^3 = -1 + j0.5$.

$$z^3 = -1 + j0.5 = 1.12\underline{/153.4°} = 1.12e^{j(153.4)(\pi)/180}$$

$$z = (1.12e^{j2.68})^{1/3} = \sqrt[3]{1.12}\,e^{j2.68/3}$$

$$= \text{also}\quad \sqrt[3]{1.12}\,e^{j(2.68 + 2\pi)/3} \quad \text{and} \quad \sqrt[3]{1.12}\,e^{j(2.68 + 4\pi)/3}$$

$$= 1.04\underline{/51.1°},\ 1.04\underline{/171.1°},\ 1.04\underline{/291.1°}$$

Note that to find all three cube roots, we must add 2π and 4π (one and two additional rotations) to the angle before dividing by 3.

4.2.3 Phasor Idea

Expressing a sinusoid with a complex number.   Now that we have reviewed the mathematics of the complex plane, we will return to the main quest, that of finding the steady-state solution of the ac circuit problem described in Fig. 4.3 and Eq. (4.4). Specifically, we wish to find a way to represent sinusoidal sources, such as that appearing in Eq. (4.4), repeated below:

$$v_s(t) = V_p \cos(\omega t + \theta), \qquad V_p = 100, \quad \omega = 2\pi \times 60, \quad \theta = 30°$$

Euler's formula allows us to express a sinusoidal waveform with the function

$$V_p \cos(\omega t + \theta) = \text{Re}\{V_p e^{j(\omega t + \theta)}\} \tag{4.11}$$

where Re, the "real part of," is the complex-plane notation for the projection on the real axis. Equation (4.11) follows from Eq. (4.12) below, which expresses Euler's formula in appropriate terms:

---

† Usually marked →P and →R. Indeed, some advanced scientific calculators can process complex numbers without converting back and forth between polar and rectangular forms. If your calculator has this capability, we strongly suggest that you learn to use this feature.

$$V_p e^{j(\omega t + \theta)} = V_p[\cos(\omega t + \theta) + j \sin(\omega t + \theta)]$$

(4.12)

$$= \underbrace{V_p \cos(\omega t + \theta)}_{\text{Real part}} + j \underbrace{V_p \sin(\omega t + \theta)}_{\text{Imaginary part}}$$

Rotating point in the complex plane. The left side of Eq. (4.12) has great importance in this development and deserves more attention. We investigate it in the equation

$$V_p e^{j(\omega t + \theta)} = V_p(e^{j\omega t})(e^{j\theta}) = (V_p e^{j\theta})e^{j\omega t}$$

(4.13)

$$= (V_p\underline{/\theta})e^{j\omega t} = (100\underline{/30°})e^{j120\pi t}$$

The first line results from some juggling of terms and the rules of exponents. The second line emphasizes that this term consists of a complex constant, $100\underline{/30°}$ in this case, and a time function. The time function, $e^{j\omega t}$, describes the movement of a point in the complex plane. If we follow its progress as time increases, as shown in Fig. 4.11, we find it to be a point at unit distance from the origin rotating around the origin, starting at the real axis at $t = 0$ and passing through that point 60 times every second. The time function $e^{j\omega t}$ describes a point rotating in the complex plane at an angular frequency of ω radians per second. When we multiply this rotation function by the complex number $100\underline{/30°}$, we move the point out to a magnitude of 100 and counterclockwise by an angle of 30° at $t = 0$ as shown in Fig. 4.12 (point a). The rotation begins from that point and rotates around a

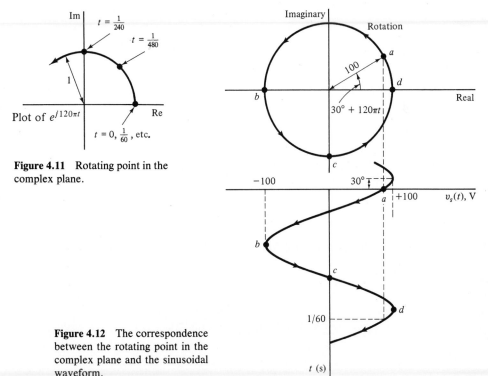

Figure 4.11 Rotating point in the complex plane.

Figure 4.12 The correspondence between the rotating point in the complex plane and the sinusoidal waveform.

circle of radius 100, 60 times a second. The projection on the real axis is shown below the complex plane (you have to turn the page on its side to look at the resulting time function). Clearly, we have the desired sinusoid—correct amplitude, phase, and frequency. Figure 4.12 is a graphical interpretation of Eq. (4.11), illustrated for the values appropriate for our specific case. We have now shown how to represent a sinusoid by a rotating point in the complex plane. To see why this is useful, we must return to the differential equation we are solving.

Using complex numbers to solve the DE. Our DE is

$$L\frac{di}{dt} + Ri = V_p \cos(\omega t + \theta) = \mathrm{Re}\,\{V_p\underline{/\theta}\,e^{j\omega t}\}$$

$$= \mathrm{Re}\,\{\underline{\mathbf{V}}_s e^{j\omega t}\} \tag{4.14}$$

We have rewritten the DE in Eq. (4.14) and used the rotating point form on the right side. The third form on the right side uses the symbol $\underline{\mathbf{V}}_s$ to represent the complex number containing the source amplitude and phase, $V_p\underline{/\theta}$ in general, $100\underline{/30°}$ in this case. We use the uppercase bold, underlined symbol to indicate that it is a constant complex number.

Recall that we know something about the unknown current, $i(t)$. Earlier we argued that the output of the system (RL circuit in this case) had to be a sinusoid of the same frequency as the input. Only the amplitude and phase of that sinusoid must be determined. Using our scheme for representing sinusoids by complex numbers, we know that $i(t)$ may be represented in the form

$$i(t) = \mathrm{Re}\{\underline{\mathbf{I}}e^{j\omega t}\}, \qquad \text{where } \underline{\mathbf{I}} = I_p\underline{/\theta_I} \tag{4.15}$$

We have thus introduced the complex number, $\underline{\mathbf{I}}$, as the unknown, everything else in the equation having been specified in the problem statement. We have numerical values for L and R, and of course ω is known from the input. We can now substitute Eq. (4.15) into Eq. (4.14):

$$L\frac{d}{dt}\,\mathrm{Re}\{\underline{\mathbf{I}}e^{j\omega t}\} + R \times \mathrm{Re}\{\underline{\mathbf{I}}e^{j\omega t}\} = \mathrm{Re}\{\underline{\mathbf{V}}_s e^{j\omega t}\} \tag{4.16}$$

We wish to avoid taking unnecessary excursions into mathematical proofs, so you will have to take our word for this assertion: The derivative can be taken inside the "real part" operation

$$\frac{d}{dt}\,\mathrm{Re}\{\underline{\mathbf{I}}e^{j\omega t}\} = \mathrm{Re}\left\{\frac{d}{dt}\,\underline{\mathbf{I}}e^{j\omega t}\right\} = \mathrm{Re}\left\{\underline{\mathbf{I}}\frac{d}{dt}\,e^{j\omega t}\right\}$$

The exponential function with base e has the special property that its derivative is proportional to the function, as shown in the equation

$$\frac{d}{dx}\,e^{\alpha x} = e^{\alpha x}\frac{d}{dx}(\alpha x) = \alpha e^{\alpha x} \tag{4.17}$$

Applying this same procedure to the $e^{j\omega t}$ function, we obtain the result

$$\frac{d}{dt}\,e^{j\omega t} = e^{j\omega t}\frac{d}{dt}(j\omega t) = j\omega e^{j\omega t} \tag{4.18}$$

We substitute the derivative back into Eq. (4.16). Because the "real part" operation distributes, we can collect terms:

$$\text{Re}\{L(j\omega\underline{I}e^{j\omega t}) + R(\underline{I}e^{j\omega t})\} = \text{Re}\{\underline{V}_s e^{j\omega t}\} \tag{4.19}$$

$$\text{Re}\{(j\omega L\underline{I} + R\underline{I} - \underline{V}_s)e^{j\omega t}\} = 0 \tag{4.20}$$

Equation (4.20) emerges from the manipulations; but what does it mean? Look first at the term, $j\omega L\underline{I} + R\underline{I} - \underline{V}_s$. This represents a complex number, a constant. Because it contains the complex number representing the unknown current amplitude and phase, $\underline{I}$, this term in Eq. (4.20) is an unknown constant, a point somewhere in the complex plane. Of course, the $e^{j\omega t}$ term rotates this unknown constant, and the Re represents the projection of the rotating complex constant on the real axis. The right side of Eq. (4.20) requires that this projection be zero at all times; hence the complex constant must itself be zero. If the point were not rotating, many complex constants could have zero projection (all points on the imaginary axis have this property), but the origin is the only point that continues to have zero projection when rotated about the origin. Equation (4.20) thus requires

$$j\omega L\underline{I} + R\underline{I} - \underline{V}_s = 0 \tag{4.21}$$

Equation (4.21) involves only complex constants; time is no longer a factor. We know all the constants except $\underline{I}$, so we can solve for the unknown $\underline{I}$:

$$\underline{I} = \frac{\underline{V}_s}{R + j\omega L} = \frac{100\underline{/30°}}{100 + j(120\pi \times 0.5)}$$

$$= \frac{100\underline{/30°}}{213\underline{/62.1°}} = 0.469\underline{/-32.1°}$$

Recall that the magnitude of $\underline{I}$ represents the peak value of the sinusoidal steady-state current flowing in the RL circuit and the angle of $\underline{I}$ represents the phase angle of the current. Hence we may write

$$i_p(t) = 0.469 \cos(120\pi t - 32.1°) \tag{4.22}$$

This at last is the steady-state solution to the DE in Eq. (4.4) and hence this is the current that flows in the circuit shown in Fig. 4.3 after the transient period passes. Before finding the transient part of the solution, we will summarize the argument.

Summary of the argument. Since the development of Eq. (4.22) has taken many pages and extensive math review, you may feel that you have been led through a complicated argument. Actually, there are but few steps, which we will now review. First we wrote the DE and focused on the steady-state solution.

$$L\frac{di}{dt} + Ri = V_p \cos(\omega t + \theta)$$

We then represented the sinusoidal voltage and the unknown sinusoidal current by the real parts of rotating complex constants and took the derivative. The result was

$$\text{Re}\{j\omega L\underline{I}e^{j\omega t}\} + \text{Re}\{R\underline{I}e^{j\omega t}\} = \text{Re}\{\underline{V}_s e^{j\omega t}\} \tag{4.23}$$

We then collected terms and reasoned that the resulting complex constant had to vanish when the equation is put into the form of Eq. (4.20). This amounts to the same thing as dropping the Re and the $e^{j\omega t}$ in Eq. (4.23). Note that we did not *cancel* these factors—that would be mathematically incorrect—but the result is the same as if we had canceled them. The resulting equation is solved for $\underline{I}$, and the result is easily interpreted in terms of the amplitude and phase of the current in the circuit.

Complete solution. We may now determine the complete solution to Eq. (4.4) using principles developed in Chapter 3. The form of the solution is

$$i(t) = Ae^{-t/\tau} + 0.469\cos(120\pi t - 32.1°)$$

where $\tau = L/R = 1/200$ s and A is an unknown constant to be determined from the initial condition. The initial current must be zero, $i(0^+) = 0$, due to the inductor; hence

$$0 = Ae^{0/\tau} + 0.469\cos(-32.1°)$$

Hence $A = -0.397$. Thus the total solution is

$$i(t) = -0.397e^{-200t} + 0.469\cos(120\pi t - 32.1°)$$

This response is shown in Fig. 4.13. Note that the current rapidly approaches the sinusoidal steady-state part of the solution.

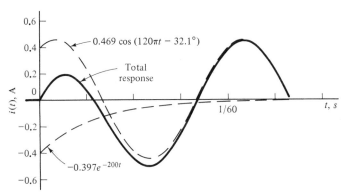

Figure 4.13 The current is the sum of the transient and steady-state responses.

Shortening the procedure. This method for determining the sinusoidal steady state admits to additional shortcuts. If we compare Eq. (4.21) with the original DE, we see a direct correlation between terms:

$$i(t) \Rightarrow \underline{I}, \qquad v_s(t) \Rightarrow \underline{V}_s, \qquad \frac{d}{dt} \Rightarrow j\omega \tag{4.24}$$

We have shown that these are legitimate transformations. These suggest a shorter method:

1. Write the DE.
2. Perform the transformations shown in Eq. (4.24) on the known sinusoidal source, the unknown voltage or current, and the d/dt.
3. Solve the resulting complex equation for the unknown.
4. Interpret the results by substituting the amplitude and phase back into the standard form for a sinusoid.

Another example, solved with the short method. Let us follow this procedure on another problem. Figure 4.14 shows an RC circuit driven by a sinusoidal current source. We wish to solve for the current in the capacitor. The relevant equations are KCL at the top node,

$$-i_s(t) + i_R(t) + i_C(t) = 0 \tag{4.25}$$

and the definitions of R and C,

$$i_R = \frac{v}{R} \quad \text{and} \quad i_C = C\frac{dv}{dt}$$

We can eliminate i_R from the KCL equation by taking the derivative of Ohm's law and substituting for dv/dt. The result is

$$\frac{di_R}{dt} = \frac{1}{R}\frac{dv}{dt} = \frac{1}{RC}i_C$$

Now we differentiate Eq. (4.25), make this substitution, and rearrange terms into the usual form:

$$\frac{di_C}{dt} + \frac{1}{RC}i_C = \frac{d}{dt}i_s(t)$$

This completes the first step of our solution. The second step is to transform this DE into a complex equation with the changes suggested by Eq. (4.24).

$$i_s \Rightarrow \underline{I}_s, \qquad i_C \Rightarrow \underline{I}_C, \qquad \frac{d}{dt} \Rightarrow j\omega$$

The resulting complex equation is

$$j\omega\underline{I}_C + \frac{1}{RC}\underline{I}_C = j\omega\underline{I}_s$$

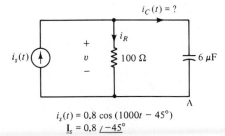

$i_C(t) = ?$

$i_s(t) = 0.8 \cos(1000t - 45°)$
$\underline{I}_s = 0.8 \,\underline{/-45°}$

Figure 4.14 Solve for the steady-state current in the capacitor.

We can solve the equation for $\underline{I}_C$, which represents the amplitude and phase of the unknown current in the capacitor. The results are

$$\underline{I}_C = \frac{j\omega\underline{I}_s}{(1/RC) + j\omega} = \frac{j1000(0.8\underline{/-45°})}{1667 + j1000}$$

$$= \frac{800\underline{/90° - 45°}}{1944\underline{/31.0°}} = 0.412\underline{/14.0°}$$

The last step consists in writing the current in the standard form for sinusoids.

$$i_C(t) = 0.412 \cos(1000t + 14.0°) \text{ A}$$

You must admit that apart from the manipulation of the complex numbers, which may still be unfamiliar to you, the derivation of the DE is now the hardest part of the solution. We will soon introduce additional shortcuts for simplifying that part of the problem. First we want to introduce some vocabulary.

One important operation is representing a sinusoidal voltage or current by a complex number, as in Eq. (4.15). These complex representations are called *phasors*. A phasor voltage is the complex number that gives the time-function sinusoid when rotated at the proper frequency and projected on the real axis. Because it is hard to see a single point, phasors are usually drawn as an arrow from the origin, as in Fig. 4.15. Because phasors add like vectors, they are often incorrectly referred to as vectors.

We use the expression *frequency domain* to refer to the form of the DE after it has been transformed into a complex equation. The DE is called the *time-domain* formulation of the problem because time is an important factor, but the transformed equation belongs to the frequency domain—time is no longer a factor. The phasor voltages and currents belong to the frequency domain. We move from the frequency domain to the time domain by rotating the phasors at the proper frequency and taking projections on the real axis. The factor $e^{j\omega t}$ represents the bridge between the time and frequency domains because it is used to move back and forth between the two domains.

Example. In Fig. 4.16a, we show a series RC circuit, with the voltage across the capacitor given as $5 \sin(500t)$ V. We are to determine the voltage across the source that produces this prescribed voltage. Considering that the current is $C \, dv_C/dt$, we write the DE for the source voltage as

$$v_s(t) = 100 \times 25 \times 10^{-6} \frac{dv_C}{dt} + v_C \tag{4.26}$$

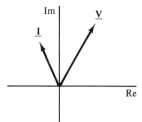

Figure 4.15 Voltage phasor and current phasor.

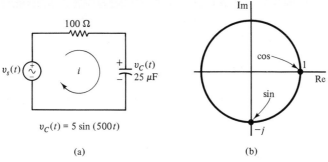

$v_C(t) = 5 \sin(500t)$

(a)

(b)

Figure 4.16 (a) The voltage across C is given; find the source voltage; (b) the frequency domain equivalent of a sine function is $-j$ or $1\underline{/-90°}$

which transforms into the frequency domain as

$$\underline{\mathbf{V}}_s = 25 \times 10^{-4}(j500)\underline{\mathbf{V}}_C + \underline{\mathbf{V}}_C = (1 + j1.25)\underline{\mathbf{V}}_C \qquad (4.27)$$

where $\underline{\mathbf{V}}_s$ and $\underline{\mathbf{V}}_C$ are phasors representing the voltages across the source and capacitor, respectively. Because $v_C(t)$ is a sine (instead of a cosine) function, we must represent it as

$$v_C(t) = 5 \sin(500t) \Rightarrow \underline{\mathbf{V}}_C = -j5 \qquad (4.28)$$

We may show that sine transforms into $-j$ by using the trigonometric identity $\sin(x) = \cos(x - 90°)$, but a clearer demonstration is shown in Fig. 4.16b. We have identified $-j$ with sine because that point produces a sine function when rotated and projected on the real axis. Note that the projection is zero at $t = 0$, but increases to unity one-quarter of a period later. Finally, we substitute Eq. (4.28) into Eq. (4.27) and transform back into the time domain.

$$\underline{\mathbf{V}}_s = (1 + j1.25)(-j5) = 1.601\underline{/51.3°} \times 5\underline{/-90°}$$

$$= 8.00\underline{/-38.7°} \qquad (4.29)$$

$$v_s(t) = 8.00 \cos(500t - 38.7°) \text{ V}$$

4.2.4 Check Your Understanding

1. Convert $10 - j12$ and $-30 + j58$ to polar form.
2. What is the rectangular form for $4\underline{/25°}$ and $0.025\underline{/-140°}$?
3. Convert $13.5e^{j0.86}$ to polar and rectangular form.
4. What is the complex conjugate of $z = -2 + j6$?
5. Give both square roots of $1 + j1$.
6. What are the phasors representing $-6 \cos(\omega t - 30°)$ and $5 \sin(\omega t + 10°)$?
7. What is the time-domain sum of the two sinusoids in the previous question?
8. In transforming a DE into the frequency domain, what replaces the d/dt?

Answers. 1. $15.6\underline{/-50.2°}$, $65.3\underline{/117.3°}$; 2. $3.63 + j1.69$, $(-1.92 - j1.61) \times 10^{-2}$; 3. $13.5\underline{/49.3°}$, $8.81 + j10.2$; 4. $-2 - j6$; 5. $1.19\underline{/22.5°}$, $1.19\underline{/202.5°}$; 6. $6\underline{/150°}$, $5\underline{/-80°}$; 7. $4.74 \cos(\omega t - 156°)$; 8. $j\omega$.

4.3.1 Basic Idea

Look at the answer.   Let us look once more at the steady-state response of the *RL* circuit we solved, shown again in Fig. 4.17. This is a simple circuit with *R* and *L* in series; we know the sinusoidal driving voltage and we have solved for the current. When expressed as phasor voltages and currents, our results were

$$\underline{I} = \frac{\underline{V}_s}{R + j\omega L} \tag{4.30}$$

We interpret Eq. (4.30) as stating that the phasor current is found by dividing the phasor voltage by the sum of two terms, the resistance and $j\omega L$. Because Eq. (4.30) closely resembles Ohm's law, every part of this expression has meaning to us except the $j\omega L$ part. Clearly, this term represents the effect of the inductor in the circuit; the inductance appears in no other part of the answer. What can it mean?

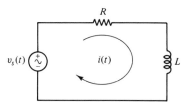

Figure 4.17   *RL* circuit again.

Impedance.   This question, or one similar to it, sparked long ago an idea that leads to the final shortcut in analyzing ac circuits—the idea of impedance. This idea is the following: Since we are dealing only with sinusoidal voltages and currents, and since these can be represented by complex numbers, why not represent *R*'s, *L*'s, and *C*'s by complex numbers also? This suggests transforming Ohm's law and the definitions of *L* and *C* into the frequency domain, that is, representing them by complex equations.

Impedance of an inductor.   Here is how it works out for the inductor: The defining equation for an inductor is

$$v_L = L\frac{di_L}{dt}$$

Because we are dealing with sinusoids, we can transform this equation into a frequency-domain equivalent:

$$v_L(t) \Rightarrow \underline{V}_L, \qquad i_L(t) \Rightarrow \underline{I}_L, \qquad \frac{d}{dt} \Rightarrow j\omega$$

The result is

$$\underline{V}_L = j\omega L\underline{I}_L \tag{4.31}$$

The phasor interpretation of Eq. (4.31) is shown in Fig. 4.18. Equation (4.31) gives both the magnitude and phase relationships between the phasor voltage and phasor current for an inductor. The magnitude of the phasor voltage is ωL times the magnitude of the phasor current. Since $j1 = 1\underline{/90°}$, the phase of the phasor voltage leads the phase of the phasor current by 90°. We can verify this interpretation by taking the derivative in the time domain. If $i_L = I_p \cos(\omega t + \theta_I)$, then $v_L(t)$ is

$$v_L = L\frac{d}{dt} I_p \cos(\omega t + \theta_I) = -\omega L I_p \sin(\omega t + \theta_I)$$

$$= \omega L I_p \cos(\omega t + \theta_I + 90°)$$

where we have used the identity that $-\sin\theta = \cos(\theta + 90°)$. We use the word *impedance* to refer to the complex number relating the phasor voltage and phasor current of an element in an ac circuit, as shown in Fig. 4.19.

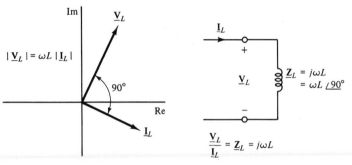

Figure 4.18 For an inductor, current lags voltage by 90°.

Figure 4.19 Representing an inductor by its impedance.

We define impedance as the phasor voltage divided by the phasor current, as in

$$\underline{\mathbf{Z}} = \frac{\mathbf{V}}{\mathbf{I}} \tag{4.32}$$

Applying this definition to Eq. (4.31), we obtain the result shown in Fig. 4.19. Hence $\underline{\mathbf{Z}}_L = j\omega L$ is the impedance of an inductor at frequency ω.

Kiirchhoff's laws in the frequency domain.   You will note that Fig. 4.19 also introduces the practice of putting the phasor voltage and current on the circuit diagram. This is legitimate and useful because, for sinusoidal voltages and currents, KVL and KCL can also be transformed directly into the frequency domain; for example,

$$-v_s(t) + v_R(t) + v_L(t) = 0 \Rightarrow -\underline{\mathbf{V}}_s + \underline{\mathbf{V}}_R + \underline{\mathbf{V}}_L = 0 \tag{4.33}$$

Equation (4.33) gives the frequency-domain version of KVL, as applied to the *RL* circuit. If there were nodes in the circuit, we could also write frequency-domain versions of KCL at these nodes.

Summary. We can represent all voltages and currents in an ac circuit as phasors. These phasors are complex numbers representing the amplitudes and phases of the waveforms in the sinusoidal steady state. Phasor voltages and currents obey KVL and KCL. We can also represent R and L (and C, when we get around to it) as complex numbers, which are called *impedances*. Figure 4.20 gives the impedance of R, which is Ohm's law for phasors. The impedance of a resistor is real because its voltage and current have the same phase.

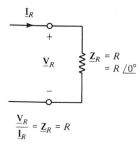

$$\frac{V_R}{I_R} = Z_R = R$$

Figure 4.20 The impedance of a resistor is real because no phase shift occurs.

Representing the circuit in the frequency domain. We can transform the entire circuit into the frequency domain, representing voltages and currents as phasors and representing the resistor and inductor as impedances. Because phasor voltages and currents obey KVL and KCL, we can use the techniques of dc circuit theory to solve the circuit, only we must now work with impedances as if they were "complex resistors." Using impedances most efficiently formulates the circuit equations (DEs in the time domain) and determines the sinusoidal steady-state response.

Benefits of the impedance concept. We can now solve our original problem by an efficient method. Figure 4.21 shows the circuit transformed into the frequency domain. Because we have a simple series circuit, we can find $i(t)$ by the method of equivalent resistance (actually, equivalent impedance), combining the impedances of R and L in series. If we were faced with a more complicated circuit, we might analyze the circuit using node voltages, loop currents, or a Thévenin equivalent circuit. Through the concept of impedance, all the techniques of dc circuits can be applied to ac circuit problems.

By introducing the impedances we avoid the task of writing the DE. In fact, writing the DE corresponds to the "nonmethod" of solving dc circuits because

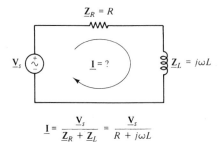

$$I = \frac{V_s}{Z_R + Z_L} = \frac{V_s}{R + j\omega L}$$

Figure 4.21 Frequency-domain version of the RL circuit.

we introduce many variables and then seek sufficient equations to eliminate unwanted variables from the equations. The impedance concept eliminates that necessity.

Let us rework the second example (Fig. 4.14) using frequency-domain techniques. Because the circuit contains a capacitor, we must derive the impedance of a capacitor, as symbolized in Fig. 4.22. This results from transforming the definition of a capacitor into the frequency domain,

$$i_C = C\frac{dv}{dt} \Rightarrow \underline{\mathbf{I}}_C = C(j\omega\underline{\mathbf{V}}_C) = j\omega C\underline{\mathbf{V}}_C$$

Thus

$$\underline{\mathbf{Z}}_C = \frac{\underline{\mathbf{V}}_C}{\underline{\mathbf{I}}_C} = \frac{1}{j\omega C} = \frac{j}{j^2\omega C} = j\frac{-1}{\omega C} = \frac{1}{\omega C}\underline{/-90^\circ} \tag{4.34}$$

For a capacitor, the voltage lags the current by 90°, as shown in Fig. 4.23.

$\underline{\mathbf{I}}_C$
+
$\underline{\mathbf{V}}_C$
$\underline{\mathbf{Z}}_C$
−
$\underline{\mathbf{Z}}_C = \frac{\underline{\mathbf{V}}_C}{\underline{\mathbf{I}}_C}$
$= \frac{1}{j\omega C}$
$= \frac{1}{\omega C}\underline{/-90^\circ}$

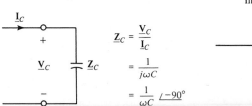

Figure 4.22 Impedance of a capacitor.

Im
$\underline{\mathbf{I}}_C$
90°
Re
$\underline{\mathbf{V}}_C$

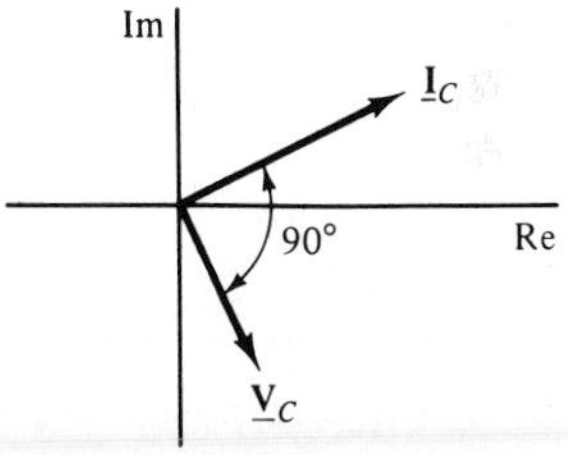

Figure 4.23 For a capacitor, the current leads the voltage by 90°.

We now transform the circuit of Fig. 4.14 into the frequency domain using phasor currents and impedances, Fig. 4.24. Because R and C are connected in parallel, we may treat them like resistors in parallel and find $\underline{\mathbf{I}}_C$ with a current-divider relationship. The results are the same as those obtained from the DE, once you do the complex algebra to show the equivalence.

Figure 4.24 We can use impedance in a current-divider relationship.

R
$\frac{1}{j\omega C}$
$\underline{\mathbf{I}}_C$
$\underline{\mathbf{Z}}_{eq} = R \parallel \frac{1}{j\omega C} = \frac{1}{(1/R) + j\omega C}$
$\underline{\mathbf{I}}_C = \frac{\underline{\mathbf{Z}}_{eq}}{\underline{\mathbf{Z}}_C} \times \underline{\mathbf{I}}_s$
$= \left[\frac{j\omega C}{(1/R) + j\omega C}\right] \times \underline{\mathbf{I}}_s$

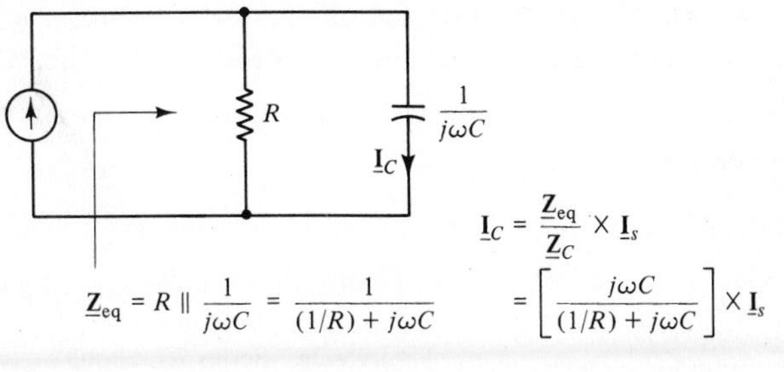

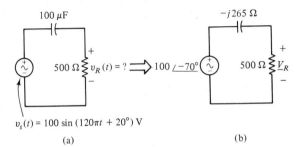

$v_s(t) = 100 \sin (120\pi t + 20°) \text{ V}$

(a)

(b)

Figure 4.25 (a) Time-domain circuit; (b) frequency-domain circuit.

Example. Figure 4.25a shows a circuit in the time domain in sinusoidal steady state; we are to determine the voltage across the resistor. Figure 4.25b shows the circuit, source, and unknown transformed into the frequency domain. Note that the sine function introduces a phase shift of $-90°$ in the phasor source voltage. The phasor voltage across the resistor can be determined from a voltage divider:

$$\mathbf{V}_R = 100\underline{/-70°} \times \frac{500}{500 - j265} = 88.3\underline{/-42.1°} \text{ V} \tag{4.35}$$

The time-domain voltage across the resistor is thus

$$v_R(t) = 88.3 \cos(120\pi t - 42.1°) \tag{4.36}$$

4.3.2 Phasor Diagrams for *RL*, *RC*, and *RLC* Circuits

You may wonder why we use the grandiose word "domain" for the techniques described above. You are correct in observing that we do not yet have much of a domain, only a technique for analyzing ac steady-state circuits at a single frequency. But there really is a domain; the exploration of this idea continues as an important theme of the remainder of this book. Extensive exploration of this concept is reserved for Chapters 8, 10, and 11.

The viewpoint in this section is that we have available a sinusoidal source, the frequency of which we vary. Here we explore the effects of frequency changes on the response of series and parallel circuits to gain insight into the properties of inductors and capacitors in ac circuits.

RL Circuits

Series *RL* circuit. Figure 4.26 shows a series *RL* circuit excited by a current source. We will investigate the effect of the inductor as frequency varies. If there were no inductor in the circuit, it would exhibit the same behavior at all

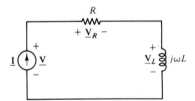

Figure 4.26 Series *RL* circuit.

The Analysis of AC Circuits   Chap. 4

frequencies; all voltages and currents would be in phase with each other and in a fixed ratio. With an inductor in the circuit, however, the properties of the circuit change as frequency is varied.

The impedance of the series circuit is

$$\underline{Z} = R + j\omega L$$

The relationship between the phasor voltage and current is thus

$$\underline{V} = \underline{Z}\underline{I} = (R + j\omega L)\underline{I} \tag{4.37}$$
$$= R\underline{I} + j\omega L\underline{I} = \underline{V}_R + \underline{V}_L$$

In Eq. (4.37) we have interpreted the two terms in the impedance as relating to the voltages across the resistor, $\underline{V}_R$, and the inductor, $\underline{V}_L$. Since the current is common to both R and L, we have drawn the phasor diagram in Fig. 4.27 with $\underline{I}$ as the phase reference, that is, $\underline{I} = I_p\underline{/0°}$. The phase of the voltage across the resistor, $\underline{V}_R$, is the same as the phase of $\underline{I}$, but the phase of the voltage across the inductor leads by 90°. The total voltage, $\underline{V}$, the phasor sum of $\underline{V}_R$ and $\underline{V}_L$, thus leads the current by a phase angle somewhere between 0° and 90°. The phase angle by which the voltage leads the current is the geometric angle of $\underline{Z}$ in the complex plane, θ, as given in Eq. (4.38) and Fig. 4.28.

$$\underline{Z} = R + j\omega L = |\underline{Z}|\underline{/\theta}$$

where

$$|\underline{Z}| = \sqrt{R^2 + (\omega L)^2}, \qquad \theta = \tan^{-1}\frac{\omega L}{R} \tag{4.38}$$

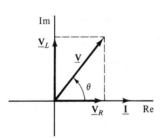

Figure 4.27 The inductor causes current to lag voltage.

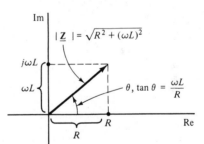

Figure 4.28 The angle of the impedance represents the phase shift.

Frequency effects on the *RL* series circuit. Now let us examine what happens to the phasor diagram when frequency is varied, other parameters being held constant. Consider first very low frequencies. By "low" we mean those frequencies where the imaginary part of $\underline{Z}$ is small compared with the real part, that is, where $\omega L \ll R$. For low frequencies, the phase angle is small, meaning that $\underline{V}$ and $\underline{I}$ are almost in phase with each other, and the magnitude of the impedance is essentially equal to R. Thus the inductor has little effect at low frequencies. We saw in Chapter 3 that inductors become invisible at dc once their initial energy requirements are met. The foregoing discussion shows inductors to be virtually invisible at low ac frequencies. Or we can put it the other way around and treat dc behavior as a limiting case of ac as ω approaches zero. The mathematics supports this approach since $e^{j\omega t} \to 1$ as $\omega \to 0$. Whichever outlook we choose, Ohm's law dominates the behavior of the circuit at low frequencies.

As frequency increases, the presence of the inductor is shown by an increase in the impedance of the inductor and hence an increase in the voltage across the inductor. This produces increased overall voltage and increased phase difference between the total voltage and the current. When $\omega L = R$, for example, the phase difference is 45° and the voltage is increased by $\sqrt{2}$ because the magnitude of the total impedance has increased by $\sqrt{2}$. As the frequency goes yet higher, the inductor comes to dominate the behavior of the circuit. The phase shift approaches 90° because the impedance approaches $j\omega L$. To get a feeling for how the impedance of the inductor gets large at high frequencies, imagine shaking a massive object in your hands. If you shake slowly (low frequencies), not much force is required, but as you attempt to shake faster, more force is required. Similarly, it takes more voltage to put the same current through an inductor as the frequency is increased.

The series *RL* circuit in Fig. 4.26 would be an appropriate circuit model for an electric motor under steady load. The resistance would represent heat and friction losses in the motor, plus the energy converted to mechanical form and applied to a mechanical load. The inductance would represent magnetic energy storage in the motor structure. If we were to draw the phasor diagram for the motor, it would look like Fig. 4.28 except that the voltage would normally be used as the phase reference, as in Fig. 4.29.

Voltage is customarily used as the phase reference in ac power circuits because lights, motors, heaters, and so on, are parallel loads requiring a common voltage. Because voltage is the phase reference, current is said to *lag* for an inductive load and, as we will soon see, *lead* for a capacitive load.

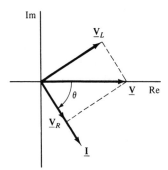

Figure 4.29 Phasor diagram redrawn with voltage as the phase reference.

Resistive and reactive parts. The foregoing discussion suggests that an impedance can represent more than a physical resistor and a physical inductor in an ac circuit. The *real part* of the impedance represents losses to the circuit, that is, energy leaving electrical form and converted to heat or some other form of energy such as mechanical work. The *imaginary part* of the impedance represents energy storage, in this instance magnetic energy storage in a motor. Because of these broader interpretations of impedance, which we will explore more fully in the next chapter, names are given to the real and imaginary parts of the impedance. The real part of $\underline{Z}$ is called the *resistive part* and the imaginary part is called the *reactive part*. The reactive part is given the symbol, X, as used in

$$\underline{Z} = R + jX = \text{(resistive)} + j\text{(reactive)} \tag{4.39}$$

The noun form of "reactive" is "reactance," so it would be correct to say: "The reactance of an inductor is ωL." Reactance is a real number having the units ohms.

Why do we need these new words? Why can't we speak in terms of inductance? One reason we have already given—that the impedance represents more than simple R and L. The other reason becomes important when frequency is varied. The reactance of a true inductor varies linearly with frequency, but the reactance of more complicated circuits or devices does not vary linearly with frequency. Reactance is a more general concept than inductance in this broader context.

Parallel *RL* load. For the parallel RL load shown in Fig. 4.30, it is natural from both practical and mathematical considerations to use voltage as a phase reference. Since both R and L are connected in parallel with an ideal voltage source, we can determine their currents independently:

$$\underline{I}_R = \frac{\underline{V}}{\underline{Z}_R} = \frac{\underline{V}}{R}, \quad \underline{I}_L = \frac{\underline{V}}{\underline{Z}_L} = \frac{\underline{V}}{j\omega L}, \quad \underline{I} = \underline{I}_R + \underline{I}_L = \underline{V}\left(\frac{1}{R} + \frac{1}{j\omega L}\right) \tag{4.40}$$

The impedance of the parallel load is calculated according to the generalized concept of impedance, $\underline{Z} = \underline{V}/\underline{I}$:

$$\underline{Z} = \frac{\underline{V}}{\underline{I}} = \frac{1}{1/R + 1/j\omega L} = R \parallel (j\omega L) = |\underline{Z}|\,\underline{/\theta} \tag{4.41}$$

where $\theta = \tan^{-1}(R/\omega L)$.

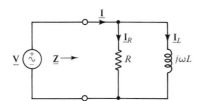

Figure 4.30 Parallel RL circuit.

Frequency effects in *RL* parallel circuit. The phasor diagram for the parallel RL circuit is shown in Fig. 4.31. Notice that the current lags the voltage. Let us now consider changes in the phasor diagram as frequency is varied. At low frequencies, $\omega L \ll R$, the current in the inductor is much larger than the current in the resistor because the reactance of the inductor approaches a short circuit at dc. This would cause the phase, θ, to approach $-90°$ (current lagging voltage).

As frequency increases, the reactance of the inductor will increase and the resistor gains importance. At $\omega L = R$, the phase is $-45°$, current lagging voltage, and the magnitude of the total current will be $\sqrt{2}$ times the current in the resistor. As higher frequencies are reached, the reactance of the inductor exceeds that of

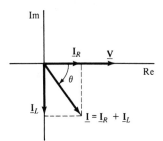

Figure 4.31 Current still lags voltage due to the inductor.

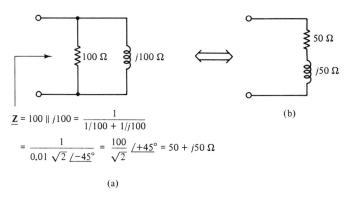

$$\underline{Z} = 100 \,\|\, j100 = \frac{1}{1/100 + 1/j100}$$

$$= \frac{1}{0.01 \sqrt{2} \,\underline{/-45^\circ}} = \frac{100}{\sqrt{2}} \,\underline{/+45^\circ} = 50 + j50 \; \Omega$$

(a)

(b)

Figure 4.32 A parallel circuit and a series circuit can be equivalent. (a) Parallel circuit; (b) series equivalent of parallel circuit.

the resistor, and the impedance of the parallel combination approaches that of the resistor.

In Fig. 4.32 we have converted a parallel circuit to an equivalent series circuit. The series equivalent of $100 \,\|\, j100$ is thus $50 + j50 \; \Omega$. The two circuits are equivalent, however, at a single frequency only; they would exhibit different characteristics if frequency were varied.

RC Circuits

Series RC circuit. Figure 4.33 shows a series RC circuit. The impedance of the series combination is

$$\underline{Z} = R + \frac{1}{j\omega C} = \sqrt{R^2 + \left(\frac{1}{\omega C}\right)^2} \; \underline{/-\tan^{-1}\frac{1}{\omega RC}} \tag{4.42}$$

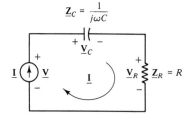

Figure 4.33 Series RC circuit.

The Analysis of AC Circuits   Chap. 4

To put Eq. (4.42) into the form of Eq. (4.39), we use the identity $1/j = j/(j)^2 = -j$:

$$\mathbf{Z} = R + j\left(-\frac{1}{\omega C}\right) \tag{4.43}$$

Hence we see that the reactance of a capacitor is negative. Figure 4.34 shows the corresponding phasor diagram. We use the current as the phase reference, so the voltages are shown lagging. However, we would normally say the current leads in a capacitive circuit because voltage would be considered the phase reference for the reasons given above.†

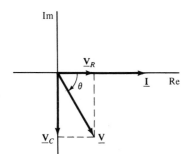

Figure 4.34 Current leads voltage in the capacitive circuit.

Frequency effects in *RC* series circuit.   In this case, we will examine the frequency behavior by starting at high frequencies. At high frequencies, the reactance of the capacitor $(-1/\omega C)$ is very small and the circuit appears resistive. The phase angle is nearly zero, with the current slightly leading the voltage. As frequency is decreased, however, the reactance of the capacitor increases. This increases $\mathbf{V}_C$, which increases the total voltage and the phase angle. At $1/\omega C = R$, the phase is 45°, leading current, and the voltage has increased by $\sqrt{2}$ from its high-frequency value. At low frequencies, the reactance of the capacitor becomes very large, approaching an open circuit at dc. Thus at low frequencies the capacitor dominates the behavior of the series combination, and the current phase approaches 90° leading the voltage.

Parallel *RC* circuit.   Figures 4.35 and 4.36 show a parallel *RC* circuit and the corresponding phasor diagram. The currents in the resistor and the capacitor are independent because they are in parallel with an ideal voltage source. Hence the current can be derived directly from consideration of the impedances of each parallel branch.

$$\mathbf{I}_R = \frac{\mathbf{V}}{\mathbf{R}}, \qquad \mathbf{I}_C = \frac{\mathbf{V}}{1/j\omega C} = j\omega C \mathbf{V} \tag{4.44}$$

$$\mathbf{I} = \mathbf{I}_R + \mathbf{I}_C = \mathbf{V}\left(\frac{1}{R} + j\omega C\right), \qquad \theta = \tan^{-1}(\omega RC) \tag{4.45}$$

† A phrase for remembering phase relationships is "ELI and ICEman." We now use V instead of E for voltage, and the iceman is only a memory, but the phrase is still useful.

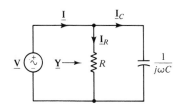

Figure 4.35 Parallel *RC* circuit.

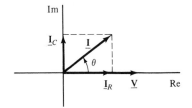

Figure 4.36 The capacitor causes the current to lead the voltage.

We can derive the impedance from Eq. (4.45), as we did for the *RL* parallel circuit in Eq. (4.41), but Eq. (4.45) leads naturally to the definition of "admittance."

Admittance.   Recall from Chapter 2 that we found it convenient to introduce the concept of conductance, $G = 1/R$, to simplify discussion of parallel combinations of resistors. There the concept of conductance brought few practical benefits, because modern calculators handle reciprocals without difficulty, but the corresponding concept in ac circuits has considerable theoretical and practical importance. *Admittance*, $\underline{Y}$, is defined as

$$\underline{Y} = \frac{\underline{I}}{\underline{V}} = \frac{1}{\underline{Z}} \tag{4.46}$$

Clearly, the admittance of the parallel *RC* circuit in Fig. 4.35 is

$$\underline{Y} = \frac{\underline{I}}{\underline{V}} = \frac{1}{R} + j\omega C = G + j\omega C \tag{4.47}$$

where we have introduced the conductance of the resistor, G. Admittance is useful for dealing with parallel circuits and has importance because parallel connections are common in practice. We have a specialized vocabulary associated with admittance. The real part of the admittance is called the *conductive part* or the *conductance*. The imaginary part of the admittance is called the *susceptive part* or the *susceptance*. Thus we say that the conductive part of the admittance of a parallel *RC* circuit is G and the susceptive part is ωC. The standard symbols for the conductance and susceptance are

$$\underline{Y} = G + jB \tag{4.48}$$

Note that B is a real number. For example, the conductive part of the parallel *RL* circuit in Fig. 4.32 is 0.01 S (siemens) and the susceptive part is $B = -0.01$ S. Note that the susceptance of the inductive circuit is negative, whereas the susceptive part of the admittance of the capacitive circuit is positive—just opposite the signs for the reactance.

RLC Circuits

Series *RLC*.   Figure 4.37 shows a series *RLC* circuit. The impedance of the circuit is

The Analysis of AC Circuits   Chap. 4

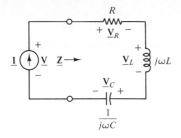

Figure 4.37 Series *RLC* circuit.

$$\mathbf{Z} = R + j\omega L + \frac{1}{j\omega C} = R + j\left(\omega L - \frac{1}{\omega C}\right) = |\mathbf{Z}|\,\underline{/\theta} \qquad (4.49)$$

where

$$|\mathbf{Z}| = \sqrt{R^2 + \left(\omega L - \frac{1}{\omega C}\right)^2} \quad \text{and} \quad \theta = \tan^{-1}\frac{\omega L - 1/\omega C}{R}$$

The reactive part of the impedance now combines the effects of the inductor and the capacitor. The frequency response of this reactive term gives this circuit interesting and useful characteristics. The phasor diagram is shown in Fig. 4.38. We have drawn the phasors showing the voltage across the inductor greater in magnitude than the voltage across the capacitor. This situation would be appropriate for high frequencies.

The frequency characteristics of the series *RLC* circuit combine those of the series *RC* and *RL* circuits. At dc the circuit acts as an open circuit because of the capacitor. At low frequencies the reactance of the capacitor dominates and the phase angle would be nearly 90°, with current leading voltage. As frequency increases, however, the inductive reactance becomes significant and at a special frequency grows to the point of canceling the negative reactance of the capacitor. This frequency occurs when $\omega_r L = 1/\omega_r C$, or $\omega_r = 1/\sqrt{LC}$, and is called the frequency of series resonance. At resonance, the inductor and capacitor combination becomes invisible and *R* is the total impedance of the circuit. As the increasing frequency passes through resonance, the phase changes from leading to lagging (current lagging voltage), and at resonance the phase is zero. At frequencies above resonance, the inductor dominates the circuit characteristics and the phasor diagram of Fig. 4.38 shows the relative voltages. At very high frequencies the phase approaches 90° lagging.

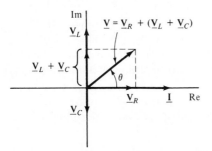

Figure 4.38 Above resonance the inductor dominates, so current lags voltage.

This circuit, with its series resonance, has importance in electronics because it can be used to select one group of frequencies from a broader group. For example, this circuit can be used as part of a radio filter that selects one station for reception, rejecting all others. We will discuss this circuit more fully in Chapter 5 as we discuss energy in ac circuits. We will see that resonance occurs when magnetic and electric energy requirements are equal, just as a mechanical system exhibits resonance when kinetic and potential energy requirements are balanced.

Example: series resonance. As an example of a circuit exhibiting series resonance, we will calculate some characteristic frequencies for the circuit shown in Fig. 4.39. The impedance is

$$\underline{Z}(\omega) = 10 + j\left(2 \times 10^{-6}\omega - \frac{10^8}{\omega}\right) \Omega$$

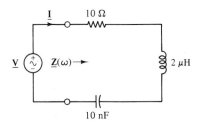

Figure 4.39 Series *RLC* circuit. We will find the resonant frequency and the width of the resonance.

The resistive part of the impedance is constant but the reactive part varies with frequency. Resonance occurs when the total reactance is zero:

$$2 \times 10^{-6}\omega = \frac{10^8}{\omega} \Rightarrow \omega_r = 7.071 \times 10^6 \ (f_r = 1125 \ \text{kHz})$$

where ω_r and f_r are the resonant frequency. At resonance, that is, at a frequency of 1125 kHz, the impedance of the circuit is 10 Ω resistive. Frequencies will exist below and above resonance where the reactance is equal to the resistance, and the phase will be $\pm 45°$. These occur at

$$\left| 2 \times 10^{-6}\omega - \frac{10^8}{\omega} \right| = 10 \qquad (4.50)$$

Equation (4.50) leads to two quadratic equations, which yield frequencies of 1592 and 796 kHz. At the lower frequency the circuit would be capacitive, and the angle of the impedance would be negative (leading current). At the higher frequency the circuit would be inductive, and the angle of the impedance would be positive (lagging current). Since these frequencies are in the AM radio band, this circuit might find application as a frequency filter in an AM radio.

Parallel *RLC* circuit. The parallel *RLC* circuit (Fig. 4.40) combines the properties on the parallel *RL* and *RC* circuits. The admittance of the circuit is

$$\underline{Y} = G + j\omega C + \frac{1}{j\omega L} = G + j\left(\omega C - \frac{1}{\omega L}\right) \qquad (4.51)$$

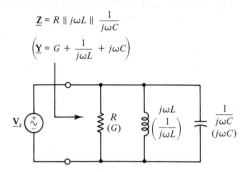

$$\underline{Z} = R \parallel j\omega L \parallel \frac{1}{j\omega C}$$

$$\left(\underline{Y} = G + \frac{1}{j\omega L} + j\omega C\right)$$

Figure 4.40 Parallel *RLC* circuit. Admittance (in parentheses) is natural for parallel circuits.

At low frequencies, the susceptance of the inductor $(-1/\omega L)$ is large and dominates the admittance expression. The admittance is large (infinite at dc) and the phase approaches $-90°$, lagging current. As frequency is increased, the inductive susceptance diminishes and the capacitive susceptance grows until they become equal. This is a parallel resonance and it occurs when $\omega_r C = 1/\omega_r L$, or $\omega_r = 1/\sqrt{LC}$, the same frequency as for series resonance. Thus series and parallel resonance occur at the same frequency for the same ideal inductor and capacitor. At resonance, the admittance is pure conductance, $\underline{Y} = G = 1/R$.

As the frequency increases above resonance, the capacitive susceptance dominates, and as the frequency approaches very high frequencies, the admittance becomes very large and the phase approaches $+90°$, leading current. Thus the admittance is minimum at resonance and becomes very large at low and high frequencies. Put differently, at low and high frequencies the impedance is very small, approaching a short circuit, but the impedance has a maximum at the frequency of parallel resonance. This contrasts with the series case, where the impedance is minimum at resonance.

4.3.3 Summary

In this chapter we have presented an efficient method for determining the steady-state response of ac circuits. We showed that sinusoidal time functions can be represented by complex numbers called phasors. Circuit elements are represented by complex numbers called impedances, and impedances can be combined in series or parallel like resistors in dc circuits. The DEs describing circuit behavior in the time domain become complex algebraic equations in the frequency domain. Finally, we investigated the effects of frequency variations in series and parallel *RL*, *RC*, and *RLC* circuits.

Table 4.1 summarizes the relationships between the time domain and the frequency domain that we have developed in this chapter.

4.3.4 Check Your Understanding

1. What is the impedance, including units, of a 0.7-H inductor at 50 Hz?
2. What is the impedance of a 80-μF capacitor at 120 Hz?
3. A capacitor of 1 kHz has an impedance with a magnitude of 20 Ω. What is the magnitude of the impedance at 2 kHz?

Table 4.1 Time-domain and Frequency-domain Transforms

Time Domain	Frequency Domain
Sinusoid	Phasor
Phase angle	Angle in complex plane
$V_p \cos(\omega t + \theta)$	$\mathbf{V} = V_p\underline{/\theta}$
DEs	Arithmetic with complex numbers
d/dt	$j\omega$
Cosine function	1
Sine function	$-j = 1\underline{/-90°}$
R, L, and C	Impedances
R	$\mathbf{Z}_R = R$
L	$\mathbf{Z}_L = j\omega L = \omega L\underline{/+90°}$
C	$\mathbf{Z}_C = 1/(j\omega C) = (1/\omega C)\underline{/-90°}$

4. What is the magnitude of the total impedance of a 20-Ω resistor in series with a 20-mH inductor at 80 Hz?

5. What is the conductive part of $\mathbf{Y} = 0.012\underline{/+12°}$?

6. What is the reactive part of $\mathbf{Z} = 18\underline{/-26°}$?

7. If the current lags the voltage in an RL circuit, what type of circuit is it: series, parallel, or either?

8. At resonance, the input impedance to a series RLC circuit is a minimum or a maximum?

Answers. 1. $j219.9\ \Omega$; 2. $-j16.6\ \Omega$; 3. $10\ \Omega$; 4. $22.4\ \Omega$; 5. 11.7×10^{-3} S; 6. -7.89; 7. either; 8. minimum.

PROBLEMS

Section 4.1: Introduction to Alternating Current

P4.1. Find the frequencies in hertz and in radians per second for the following:
(a) The rotation of the earth on its axis relative to the earth–sun line.
(b) The rotation of a bike tire (26-in diameter) at 20 mph.
(c) The second hand of a watch.
(d) A dentist's drill rotating at 200,000 rpm.
(e) A 33.3 rpm LP phonograph record.

P4.2. The sidereal day measures the earth's rotation relative to the fixed stars and is 3 minutes, 56 seconds shorter than the mean solar day. What is the earth's angular velocity on its axis relative to absolute space in radians/hour? Give to six-place precision.

Answer. $= 0.262516$ rad/h

P4.3. The maximum elevation angle of the sun in Austin, Texas, is $E(t) = 60° + 23.5° \cos(\omega t - \theta)$.
(a) If t is in months, what is ω?
(b) Estimate θ in radians if $t = 0$ on January 1.

P4.4. A sinusoidal function is shown in Fig. P4.4. Determine the frequency, phase, and amplitude for expressing this sinusoid in the standard form:

$$i(t) = I_p \cos(\omega t + \theta)$$

Answer. $i(t) = 9 \cos(5.24 \times 10^6 t - 70.5°)$ mA

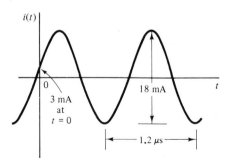

Figure P4.4

P4.5. Sketch the sinusoidal voltage $v(t) = 60 \cos(100\pi t - 120°)$ V.

Section 4.2: Representing Sinusoids with Phasors

P4.6. Given three complex numbers: $z = 2 - j6$, $w = 5\underline{/-105°}$, and $u = 8e^{j2}$:
   (a) Find $|z|$.
   (b) Find u^* in rectangular form.
   (c) Evaluate $z/(w - u)$ and place the result in exponential form.
   (d) Solve for s (a complex number) if $z(s - w) = u$ and express s in polar form.

   Answers. (a) 6.32; (b) $-3.33 - j7.27$; (c) $0.515e^{j0.155}$; (d) $5.58\underline{/-117.2°}$

P4.7. For the complex numbers $z_1 = 2 + j3$ and $z_2 = -1 + j2$:
   (a) Show that $|z_1 \times z_2| = |z_1| \times |z_2|$.
   (b) Show that $|z_1/z_2| = |z_1|/|z_2|$.
   (c) Show that $|z_1| + |z_2| \neq |z_1 + z_2|$.

P4.8. Given that $z = x + jy$ is a general complex number:
   (a) Show that $\mathrm{Re}\{z\} = (z + z^*)/2$.
   (b) Show that

$$\frac{1}{z} = \frac{x}{x^2 + y^2} - \frac{jy}{x^2 + y^2}$$

   (c) Solve for the first time when $\mathrm{Re}\{ze^{j\omega t}\} = 0$ if z is $2 - j6$ and $\omega = 100$.

P4.9. Evaluate the following expressions:
   (a) $(1.2 + j6) \, 5\underline{/-118°} + 6\underline{/+12°}$
   (b) $\dfrac{0.02 + j0.015}{3 + j7} - \dfrac{1}{1000\underline{/+66°}}$

   (c) $\dfrac{1}{\dfrac{1}{1 + j2} + \dfrac{1}{5\underline{/+20°}}}$

P4.10. (a) What is the phasor for $v(t) = 5.2 \cos(100t - 90°)$?

     (b) Sketch one cycle of the time function, $v(t)$, represented by the phasor $\mathbf{V} = 60e^{j\pi/3}$, $f = 60$ Hz.

Answer. (a) $5.2\underline{/-90°}$

P4.11. Use phasor techniques in the following.

     (a) Find $2 \cos(100t - 45°) - 3 \cos(100t + 60°)$.

     (b) Find $50 \sin(100t) + d \cos(100t - 30°)/dt$.

     (c) Use phasor techniques to evaluate the derivative of $i(t) = 20 \sin(500\,t)$ at $t = 2$ ms. *Hint:* Write the formula in the time domain and transform into the frequency domain, using $j\omega$ for d/dt. Then put the specific ω and t in the $e^{j\omega t}$ and take the real part.

Answers. (a) $4.01 \cos(100t - 91.2°)$; (c) 5400 A/s

P4.12. On page 125 it was argued that if

$$\text{Re}\{\underline{\mathbf{Z}}e^{j\omega t}\} = \text{Re}\{\underline{\mathbf{W}}e^{j\omega t}\}$$

for all t, then $\underline{\mathbf{Z}} = \underline{\mathbf{W}}$. In effect the Re and the $e^{j\omega t}$ can be dropped. Prove this by letting $\underline{\mathbf{Z}} = Z_r + jZ_i$ and $\underline{\mathbf{W}} = W_r + jW_i$ and evaluating the equation at the times when $\omega t = 0$ and $\omega t = \pi/2$.

P4.13. Solve for $v(t)$ in the circuit shown in Fig. P4.13 using the phasor methods described in Section 4.2.3. You may use the "shortened method" if you wish.

Answer. $v(t) = 71.6 \cos(1000t + 46.6°)$ V

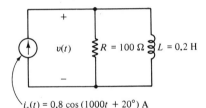

$i_s(t) = 0.8 \cos(1000t + 20°)$ A   **Figure P4.13**

P4.14. Solve for $v_C(t)$ in the circuit shown in Fig. P4.14 using the phasor methods described in Section 4.2.3. You may use the shortened method if you wish.

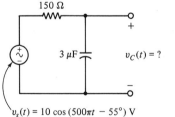

$v_s(t) = 10 \cos(500\pi t - 55°)$ V   **Figure P4.14**

P4.15. Rework Problem P4.14 for the total response if the voltage is applied with a switch closure at $t = 0$.

P4.16. The mechanical system shown in Fig. P4.16 is excited by a rotating wheel that gives approximately a sinusoidal displacement for x_1. The differential equation of the displacement x_2 is

$$m\frac{d^2x_2}{dt^2} + D\frac{dx_2}{dt} + Kx_2 = D\frac{dx_1}{dt}$$

(a) Find ω.

(b) Transform the DE to the frequency domain: $x_2(t) \Rightarrow \mathbf{X}_2$.

(c) Solve for $\mathbf{X}_2$.

(d) Write $x_2(t)$.

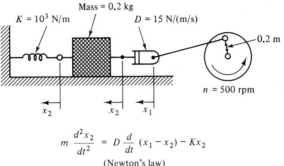

$$m \frac{d^2 x_2}{dt^2} = D \frac{d}{dt}(x_1 - x_2) - Kx_2$$

(Newton's law)

Figure P4.16

Section 4.3: Impedance: Representing the Circuit in the Frequency Domain

P4.17. Make a chart for resistors, capacitors, and inductors with the following columns: name, symbol, time-domain equation, frequency-domain equation, impedance in rectangular form and impedance in polar form.

P4.18. (a) What values of capacitance and inductance have an impedance with a magnitude of 10 Ω at a frequency of 1 kHz?

(b) What would be the magnitudes of the impedance of this C and this L at 2 kHz?

P4.19. Using the frequency-domain versions of KVL and KCL, show that two impedances in series add like resistors in series, that is, $\mathbf{Z}_{eq} = \mathbf{Z}_1 + \mathbf{Z}_2$. Show also that two impedances in parallel add like resistors in parallel, that is,

$$\mathbf{Z}_{eq} = \mathbf{Z}_1 \| \mathbf{Z}_2 = \frac{1}{(1/\mathbf{Z}_1) + (1/\mathbf{Z}_2)}$$

P4.20. Find the impedance in Fig. P4.20.

Answer. $22.4\underline{/-63.4°}$

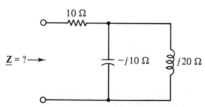

Figure P4.20

P4.21. Determine the input impedance of the circuits shown in Fig. P4.21.

Figure P4.21

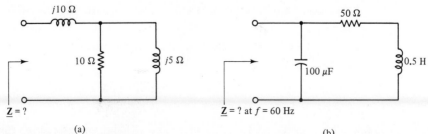

$\mathbf{Z} = ?$

(a)

$\mathbf{Z} = ?$ at $f = 60$ Hz

(b)

147

P4.22. Use the techniques of the frequency domain to solve for $i(t)$ in the circuit shown in Fig. P4.22.

    (a) Find the frequency-domain version of the circuit, using phasors to represent sinusoidal functions, known and unknown, and impedances to represent circuit components.

    (b) Using parallel and series combinations, find $\underline{Z}_{eq}$ as seen by the voltage source.

    (c) Solve for $\underline{I}$.

    (d) Convert back to the time domain.

Answer. $i(t) = 0.707 \cos(1000t - 81.9°)$ A

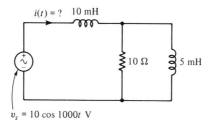

$v_s = 10 \cos 1000t$ V   **Figure P4.22**

P4.23. Consider a 5-μF capacitor and a 8-Ω resistor.

    (a) They are connected in series. At what frequency in hertz is their series impedance 20 Ω in magnitude?

    (b) If the resistor and capacitor are now placed in parallel, find the frequency at which their combined impedance is 7 Ω in magnitude.

P4.24. For the circuit shown in Fig. P4.24, determine $v(t)$ using phasor techniques. Sketch $v(t)$ in the time domain.

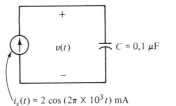

$i_s(t) = 2 \cos (2\pi \times 10^3 t)$ mA   **Figure P4.24**

P4.25. Figure P4.25a shows a circuit in sinusoidal steady state, with the phasor diagram in Fig. 4.25b. Assume the load consists of a resistor in series with a reactive component. The frequency is 60 Hz.

    (a) What is the voltage at $t = 0$?

    (b) What is the magnitude of the impedance?

    (c) What is the resistance of the circuit?

    (d) What is the reactive component (type and value)?

Answers. (a) 50 V; (b) 24.1 Ω; (c) 16 Ω; (d) inductor, 47.7 mH

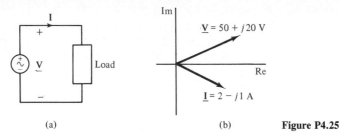

(a) (b) **Figure P4.25**

P4.26. The circuit shown in Fig. P4.26 is in sinusoidal steady state. Determine the maximum value of the current and the first time after $t = 0$ at which the maximum current occurs.

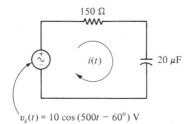

Figure P4.26

P4.27. A 60-Hz ac source and load are connected as indicated in Fig. P4.27. The phasor voltage is $\underline{V} = 100 + j0$ V and the phasor current is $\underline{I} = 6 - j5$ A.
(a) Find the instantaneous current at $t = 0$.
(b) What is the impedance of the ac load in rectangular form?
(c) Assuming that the load is a series resistor and inductor, find the value of the inductance.

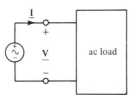

Figure P4.27

P4.28. For the circuit shown in Fig. P4.28:
(a) Draw a phasor diagram showing $\underline{V}_s$, $\underline{I}$, $\underline{V}_R$, and $\underline{V}_C$. The voltages must be drawn to consistent scale and shown to add in accordance with the phasor KVL.
(b) Find $v_R(t)$ and sketch along with the source voltage.

Answer. $v_R(t) = 14.1 \cos(800\pi t + 38.5°)$

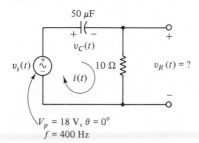

Figure P4.28

P4.29. For the circuit shown in Fig. P4.29:
(a) At what frequency would the magnitude of the input impedance be 200 Ω?
(b) What is the angle of the impedance of this frequency?
(c) What value of C should be added in series to make the circuit appear pure resistive at this frequency? In parallel? (The answers to these two questions differ.)

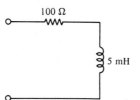

100 Ω

5 mH

Figure P4.29

P4.30. Convert the circuit shown in Fig. P4.30 to the equivalent parallel circuit at $f = 1$ kHz.

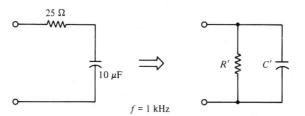

25 Ω

10 μF

$f = 1$ kHz

R' C'

Figure P4.30

P4.31. The circuit in Fig. P4.31 is to be represented by a Norton equivalent circuit. Determine $\underline{I}_N$ and $\underline{Z}_{eq}$.

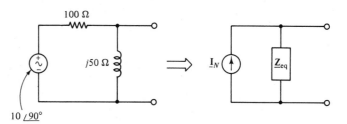

100 Ω

$j50\ \Omega$

$\underline{I}_N$ $\underline{Z}_{eq}$

10 $\angle 90°$

Figure P4.31

P4.32. What is the total admittance of $\underline{Y}_1 = 1 + j2\mho$ and $\underline{Y}_2 = 1.5 - j2.5\mho$ connected in parallel? What is the magnitude of the input impedance of this parallel combination?

Answers. $= 2.5 - j0.5\mho$, 0.392 Ω

General Problems for Chapter 4

P4.33. By proper choice of X_C and X_L, the 10-Ω resistance in Fig. P4.33 can be transformed to "look" like a 100-Ω resistor at a specified frequency, as indicated in the figure. Find X_C and X_L and, from them, C and L to accomplish this transformation at 1 kHz.

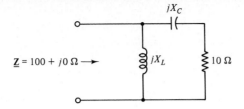

Figure P4.33

P4.34. For the circuit shown in Fig. P4.34, find:

(a) $v_R(t)$ for $\omega = 0$

(b) $v_R(t)$ for $\omega = 4000$ rad/s

(c) Frequency in hertz for which the amplitude of $v_R(t)$ is maximum

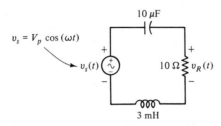

Figure P4.34

P4.35. For the circuit in Fig. P4.35, $\underline{V} = 30\underline{/0°}$.

(a) Show $\underline{V}$ and $\underline{I}$ on a phasor diagram.

(b) With $\underline{V}$ unchanged except that the frequency is doubled, show the phasor diagram. Use primed $\underline{V}$ and $\underline{I}$ for this case.

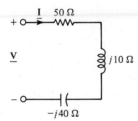

Figure P4.35

P4.36. In Fig. P4.36, the design goal is to have $i(t)$ lead $v_s(t)$ by 65° of phase.

(a) What is in the box: R, L, or C?

(b) What is its numerical value?

(c) What is the peak value of $i(t)$?

(d) If the frequency were doubled, what would be the phase difference between $i(t)$ and $v_s(t)$? Consider phase positive if $i(t)$ leads $v_s(t)$.

(e) At extremely low frequency, what would be the phase shift between $i(t)$ and $v_s(t)$?

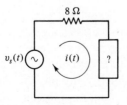

$v_s(t) = 120\sqrt{2}\cos(120\pi t)$ **Figure P4.36**

5 Power in AC Circuits

5.1 AC POWER AND ENERGY STORAGE: THE TIME-DOMAIN PICTURE

5.1.1 Importance of Power and Energy

Power and energy are important in ac problems for several reasons. Energy concerns you as a user of electricity because the local electric utility charges you for the energy you use. More important, energy plays a vital role in describing the behavior of physical systems. The roller coaster problem of freshman physics—where you compute the speed at some point on the track from a difference in height—exemplifies how energy considerations often sweep away many details of a problem and lead directly to a useful result. Indeed, the more experience you gain in analyzing physical systems, the more you should become impressed with the importance of energy in revealing the true workings of a system. Some feel, as does your author, that any analysis is incomplete until energy relationships are explored and understood.

In this chapter we investigate energy and power relationships in electrical circuits. We begin by looking at averages because average power is frequently our focus. Then we will examine energy and power relationships in ac circuits from the viewpoint of the time domain. Next comes the frequency-domain viewpoint, and we succeed in expressing energy and power relationships with phasors. Finally, we introduce three topics of eminent practical value: transformers, three-phase circuits, and electrical safety.

5.1.2 Average Values of Electrical Signals

What is an average?   Everybody knows how to calculate a numerical average. We compute the average of 12, 9, and 15 by adding the numbers (36), dividing by the number of values we are averaging (3), and getting the average (12). In general, the average of n numbers, x_i, $i = 1, 2, \ldots, n$, is

$$X_{\text{average}} = \frac{\sum\limits_{i=1}^{n} x_i}{n} \quad \text{or} \quad nX_{\text{average}} = \sum_{i=1}^{n} x_i \tag{5.1}$$

In the second form of Eq. (5.1), we see that the average value multiplied by the number of samples is equal to the sum of the numbers.

That is how to compute an arithmetic average, but what is the definition of an average? An average is a number that characterizes in some aspect a body of information. The arithmetic average, for example, gives us some idea of the size of the numbers, such as the average price of gasoline in Kansas. There are many kinds of averages. The grade-point average provides an example near to the heart of most college students. This average characterizes the academic performance of a student even though many important factors, such as course load and difficulty, are ignored. Nevertheless, if we insisted on one indication of excellence, the GPA would be used. Thus an average is a number that characterizes a body of information in one particular way, omitting all the rest of the information.

Computing time averages.   The time average of a periodic function can be defined as a generalization of the arithmetic average. The periodic function in Fig. 5.1 provides an example. Because the function is periodic, the average over all time will be the same as the average over one period; hence we can limit our attention to the time from $t = 0$ to $t = T$, as shown in Fig. 5.2. We can define the time average by analogy with the second form of Eq. (5.1): n becomes T, the period; X_{average} becomes V_{avg}; and the summation of the numbers becomes the summation of all the heights of the function, which is the integral over the time period from 0 to T. Thus we have the definition of time-average voltage:

$$TV_{\text{avg}} = \int_0^T v(t)\, dt \Rightarrow V_{\text{avg}} = \frac{1}{T} \int_0^T v(t)\, dt \tag{5.2}$$

Figure 5.1 Find the time-average voltage.   **Figure 5.2** One period of the waveform.

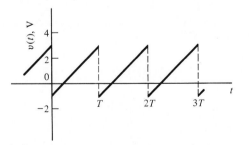

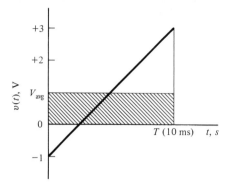

Figure 5.3 The average multiplied by the period has the same area as the original waveform, counting area below the axis as negative.

The first form of Eq. (5.2) can be interpreted in terms of area, as shown in Fig. 5.3. Because V_{avg} is a constant, the product on the left side represents the area on the $v(t)$ graph of a rectangle having base T and height V_{avg}. The right side is the area under the $v(t)$ curve, counting area above the axis as positive and area below the axis as negative. This geometric interpretation is shown in Fig. 5.3. Equation (5.2) requires the two areas to be equal. The second form of Eq. (5.2) serves for computing averages. The present example can be handled as the areas of triangles, but we will use calculus to illustrate the more general method. First we derive the equation of $v(t)$ in the slope–intercept form: the intercept is -1; the slope is $3 - (-1)$ divided by 10 ms, or 400 V/s. Thus the equation is

$$v(t) = -1 + 400t \qquad 0 < t < 10 \text{ ms} \tag{5.3}$$

We now subsitute into Eq. (5.2), with the result

$$V_{avg} = \frac{1}{10 \text{ ms}} \int_0^{10 \text{ ms}} (-1 + 400t)\, dt = 10^2 \left(-t + \frac{400t^2}{2} \right)_0^{10^{-2}} = +1 \tag{5.4}$$

This is the value indicated on Fig. 5.3 and its validity is apparent.

Some special averages.   Several results follow from the definition of time average. The time average of a dc (constant) voltage or current is the dc value. For this reason the time-average value of a signal, such as that shown in Fig. 5.3, is often called the dc value or dc component of the signal. Another result is that the average value of a sinusoidal waveform is zero. The sinusoidal function has equal areas above and below the time axis and thus has zero average (or dc) value.

If we have the sum of two signals, say, two voltage sources connected in series, the average of the sum is the sum of the averages of the component signals. This follows from Eq. (5.2) because the integral distributes to the two functions.

$$(v_1 + v_2)_{avg} = \frac{1}{T} \int_0^T [v_1(t) + v_2(t)]\, dt$$

$$= \frac{1}{T} \left[\int_0^T v_1(t)\, dt + \int_0^T v_2(t)\, dt \right] = V_{1\,avg} + V_{2\,avg}$$

One example would be the sum of a dc and a sinusoidal signal; the average would be the dc value because the sinusoidal part would average to zero.

5.1.3 Effective or Root-Mean-Square Value

Time-average power. We will now consider the time-average power in a dc circuit, Fig. 5.4. We have indicated that the resistor is hot because the electrical energy into the resistor appears as heat. From Chapter 2 we know that the power into the resistor is $(V_{dc})^2/R$, but we will derive this result here from more general considerations, which we then apply to the heating of a resistor with an ac source. Voltage is energy/charge and current is charge flow/time; it follows that the instantaneous power (energy/time) into a circuit element is

$$p(t) = v(t)i(t)$$

For our dc circuit in Fig. 5.4 both voltage and current are constant, so the instantaneous power into the resistor is $v \times i = (V_{dc})^2/R$, a constant. The time-average power (P) into R is thus

$$P = \frac{1}{T}\int_0^T p(t)\, dt = \frac{1}{T}\int_0^T \frac{(V_{dc})^2}{R}\, dt = \frac{(V_{dc})^2}{R} \tag{5.5}$$

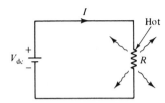

Figure 5.4 The heat is a measure of the average power.

The time-average power determines how hot the resistor will become. The movement of charge through the resistor imparts thermal energy to the material. The input electrical power appears as a heat source internal to the resistor; the temperature of the resistor depends on this input and on its thermal coupling to the environment. The more power into the resistor, the hotter it will become. Physical resistors are available for $\frac{1}{4}$, $\frac{1}{2}$, 1, 2 watts, and so on. The power rating indicates how much power the resistor can accept without burning out or changing its resistance value significantly.

Power in AC circuits. Figure 5.5 shows the same circuit with an ac source. The instantaneous power is

$$p(t) = v(t)i(t) = V_p \cos \omega t \times \left(\frac{V_p}{R}\cos \omega t\right) = \frac{V_p^2}{R}\cos^2 \omega t, \qquad \omega = \frac{2\pi}{T} \tag{5.6}$$

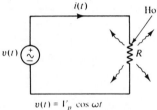

$$v(t) = V_p \cos \omega t$$
$$p(t) = \frac{v^2(t)}{R} = i^2(t)R$$

Figure 5.5 The heat is a measure of the time-average power.

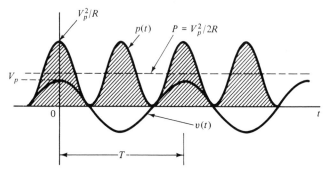

Figure 5.6 The energy flows into the resistor in spurts. The power is nonnegative at all times.

Figure 5.6 shows a plot of the instantaneous power [Eq. (5.6)]. Although the charges move back and forth in the resistor, the power is always nonnegative; this is shown mathematically from the squaring of the voltage or current in the power relationship. The nonnegative power results physically because the moving electrons heat up the material regardless of their direction of flow. The energy does not flow smoothly into the resistor but flows in spurts, twice each cycle of the ac waveform, as shown in Fig. 5.6. Thus the resistor is heated cyclically. In most resistors this time variation is unimportant because thermal inertia smooths out the heating variations and the temperature remains essentially constant. However, these variations can be a problem in incandescent lighting, where the thermal inertia of the filament is small. The power frequencies of 50 or 60 Hz were established high enough to make the flickering of the light (at 100 or 120 Hz) barely noticeable.

Effective or root-mean-square value of a sinusoid.   The peak value for the power function in Fig. 5.6 is V_p^2/R. From the symmetry of the $p(t)$ function, it is apparent that the average value is half that peak, or $V_p^2/2R$, as indicated on Fig. 5.6. The analytic computation of this result appears in Eq. (5.7). The integral uses the trigonometric identity $\cos^2 \omega t = (1 + \cos 2\omega t)/2$:

$$P = \frac{1}{T} \int_0^T \frac{V_p^2}{R} \cos^2 \omega t \, dt = \frac{V_p^2}{RT} \int_0^T \left(\frac{1}{2} + \frac{1}{2} \cos 2\omega t \right) dt$$

$$= \frac{V_p^2}{RT} \left(\frac{t}{2} + \frac{\sin 2\omega t}{4\omega} \right)_0^T = \frac{V_p^2}{2R} \tag{5.7}$$

Equation (5.7) leads to the *effective value* of the ac voltage. When we speak of "effective," the effect to which we refer is the heating effect, or more generally the time-average energy conversion from electrical to nonelectrical form. The effective value of an ac waveform is the equivalent dc value that would heat the resistor as hot as the ac waveform heats it.

$$P = \frac{V_e^2}{R} = \frac{V_p^2}{2R} \Rightarrow V_e = \frac{V_p}{\sqrt{2}} \tag{5.8}$$

where V_e is the effective voltage. Equation (5.8) equates the time-average power from an equivalent dc source with magnitude V_e, the effective value, to the time average of power due to the ac source. Equation (5.8) defines the effective value of an ac source, that the effective value of the sinusoidal voltage (or current, if we were dealing with current) is $1/\sqrt{2}$ or 0.707 times the peak value. For example, an ac voltage with a peak value of 20 V would be equivalent in heating effect to a $20/\sqrt{2} = 14.1$-V battery. But we warn you that Eq. (5.8) applies only for the sinusoidal waveform, as we will illustrate shortly.

Effective value in general.   The effective value is often referred to as the root-mean-square value (rms). The general definition of effective value of a periodic function is

$$ P = \frac{V_e^2}{R} = \frac{1}{T}\int_0^T \frac{v^2(t)}{R}\, dt \Rightarrow V_e = \sqrt{\frac{1}{T}\int_0^T v^2(t)\, dt} \qquad (5.9) $$

where T is the period. Here we have illustrated the definition of rms for voltage, $v(t)$, but a similar expression would apply for the effective value of a current. The effective value, or rms, is the square *root* of the *mean* (i.e., the average, as in "mean sea level") of the *square* of the function.

The practical importance of rms values is suggested by the way voltmeters and ammeters are calibrated. If you measured the voltage at the wall outlet in your room with an ac voltmeter, you would expect something around 120 V. Would that be peak or rms or what? If you "looked" at the voltage with an oscilloscope, you would observe the peak-to-peak voltage of the sinusoid to be about 340 V. Consequently, the peak value is 170 V and the rms is about 120 V. Generally, ac meters are calibrated to indicate rms for a sinusoid.

Example.   To illustrate the computation of the rms value of a nonsinusoidal waveform, we will compute the rms value of the waveform in Fig. 5.1. We derived the equation of the voltage during the first period to be

$$ v(t) = -1 + 400t, \qquad 0 < t < 10 \text{ ms} $$

Substituting into Eq. (5.9) and integrating the function, we obtain

$$ V_e = \sqrt{\frac{1}{10 \text{ ms}}\int_0^{10 \text{ ms}} (-1 + 400t)^2\, dt} = 1.53 \text{ V} $$

Measuring AC voltage.   Many ac voltmeters would not indicate the true rms (1.53 V) of the waveform in Fig. 5.1. A meter designed to square and average the instantaneous waveform could be complicated and expensive, so for simple voltmeters some other property of the waveform is measured and the rms is inferred from that, assuming a sinusoidal shape. For example, a common type of meter actually responds to the peak-to-peak value of the waveform, but the meter scale is marked to indicate the peak-to-peak value divided by $2\sqrt{2}$, which would be the rms for a sinusoid. Such a meter would indicate $4/2\sqrt{2} = 1.41$ V for the waveform in Fig. 5.1, clearly not the true rms value for this (nonsinusoidal) waveform. Thus meter readings require careful interpretation when measuring non-sinusoidals.

Summary. We can represent the power-producing capability of a waveform with an average value called the effective (or rms) value of the waveform. This is defined as the value that, when squared, is equal to the time average of the square of the waveform. For a dc waveform, the effective value is equal to the dc value. For a sinusoidal waveform, the effective value is $1/\sqrt{2}$ times the peak value. For other waveforms, the effective value may be calculated by squaring and averaging. Electrical meters are designed to indicate the effective value of a sinusoidal waveform but may not indicate the effective value for other wave shapes.

5.1.4 Power and Energy Relations for *R*, *L*, and *C*

Resistance. We have discussed the power relationship for resistance in deriving the effective value of a sinusoid. As shown in Fig. 5.6, the energy flows unilaterally into the resistor, not smoothly but in spurts. The time-average power into the resistor is

$$P_R = \frac{V_p^2}{2R} = \frac{V_e^2}{R} = I_e^2 R \tag{5.10}$$

where I_e is the effective current. Consider a 120-V, 100-W light bulb. According to Eq. (5.10), this indicates a resistance value of $(120)^2/100 = 144\ \Omega$ (recall that the 120 V is the rms value). The energy consumed by the bulb, operated for 24 hours, would be 24 hours $\times$ 0.100 kW or 2.4 kWh (kilowatthours). This represents the total energy consumed, and at 6 cents/kWh the bulb would operate for about 15 cents per day.

Inductance. We will calculate the instantaneous and time-average magnetic energy stored by an inductor. The current and voltage for an inductor are

$$i_L(t) = I_p \cos \omega t$$

$$v_L(t) = L\frac{di}{dt} = -\omega L I_p \sin \omega t$$

These are shown in Fig. 5.7. The instantaneous power into the inductor is the product of voltage and current:

$$p_L(t) = v_L(t)i_L(t) = (-\omega L I_p \sin \omega t)(I_p \cos \omega t) \tag{5.11}$$

$$= -\frac{\omega L I_p^2}{2} \sin 2\omega t$$

where we have used the trigonometric identity $(\sin \omega t)(\cos \omega t) = (\sin 2\omega t)/2$. The power into the inductor is shown in Fig. 5.7 as the lightly shaded area. We notice that the average power is zero, for the inductor loses no energy. Thus an inductor gives back on the average as much energy as it receives. (This is true only for an ideal inductor; a real inductor would have resistive losses.) Being the

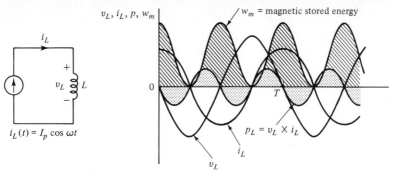

Figure 5.7 Voltage, current, stored energy, and power into the inductor.

product of v_L and i_L, the power is zero four times each cycle, when either the voltage or the current is zero. For a 60-Hz source, the energy would pulsate in and out of the inductor 120 times per second because it makes two round trips from source to inductor each cycle. For this reason, heavy electrical equipment, such as transformers and motors, often hums audibly at 120 Hz.

The magnetic energy stored in an inductor is given in Eq. (3.3) as

$$w_m(t) = \tfrac{1}{2}Li_L^2(t)$$

For the sinusoidal source, we find the instantaneous stored energy to be

$$w_m(t) = \tfrac{1}{2}L(I_p \cos \omega t)^2 = \frac{LI_p^2}{4}(1 + \cos 2\omega t) \qquad (5.12)$$

This stored energy is shown as the heavily shaded area in Fig. 5.7. The stored energy is nonnegative and pulsates at twice the ac source frequency. This is shown mathematically by the squaring of the negative current peaks. While the stored energy is increasing, the power into the inductor is positive. During this period of time the ac source supplies energy and the inductor acts as a load. While the stored energy is decreasing, the power into the inductor is negative, indicating that the inductor now acts as a source, returning energy to the ac source.

The time-average stored energy is one-half of $\tfrac{1}{2}LI_p^2$, the peak stored energy shown by Eq. (5.12). This can be written

$$W_m = \tfrac{1}{2}LI_e^2 \qquad (5.13)$$

where W_m is the average stored magnetic energy and I_e is the effective value of the current. The time-average energy is important for two reasons. Although the time-average power into the inductor is zero, the time-average stored energy must be supplied to the inductor when the ac source is originally connected to the inductor. Additionally, the time-average stored energy has importance because it indicates the magnitude of the energy pulsations in the inductor. Specifically, we know that the ac source must lend twice this amount of energy to the inductor twice each cycle.

Capacitance. We will calculate the instantaneous and time-average electric energy stored by a capacitor. The voltage and current for a capacitor are

$$v_C(t) = V_p \cos \omega t$$

$$i_C(t) = C\frac{dv}{dt} = -\omega C V_p \sin \omega t$$

These are plotted in Fig. 5.8. The instantaneous power into the capacitor is the product of voltage and current:

$$p_C(t) = v_C(t)i_C(t) = (V_p \cos \omega t)(-\omega C V_p \sin \omega t) \qquad (5.14)$$

$$= -\frac{\omega C V_p^2}{2}\sin 2\omega t$$

where we have again used the trigonometric identity $(\sin \omega t)(\cos \omega t) = (\sin 2\omega t)/2$. The power into the capacitor is shown in Fig. 5.8 as the lightly shaded area. We notice that the average power is zero, which indicates that the capacitor is lossless. (This is true only for an ideal capacitor; a real capacitor would have some loss, though not as much as a real inductor.) Being the product of v_C and i_C, the power is zero four times each cycle, when either voltage or current is zero.

The pulsation of the power at twice the ac frequency corresponds to the shuttling of electric energy between source and capacitor. The stored electric energy in a capacitor is given in Eq. (3.8) as

$$w_e(t) = \tfrac{1}{2}Cv_C^2(t) \qquad (5.15)$$

For the sinusoidal source, we find the instantaneous stored energy to be

$$w_e(t) = \tfrac{1}{2}C(V_p \cos \omega t)^2 = \frac{CV_p^2}{4}(1 + \cos 2\omega t) \qquad (5.16)$$

This stored electric energy is shown as the heavily shaded area in Fig. 5.8. The stored energy remains nonnegative and pulsates at twice the ac source frequency. This is shown mathematically by the squaring of the negative half-cycle of the voltage. While the stored energy is increasing, the power into the capacitor is positive. During this period the ac source supplies energy and the capacitor acts as a load. While the stored energy is decreasing, the power into the capacitor goes negative, indicating that the capacitor is acting as a source, returning energy to the ac voltage source.

Figure 5.8 Voltage, current, stored energy, and power into the capacitor.

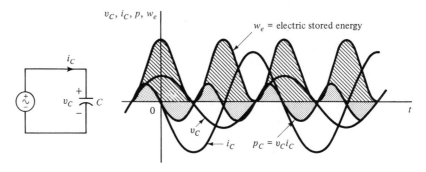

The time-average stored energy is one-half of $\frac{1}{2}CV_p^2$, the peak stored energy. This can be written

$$W_e = \tfrac{1}{2}CV_e^2 \qquad (5.17)$$

where W_e is the average stored electric energy and V_e is the effective value of the voltage.

5.1.5 General Case for Power in an AC Circuit

We have dealt with resistance, inductance, and capacitance separately and shown the role of each in power and energy relationships. We will now consider the general case of circuits containing combinations of resistors, inductors, and capacitors. We will think in terms of the circuit shown in Fig. 5.9, a parallel RLC circuit, although our results and interpretations will apply to all ac circuits.

This circuit is analyzed with frequency-domain techniques on page 142. The impedance, $\mathbf{Z} = |\mathbf{Z}|\underline{/\theta}$, is the reciprocal of the admittance given in Eq. (4.51). We are dealing in this section with power and energy in the time domain, so we will express the relationship between voltage and current generally

$$i(t) = I_p \cos(\omega t - \theta) \quad \text{and} \quad v(t) = V_p \cos(\omega t) \qquad (5.18)$$

where $V_p = |\mathbf{Z}|\,I_p$. Although impedance is a frequency-domain concept and is computed in the frequency domain, the magnitude of the impedance is a real number and hence the previous is a legitimate time-domain equation.

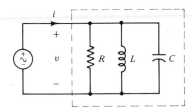

Figure 5.9 General RLC load.

Of immediate interest is the time-average power into the RLC load. The instantaneous power is

$$p(t) = v(t)i(t) = V_p \cos(\omega t)I_p \cos(\omega t - \theta)$$
$$= \frac{V_p I_p}{2}[\cos\theta + \cos(2\omega t - \theta)] \qquad (5.19)$$

The second form of Eq. (5.19) follows the application of the trigonometric identity

$$(\cos A)(\cos B) = \frac{\cos(A - B) + \cos(A + B)}{2}$$

with A identified with ωt and B with $\omega t - \theta$. We have plotted Eqs. (5.18) and (5.19) for $\theta = 56°$ in Fig. 5.10. The power function still must go through zero when either v or i is zero, but here the picture lies intermediate between those for pure resistance and for a lossless energy storage element. As Eq. (5.19) indicates in the second form, the power can be considered a combination of a

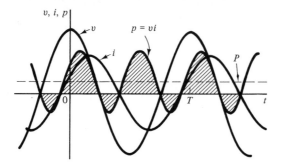

Figure 5.10 Voltage, current, and power into a general load.

constant term plus a fluctuating term at twice the frequency of the ac source. Because the average of the fluctuating term is zero, the time-average power is the constant term

$$P = \frac{V_p I_p}{2} \cos \theta = V_e I_e \cos \theta \tag{5.20}$$

The second form of Eq. (5.20) is preferred because meters indicate effective values, and because we usually speak in terms of effective values when discussing power relationships. Equation (5.20) reminds us of the power in the dc case, with effective values for voltage and current, except that we now have the cos θ factor. Figure 5.11 shows the effect of the cos θ factor, which is called the *power factor*. When voltage and current have the same phase, that is, when θ is zero, the power factor is unity and the power into the load is maximum, $V_e I_e$. When the voltage has a different phase from the current, the power factor is less than unity and the time-average power is decreased proportionally. This occurs symmetrically, whether the circuit is inductive (positive θ) or capacitive (negative θ). When the phase is ±90°, no time-average power is transferred to the load.

The fluctuating term in Eq. (5.19) shares aspects of the resistive load shown in Fig. 5.6, where we interpreted the fluctuations in terms of pulsating energy flow, and the lossless load (*L* or *C*), where we interpreted the fluctuations as energy being shuttled back and forth between source and load. That is, if the power factor were near unity (small θ), the fluctuations would indicate pulsating energy flow, but if the power factor were low (θ near ±90°), the fluctuations would represent shuttled energy. For all power factors less than unity, there exists a part of each cycle during which the instantaneous power is negative and the load acts as a source for a period of time.

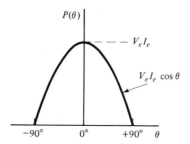

Figure 5.11 Average power versus θ.

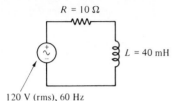

$R = 10\ \Omega$

$L = 40$ mH

120 V (rms), 60 Hz

Figure 5.12 Series RL circuit.

The analysis of the RLC load pictured in Fig. 5.9 will be resumed in the next section. Here for simplicity we will work out power and energy calculations for the RL circuit pictured in Fig. 5.12. We are working in the time domain for our power calculations, but frequency-domain techniques are appropriate for finding the voltage and current that we need in the power calculations. We will use the voltage source as our phase reference, so it would be represented by the phasor $120\sqrt{2}\underline{/0°}$. To calculate the current, we need the impedance,

$$\underline{Z} = 10 + j120\pi \times 0.04 = 10 + j15.1 = 18.1\underline{/56.4°}\ \Omega$$

The magnitude of the current is thus $120\sqrt{2}/18.1 = 6.63\sqrt{2}$ and the phase angle of the current is $-56.4°$. Thus the time-domain voltage and current are

$$v(t) = 120\sqrt{2}\cos(120\pi t)$$

$$i(t) = 6.63\sqrt{2}\cos(120\pi t - 56.4°)$$

The numbers in this example correspond to the plots in Fig. 5.10. From Eq. (5.20) we compute the time-average power delivered to the RL load by the ac voltage source to be

$$P = 120 \times 6.63\cos(56.4°) = 440\ \text{W}$$

We know from physical considerations that this power represents electrical energy converted to thermal energy in the resistor. This interpretation is confirmed by direct calculation of the power into the resistor.

$$P_R = I_e^2 R = (6.63)^2(10) = 440\ \text{W}$$

The magnetic energy storage represented by the inductance has an average value of

$$W_m = \tfrac{1}{2}LI_e^2 = \tfrac{1}{2}(0.04)(6.63)^2 = 0.880\ \text{J} \tag{5.21}$$

The instantaneous magnetic energy storage fluctuates between zero and twice the average energy calculated in Eq. (5.21). This energy must be loaned twice each cycle to the load by the source.

You may have noticed that we used rms values in every power and energy calculation in the foregoing example, and we started out with the rms value of the source voltage because that is what a meter would indicate. However, we were careful to use peak values for phasors and for time functions. This required inserting and taking out a lot of $\sqrt{2}$'s that never entered in the calculation. To be consistent with our definition of phasors, we must carry these $\sqrt{2}$'s in the notation. Indeed, many texts and most practitioners drop them out and use rms values for phasors, but we favor the more explicit approach and will maintain it in this chapter.

We could push our investigation of power and energy in the time domain a little further but we would rather move on to the frequency domain. The important question we ask is: Can such power calculations be made in the frequency domain without explicitly considering the time functions? The next section shows the answer to be yes. As before, the frequency-domain viewpoint yields efficiency in calculation and suggests new insights.

5.1.6 Check Your Understanding

1. What is the peak value of a sinusoidal current if a standard ac ammeter indicates 5 A?
2. An electric iron (for ironing clothes) converts approximately 1200 W of electrical power to heat. Estimate the rms current to the iron. What about a 100-W light bulb? Assume 120 V.
3. For a resistor, the time-average power is one-half the peak instantaneous power. (T/F?)
4. In an ac circuit the peak stored energy in a capacitor is 10 μJ. What is the time-average stored energy?
5. What is the time-average value of $v(t) = 10 + 5 \cos(100t)$ V?
6. If the power factor is 0.75 (lagging), by what angle does the current lag the voltage?
7. For a dc voltage, a dc voltmeter measures 10 V. What is the time-average voltage? What is the rms value of the voltage? For an ac voltage, an ac voltmeter measures 10 V. What is the time-average value of the voltage? What is the rms value of the voltage?

Answers. 1. 7.07 A; 2. 10 A, 0.83 A; 3. true; 4. 5 μJ; 5. 10 V; 6. 41.4°; 7. 10 V, 10 V, 0 V, 10 V.

5.2 POWER AND ENERGY IN THE FREQUENCY DOMAIN

5.2.1 Time-average Power

The time-average power in an ac circuit is given in Eq. (5.20). This involves the peak (or rms) values of the voltage and current and the power factor, which is the cosine of the phase angle between the voltage and current. All these quantities can be expressed through frequency-domain concepts; indeed, the frequency-domain concept of impedance provides the most sensible way to determine the magnitude of the current (given the voltage) and the phase angle. Thus in calculating time-average power it is unnecessary to consider the time-domain voltage and current.

Let us reconsider the *RL* circuit in Fig. 5.12 wholly in the frequency domain (Fig. 5.13). The time-average power into the load is

$$P = \frac{V_p I_p}{2} \cos \theta = \tfrac{1}{2} | \underline{V} | | \underline{I} | \cos \theta \qquad (5.22)$$

"in-phase" current

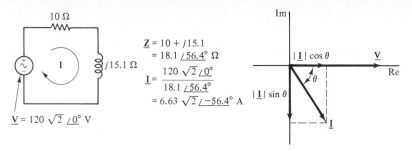

Figure 5.13 Solution in the frequency domain. The phasor diagram shows the "in-phase" and "out-of-phase" components of the current.

Figure 5.13 suggests an interpretation of the power factor. If we associate the $\cos \theta$ with the magnitude of $\mathbf{I}$, the term $|\mathbf{I}| \cos \theta$ can be interpreted as the projection of the current phasor on the voltage phasor. This part of the current is "in phase" with the voltage, and we conclude that the time-average power is given by the product of the voltage and the current in phase with the voltage.

The power calculated by Eq. (5.22) is called the *real power* for two reasons. It is real because of the geometric interpretation of the phasor diagram, because it is the product of the phasor voltage and the real component of the current (with the voltage considered the phase reference, of course). But it is also called real because this is the time-average power registered by a wattmeter—the power that makes your "electric meter" revolve and affects your electric bill. We give this power a special name to distinguish it from other kinds of ac power that we define below.

5.2.2 Reactive Power

Definition of reactive power. We have an interpretation of the product of the voltage with the real component of the current in Fig. 5.13. Can we place a meaningful interpretation on the product of the voltage with the "out-of-phase" component of the current?

$$Q = \tfrac{1}{2} |\mathbf{V}| \times \text{"out-of-phase" current}$$

$$= \tfrac{1}{2} |\mathbf{V}| |\mathbf{I}| \sin \theta = ?$$

where Q is called the reactive power. To interpret the product of the voltage and the out-of-phase current, we will return to the parallel RLC circuit, repeated in Fig. 5.14. We can calculate the current by multiplying the voltage by the admittance of the circuit, given in Eq. (4.51).

$$\mathbf{I} = \underbrace{\frac{\mathbf{V}}{R}}_{\text{in phase}} + \underbrace{j\mathbf{V} \left(\omega C - \frac{1}{\omega L} \right)}_{\text{out of phase}} = |\mathbf{I}| \cos \theta + j |\mathbf{I}| \sin \theta \qquad (5.23)$$

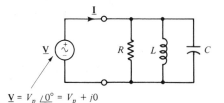

$\underline{\mathbf{V}} = V_p \underline{/0^\circ} = V_p + j0$

Figure 5.14 General *RLC* load.

Because $\underline{\mathbf{V}}$ is a real quantity, we can express both the real power and the reactive power mathematically by multiplying Eq. (5.23) by $\frac{1}{2}\underline{\mathbf{V}}$. We will seek an interpretation of the out-of-phase term by juggling the resulting mathematical expressions until we have everything expressed in terms of stored energy. Hence we multiply both sides of Eq. (5.23) by $\frac{1}{2}\underline{\mathbf{V}}$ and work on the result.

$$\frac{1}{2}\underline{\mathbf{V}}\,\mathbf{I} = \frac{|\underline{\mathbf{V}}|^2}{2R} + j\frac{|\underline{\mathbf{V}}|^2}{2}\left(\omega C - \frac{1}{\omega L}\right)$$

$$= \underbrace{\frac{|\underline{\mathbf{V}}|^2}{2R}}_{\text{I}} + j\left(\underbrace{\frac{|\underline{\mathbf{V}}|^2\omega C}{2}}_{\text{II}} - \underbrace{\frac{|\underline{\mathbf{V}}|^2}{2\omega L}}_{\text{III}}\right) \tag{5.24}$$

Although $\underline{\mathbf{V}}$ is a real quantity, we have used $|\underline{\mathbf{V}}|^2 = \underline{\mathbf{V}}^2$ for generality. The first term (I) we have already interpreted; this is the real power, P, the power that makes the electric meter revolve. The second term (II), from the capacitor, can be interpreted in terms of the peak stored energy in the capacitor. Equation (5.16) can be introduced to yield

$$\frac{|\underline{\mathbf{V}}|^2\omega C}{2} = \omega \times W_{ep}$$

where W_{ep} is the peak electric energy in the capacitor. The third term (III), from the inductor, requires some juggling. To express the magnetic stored energy, we need the current in the inductor. Recall that the peak magnetic energy storage in the inductor is derived from Eq. (5.12) as

$$W_{mp} = \tfrac{1}{2}L\,|\underline{\mathbf{I}}_{\text{L}}|^2 \tag{5.25}$$

To introduce the current, we require the impedance of the inductor:

$$|\underline{\mathbf{V}}| = |j\omega L\underline{\mathbf{I}}_{\text{L}}| = \omega L\,|\underline{\mathbf{I}}_{\text{L}}| \tag{5.26}$$

Note that j goes away because $|j| = |1\underline{/90^\circ}| = 1$. We substitute Eq. (5.26) into term (III) to obtain

$$\frac{|\underline{\mathbf{V}}|^2}{2\omega L} = \omega\left(\frac{|\underline{\mathbf{I}}_{\text{L}}|^2 L}{2}\right) = \omega \times W_{mp}$$

where W_{mp} is the peak magnetic energy. Putting the two energy terms back into Eq. (5.24), we have

$$\tfrac{1}{2}\underline{\mathbf{V}}\,\mathbf{I} = P + j\omega(W_{ep} - W_{mp}) \tag{5.27}$$

Power in AC Circuits   Chap. 5

Equation (5.27) is what we sought because it expresses the reactive power in terms of energy relationships, but we will make one change. We arbitrarily change the sign of the imaginary term, for reasons to be explained shortly. The results are

$$\tfrac{1}{2}\underline{V}\,\underline{I} = P + jQ \tag{5.28}$$

where

$$P = \frac{|\,\mathbf{V}\,|^2}{2R}, \qquad \text{the real power} \tag{5.29}$$

and

$$Q = \omega(W_{mp} - W_{ep}), \qquad \text{the reactive power} \tag{5.30}$$

The reactive power, Q, is seen to indicate the imbalance between peak magnetic energy storage and the peak electric energy storage in the circuit.

Meaning of reactive power. We may understand the significance of the reactive power by considering the stored energy requirements of the capacitor and inductor. The capacitor stores its maximum energy when the voltage is maximum. The inductor stores its maximum energy when its current is maximum, which occurs when the voltage is zero because of the 90° phase shift. Thus the capacitor and inductor require energy at different times in the ac cycle. When magnetic and electric energy requirements are balanced internal to the load, the source does not have to supply any stored energy externally because the load takes care of its own requirements for stored energy. But when these energies are not balanced internally, the source must lend energy cyclically to the load. The reactive power represents this energy lent twice each cycle to the load.

Sign of reactive power. Why did we arbitrarily change the sign of the reactive power term? This change is traditional for historical and practical reasons. When the load stores more magnetic energy than electric energy, the reactive power is positive (after the sign is changed); and when electric energy storage dominates, the reactive power is negative. In a practical power system, magnetic energy requirements always dominate because almost all power equipment is magnetic in its operations—motors, transformers, induction furnaces. Thus the power company usually supplies magnetic energy; that is, the current normally lags the voltage in a power system. Hence we call magnetic reactive power positive because it's normal, and we change the sign in the mathematical expression to conform with this preference.

The reactive power, Q, is so named because it is associated with the reactance, the imaginary part of the impedance in the circuit. To help distinguish between the real power, P, and the reactive power, Q, we use different units for these quantities. The unit of real power is watts. The unit of reactive power is *volt-amperes reactive*, or VAR (rhymes with "jar"). The kVAR (kiloVAR) is used for large amounts of reactive power, as kW is used for large quantities of real power.

5.2.3 Complex Power

Definition of complex power. The complex sum of the real and reactive power in Eq. (5.28) is called the *complex power* and can be written

$$\underline{S} = \tfrac{1}{2}\underline{V}\underline{I}^* \tag{5.31}$$

where $\underline{S}$ is the complex power. The complex conjugate of the current changes the sign of the imaginary part and thus introduces mathematically the customary change of sign. The expression for the complex power in Eq. (5.31) also permits us to relax the requirement that the phasor voltage, $\underline{V}$, be a real quantity. Note that if

$$\underline{V} = |\underline{V}|\underline{/\theta_V} \quad \text{and} \quad \underline{Z} = |\underline{Z}|\underline{/\theta}$$

then

$$\underline{I} = \frac{\underline{V}}{\underline{Z}} = |\underline{I}|\underline{/\theta_V - \theta}$$

where θ_V is the phase of the voltage, no longer assumed zero. The complex power, as defined in Eq. (5.31), would be

$$\underline{S} = \tfrac{1}{2}\underline{V}\underline{I}^* = \tfrac{1}{2}|\underline{V}|\underline{/\theta_V} \times |\underline{I}|\underline{/-\theta_V + \theta}$$

$$= \tfrac{1}{2}|\underline{V}||\underline{I}|\underline{/\theta} = \tfrac{1}{2}|\underline{V}||\underline{I}|(\cos\theta + j\sin\theta)$$

because the complex conjugate changes the sign of the current phase angle. Thus the phase of the voltage, or the phase reference generally, drops out with the expression of complex power in Eq. (5.31) and only the phase difference between voltage and current remains.

Example: complex power. To illustrate these ideas, we will continue the example of Fig. 5.13. The voltage was $\underline{V} = 120\sqrt{2}\underline{/0°}$ and the current we found to be $\underline{I} = 6.63\sqrt{2}\underline{/-56.4°}$ in polar form or $\underline{I} = 3.67\sqrt{2} - j5.53\sqrt{2}$ in rectangular form. The complex power is

$$\underline{S} = \tfrac{1}{2}\underline{V}\underline{I}^* = \tfrac{1}{2}(120\sqrt{2} + j0)(3.67\sqrt{2} + j5.53\sqrt{2})$$

$$= 120 \times 3.67 + j120 \times 5.53 = 440 + j663$$

Thus the real power is 440 W and the reactive power is $+663$ VAR. The positive sign indicates predominant magnetic energy storage. Note that the $\sqrt{2}$'s all dropped out, as they always do in power calculations.

Apparent power. The complex power is a complex number yielding information about both the flow of time-average power and the shuttling of loaned energy between source and load in an ac circuit. The magnitude of the complex power, $|\underline{S}|$, also has significance. This magnitude, which is a real number, is called the *apparent power*. The apparent power results when you measure the voltage and current with meters and multiply the measured values without regard for phase.

$$|\mathbf{S}| = \tfrac{1}{2}|\mathbf{V}||\mathbf{I}| = V_e I_e \quad \text{volt-amperes} \tag{5.32}$$

In Eq. (5.32) we need not use $|\mathbf{I}^*|$ because the magnitude is the same for $\mathbf{I}$ and $\mathbf{I}^*$. The units of apparent power are *volt-amperes* (VA). Apparent power is important as a measure of the operating limits in large electrical equipment such as transformers, motors, and generators. Losses in the wires are proportional to the square of the current in the machine, whereas losses in the magnetic materials are roughly proportional to the square of the operating voltage. Because electrical machinery is operated with the voltage more or less constant, apparent power limits imply current limits. In the present example the apparent power is $120 \times 6.63 = 796$ VA.

Summary.   The complex power gives concise information about power and energy flow in an ac circuit. The real part of the complex power is the time-average power in watts. The real power heats resistors, turns motors, and makes the electric meter revolve. The imaginary part of the complex power is the reactive power in VARs, which is proportional to the electrical energy lent to the load by the ac source twice each cycle. The reactive power is considered positive when the load is inductive and negative when the load is capacitive, although the latter case is rare in power systems. The magnitude of the complex power is the apparent power in volt-amperes and indicates the operating level of a power system.

The three units for ac power—watts, volt-amperes reactive, and volt-amperes—all have the proper scientific units for power (J/s). We use different names to clarify communication when speaking of the various kinds of "power" in ac circuits. These distinctions between watts, VARs, and VAs will aid your thought and communication about these related concepts.

Power triangles.   The apparent power, real power, and reactive power form a right triangle. Figure 5.15 shows this *power triangle* for the present example. The angle at the origin is θ, the angle of the impedance ($+56.4°$ in this case). Notice that when the current lags the voltage, the power triangle is drawn above the axis. This reversal is created by changing the sign on the reactive power term, as discussed above.

Because the various kinds of powers form a triangle, and because the power factor is the cosine of an angle of that triangle, there follow a host of formulas relating power factor (PF) with the various types of powers. Below we list several

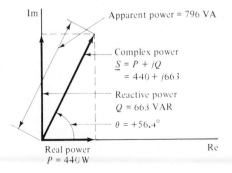

Figure 5.15   The power triangle pictures complex, apparent, real, and reactive power.

of these, which you can verify from the definitions and common trigonometric identities.

$$P = |\underline{S}| \, \text{PF} \tag{5.33}$$

$$|\underline{S}| = \sqrt{P^2 + Q^2} \tag{5.34}$$

$$Q = \pm |\underline{S}| \sqrt{1 - (\text{PF})^2} \; (+ \text{ for lagging PF}) \tag{5.35}$$

Based on these formulas, and others that can easily be derived, a variety of problems in ac power systems can be formulated, of which the following is typical.

Motor example. A 230-V motor has an output of 3 horsepower (hp). The motor efficiency is 78% and the power factor is 0.83 lagging. Find the input current, the apparent power, and the reactive power. Draw a phasor diagram.

The first step in the solution is to convert horsepower to watts. Of course we could look up the conversion factor in a table, but think about Christopher Columbus (1492), divide by 2, and you have it (746 W/hp). Thus the mechanical output is 3 × 746 = 2.24 kW. The efficiency would be the mechanical power output divided by the electrical power input; hence the input power is 2.24/0.78 = 2.87 kW. This would be the real power, so we have stated the units as watts. We know the real power, the voltage, and the power factor, so we can compute the current from Eq. (5.20).

$$P = V_e I_e \times \text{PF} \Rightarrow I_e = \frac{2.87 \text{ kW}}{230 \times 0.83} = 15.0 \text{ A (rms)}$$

From this result follows the apparent power, $|\underline{S}| = 230 \times 15.0 = 3.46$ kVA (kilovolt-amperes). The reactive power can be computed from Eq. (5.34).

$$Q = \sqrt{|\underline{S}|^2 - P^2} = \sqrt{(3.46)^2 - (2.87)^2} = +1.93 \text{ kVAR}†$$

Because the current is lagging, the reactive power is positive.

For the phasor diagram in Fig. 5.16 we have used the voltage for the phase reference, and shown the current lagging. The phase angle of the current follows from the power factor, $\theta = \cos^{-1}(0.83) = 33.9°$. The power triangle showing the real, reactive, and apparent powers would have the same angle but would be drawn above the real axis, similar to Fig. 5.15.

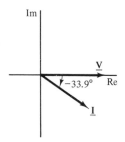

Figure 5.16 Motor voltage and current.

Power-factor correction in a transmission system. The power company uses capacitors in their transmission and distribution systems to improve efficiency and uniformity of voltage. Figure 5.17a models a source, load, and distribution line, which is inductive. By "distribution line" we refer to the power lines on poles that one sees lining streets and alleys in the industrial world. Figure 5.17c shows the phasor diagram for the system with 0.9 power factor, lagging current. The phase of the voltage drop in the line causes a difference between load and source voltages, and hence voltage regulation would be poor. Figure 5.17d shows the phasor diagram with unity power factor and the same load. Here

† You may not get this exact answer unless you start at the beginning of the problem and retain full accuracy in your calculator. In this book we write only three significant digits, but we calculate results with full precision and then round.

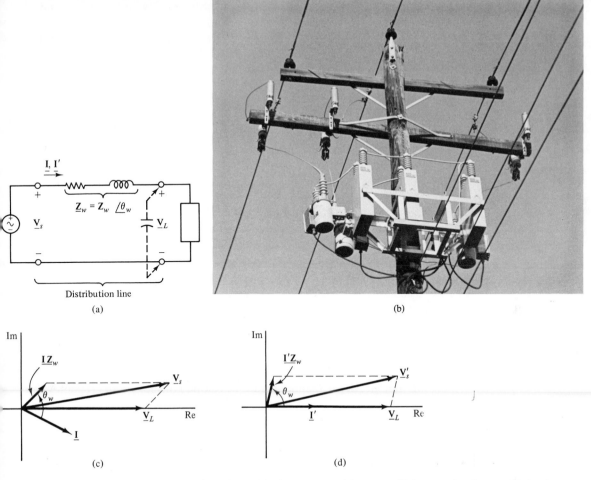

Figure 5.17 (a) Capacitors placed near the load improve efficiency and voltage regulation because the transmission system is inductive; (b) a 12.5-kV three-phase distribution power line with six capacitors between the lines and neutral; (c) phasor diagram for the system with 0.9 power factor and no capacitors; (d) phasor diagram for the systems with same load and unity power factor (with capacitors). Note that line current is smaller and $\underline{V}_L$ and $\underline{V}_s$ are more nearly equal than in (c).

line current is smaller, and load and source voltages are more nearly equal. For this reason the power company adds capacitors (Fig. 5.17b) to their distribution system.

Importance of reactive power. Power companies have to be careful about the reactive power load on their systems. As stated above, the limits of larger power equipment such as generators and transformers are described by the apparent power, the Pythagorean sum of the real and reactive powers. Thus if the reactive power becomes large, a piece of equipment may become overloaded even though the load of real power is moderate. Also, the reactive energy must be transported from generator to user, often over great distances. Reactive power increases line current, and hence increases line losses. The losses in the transmission system are not charged directly to the consumer, since the watthour meter is placed on the load. For this reason, the power company might penalize customers whose requirements for reactive power are great. Often the industrial consumer can save money by placing a bank of capacitors in parallel with an inductive load to store energy locally. In effect, they receive the stored energy from the power company only once and then keep it "in house" with the capacitors. The industrial customer thus "corrects" his power factor by creating a resonance between the electric energy stored by the added capacitance and the magnetic energy used by motors or other heavy equipment. As shown above, the power company adds capacitors to the power lines to improve the characteristics of their distribution system.

5.2.4 Reactive Power in Electronics

Reactive power flow also plays an important role in electronics. Here we normally do not deal with large amounts of power; indeed, the goal often is to make full use of the small amounts of power that are available. To see the role of reactive power, let us examine the conditions for maximum power transfer in an ac circuit.

Thévenin and Norton equivalent circuits for AC. In Chapter 2 we justified the Thévenin equivalent circuit by using the concepts of linearity and superposition. In Chapter 4 we reduced ac problems to equivalent dc problems through the use of phasors, through which sources and circuits are represented by complex numbers. All the techniques we developed for dc circuits remain valid for solving ac circuits, including Thévenin and Norton equivalent circuits. In Fig. 5.18 we show a Thévenin equivalent circuit with a phasor voltage source and an output impedance, $\underline{Z}_{eq}$, which we have expressed in terms of a resistive and reactive part for benefit of the derivation that follows. Recall that the circuit replaced by the Thévenin equivalent circuit can be arbitrarily complicated.

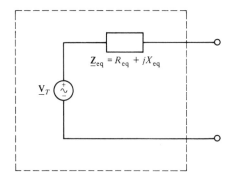

Figure 5.18 For an ac circuit the Thévenin voltage is a phasor and the output resistance becomes an impedance.

Maximum power transfer for AC sources. Let us consider a typical situation that might arise in electronics, that of getting maximum power out of a radio antenna. Specifically, consider the telescoping AM radio antenna on an automobile, as suggested in Fig. 5.19. Radio waves are radiated by a commercial station, perhaps at considerable distance, and these waves interact with the an-

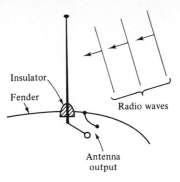

Figure 5.19 The antenna plus the radio station can be represented by a Thévenin equivalent circuit.

tenna to give a small voltage, typically 10 mV rms, between the fender and the end of the antenna. The derivation of the output impedance of such an antenna lies beyond the scope of this text, but we can say a little based on our understanding of electric and magnetic energy. As a circuit element, the antenna represents a wire that leads nowhere, that is, an open circuit. The ratio waves tend to make current flow on the wire, but a small current produces a buildup of charge on the wire and the current stops. Thus we anticipate that an antenna of this type would build up charge but carry little current. This suggests that electric energy storage would dominate magnetic energy storage and that the output impedance would be capacitive. We have shown typical values in Fig. 5.20. The 10-Ω resistive part of the output impedance represents the resistance of the antenna and the energy reradiated by the antenna currents. Our task is to specify the input impedance of the radio, $\underline{Z}_L = R + jX$, to receive maximum power out of the antenna. The average power into the load would be $|\underline{I}|^2 R/2$, where $\underline{I}$ is the current in the load. This current can be calculated as the voltage divided by the total impedance in the circuit, the sum of $\underline{Z}_{eq}$ and $\underline{Z}_L$.

$$\underline{I} = \frac{\underline{V}_T}{\underline{Z}_{eq} + \underline{Z}_L} = \frac{10\sqrt{2}\underline{/0°}}{(10 + R) + j(X - 150)} \quad \text{mA}$$

Because the phase does not matter in this power calculation, we will consider only the magnitude

$$|\underline{I}| = \frac{10\sqrt{2}}{\sqrt{(10 + R)^2 + (X - 150)^2}} \quad \text{mA}$$

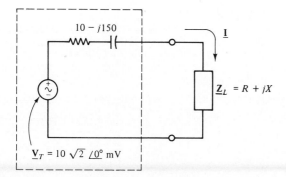

Figure 5.20 Find R and X to maximize power in R.

Consequently, the average power delivered to the radio by the antenna would be

$$P(R, X) = \frac{|\mathbf{I}|^2 R}{2} = \frac{2 \times (10^{-2})^2 R/2}{(10 + R)^2 + (X - 150)^2}$$

This is the function to be maximized, so we might make an assault using the methods of differential calculus. Before taking derivatives, however, we should note the effect of the load reactance, X. Being in the denominator and being squared, the total reactance can only decrease the power. Clearly, the best we can do is set X to $+150$ Ω. Once we do that, the power is a function of R only:

$$P(R, +150) = \frac{10^{-4}R}{(10 + R)^2} \Rightarrow R = 10 \ \Omega \text{ for maximum power}$$

We have written by inspection the value of R that gives maximum power because we recognize that the problem has been reduced to that solved back in Chapter 2. That is, once the load reactance (X) is adjusted to balance the reactance in the source, the maximum power follows from setting the resistance of the load equal to the resistive part of the output impedance of the source.

In general, the maximum power transfer will occur when the load has the same resistance as the source but the opposite reactance. We therefore create a local resonance from the viewpoint of the load, in effect balancing the equivalent stored energies. In the case solved above, you can confirm that the power delivered to the radio by the antenna is 2.5×10^{-6} W. Not much power, but we know that it must be adequate since AM radios work. In Section 6.4.4 we will show how transistors are used to amplify these small signals to a level where loudspeakers produce audible sound.

A power system is designed to provide uniform voltage, independent of load. For this reason, the source impedance is always small relative to load impedance. Maximum power transfer is irrelevant to a power system.

5.2.5 Check Your Understanding

1. If the current leads the voltage in an ac circuit, the reactive power is positive or negative?
2. If the real power is 600 W and the reactive power is -300 VAR, what is the apparent power?
3. The complex power depends only on the relative phase between voltage and current, not the absolute phase. (T/F?)
4. An electronic circuit has an open-circuit voltage of 100 mV (rms) and an output impedance of $20 + j10$ Ω. What load will draw maximum power from this source? What will be the power in this load?
5. To exhibit a resonance, a circuit must contain both inductors and capacitors. (T/F?)

Answers.   1. Negative; 2. 671 VA; 3. true; 4. 20 $-j10$ Ω, 1.25 $\times$ 10^{-4} W; 5. true.

Power in AC Circuits   Chap. 5

5.3.1 Introduction

A transformer is a highly efficient device for changing ac voltage from one value to another, for example, from 120 V to 6 V. Transformers come in all sizes, from the enormous transformers used in power substations to the small transformers used for doorbells. [Sometimes the word "transformer" is used for a device that employs a transformer (in the first sense) but includes other controls or devices, such as the "transformers" used to control model railroad trains.]

The transformer gives to ac a feature lacking in dc power systems. Using a transformer, we can efficiently change ac voltage from small amplitudes to large amplitudes, or vice versa. Such changes are not simply accomplished with dc voltage.

Figure 5.21 shows a simple transformer. It consists of two coils and an iron core, which enhances the magnetic coupling between the coils. Let us say that we construct such a device, connect one coil to an ac voltage source, and connect the other coil to a resistive load. In this case, we find that the resistive load gets hot, suggesting the flow of electrical energy from the ac source through the transformer and into the load. Furthermore, we find that the transformer does not get very hot, suggesting that the transformer is an efficient device for coupling load and source.

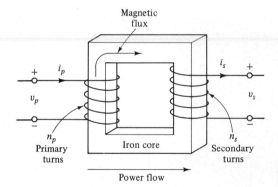

Figure 5.21 Simple electrical transformer.

The coil connected to the ac source is called the *primary*, and that connected to the load is called the *secondary*. There is nothing special about the two sides, for the transformer can convey power either way. In most applications the power flows in only one direction and hence these names are useful.

The ideal transformer. In Chapter 13 we will describe physical construction, circuit models, and applications of transformers in power systems. Here we define the *ideal transformer* as a circuit element and explore its properties in voltage, current, and impedance transformation. Primary and secondary voltage and current variables are defined in Fig. 5.22, which shows the circuit-theory

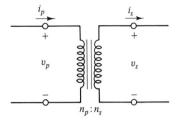

Figure 5.22 Circuit symbol for an ideal transformer. The primary voltage and current form a load set, and the secondary voltage and current form a source set.

symbol for an ideal transformer. Note that the primary variables form a load set and the secondary variables form a source set. These definitions are customary and indicate that the primary acts as a *load* to the power system supplying power to the transformer but the secondary acts as a *source* to the loads connected to it. We also define n_p and n_s, the turns on the primary and secondary.

The primary and secondary voltages in the ideal transformer are related by

$$\frac{v_p}{n_p} = \frac{v_s}{n_s} \tag{5.36}$$

Thus the side of the transformer with the larger number of turns has the larger voltage; indeed, the voltage per turn is constant for a given transformer. The primary and secondary currents in the ideal transformer are related by

$$n_p i_p = n_s i_s \tag{5.37}$$

Thus the side of the transformer with the larger number of turns has the *smaller* current. For example, a transformer to increase the voltage would have a primary with few turns of large wire (small voltage, large current) and the secondary would have many turns of small wire (large voltage, small current).

Equations (5.36) and (5.37) define the ideal transformer as a circuit element, with one restriction—no dc. The primary of the transformer is, like an inductor, a short circuit to dc and provides no coupling to the secondary unless voltage and current are changing.

No losses.   Multiplication of the left and right sides of Eqs. (5.36) and (5.37) shows that the instantaneous power into the primary (p_{in}) is equal to the instantaneous power out of the secondary (p_{out})

$$\frac{v_p}{n_p} \times n_p i_p = \frac{v_s}{n_s} \times n_s i_s \Rightarrow p_{in} = p_{out} \tag{5.38}$$

Thus the ideal transformer has no losses and stores no energy.

Impedance transformer.   Equations (5.36) and (5.37) reveal a very useful property of transformers—resistance (or impedance) transformation. Figure 5.23 shows an ideal transformer, a load resistor (R_L) connected to the secondary, and an equivalent resistance (R_{eq}) defined at the primary. This section shows that such an equivalent resistance is meaningful and that its value depends on the load resistance and the turns ratio of the transformer. The equations of the circuit are those of the ideal transformer in Eqs. (5.36) and (5.37) plus Ohm's law for R_L. If we divide Eq. (5.36) by Eq. (5.37), we obtain

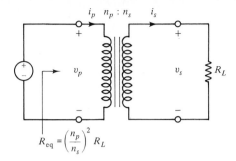

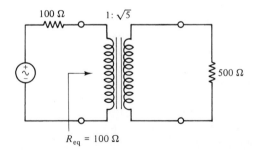

Figure 5.23 The transformer will change R_L to an equivalent resistance R_{eq}.

$$\frac{1}{n_p^2}\frac{v_p}{i_p} = \frac{1}{n_s^2}\frac{v_s}{i_s} \tag{5.39}$$

But v_s/i_s is the load resistor (R_L), and v_p/i_p defines the equivalent resistance (R_{eq}) into the primary, so Eq. (5.39) leads to the value of the equivalent resistance:

$$R_{eq} = \left(\frac{n_p}{n_s}\right)^2 R_L \tag{5.40}$$

Thus the load resistance is transformed by the square of the turns ratio†. By using a transformer, we can make a large resistor appear small, or we can make a small resistor appear large. For example, if we wish to derive maximum power out of a source with an output resistance of 100 Ω, we must use a load resistance equal to the output resistance of the source (page 174). If we are required to furnish this power to a load resistance of 500 Ω, as shown in Fig. 5.24, we can accomplish the task with a transformer having $\sqrt{100/500}$ for a turns ratio. Because we wish to make the 500-Ω resistor look smaller, we must connect it to the high-voltage side of the transformer, the side with the more turns, and look into the low-voltage side to see the smaller resistance.

Transforming impedances is a useful technique in the analysis of circuits containing transformers. Figure 5.25a shows an ac circuit: we are to solve for $\underline{I}_L$. One approach would be to write the circuit equations, plus those of the ideal transformer, and proceed to eliminate variables. Without defining transformer primary and secondary variables, however, we may transform the load impedance into the primary and solve directly for the primary current. Figure 5.25b shows

Figure 5.24 The transformer turns ratio maximizes the power to the 500-Ω load.

† Note that the ratio of primary to secondary turns determines voltage, current, and impedance transformation properties. For this reason, the *turns ratio* is often stated as $1:n$ (or $n:1$), where n is not necessarily an integer and can be less than unity.

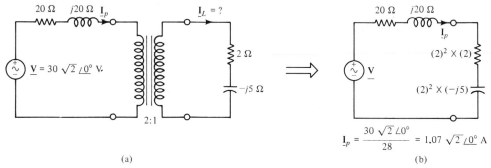

Figure 5.25 (a) Solve for I_L; (b) equivalent circuit.

the result of impedance transformation, and the primary current is easily determined as shown in the figure. The secondary current, which is the load current, is greater by the turns ratio, twice as great in this case. Note that we have twice as many primary turns as secondary turns; hence the secondary voltage is half the primary voltage and the secondary current is twice the primary current, the product (the power) remaining constant. Thus the secondary current is $2.14\sqrt{2}\underline{/0°}$ A.

5.3.2 Transformer Applications

Voltage and current transformation. We have shown that transformers change voltage levels, current levels, and impedance levels. Each of these properties leads to important applications. Voltage changes are frequently required to convert the standard power distribution voltages, usually 120 V, into higher or lower ac voltages as required by specific devices. A TV set normally contains one or more transformers to furnish ac voltage to several internal "power supplies," a subject we will study in Chapter 6. The "coil" in an automotive ignition system is a transformer whose function is to provide high voltage to the spark plugs, where we want sparks, and to keep the voltage low at the points, where we do not want sparks. Transformers are also used to reduce 120 V to lower values for safety. Doorbell and thermostat circuits provide examples.

A few transformer applications arise out of need for high current levels. Arc welders and ac electromagnets are applications where transformers are used to produce large currents. One is also tempted to list the hand-held electric soldering gun in this category, but it probably belongs in the third category.

Impedance transformation. The third, and probably the most important, class of transformer applications involves impedance transformation. We have already illustrated the use of transformers in effecting maximum power transfer, a common application in electronics. The most important transformer applications of this class, however, lie in power distribution. Figure 5.26 suggests the generation and delivery of electrical power to a distant user over a distribution line. Although we have shown identical transformers at each end to simplify the anal-

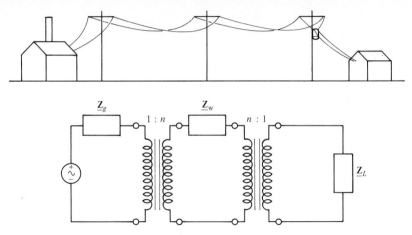

Figure 5.26 Simple power distribution system and its equivalent circuit.

ysis, the two transformers would not be identical, for the power would not be generated at the low voltage required by the user.

A power distribution system should offer constant voltage and high efficiency. The user wants constant voltage, independent of load, because his equipment is designed to operate at standard voltage. He requires, for example, that the voltage remain resonably near 120 V whether he uses 1 kW or 10 kW of power. This requires, in turn, that the equivalent impedance of the source, as seen from the user's point of view, be as low as possible. Of course, an ideal voltage source has zero output impedance, but the generator has an inherent output impedance, $\mathbf{Z}_g$, and the distribution line has an impedance, $\mathbf{Z}_w$, as shown in Fig. 5.26. The generator output impedance can be made quite small by good design, but the resistance and reactance of the transmission line can only be reduced within limits due to the large distances.

The transmission system is required to be efficient to reduce costs and energy waste. We shall use the impedance-transforming properties of the transformers to improve system characteristics. First we will transform $\mathbf{Z}_L$ with the transformer at the load. This raises the load impedance by a factor of n^2 and places it in series with $\mathbf{Z}_w$. We now can transform this series combination with the transformer at the generator, which lowers the impedances by n^2. The resulting equivalent circuit is shown in Fig. 5.27. Note that the load impedance appears with its true value in the final equivalent circuit, as does the generator impedance, but the impedance of the transmission line is reduced by the factor n^2. This will reduce its losses accordingly.

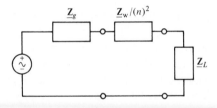

Figure 5.27 Simplified equivalent circuit for the power distribution system.

This reduction in transmission line losses can be understood by considering the current in the transmission line. The transformer at the generator increases the voltage and decreases the current by its turns ratio, the total power being unchanged. The transmission-line losses will be $I^2 R_w$, where I is the rms current in the line and R_w is the wire resistance. Reduction of the line current by n therefore reduces the losses by n^2, and this reduction is reflected in the equivalent circuit in Fig. 5.27. For this reason, electrical power is distributed at extremely high voltages, up to 750 kV.

Figure 5.27 also suggests that the voltage regulation at the load (voltage relatively independent of load current) is improved by the use of a high-voltage transmission line. The load impedance appears with its true value, which shows that the same equivalent circuit would have resulted had we transformed all impedances to the load end of the circuit. Thus the output impedance of the generator–transmission line system is $\underline{Z}_g + \underline{Z}_w/n^2$. The effect of the transmission-line impedance is therefore reduced by the square of the turns ratio, to the improvement of load-voltage regulation.

Example. Power is generated at 24 kV, 100 miles distant from a town that uses 50 MW at 12 kV, as shown in Fig. 5.28. The transmission line has an impedance of $0.1 + j0.8$ Ω/mile. What should be the transmission voltage for an efficiency of 98.5% for the transmission system?

The load current is $50 \times 10^6/12 \times 10^3 = 4167$ A, and hence the transmission line current is

$$I = 4167 \times \frac{12}{V} \tag{5.41}$$

where V (in kV) is the line voltage to be determined. The line resistance is 10 Ω and the allowed loss is 1.5% of the 50-MW load; hence

$$\left(4167 \times \frac{12}{V}\right)^2 \times 10 = 0.015 \times 50 \times 10^6 \tag{5.42}$$

Thus the required voltage is 183 kV.

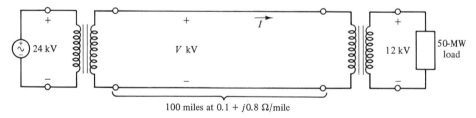

Figure 5.28 The transmission system is required to have 98.5% efficiency.

Summary. Transformers are used for voltage, current, and impedance transformation. Applications abound in both electronics and power systems engineering. Realistic power systems, however, would use three-phase voltages for generation and distribution of power, as we discuss in the following section.

5.3.3 Check Your Understanding

1. An ideal transformer has no losses. (T/F?)
2. In a transformer, a large number of turns on a winding goes with high or low voltage for that winding? What about current?
3. Transformers can be used to change voltage levels, current levels, impedance levels, or all three?
4. If we want to make a load impedance look smaller, which winding (primary or secondary) should have more turns?
5. To reduce losses in a electrical power transmission system, the voltage level should be raised or lowered?
6. If a larger voltage needs to be transformed to a smaller voltage, which side of the transformer should have the larger wire size?

Answers. 1. true; 2. more turns go with the larger voltage and the smaller current; 3. all three; 4. secondary; 5. raised; 6. the low-voltage side.

5.4 THREE-PHASE POWER SYSTEMS

5.4.1 Introduction

Importance of three-phase systems. If you looked out your window at this moment, you would probably see some power lines. Count the wires and you will likely find there to be four. Go examine a pole closely and you will see that at each pole one of the four wires is connected to a conductor that comes down the pole and enters the ground.

When you are driving cross country and see a large electrical transmission line, you will again see four wires. One of them, usually running along the top of the towers, will be noticeably smaller than the other three. If you look closely, you will again see that the small wire is grounded at every tower. Four wires—what can it mean?

We have hitherto been discussing single-phase circuits, one ac generator connected to a load with two wires. The three ungrounded wires in the transmission systems we have just described are driven by three ac generators. The grounded wire increases safety and protection from lightning. Power is conveyed by the three larger wires in the form of three-phase electric power. The overwhelming majority of the world's electric power is generated and distributed as three-phase power. If you were to examine a catalogue of industrial-grade motors, you will discover that all the larger motors, bigger than a few horsepower, would be three-phase motors.

What is three-phase power? Physically, there are three wires that carry the power, and often a fourth wire, called the *neutral*, which is grounded. In enclosed cables, the active wires are normally colored red, black, and blue; and the neutral, if present, will be white or gray. The phases are traditionally designated A, B, and C, and the time-domain voltages between them are as shown in Fig. 5.29. The Greek lowercase letter ϕ (phi) is a common abbreviation for

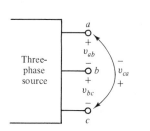

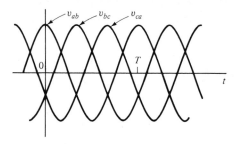

Figure 5.29 Voltages of a three-phase system in the time domain.

"phase"; hence 3φ means three phase and 1φ means single phase. The voltages are expressed mathematically as

$$v_{ab}(t) = V_p \cos \omega t$$

$$v_{bc}(t) = V_p \cos(\omega t - 120°)$$

$$v_{ca}(t) = V_p \cos(\omega t - 240°)$$

(5.43)

The frequency-domain picture for a three-phase system is shown in Fig. 5.30. We have used $\underline{V}_{ab}$ as our phase reference and shown $\underline{V}_{bc}$ following by 120°, then $\underline{V}_{ca}$. This is known as an *ABC* phase sequence and corresponds to the time-domain representation in Fig. 5.29 and Eqs. (5.43).

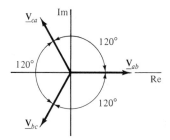

Figure 5.30 Voltages of a three-phase system in the frequency domain.

Advantages of three-phase power. We have shown in Fig. 5.6 that single-phase power produces a pulsating flow of energy. A smooth flow of energy from source to load is achieved by a balanced three-phase system. If we have identical resistive (R) loads connected between the three phases, the instantaneous flow of power would be given by Eq. (5.44). We use the trigonometric identity $\cos^2 \alpha = (1 + \cos 2\alpha)/2$ to derive the second and third forms.

$$p(t) = \frac{v_{ab}^2(t)}{R} + \frac{v_{bc}^2(t)}{R} + \frac{v_{ca}^2(t)}{R}$$

$$= \frac{V_p^2}{2R} [1 + \cos 2(\omega t) + 1 + \cos 2(\omega t - 120°) + 1 + \cos 2(\omega t - 240°)]$$

$$= \frac{3V_p^2}{2R} + \frac{V_p^2}{2R} [\cos(2\omega t) + \cos(2\omega t - 240°) + \cos(2\omega t - 480°)]$$

(5.44)

We see a constant term and a term that appears to be time varying at twice the source frequency. Actually, the second term adds to zero at all times. This is easily shown by a phasor diagram; indeed, the phasors representing these terms give the same phasor diagram as Fig. 5.30, since a phase of $-480°$ is the same as $-120°$. Clearly, the phasor sum of the three symmetrical phasors is zero, and hence the fluctuating power term is also zero at all times. Thus Eq. (5.44) reduces to

$$p(t) = \frac{3V_p^2}{2R} \quad \text{(a constant)}$$

This constant flow of energy effects general smoothness of operation in three-phase electrical equipment. A rough analogy is suggested by comparing an engine having one cylinder with an engine having many cylinders—clearly, the multi-cylinder engine will run smoother.

Compared with a single-phase system, distribution losses are proportionally less for a three-phase system. Additionally, three-phase motors offer advantages over single-phase motors in both startup and run characteristics. In short, three-phase systems are supremely important for the generation, distribution, and use of electrical power, particularly in industrial settings.

5.4.2 Three-Phase Power Sources

Three single-phase sources.   Three-phase generators produce three single-phase voltages with the required 120° phase shifts, which are internally connected to produce a three-phase source. In this section, we consider that the three single-phase voltages are brought out of the generator to a terminal board, and our job is to connect the resulting six terminals together to produce a three-wire, three-phase source. The terminal board is shown in Fig. 5.31, and the phasor diagram of the available voltages is shown in Fig. 5.32.

Figure 5.31 External terminals for the three separate phases. These must be connected to produce a three-phase system.

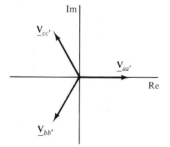

Figure 5.32 Phasor voltages in the three coils. These must be connected externally to make a true three-phase system.

Delta (Δ) and wye (Y) connections.   The symmetry of the desired phasor voltages suggests that we require some sort of symmetrical connection for our three voltages represented in Fig. 5.31. There are only two symmetrical configurations involving three elements connected end-to-end, and these are shown in Fig. 5.33. The closed ring is usually called a *delta* (for the Greek letter Δ), even when it is drawn upside down or on its side. The configuration with the common point is sometimes called a star configuration, but in three-phase terminology it

Figure 5.33 The only two symmetrical configurations of three elements: (a) delta; (b) wye.

is usually called a *wye configuration* (a phonetic spelling of the letter Y), regardless of orientation.

These symmetric configurations suggest two solutions of our question posed above. We require three terminals for a source of three-phase power. The delta has three terminals, and hence its application as a possible configuration for the three-phase generators is clear. The wye, on the other hand, has four terminals, counting the common connection in the center. This gives us the fourth wire mentioned earlier, the neutral wire that is grounded. We have to connect the terminals in such a way that they have the geometric symmetry of the delta or wye configurations in Fig. 5.33, but we have also retained the electrical symmetry indicated by the phasor diagram in Fig. 5.30.

Delta connection. The delta configuration places the three generators in a closed ring. Figure 5.34b shows one possible connection for the delta. We must be careful, however, if we are to connect c' to a, for we then will have the generator connected in a closed ring. It would be like jump-starting a car having a weak battery, as shown in Fig. 5.34c. The circuit can be closed if the polarities are correct; for with the connection marked "Yes" there will be at most a small voltage across the gap and only small currents will flow through the batteries. But if the polarities are wrong, there will be approximately 24 V across the gap and a huge current will flow if the connection is made.

Similarly, if we are to close the ring of generators in Fig. 5.34b, we require the voltage across the gap to be small. This requires that

$$\underline{V}_{ac'} = \underline{V}_{aa'} + \underline{V}_{bb'} + \underline{V}_{cc'} = \text{small} \tag{5.45}$$

Notice that Eq. (5.45) follows from the rule for adding subscripted voltages given on page 15 because a' is connected to b and b' is connected to c. That is, the general rule is

$$V_{ab} = V_{ax} + V_{xb}$$

Figure 5.34 Potential delta connection. The voltage across a and c' must be zero if the ring is to be closed.

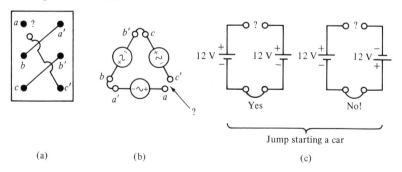

(a)   (b)   (c)

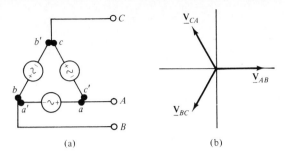

(a)

(b)

Figure 5.35 Delta connection with phasor diagram.

We can extend this to $V_{ab} = V_{ax} + V_{yb}$, if x and y are connected. The phasor diagram for the sum in Eq. (5.45) is easily derived from Fig. 5.32, clearly, the sum is small, ideally zero. Thus it is safe to close the ring of generators and bring out the connected terminals as a three-phase source. Figure 5.35b shows the phasor diagram of the final connection. Another possible delta connection would result with a connected to b', b connected to c', and c connected to a'. We leave the investigation of this possibility for a homework problem.

Wye connection. A wye connection results from connecting a', b', and c', as shown in Fig. 5.36a. With this connection the magnitudes of $\underline{V}_{ac}$, $\underline{V}_{ba}$, and $\underline{V}_{cb}$ are $\sqrt{3}$ greater than those of the component voltages, $\underline{V}_{aa'}$ and $\underline{V}_{bb'}$, and $\underline{V}_{cc'}$ and the phase of $\underline{V}_{ac}$ lies at $-30°$ relative to $\underline{V}_{aa'}$. These combinations are illustrated in Fig. 5.36b by the forming of $\underline{V}_{ac}$ through addition of $\underline{V}_{aa'}$ and $\underline{V}_{c'c}$, which is the negative of $\underline{V}_{cc'}$. The $\sqrt{3}$ comes from the 30°–60°–90° triangle, as we have shown in Fig. 5.36c. Unlike the delta connection, the wye connection has all three generators connected to a common point. In Fig. 5.37, we label the three lines

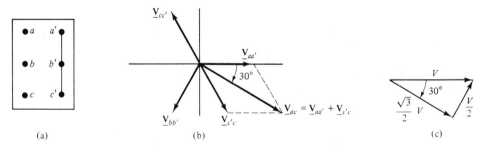

(a)

(b)

(c)

Figure 5.36 Possible wye connection.

Figure 5.37 Wye connection. The line voltages are $\sqrt{3}$ larger than the phase voltages.

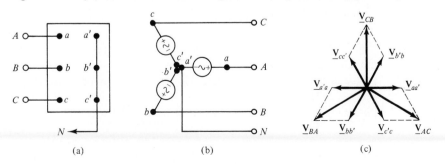

(a)

(b)

(c)

A, *B*, and *C*, and the neutral we label *N*. This is a wye connection even though our Y comes out lying down. The wye connection of the generators leads to the four-wire system which we described at the beginning of this section. The wye connection is important in the distribution of electric power.

Line voltage and phase voltage.　We must distinguish between the line-to-line voltages, usually called *line voltages*, and the line-to-neutral voltages, usually called *phase voltages* for the wye connection. As we have shown, the line voltages are $\sqrt{3}$ times the phase voltages. The line voltages are displaced by 30° from the phase voltages, so we would have to make $\underline{V}_{AC}$ our phase reference to have the line voltages in Fig. 5.37 conform to the picture in Fig. 5.30.

Phase rotation.　The phase rotation came out *ABC* in both systems we developed. This is a mere accident of labeling; we could easily relabel the connections to have *ACB*. That is, we can erase *B* and *C* (actually, any two) and switch them to achieve an *ACB* phase sequence. The notation is arbitrary, but the physical phase rotation is very important. The rotational direction of a three-phase motor, for example, depends on the phase rotation of the input voltages. Because phase rotation is important, several techniques exist for determining the phase rotation of a three-phase power system.

Other possible connections.　We have now shown the two ways for connecting the voltages in Fig. 5.32 to give a three-wire, symmetrical power system. Actually, we have shown one version of each way, for there are minor variations on the procedures. For example, we can make *a*, *b*, and *c* the neutral for a wye. But we have shown the two basic connections. Each is important; the delta fits some applications of three-phase power, the wye excels for others. We will postpone our discussion of typical applications until we have explored the ways for connecting three-phase loads.

5.4.3 Three-phase Loads

Delta-connected loads.　Like three-phase generators, three-phase loads can be connected in delta or wye. Figure 5.38 shows a balanced three-phase resistive load connected as a delta. The source of the three-phase power is not shown; we will assume *ABC* phase rotation and a line voltage $\sqrt{2}\, V_L = |\underline{V}_{AB}|$ $= |\underline{V}_{BC}| = |\underline{V}_{CA}|$ where V_L is the effective line voltage. We have shown no neutral because the load offers no place for connecting a neutral. In practice, however, a three-phase load, such as a three-phase motor, would be housed in a physical structure of some kind, and this structure would normally be grounded directly to earth ground and through the neutral.

With a load connected as a delta, we must distinguish between the line currents and the phase currents flowing in the resistors. Figure 5.39 shows the phase currents, $\underline{I}_{AB}$ and $\underline{I}_{BC}$ and $\underline{I}_{CA}$. These currents are in phase with the line voltages. The line current, say, $\underline{I}_A$, can be determined by phasor addition of the phase currents. Kirchhoff's current law at the top node is

$$\underline{I}_A = \underline{I}_{AB} + \underline{I}_{AC}$$

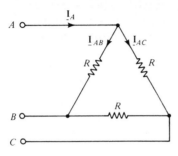

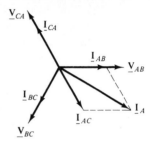

Figure 5.38 Delta-connected load.

Figure 5.39 Phase current addition to yield line current.

but

$$\underline{I}_{AC} = -\underline{I}_{CA}$$

and hence the currents add as in Fig. 5.39. The other line currents could be determined similarly; indeed, the picture develops like that of the wye generator connection in Fig. 5.37c. We see that the line currents are $\sqrt{3}$ greater than the phase currents. For the resistive load, the phase of the line current in A is in phase with the average of the phases of $\underline{V}_{AB}$ and $\underline{V}_{AC}$. The phase relationships are shown in Fig. 5.40.

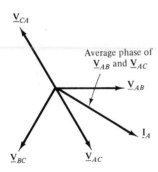

Figure 5.40 For a resistive load the line current is in phase with the average phase of the voltages to the other two lines.

With balanced loads that are not resistive, the phasor diagram shown in Fig. 5.39 changes only slightly. The magnitude of the phase currents is computed by dividing the line voltages (which are also the phase voltages) by the magnitude of the phase impedance. The phase currents will lead or lag the line voltages according to the angle of the impedance. Hence the line currents will also be shifted in phase by the angle of the impedance. We give an example below.

Power in three-phase delta connections. The total power to a resistive load would be the sum of the powers delivered to the three resistors; and this would be

$$P = 3P_R = 3(V_\phi I_\phi) \tag{5.46}$$

because the power factor is unity for a resistor. In Eq. (5.46), V_ϕ and I_ϕ represent the rms values of the phase voltage and current. We desire, however, to express the total power in terms of line voltage and current. The phase currents are often inaccessible for measurements, but the line voltage and current can always be measured. Consequently, we introduce the line voltage and current

$$P = 3V_L \frac{I_L}{\sqrt{3}} = \sqrt{3}\ V_L I_L \qquad (5.47)$$

where V_L and I_L are the rms line voltage and current. Hence with three identical resistors connected as a delta, you measure the line voltage and line current, take their product and then, because it is three phase, you multiple by $\sqrt{3}$.

If the load were not resistive, you would multiply also by the power factor. The more general formula is thus

$$P = \sqrt{3}\ V_L I_L \times \text{PF} \qquad (5.48)$$

when PF is the power factor. In applying Eq. (5.48), the power factor is the cosine of the angle of the phase impedance and is not the phase angle between line current and line voltage. The power factor angle is, however, the angle between the phase of the line current and the average phase of the voltages to the two other lines, as defined in Fig. 5.40.

Example. A 230-V three-phase power system supplies 2000 W to a delta-connected, balanced load with a power factor of 0.9, lagging. Determine the line currents, the phase currents in each phase of the load, and the phase impedance. Draw a phasor diagram.

First we calculate the magnitude of the line currents from Eq. (5.48).

$$P = \sqrt{3}\ V_L I_L \times \text{PF} \Rightarrow I_L = \frac{2000}{\sqrt{3}\ (230)(0.9)} = 5.58\ \text{A (rms)}$$

The phase currents are smaller by $\sqrt{3}$, so

$$I_\phi = \frac{I_L}{\sqrt{3}} = \frac{5.58}{\sqrt{3}} = 3.22\ \text{A}$$

This allows us to calculate the impedance in each phase of the delta. The angle of the impedance is implied by the power factor: $\theta = \cos^{-1}(0.9) = +25.8°$, $+$ because the current is lagging (inductive).

$$\mathbf{Z}_\phi = \frac{V_\phi}{I_\phi}\underline{/\cos^{-1}(\text{PF})} = \frac{230}{3.22}\underline{/\cos^{-1}(0.9)} = 71.4\underline{/+25.8°}$$

We can now draw the phasor diagram (Fig. 5.41). We will use $\underline{\mathbf{V}}_{AB}$ for the phase reference, with the other line voltages placed symmetrically in ABC sequence. The phase currents lag by 25.8°, as shown. The line currents can be computed by phasor addition of the phase currents, as we did in Fig. 5.39, but another approach is to use our earlier results to place the line currents behind the phase currents by 30° and greater by $\sqrt{3}$. Whichever way is chosen, only one line current need be determined, $\underline{\mathbf{I}}_A$ for example, and the other two can be constructed by symmetry.

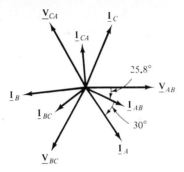

Figure 5.41 Line voltages, phase currents, and line currents. The line currents are $\sqrt{3}$ greater than the phase currents.

Wye-connected loads.   A load is said to be connected in a wye when it is connected into a neutral, as shown in Fig. 5.42, where we have labeled the input connections for the three-phase power a, b, and c and the neutral n. We have shown no connection to the neutral, but there would often be a connection between the load neutral and the source neutral, if such existed. For a perfectly balanced load no current would flow in the neutral because the three line currents add to zero. Otherwise, the type of connection of the generators producing the three-phase power (A, B, and C) is unimportant. We will calculate the line-to-neutral voltages $\mathbf{V}_{an}$, $\mathbf{V}_{bn}$, and $\mathbf{V}_{cn}$ and the line currents $\mathbf{I}_A$, $\mathbf{I}_B$, and $\mathbf{I}_C$.

First we will solve for the line-to-neutral voltages. The phase impedances are given, $\mathbf{Z}_\phi$; hence it is clear that the line currents can be determined once the line-to-neutral voltages are known; for example,

$$\mathbf{I}_A = \frac{\mathbf{V}_{an}}{\mathbf{Z}_\phi} \qquad (5.49)$$

The line-to-neutral voltages can be determined from consideration of the symmetry of the circuit. It is convenient to reason as if we know the line-to-neutral voltages (which we do not) and wish to determine from them the line-to-line voltages. The relationship between these two sets of three-phase voltages then becomes known and we henceforth can deduce either set of voltages from

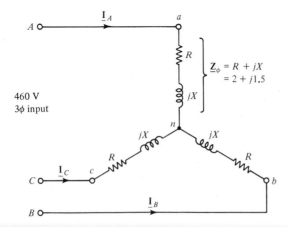

Figure 5.42 Wye-connected load. To find the line currents we must determine the line-to-neutral voltages.

the other. We assume the *ABC* phase sequence; hence the line-to-neutral voltages must appear as in Fig. 5.43, assuming that we make $\underline{V}_{an}$ the phase reference. First we will determine $\underline{V}_{AB}$. We can express the $\underline{V}_{AB}$ in terms of $\underline{V}_{an}$ and $\underline{V}_{nb}$:

$$\underline{V}_{AB} = \underline{V}_{an} + \underline{V}_{nb} \tag{5.50}$$

We can use *A* and *a*, and so on, interchangeably, because these indicate the same points in the circuit. Equation (5.50) becomes more useful when we reverse the subscripts on $\underline{V}_{nb}$, which we may do by changing the sign of that term:

$$\underline{V}_{bn} = -\underline{V}_{nb} \Rightarrow \underline{V}_{AB} = \underline{V}_{an} - \underline{V}_{bn} \tag{5.51}$$

Equation (5.51) is represented in Fig. 5.44, with the negative of $\underline{V}_{bn}$ drawn and added to $\underline{V}_{an}$. We note that $\underline{V}_{AB}$ leads $\underline{V}_{an}$ by 30°, and is somewhat greater in magnitude. The phasor addition is identical to that as shown in Fig. 5.37 and the magnitudes of phase and line voltages have the ratio $\sqrt{3}$, just as the currents in the delta-connected load. The remaining line-to-line voltages, $\underline{V}_{BC}$ and $\underline{V}_{CA}$ may be determined by similar reasoning, or more directly by arranging them in *ABC* sequence, each 120° from $\underline{V}_{AB}$. Hence we arrive at the picture shown in Fig. 5.44. The line-to-neutral voltages lag the corresponding line-to-line voltages by 30°, when we consider the two voltages with, say, *A* (and *a*) written first, like $\underline{V}_{AB}$ and $\underline{V}_{an}$. But a better way to think about the phase is to realize that the phase of the corresponding line-to-neutral voltage lies between the phases of the two line-to-line voltages that connect to the same point. That is, $\underline{V}_{an}$ will lie halfway between $\underline{V}_{AB}$ and $\underline{V}_{AC}$. This phase relation, together with the magnitude ratio of $1/\sqrt{3}$, allows us to determine easily the line-to-neutral voltages from the set of line-to-line voltages, or vice versa.

We now can solve the problem pictured in Fig. 5.42. We will assume that $\underline{V}_{AB}$ is the phase reference, that is, $\underline{V}_{AB} = 460\sqrt{2}\underline{/0°}$. From Fig. 5.44 we see that $\underline{V}_{an}$ will lie at $-30°$ and be smaller by $\sqrt{3}$, or $\underline{V}_{an} = 266\sqrt{2}\underline{/-30°}$. Hence the current in line *A* will be

$$\underline{I}_A = \frac{\underline{V}_{an}}{\underline{Z}_\phi} = \frac{266\sqrt{2}\underline{/-30°}}{2 + j1.5} = \frac{266\sqrt{2}\underline{/-30°}}{2.50\underline{/36.9°}} = 106.2\sqrt{2}\underline{/-66.9°} \quad A$$

The other line currents can be determined similarly or by symmetry from $\underline{I}_A$.

Figure 5.43 Determining line-to-line voltages from line-to-neutral voltages.

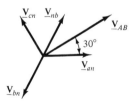

Figure 5.44 The line-to-neutral voltages are smaller by $\sqrt{3}$ and lag line-to-line voltages by 30°.

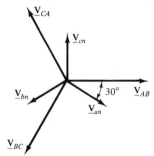

Power in AC Circuits   Chap. 5

Power in wye-connected loads.   The total power to the wye-connected load will be three times the power to each phase of the load (P_ϕ). Thus we can compute the total power with

$$P_{3\phi} = 3P_\phi = 3V_\phi I_\phi \times \text{PF} \tag{5.52}$$

where V_ϕ is the phase rms voltage, the line-to-neutral voltage in this instance, I_ϕ is the phase rms current, also the line current in this instance, and PF is the power factor of the phase impedance. Equation (5.52) is similar to Eq. (5.46), derived for the delta-connected load, but is applied differently. For the delta-connected load, the phase voltage is identical to the line-to-line voltage, but the phase current is smaller than the line current by $1/\sqrt{3}$. For the wye-connected load, the phase current is identical to the line current, but the phase voltage is smaller than the line voltage by $1/\sqrt{3}$. For the present example, the total power to the three-phase load would be

$$P_{3\phi} = 3(266)(106.2)(\cos 36.9°) = 67.7 \text{ kW}$$

The neutral of the wye-connected load might not be accessible for voltage measurement; hence it is desirable to express the total power in terms of the line voltage and current. Using the $\sqrt{3}$ ratio between phase voltage and line voltage, we may convert Eq. (5.52) to

$$P_{3\phi} = 3\left(\frac{V_L}{\sqrt{3}}\right)(I_L) \times \text{PF} = \sqrt{3}\ V_L I_L \times \text{PF} \tag{5.53}$$

where the L subscript indicates a line rms current or a line-to-line rms voltage. In Eq. (5.53) the power factor is the cosine of the angle between the line current and the average phase angle of the two line voltages involving that line. Thus Fig. 5.40 applies to this case as well as the delta-connected load because the phase voltage, $\underline{V}_{an}$ in this case, would lie symmetrically between $\underline{V}_{AB}$ and $\underline{V}_{AC}$. Of course, the power factor can also be determined from the angle of the phase impedances if they are known.

Equation (5.53) is identical to Eq. (5.48), which was presented for the delta-connected load. Clearly, the two load configurations are indistinguishable to external measurement and can only be identified in practice by examining the internal connections in the three-phase load.

Delta–wye conversions.   This does not mean, however, that the *same* set of phase impedances are equivalent when connected first in delta and then in wye. Indeed, the appearance of the circuits suggests that the delta gives parallel paths whereas the wye gives series paths. This appearance suggests that the line current for the delta would be larger than for the wye if the same phase impedance were used for each connection. It can be shown that the ratio is 3:1; that is, three identical resistors will draw three times the current (and three times the power) when connected in delta, as compared to when they are connected in wye.

From the above it follows that the delta and wye are equivalent if the phase impedances differ by a factor of 3, with the delta connection having the higher impedance. This equivalence, shown in Fig. 5.45, is often useful in solving three-phase problems. We leave proof of this equivalence for a problem at the end of this chapter.

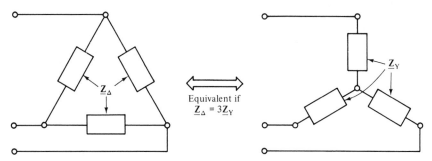

Figure 5.45 Delta and wye equivalence.

To illustrate the usefulness of this delta–wye conversion, we will determine the power to the delta-connected load in Fig. 5.46. The 10-Ω resistors represent the load, but we must consider the resistance of the wires leading to the load, which is represented by the 0.5-Ω resistors. The presence of the wire resistance undermines our previous approach for solving delta-connected loads, but if we convert the delta to an equivalent wye load, we can solve the problem. Figure 5.47 shows the circuit after conversion to wye, and it now should be clear how to proceed. The wire resistance can now be combined with the load resistance to yield a phase resistance of 3.83 Ω, and the rms line-to-neutral voltage would be $240/\sqrt{3} = 139$ V. The line current would thus be $139/3.83 = 36.1$ A, and the total power to the wires plus load would be $\sqrt{3}(240)(36.1) = 15.0$ kW. The wire losses would be $3(36.1)^2(0.5) = 1960$ W, the rest of the power going to the delta-connected load.

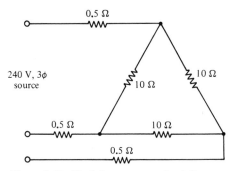

Figure 5.46 Find the power to the delta-connected load.

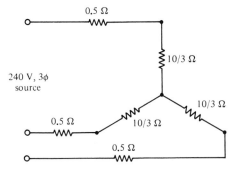

Figure 5.47 After converting the delta to a wye, we can determine the line currents.

Unsymmetric loads. A three-phase load is said to be unbalanced when the phase impedances are not identical. This is considered an undesirable situation and is avoided in practice if possible. When unbalanced loads are connected in delta, calculation of the phase and line currents becomes tedious, though straightforward. All phase and line currents must be calculated individually because symmetry has been lost.

When unbalanced loads are connected in wye, the analysis is straightforward only when the neutral of the load is connected to the neutral of the three-phase source. With the neutral connected, the three loads operate in effect as single-phase loads that share one common wire (the neutral connection). Current will flow in the neutral wire for an unbalanced load.

When there exists no neutral wire in the unbalanced wye connection, complications arise in the calculation of the line-to-neutral voltages and the line currents. Because the neutral of the load is no longer at the same voltage as the neutral of the source, the first step in solving the problem is to calculate the voltage of the neutral of the load. Then one can proceed to solve for the line currents, a tedious though straightforward calculation. This problem lies beyond the scope of this book.

5.4.4 Some Practical Matters

Applications of delta and wye sources.   We have discussed both delta and wye source connections because both find important applications. The wye source connection is used for long-distance transport of electrical power, where the resistive losses of the wires degrade system efficiency. Equation (5.53) suggests one reason: the total power from a source is the product of the line current, the line-to-line voltage, and the power factor. The power required and the power factor are determined by the load on the system, but the levels of the line voltage and current are influenced by the source connections. The wye connection gives a line-to-line voltage that is $\sqrt{3}$ greater than the delta connection and hence, for the same power, the line current will be $\sqrt{3}$ smaller. The losses in the wire vary with the square of the line current and, as a consequence, losses will be one-third as great with a wye source connection, as compared with a delta connection. Thus the high-voltage transformer windings are connected in a wye configuration for power transmission and distribution systems.

The delta source connection finds wide application when three single-phase circuits must be derived from a three-phase source. This three-phase to single-phase conversion is required in residential areas, because single-phase power is required for household lighting and appliances. The power would be brought into the neighborhood at a line voltage of at least 12,000 V, and would be stepped down to 240/120 V with a transformer (more details below). The low-voltage secondaries of the distribution transformer would be connected in a delta, and several households would be served with single-phase power from each phase of the delta. This arrangement is advantageous because it minimizes interaction between the three single-phase circuits as compared with the wye connection.

How a house is wired.   We do not recommend that you do it, but *if* you sought out the circuit breaker box (or the fuse box in older structures) for your dwelling, and *if* you removed the safety cover, you would discover three wires coming into the box from the transformer on the pole in the alley. One wire would be red, another black, and the third white. Does this mean that you have three-phase power coming into your residence? No, this is not three-phase power; this

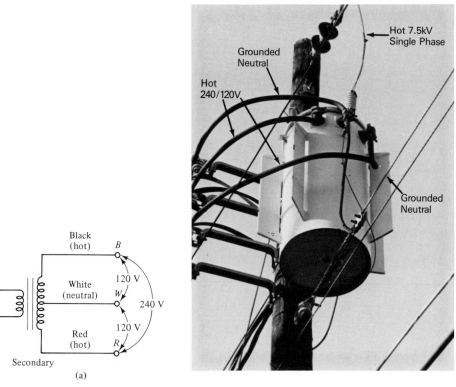

Figure 5.48 (a) A 120/240-V household power system; (b) pole transformer.

is 240/120-V single-phase power. Figure 5.48a shows the usual arrangement at the transformer secondary. The white wire is the neutral and is grounded at the transformer by a wire that enters the moist earth and should also be grounded through the household plumbing. The black and red wires are "hot," each carrying 120 V relative to the neutral. The two 120-V voltages are opposite in phase and hence add to 240 V. That is, the neutral is at the midpoint between the two lines that carry 240 V and hence 120 V is developed between each hot line and the neutral.

The 240-V power is used for heavy equipment such as air conditioners and certain power tools. Most appliances operate with 120 V, so the lighting and appliance circuits in the house are connected between the hot wires and neutral. The red color is not used in the household 120-V wiring: black means hot and white neutral. Modern wiring codes require a third wire for a separate ground. The ground wires does not carry power like the neutral wire but enables equipment to have a separate ground connection independent of the power circuit.

Figure 5.49 shows a modern appliance outlet, which is both *polarized* and grounded with a three-wire system. The outlet and plug are said to be polarized because the hot and neutral connections differ in size. Some loads, such as table lamps, may work equally well, and be equally as safe, plugged in either way. Such a load would have an unpolarized plug, which would fit into the outlet either way. Other loads are equipped with polarized plugs that fit in only one way, thus

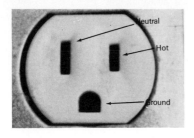

Figure 5.49 Grounded and polarized household outlet.

controlling which wire is neutral. Note that for safety the neutral of the load (the male) cannot make connection with the hot of the source. Incidentally, if you are wiring an appliance outlet, pay close attention to polarity. Usually, the screw to which you should connect the hot (black) wire has a copper color and the screw to which you should connect the neutral (white) wire has a silver color. If you are installing a lighting fixture in a ceiling, the hot goes through the wall switch to the center of the receptacle and the neutral connects to the screw threads.

5.4.5 Electrical Safety

Although use of electrical power underlies much of our modern way of life, the average citizen is often uninformed of the dangers of electrical power. One person might fear to touch the terminals of a 12-V auto battery, while another might think nothing of sticking a finger into a light socket to see if there is any power. (The battery is safe, but please keep your finger out of the light socket.) The purpose of this section is to present some basic information about electrical safety so that the reader will be able to recognize a dangerous situation and hopefully stay out of trouble.

The circuit theory of this subject is simple enough: Ohm's law is the key,

$$I = \frac{V}{R}$$

where R is you. Although we are accustomed to signs warning "DANGER, HIGH VOLTAGE," it is actually the current that affects our bodies. Many thousands of volts will do no more than startle provided the current is small, as when we feel a small spark of static electricity after sliding across an auto seat. In the following we speak first of the physiological effects of electric current on the human body. We then discuss body resistance and the resistance of typical surroundings. Finally, we will offer some advice about electrical safety.

Physiological Effects of Current. Figure 5.50 shows the variety of effects that electrical current may have on the human body and the current levels at which they occur. Injury could be caused indirectly through being startled and losing muscular control or directly through burns. Death could result from suffocation, through loss of proper function of the heart, or through severe burns. The large ranges in Fig. 5.50 are represented because of wide variation between individuals in body size, condition, and tolerance to electrical shock. Clearly of importance is the region of the body through which the current passes; current

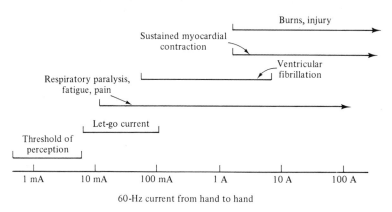

Figure 5.50 Physiological effects of electricity. (Adapted from Webster, *Medical Instrumentation, Application and Design*, Copyright © 1978 Houghton Mifflin Company. Adapted with permission.)

passing through the lower part of the leg, for example, might be painful but would be unlikely to affect heart action.

The surprising aspect of Fig. 5.50 is that small currents can have serious effects. This occurs because the communication system of the human body is electrical in nature and misbehaves under external electrical influence.

When a person is experiencing electrical shock, time becomes an important factor, for the damaging effects are progressive. Hence it is important to remove the source of electrical energy from a shock victim before other action is taken. Particularly serious is the condition of ventricular fibrillation, where the heart loses its synchronized pumping action and circulation ceases. This condition is very dangerous because the heart may not resume normal action when the source of electrical power is removed; sophisticated medical equipment is required to restore coherent heart action.

Resistance. Earlier we stated that the resistance is you. This would be true if you came into simultaneous contact with both wires of an electrical circuit, say, by grabbing a wire with each hand, but most serious electrical shocks occur through the situation portrayed in Fig. 5.51. Here we have shown the victim in simultaneous contact with the hot wire and with "ground." Ground may be the moist earth, the plumbing of a house, or even a concrete floor that is in contact with the plumbing. In this case the resistance that influences the amount of current through the body includes not only the resistance of the body but also the resistance of the shoes and the resistance between the shoes and earth ground.

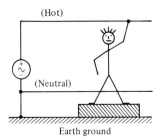

Figure 5.51 Most shocks occur between the hot wire and ground.

To assess the danger of a given situation, we must estimate the resistance of the "circuit" of which our body might become an unhappy part. Table 5.1 shows some basic information that would allow such an estimate of the total resistance to ground. What, for example, would you experience if you were standing on moist ground with leather-soled shoes and you unwittingly grab hold of a 120-V wire? Taking the lowest values in Table 5.1, we estimate the following resistances: 3 kΩ for the grasp, 200 Ω for the body, and 5 kΩ for the feet–shoes. Thus the largest current you might carry would be 120 V/(8.2 kΩ), about 15 mA. From Fig. 5.50 we judge that there is a fair chance that you will be unable to release your grasp and that you might hence be unable to breathe. This is therefore a dangerous situation.

One factor that Table 5.1 does not contain concerns the breakdown voltage of our skin resistance. At approximately 700 V (ac) the resistance of the skin drops to near zero: in effect, a spark burns a hole in the skin. Thus the resistance of the skin, which might save your life for a lower voltage, becomes ineffective at such high voltages. For this reason, high voltages are seldom used for distribution of power in industrial applications, and 240 V is the highest voltage used in residential wiring.

An interesting, and alarming, calculation the reader might wish to perform is the following: Making the most pessimistic assumptions about body resistance,

TABLE 5.1 RESISTANCE

(a) For Various Skin-Contact Conditions

Condition (area to situ)	Resistance	
	Dry	Wet
Finger touch	40 kΩ–1 MΩ	4–15 kΩ
Hand holding wire	15–50 kΩ	3–6 kΩ
Finger–thumb grasp	10–30 kΩ	2–5 kΩ
Hand holding pliers	5–10 kΩ	1–3 kΩ
Palm touch	3–8 kΩ	1–2 kΩ
Hand around 1½-in. pipe (or drill handle)	1–3 kΩ	0.1–1.5 kΩ
Two hands around 1½-in. pipe	0.5–1.5 kΩ	250–750 Ω
Hand immersed	—	200–500 Ω
Foot immersed	—	100–300 Ω
Human body, internal, excluding skin = 200–1000 Ω		

(b) For Equal Areas (130 cm^2) of Various Materials

Material	Resistance
Rubber gloves or soles	More than 20 MΩ
Dry concrete above grade	1–5 MΩ
Dry concrete on grade	0.2–1 MΩ
Leather sole, dry, including foot	0.1–0.5 MΩ
Leather sole, damp, including foot	5–20 kΩ
Wet concrete on grade	1–5 kΩ

Source: Adapted from Ralph Lee, "Electrical Safety in Industrial Plants," *IEEE Spectrum*, June 1971. © 1971, IEEE.

resistance to ground, and loss of skin resistance (say, cuts or blisters on the hands and feet), calculate the least voltage that might prove fatal. Such a calculation would have you standing in water or on a metal floor without shoes. Although we are describing an unusual situation, we still urge you to make the calculation.

Of more importance are the factors that increase your safety when working around electrical power. Make sure that the power is off before working on any electrical wiring or electrical equipment. Wear gloves and rubber-soled shoes. Avoid standing on a wet surface or on moist ground. Do not work alone around exposed electrical power.

Grounding. From our discussion of safety, you might notice that danger increases because the electrical power system is grounded. If the circuit were *floating*, that is, not grounded, the only way to get shocked would be to come into contact with both wires simultaneously, an unlikely event. How can we reconcile this viewpoint with the common idea that electrical circuits are grounded for safety? Actually, both ideas are valid, and there is no contradiction; there is more involved in the issue than we have discussed thus far. Suppose that the electrical power system in your building were floating. Everything would function correctly and safely until something went wrong with the equipment. But if, say, the wiring of the transformer on the pole became defective and a connection developed between primary and secondary, in this event the 120-V circuits could float at 12,000 V or even 24,000 V. This would endanger virtually every piece of equipment on the line. Also, if the voltage suddenly were floated at 12,000 V, the danger to you becomes much greater. For this reason, the secondary of the transformer is grounded for protection of life and property. For if the secondary is grounded and a fault develops in the transformer, a large current flows immediately, a fuse or circuit breaker opens the circuit, and the source of power becomes disconnected from the offending part of the circuit.

Given that the power system in the building should be grounded, the necessity for grounding the equipment with the three-wire system in Fig. 5.48 becomes apparent. On a piece of equipment, such as a washing machine, if a fault develops between the hot side of the power and the metal chassis, again a fuse or circuit breaker will respond to the large current flowing through the ground connection and will remove power from the circuit containing the faulty connection. Hence the safest way to install the power system involves grounding the power system and the equipment with which you might come into contact.

5.4.6 Check Your Understanding

1. A three-phase circuit measures 266 V (rms) between earth ground and line A. What would be the voltage between lines B and C?
2. For a delta-connected load, the phase and line currents or voltages are the same?
3. A three-phase load uses 50 kW at 480 V. The current is 64 A. Find the reactive power required by the inductive load.
4. In a 120/240-V single-phase system, what is the color of the wire that should have a very small voltage to the plumbing of the building?

5. What range of current through the human body is most likely to result in death, $i <$ 1 mA, 100 mA $< i <$ 1 A, or 10 A $> i$?

6. In most electrocutions, current passes from the hot wire through the victim to the neutral wire or to earth ground?

7. When working on electrical devices, your body should be grounded for safety. (T/F?)

Answers. 1. 460 V; 2. voltages; 3. +18.2 kVAR; 4. the neutral wire should be white; 5. 100 mA $< i <$ 1 A; 6. to earth ground; 7. false.

PROBLEMS

Section 5.1: AC Power and Energy Storage: The Time-Domain Picture

P5.1. Compute the time average and the rms value for the waveforms shown in Fig. P5.1.

Answers. (a) 8.0 V average and 8.72 V rms; (b) $2 V_p/\pi$ average and $V_p/\sqrt{2}$ rms, just like a pure sinusoid, because the reversal of the bottom half does not matter after the squaring.

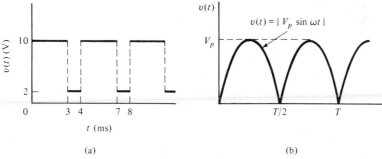

(a)

$$v(t) = | V_p \sin \omega t |$$

(b)

Figure P5.1

P5.2. Show that ac and dc power are additive. That is, if $i(t) = I_{dc} + I_p \cos \omega t$, then the time average of $i^2(t)$ is $I_{dc}^2 + (I_p/\sqrt{2})^2$.

P5.3. For the sawtooth waveform in Fig. P5.3:
   (a) Find the average value, V_{dc}.
   (b) Find the effective value, V_e.
   (c) What would be the average power if this were the voltage across a 10-Ω resistor?

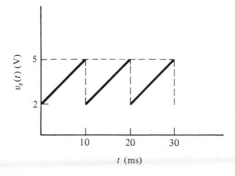

Figure P5.3

P5.4. **(a)** A current source having an rms value of 0.75 A is connected to a 10-Ω resistor. What is the power in the resistor?

(b) If the current source in part (a) is sinusoidal with a period of 10 ms, what is the equation of the current as a function of time? (Assume the cosine form and zero phase.)

(c) If the current in part (b) is put through an impedance of $10/\!-\!60°$ rather than a pure resistance, what now is the time-average power?

P5.5. Figure P5.5 shows a periodic voltage waveform.

(a) What is the time-average voltage?

(b) If this voltage is applied to a 2-μF capacitor, what would be the effective value of the current in the capacitor?

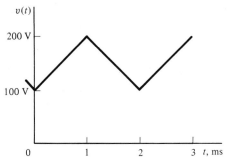

Figure P5.5

P5.6. For the circuit in Fig. P5.6, find the:

(a) Minimum instantaneous power into R

(b) Time-average power into R

(c) Maximum instantaneous power into R

(d) Time between maxima in the instantaneous power into R

Answers. (a) 0 W; (b) 1440 W; (c) 2.880 kW; (d) 3.14 ms

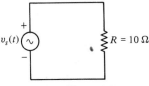

$v_s(t) = 120\ \sqrt{2}\cos{(1000t)}$ V   **Figure P5.6**

P5.7. For the circuit shown in Fig. P5.7, $v(t) = 180\cos(\omega t)$ V and $i(t) = 8\cos(\omega t)$ A. Find:

(a) Instantaneous power into the load at $t = 0$

(b) Time-average power into the load

(c) Effective value of the voltage

(d) Impedance of the load

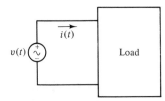

Figure P5.7

P5.8. For the circuit shown in Fig. P5.8, $v(t) = 140 \cos(377t)$ V and $i(t) = 6.2$ $\cos(377t + 30°)$ A. Find:
  (a) Effective current
  (b) Time-average power out of the source
  (c) Resistance, R
  (d) Peak stored energy in the capacitor

Answers. (a) 4.38 A; (b) 376 W; (c) 19.6 Ω; (d) 0.576 J

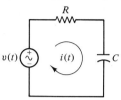

Figure P5.8

P5.9. A 117-V, 60-Hz household circuit has two 60-W lights (assume unity power factor) and a fan using 250 VA at 0.92 PF (lagging).
  (a) Draw the circuit, representing each load by its impedance. Include the switches for each load.
  (b) Determine the total current required for all loads operating simultaneously.
  (c) What capacitor in parallel with the loads will give unity power factor?

P5.10. For the circuit shown in Fig. P5.10:
  (a) Solve for $v(t)$. Let the phase of $i(t)$ be zero.
  (b) Compute the time-average power into the circuit using the power factor.
  (c) Show that the time-average power into the entire circuit, P, is equal to the time-average power dissipated in the resistor, P_R.
  (d) Find the time-average electric stored energy.

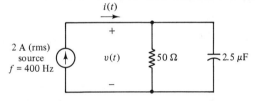

Figure P5.10

P5.11. For the circuit in Fig. P5.11:
  (a) Find $i(t)$. Use time- or frequency-domain techniques.
  (b) What is the power factor for the load?
  (c) Compute the time-average power into the load.
  (d) Show that the time-average power into the load, P, is equal to the power dissipated in the resistor, P_R.
  (e) Calculate the peak and time-average magnetic energy stored in L.

Answers. (a) $= 1.756 \cos(120\pi t - 14.8°)$; (b) 0.967; (c) 144 W; (e) 0.101 J peak

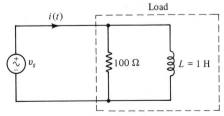

$v_s(t) = 120 \sqrt{2} \cos(120\pi t)$ V

Figure P5.11

Section 5.2: Power and Energy in the Frequency Domain

P5.12. For the circuit shown in Fig. P5.12:

$$v(t) = 600\sqrt{2} \cos(120\pi t + 30°)$$

$$i(t) = 20\sqrt{2} \cos(120\pi t + 60°)$$

(a) What is the real power?
(b) What is the reactive power?
(c) What is the apparent power?
(d) What is the impedance of the load in polar form?

Answers. (a) 10.4 kW; (b) −6 kVAR; (c) 12 kVA; (d) $30\underline{/-30°}$ Ω

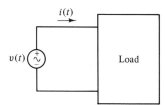

Figure P5.12

P5.13. For the 60-Hz load shown in Fig. P5.13, the voltmeter measures 120 V, the ammeter measures 5 A, and the wattmeter measures 500 W. The load consists of a resistor and inductor. Find the:
(a) Power factor
(b) Leading or lagging?
(c) Real power
(d) Apparent power
(e) Reactive power
(f) Average stored energy
(g) Draw a phasor diagram.

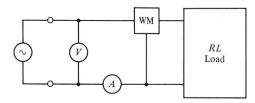

Figure P5.13

P5.14. An impedance at 60 V requires 15 A as measured by standard ac meters. Figure P5.14 shows the phasor diagram. Find the:
(a) Apparent power
(b) Power factor
(c) Real power
(d) Reactive power (give units for all powers)
(e) Impedance (rectangular form)

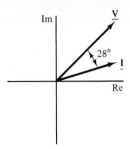

Figure P5.14

P5.15. The complex power in a load is $1000\underline{/+30°}$ volt-amperes. What are the following?

(a) Apparent power (give units)

(b) Real power (give units)

(c) Reactive power (give units)

(d) Power factor, leading or lagging

P5.16. For the circuit shown in Fig. P5.10, find:

(a) The complex power.

(b) The real and reactive power, giving correct units.

(c) The apparent power, giving correct units.

(d) Draw a power triangle for this circuit.

(e) Verify Eq. (5.30) for this circuit.

P5.17. A single-phase electric motor is monitored with an ammeter, voltmeter, and watt-meter, which indicate 5.5 A, 120 V, and 623 W. Assume 60 Hz.

(a) Draw a phasor diagram of the voltage and current, assuming the voltage at zero phase and lagging current.

(b) What is the reactive power to the motor, including the units?

(c) To improve the power factor of the motor, a capacitor is hung directly across the motor terminals. This capacitor will draw leading current and can neutralize the lagging component of the motor current. What value of capacitance will give unity power factor?

Answers. (b) +218 VAR; (c) 40.1 μF

P5.18. A 60-Hz single-phase, 1-hp motor in a washing machine (120 V) has an efficiency of 78% at full load (rated output power) and a power factor of 0.82 lagging. Find the line current, the reactive, and the apparent power to the motor. Draw a phasor diagram.

P5.19. For the circuit shown in Fig. P5.19, the wattmeter reads 1090 W, the ammeter reads 9.43 A, and the voltmeter reads 40 V.

(a) What is R?

(b) What would be the peak value of the source voltage, V_s?

Answers. (a) 7.97 Ω; (b) 170.5 V

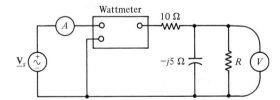

Figure P5.19

P5.20. For the circuit shown in Fig. P5.20, the voltage and current are

$$v(t) = 120\sqrt{2} \cos(120\pi t + 30°)$$

$$i(t) = 5\sqrt{2} \cos(120\pi t - 15°)$$

(a) Find the complex power into the load, $\frac{1}{2}\mathbf{V}\mathbf{I}^*$.

(b) Draw a power triangle showing the numerical values of the real, apparent, and reactive powers.

(c) Assuming that the load contains no electric energy storage, what is the peak value of the magnetic energy stored in the load?

(d) What value of capacitor connected between a and b will make the power factor to be unity, as seen by the source?

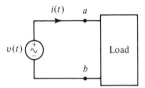

Figure P5.20

P5.21. For the circuit shown in Fig. P5.21, what should be the load impedance to draw maximum power from the circuit in the box? Give component values, not just impedance values. Assume a series circuit for the load. *Hint:* Convert the source to a Thévenin equivalent circuit.

Answer.   A 5-Ω resistor in series with a 2-mH inductor

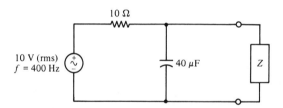

Figure P5.21

P5.22. The output impedance of a typical 120-V appliance wall outlet might be $0.5 + j0.5$. What is the theoretical available power from the outlet? Why would it be a bad idea to actually try to get that much power out?

Section 5.3: Introduction to Transformers

P5.23. An ideal transformer has 100 turns on the primary and 200 turns on the secondary. Find the primary and secondary currents for the circuit shown in Fig. P5.23.

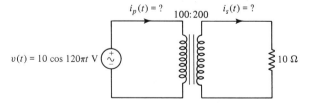

Figure P5.23

P5.24. For the circuit in Fig. P5.24, containing the ideal transformer, find:
  (a) Primary current
  (b) Load voltage
  (c) Power out of the source

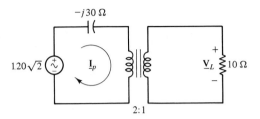

2:1

Figure P5.24

P5.25. For the circuit in Fig. P5.25 containing an ideal transformer, find:
  (a) Secondary voltage
  (b) Primary current
  (c) Secondary current
  (d) Power out of the source

Answers. (a) 15 V (rms); (b) $15/4\sqrt{5}$ A (rms); (c) $15/\sqrt{5}$ A (rms); (d) 90.0 W

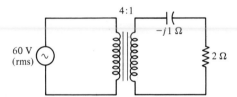

Figure P5.25

P5.26. The circuit shown in Fig. P5.26 contains a step-up transformer, a transmission line with resistance of 1.0 Ω, a step-down transformer, and a load of 10 Ω. Find the value of n to make the efficiency of the transmission circuit 99.5%.

Answer. 4.46

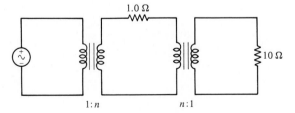

Figure P5.26

P5.27. In the circuit in Fig. P5.27, the voltmeter measures 100 V. What would be measured by:

(a) A voltmeter in the primary?
(b) An ammeter in the secondary?
(c) An ammeter in the primary?
(d) A wattmeter in the secondary?
(e) A wattmeter in the primary?
(f) A dc ohmmeter connected between primary and secondary circuits?

Answers. (a) 200 V; (b) 10 A; (c) 5 A; (d) 1 kW; (e) 1 kW; (f) ∞

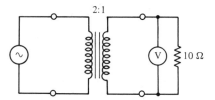

Figure P5.27

P5.28. For the ideal transformer circuit shown in Fig. P5.28 a voltmeter measures 30 volts across the primary. What would the following measure?

(a) A voltmeter across the secondary
(b) An ammeter in the primary
(c) An ammeter in the secondary
(d) A wattmeter in the primary
(e) A wattmeter in the secondary

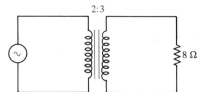

Figure P5.28

P5.29. Power is generated at 480 V and utilized at 120 V. The generator is ideal, but the load is at some distance from the generator and hence line losses are appreciable, 1 Ω total for both wires. The equivalent load resistance would draw 10 kW at unity PF if 120 V were provided. Calculate the load voltage, the load power, and the efficiency of the transmission system under the following schemes:

(a) 4:1 transformer at the generator
(b) 4:1 transformer at the load
(c) 1:4 transformer at the generator and a 16:1 transformer at the load

P5.30. For the circuit shown in Fig. P5.30, the load impedance is fixed as 1000 Ω $\| -j3000$ Ω, and the source output impedance is fixed at 10 Ω $\| j10$ Ω. Maximum power transfer is to be achieved to the load with an ideal transformer with a turns ratio of $1:n$ and a "tuning capacitor" on the source side, represented by its reactance $+jX_C$ ($X_C < 0$). Find the values for n and X_C that achieve maximum power transfer.

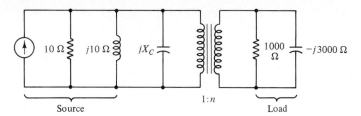

Source 1:n Load **Figure P5.30**

Section 5.4: Three-Phase Power Systems

P5.31. Pick two values of ωt in Eq. (5.44) and show by direct calculation that the time-varying terms cancel.

P5.32. On Fig. 5.34, develop a delta connection by first connecting a to b'. Draw a phasor diagram of the resulting system. Is the phase rotation *ABC* or *ACB*?

P5.33. Three 230-V (rms) generators are connected in a three-phase wye configuration to generate three-phase power. The load consists of three balanced impedances, $\underline{Z}_L = 2.8 + j1.6\ \Omega$, connected in delta. Find the line current an ammeter would measure, the apparent power, and the real power to the load. What is the phase angle between $\underline{I}_A$ and $\underline{V}_{AB}$, assuming *ABC* rotation?

P5.34. For the three-phase power system shown in Fig. P5.34 a voltmeter measures 146 V between line A and the neutral N. Find:
  (a) Line voltage
  (b) Line current
  (c) Load power factor
  (d) Apparent power
  (e) Real power

Answers. (a) 253 V (rms); (b) 65.3 A (rms); (c) 0.894; (d) 28.6 kVA; (e) 25.6 kW

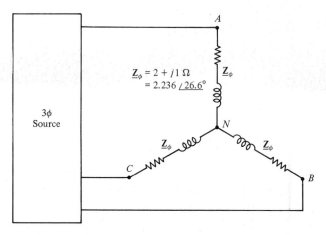

$\underline{Z}_\phi = 2 + j1\ \Omega$
$= 2.236\ \underline{/26.6°}$

Figure P5.34

P5.35. In Fig. P5.35, the voltage between A and N is 150 V as measured by a standard meter. Let this voltage be the phase reference.

    (a) What is $\underline{V}_{AB}$ as a phasor?

    (b) What would an ammeter measure as the load current?

    (c) What is the apparent power?

    (d) What is the real power?

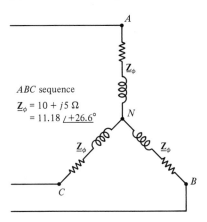

ABC sequence

$\underline{Z}_\phi = 10 + j5\ \Omega$
$= 11.18\ \underline{/+26.6°}$

Figure P5.35

P5.36. Figure P5.36 shows a three-phase source and load.

    (a) Draw appropriate connections.

    (b) Find the line current.

    (c) Find the phase voltage.

    (d) Determine the power factor.

    (e) What is the power in the load?

    (f) Assuming $\underline{V}_{AB} = 460\sqrt{2}\underline{/0°}$, find $\underline{I}_A$ as a phasor.

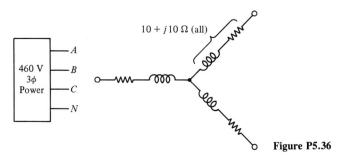

$10 + j10\ \Omega$ (all)

460 V
3ϕ
Power

A
B
C
N

Figure P5.36

P5.37. For the three-phase system shown in Fig. P5.37, find:

    (a) Phase voltage

    (b) Line voltage

    (c) Phase current

    (d) Line current

    (e) Phase impedance

    (f) Apparent power

    (g) Power factor

    (h) Real power to the load.

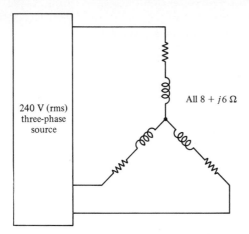

240 V (rms)
three-phase
source

All $8 + j6 \ \Omega$

Figure P5.37

P5.38. A balanced, wye-connected three-phase load is shown in Fig. P5.38. The current in line A is $\underline{\mathbf{I}}_A = 8\sqrt{2}\underline{/0°}$. The voltage from b to the neutral point is $\underline{\mathbf{V}}_{bn} = 120\sqrt{2}\underline{/-90°}$.

(a) Find the line-to-line voltage that a voltmeter would read.
(b) Find the real and reactive power into the entire three-phase load.
(c) Determine R and L, assuming 60-Hz operation.

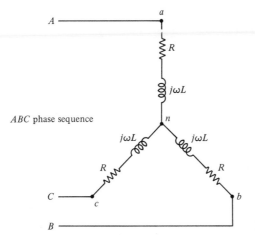

ABC phase sequence

Figure P5.38

P5.39. Demonstrate the equivalence of the delta and wye circuits in Fig. 5.45. Assume an impedance $Z_\Delta\underline{/\theta}$ for the delta and an impedance $Z_Y\underline{/\theta}$ for the wye and compute the complex power for each. Equate these powers and confirm the 3:1 ratio shown in Fig. 5.45.

P5.40. Three identical resistors are placed in a wye configuration and draw a total of 100 W from a three-phase source. What power would the same resistors draw if placed in delta?

P5.41. For the three-phase circuit shown in Fig. P5.41, the 0.2-Ω resistors represent the resistance of the distribution system. Find:
  (a) Total power out of the source, including line and load
  (b) Line losses
  (c) Distribution system efficiency.

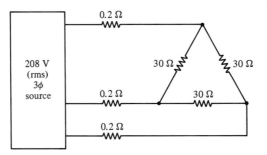

Figure P5.41

P5.42. Using delta–wye transformations, determine the total power given to the delta and wye loads in Fig. P5.42, not counting the losses in the 0.5-Ω resistors that represent losses in the connecting wires. *Hint:* The neutrals of two balanced wye loads will have the same voltage and hence may be considered as connected.

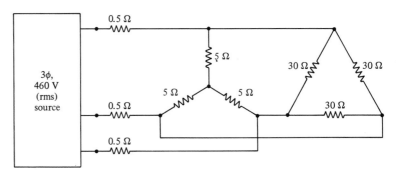

Figure P5.42

General Problems for Chapter 5

P5.43. With the circuit shown in Fig. P5.43, there is no value of the turns ratio, n, that will perfectly "match" the load to the source impedance, in the sense of eliminating all reactance and at the same time making the resistors match. However, there is still an optimum value of n that maximizes the power in the load. Find that value of n.

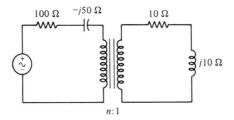

Figure P5.43

P5.44. The circuit shown in Fig. P5.44 is the Thévenin equivalent circuit of a loop antenna operating at a frequency of 570 kHz. Assuming that the input circuit of the radio consists of a capacitor in parallel with a resistor, what values of R and C will extract maximum power from the antenna. *Hint:* Convert the Thévenin to a Norton equivalent circuit. Then create a resonance and match the resistors, that is, make them equal for maximum power to the load.

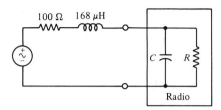

Figure P5.44

P5.45. The City of Austin distributes power with a three-phase system with 12.5 kV between the power-carrying wires. But each group of houses is served from a single wire and ground, transformed to 240/120 V by a pole transformer. This is shown in Fig. P5.45.

(a) What is the turns ratio, primary/secondary turns, of the pole transformer to give 240 V, center tapped?

(b) When a 60-W light is turned on, how much does the current increase in the high-voltage wire? Assume the power factor is unity and the transformer is 90% efficient.

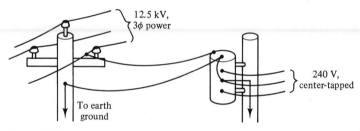

Figure P5.45

6 Introduction to Electronics

6.1 WHAT IS ELECTRONICS?

Electronics and information. In Chapter 1 we distinguished between two branches of electrical engineering, power and electronics, through their differing uses of electrical energy. In the power industry, electrical energy is used for its own sake: energy is generated from a primary source such as coal or water power, transported long distances with transmission lines, and converted into heat, illumination, mechanical work, or some other useful form. In electronics, electrical energy is used to symbolize, transport, and process information. Think of a telephone system, a radio, a computer, or a traffic control system—all use electrical energy to convey, process, and use information.

In electronics, voltage and current become electrical signals; that is, they signify something else, as the root word "sign" suggests. The voltage generated when you speak into a telephone is a signal because it reproduces the acoustic vibrations in the air, these sounds have meaning and information is exchanged in the conversation.

Chapters 6 through 11 address the main ideas that enliven electronics at the present time. As you will see from the historical survey to follow, electronics is a young enterprise. The current rate of progress in the field makes it difficult to anticipate what the future holds. But there are some major themes that, once understood, will give you a good grasp of the nature of electronics. Some of the important factors we address are (1) expanded use of the frequency domain, (2) application of the electrical properties of semiconductor materials, (3) utilization of nonlinear effects in circuits, (4) feedback, and (5) analog and digital representation and processing of information.

History of electronics. Before World War II, electronics had commercial importance primarily in radio broadcasting and in the telephone and telegraph industries. Most people who were active in these fields were educated through experience, including many physicists and electrical engineers. In those days, to study electrical engineering in college meant to study about power: motors, generators, lighting, transformers, and transmission lines.

World War II changed electronics profoundly. In addition to the obvious need to improve radio communication, one of the Allies' top-secret war projects focused on electronics. Everybody knows about the Manhattan Project and the atomic bomb; through the work of the Radiation Laboratory in developing microwave radar, scientists and engineers made an equally important contribution toward ending the conflict. The best technical minds in the country were employed in these two projects and spectacular success crowned both efforts.

The postwar fruit of the Manhattan Project was more and bigger bombs, and even today the peaceful application of nuclear technology excites continuing controversy. But widespread and, for the most part, benevolent have been the postwar fruits of the work of the Radiation Laboratory. The cathode ray tubes (CRTs) that were developed as radar displays became TV picture tubes, and soon there were high-fidelity recordings to be enjoyed. Radar techniques developed to detect enemy aircraft allowed commercial aviation to fly in all weather, and you could dial long distance directly and hear the other party without straining.

In the early 1950s the development of the transistor inaugurated the first of several "solid-state revolutions." The integrated circuit followed and later the microcomputer on a "chip." The limits of the techniques of microelectronics continue to defy the imagination of the most far-seeing sages of the electronics industry.

The idea of electronic computers was conceived before the war, but only modest applications were made, even during the war. An analog computer was developed to control naval gunnery, and some simple "logic" circuits emerged, but practical computers came after the war. The first digital computers were based on vacuum-tube technology and by present standards were huge, slow, and awkward to program. But when the idea of the digital computer merged with that of the transistor and later the integrated circuit, the development of computers became spectacular and continues so to this day.

Nature of electronics. Electronics is a big bag of tricks. Unlike circuit theory, which submits to an orderly and logical development, electronics employs a diversity of devices, techniques, and processes. It is hard to learn a little bit about electronics, because electronic circuits are complicated. Did you ever, for example, examine the circuit diagram (the schematic) for a TV set? If one fell into your hands at this moment and you set about to apply your newly gained knowledge of Ohm's law, Kirchhoff's laws, and concepts such as impedance and Thévenin equivalent circuits, you would not make much progress in understanding such a circuit diagram.

Our goal in this second part of the book is to investigate the major themes currently active in electronics. We will examine a few of the tricks in the electronics bag at the present time, but only enough to impart a beginning understanding of the nature of the subject. Do not expect to be able to fix your radio or interface a microcomputer after you master this part of the book. But you should understand how these systems work.

6.2 RECTIFIERS AND POWER SUPPLIES

6.2.1 Ideal Diode

Nonlinear devices.   The circuit theory we developed in Chapters 1 through 5 describes the properties of circuits that are linear. The defining equations for resistors, inductors, and capacitors describe linear relationships between voltage and current. Also linear are Kirchhoff's laws describing conservation of energy and charge in electrical circuits. Many of the techniques that we developed are based on the linear properties of the defining equations and Kirchhoff's laws—superposition and Thévenin equivalent circuits being obvious examples.

Electronics also uses resistors, inductors, and capacitors, and certainly Kirchhoff's laws are still linear and still important. But electronic circuits employ many devices that are nonlinear in their characteristics. Diodes, transistors, and silicon-controlled rectifiers offer examples of such nonlinear electronic devices. The nonlinearities are not undesirable hindrances to the use of these and other devices; rather, they are useful because of their nonlinear properties.

Because of the importance of nonlinear devices in electronics, we employ more graphical analysis than in circuit theory. Many of the circuit solution techniques we develop are based on graphical methods, and much of the information about device characteristics is given in graphical form. In Fig. 6.1 we show the graphical characteristics of a resistor, together with the extremes of an open circuit ($R = \infty$) and a short circuit ($R = 0$). Of course, the graphical form of Ohm's law is a straight line.

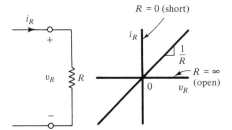

Figure 6.1 Symbol and graphical definition of a resistor.

Ideal diode characteristics.   In Fig. 6.2 we show the circuit symbol for a diode, with the associated graphical characteristic, which is nonlinear. The diode characteristic divides into two regions: the vertical region is called the forward-bias region and the horizontal region is called the reverse-bias region. Comparison with Fig. 6.1 suggests that the diode acts as a short circuit in the forward-bias region and an open circuit in the reverse-bias region. The behavior of the device

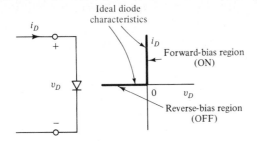

Figure 6.2 Symbol and graphical definition of an ideal diode.

in terms of current flow is as follows: As long as the current is positive in the direction of the arrow of the circuit symbol, the diode acts as a short circuit and the current flows without hindrance. The diode is ON. When the voltage is positive in the direction opposite to the diode arrow, however, the diode acts as an open circuit and no current can flow. The diode is OFF. Thus the current can flow in the direction of the arrow but cannot flow against the arrow.

The diode can be considered as the electrical equivalent of the mechanical ratchet, such as is used to tighten the net on a tennis court. The mechanical ratchet allows motion or rotation in one direction only. Similarly, the diode allows charge motion in one direction only. Another analog would be a check valve that allows fluid flow in only one direction.

The diode can also be considered a voltage-actuated switch. As long as the input voltage is positive (with the + at the top of the arrow) the switch is closed, current flows, and the diode is ON. But once the polarity of the voltage reverses, the switch opens, no current flows, and the diode is OFF.

The diode characteristic we have described is that of an ideal diode. Real diodes depart from this ideal characteristic, as will be detailed in Section 6.3.4. But many of the common applications of the diode can be understood in terms of this ideal characteristic; hence we will move to some applications before investigating the physical processes in semiconductor diodes.

6.2.2 Rectifier Circuits

Electronics and DC.   At the beginning of Chapter 4 we mentioned the struggle between dc and ac to see which form of electrical power would dominate the fledgling electrical power industry. We saw that Edison was wrong, Tesla was right, and ac came to be the common mode for the generation, distribution, and consumption of electrical power.

With the ascendency of electronics, however, dc has made a comeback, for almost all electronics circuits require dc power. You might suppose, therefore, that every electronic device would contain a battery, but this is untrue. Batteries are expensive, heavy, short-lived, bulky, and filled with chemicals likely to leak out and damage flashlights, tape recorders, and the like. Batteries are thus undesirable components and are avoided by designers except where portability is essential.

For this reason, most electronic equipment contains a *power supply* circuit. The function of the power supply is shown in Fig. 6.3. When you plug in and turn on your TV set, the ac power enters the power supply section of the electronic

Figure 6.3 Most electronic circuits require a power supply.

circuit, where it is converted to dc power. From there the dc power flows to other parts of the circuit. Such power supply circuits use the nonlinear properties of diodes.

Half-wave rectifier. Power supplies use diodes to convert ac to dc in *rectifier* circuits; the diodes are said to "rectify" the ac. Figure 6.4 shows a basic rectifier circuit. The ac voltage is represented as a sine function, the rectifier is an ideal diode, and the load is represented as a resistor, although in practice the load would be the electronic circuits that require dc power.

The rectifying effect of the diode is shown in Fig. 6.5. When the input voltage is positive, the diode turns ON and current flows with a sinusoidal shape, but when the input voltage is negative the diode turns OFF and no current flows. The resulting current flows in spurts, and the voltage across the load is the portion of the sinusoidal input that is positive. The diode blocks the negative part of the ac waveform.

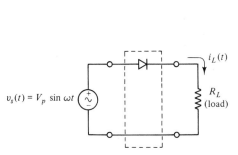

Figure 6.4 Half-wave rectifier circuit.

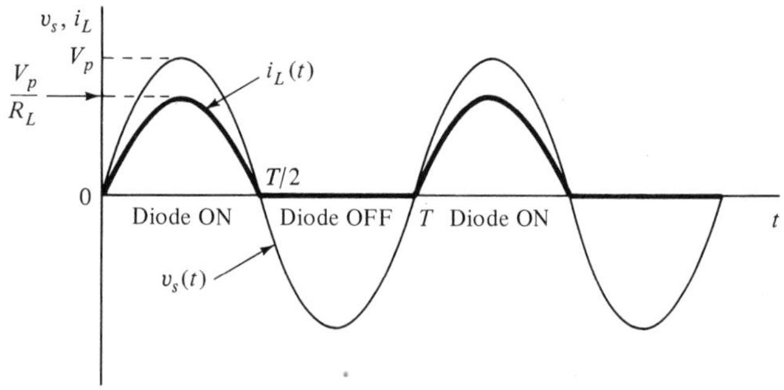

Figure 6.5 Half-wave rectifier waveforms.

Admittedly the output of this simple power supply is not pure dc power. The output, however, does contain a dc component. Indeed, if we define the dc portion of the output as the time average (see Section 5.1.2), we have a dc component of

$$I_{dc} = \frac{1}{T} \int_0^T i_L(t)\, dt = \frac{1}{T} \int_0^{T/2} \frac{V_p}{R_L} \sin \omega t\, dt$$

$$= \frac{V_p}{\pi R_L}$$

where $\omega = 2\pi/T$. This current would be indicated by a dc ammeter in series with the load.

Full-wave rectifier. It is possible to clean up or *filter* the output of the half-wave rectifier in Fig. 6.4 to better approximate true dc. Before presenting

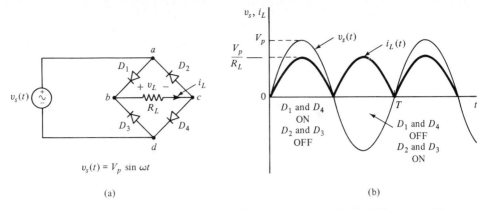

$$v_s(t) = V_p \sin \omega t$$

(a)

(b)

Figure 6.6 (a) Bridge full-wave rectifier circuit; (b) waveforms for the bridge full-wave rectifier.

filtering circuits, however, we will discuss another rectifier circuit. The circuit in Fig. 6.6a is called a full-wave bridge rectifier. This circuit uses four diodes to accomplish rectification in the following way. When the source voltage is positive, the current tends to flow through the rectifier from a to d. Diode D_1 comes ON but D_2 cannot accommodate current from a to c and will turn OFF. From b the current must flow through the resistor because D_3 will not permit current flowing directly from b to d. Finally, D_4 will turn ON and the current will flow from c to d. Thus when the source voltage is positive, positive current flows out of the $+$ of the source through D_1, through the load resistor, and finally through D_4. Diodes D_2 and D_3 are OFF during this part of the cycle. When the source voltage is negative, D_2 and D_3 turn ON while D_1 and D_4 turn OFF. Thus during the second half of the cycle, positive current flows out of the $-$ terminal of the source, through D_3 from d to b, through the load resistance, and back to the source through D_2. The two paths for the current are shown in Fig. 6.7. The current flows through the load from b to c during both parts of the cycle, thus the current through the load is as shown in Fig. 6.6b. The current again flows in spurts, but the full-wave rectifier leaves no idle time between spurts as does the half-wave rectifier.

The diodes act as switches that are activated by the voltage. When the source voltage is positive, D_1 and D_4 are turned ON and thus b and c are connected to the $+$ and $-$ terminals of the source, respectively. When the source voltage becomes negative, D_2 and D_3 are turned ON and hence b and c are again connected

Figure 6.7 Current paths for bridge rectifier: (a) v_s positive; (b) v_s negative.

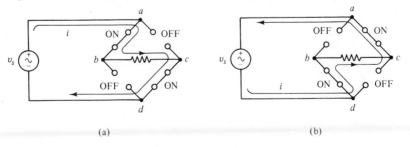

(a)

(b)

to the source, this time with b connected to the $-$ and c to the $+$ terminal. Thus b is automatically connected to the physically positive terminal of the source and c is automatically connected to the physically negative terminal. This automatic switching action occurs regardless of the shape of the source voltage—the shape can be sinusoidal, triangular, or an unpredictable communication signal in a radio circuit. The effect of the diode bridge is thus to produce across the load an output voltage that is the absolute value of the input voltage: $v_L(t) = |v_s(t)|$.

As a power supply circuit the full-wave rectifier does better than the half-wave rectifier, but the output is still a poor approximation to a pure dc voltage. By inverting the negative portions on the input voltage, the circuit doubles the dc component in the output; hence the dc component in the current is

$$I_{dc} = \frac{1}{T} \int_0^T i_L(t) \, dt = \frac{2}{T} \int_0^{T/2} \frac{V_p}{R_L} \sin \omega t \, dt$$

$$= \frac{2V_p}{\pi R_L}$$

Figure 6.8 shows a full-wave rectifier circuit that uses a transformer and two diodes. The transformer secondary is center-tapped to supply identical but opposite voltages to the two diodes. Each diode acts as a half-wave rectifier: D_1 supplies the positive part of v_{de} and D_2 supplies the positive part of v_{fe}, as shown in Fig. 6.9. The transformer gives this circuit the versatility to produce any desired dc voltage, depending on the turns ratio.

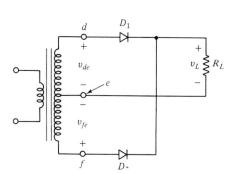

Figure 6.8 Full-wave rectifier using a center-tapped transformer.

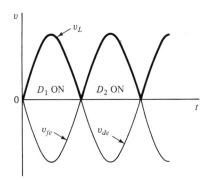

Figure 6.9 Waveforms for the full-wave rectifier.

Filtering out the ripple. The three rectifier circuits we have described produce a dc component in their outputs from an ac input. We may describe their outputs as a desired dc component plus an undesired *ripple*, as shown in Fig. 6.10. We need to eliminate, or at least greatly reduce the ripple. We need a filter, as in Fig. 6.11, to remove the undesirable ripple component from the output of the rectifier. In a later chapter we will consider filters generally; here we introduce the simplest of filters—we merely connect a capacitor across the load, as shown in Fig. 6.12. The purpose of the capacitor is to stabilize the voltage across the load resistor.

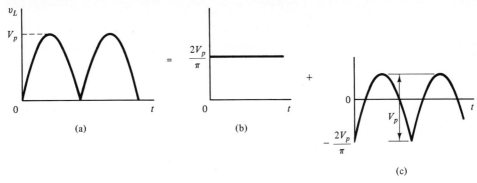

Figure 6.10 The ripple is the undesirable portion of the output. (a) Rectifier output; (b) dc component; (c) ripple component.

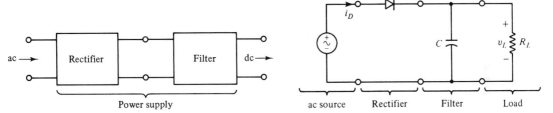

Figure 6.11 A filter is used to reduce the ripple.

Figure 6.12 Half-wave rectifier with a capacitor filter.

Let us consider that the source is $V_p \sin \omega t$, and thus passes through zero volts at $t = 0$, as shown in Fig. 6.13. As the voltage increases, the diode turns ON and current flows through the load as before. Current also flows through the capacitor and charges it to the peak value of the input ac voltage. The first spurt of current is relatively large in Fig. 6.13 due to the initial charging of the capacitor. After its peak, the input voltage drops rapidly. If the voltage of the capacitor were to follow this voltage, a rapid discharge would have to occur. However, the diode prevents discharge through the input source and turns OFF when the input voltage drops below the voltage on the capacitor because at that moment the diode be-

Figure 6.13 Waveforms for the half-wave rectifier with a capacitor filter.

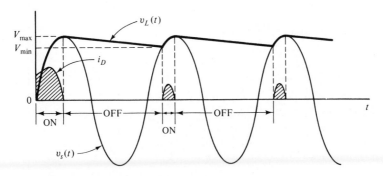

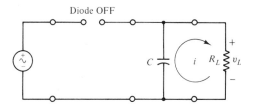

Diode OFF

Figure 6.14 When the diode is OFF, the capacitor discharges through the load.

comes reverse biased. Hence, the filter capacitor and the load resistance become disconnected from the source, which continues with the negative part of its waveform. The capacitor thus discharges through the load resistor, as shown in Fig. 6.14.

The discharge of a capacitor through a resistor is a problem we learned to solve in Chapter 3. To write the voltage as a function of time, we need to know the time constant, the initial value, and the final value. The time constant is R_LC, the initial value is V_p, and the final value would be zero if the discharge were allowed to go on forever. Thus the voltage of the load (and the capacitor) during the discharge period, when the diode is OFF, would be

$$v_L(t) = 0 + (V_p - 0)e^{-(t'/R_LC)} = V_p e^{-(t'/R_LC)}$$

where t' is measured from the peak.

To function well as a filter, the circuit should have a time constant (R_LC) much longer than the period of the input ac voltage (usually $\frac{1}{60}$ s). The load voltage thus decreases only slightly as the input voltage swings negative and again approaches its peak. Figure 6.13 exaggerates the decrease due to the discharge from what it would be in practice. During this time the diode remains OFF until the increasing input voltage becomes equal to the decreasing load voltage. As the input voltage again exceeds the load voltage, the diode turns ON and current again flows through the diode. Most of the current goes to the capacitor, replenishing the charge lost during the discharge part of the cycle. After the initial pulse of current that charges the capacitor, current flows through the diode only during these brief recharging periods, as shown in Fig. 6.13.

If the load voltage decreases only slightly during the period of the ac waveform, as shown in Fig. 6.13, the output voltage of the power supply remains approximately equal to the peak value of the input ac waveform. Thus the first benefit of the filter capacitor is to increase the dc output from V_p/π to V_p. Second, the filter capacitor greatly reduces the ripple voltage. For the case we are considering $(R_LC \gg \text{period})$, the exponential decrease of the load voltage is well approximated by a straight line, the two leading terms in a series expansion of the exponential.

$$v_L(t') = V_p e^{-(t'/R_LC)} = V_p \left(1 - \frac{t'}{R_LC} + \frac{(t'/R_LC)^2}{2} - \cdots \right) \qquad (6.1)$$

where t' is measured from the peak. Thus from its maximum value of V_p the voltage decreases to a minimum value of approximately

$$V_{\min} \approx V_p \left(1 - \frac{T}{R_LC} \right) = V_p \left(1 - \frac{1}{fR_LC} \right) \qquad (6.2)$$

at the moment when the diode turns ON and permits recharging of the capacitor. In Eq. (6.2), f represents the ac frequency, the reciprocal of the period. Consequently, the peak-to-peak ripple is

$$V_r = V_{max} - V_{min} = \frac{V_p}{fR_LC} \qquad (6.3)$$

We note from Fig. 6.13 that a more accurate approximation to the dc component of the filtered output would be the average between the maximum and minimum voltages:

$$V_{dc} \approx \frac{V_{max} + V_{min}}{2} = V_p \left(1 - \frac{1}{2fR_LC} \right) \qquad (6.4)$$

Example. We illustrate with the circuit in Fig. 6.15. The input ac has an rms voltage of 12 V at 60 Hz. Thus the peak value, and hence the maximum of the output voltage, is $12\sqrt{2} = 17.0$ V. The time constant is

$$\tau = R_LC = 100 \times 2000 \times 10^{-6} = 200 \text{ ms}$$

which is long compared to the period of 16.7 ms. The exponential decrease of the output voltage during the period when the diode is OFF can be written

$$v_L(t') = 17.0e^{-(t'/0.2)} \approx 17.0 \left(1 - \frac{t'}{0.2} \right)$$

where we have measured time from the peak. At the time of the next peak, the output voltage will have decreased to approximately

$$v_{min} \approx 17.0 \left[1 - \frac{1}{60(0.2)} \right] = 15.6 \text{ V}$$

Thus the ripple voltage is 1.4 V peak-to-peak and the dc (time-average) voltage at the load is the average between the maximum and minimum, 16.3 V. The dc current in the load is 16.3 V divided by the load resistance, or 163 mA.

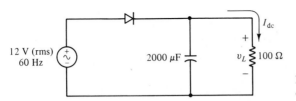

12 V (rms)
60 Hz

$2000\ \mu F$

I_{dc}

v_L $100\ \Omega$

Figure 6.15   Half-wave rectifier circuit with a capacitor filter.

Better filters. We have investigated the benefits of the simplest possible filter. The performance of this filter is adequate for many applications, but high-quality power supplies employ more sophisticated filters. Some filters add inductors and additional capacitors to reduce the ripple; others employ electronic circuits to cancel the ripple.

6.2.3 Check Your Understanding

1. An ideal diode uses no power because either its voltage or current is zero. (T/F?)
2. A half-wave rectifier circuit requires at least two diodes. (T/F?)
3. What is the average value of a full-wave rectified sinusoid having a peak value of 10 V before rectification?
4. A battery may be used as a filter in a power supply. (T/F?)
5. A rectifier circuit produces pure dc. (T/F?)
6. What is "filtered" by a filter circuit in a power supply?
7. An unfiltered full-wave rectifier puts out 12 V dc. If a large capacitor is added across the load, what would be the new voltage, assuming ideal diodes?
8. Explain why, in a good stereo, the sound does not go away immediately when you unplug the amplifier but fades out over a period of several seconds. (Try it.)

Answers. 1. true; 2. false; 3. 6.37 V; 4. true—the battery acts like an infinite capacitor; 5. false; 6. the ripple, everything but the dc; 7. 18.8 V; 8. the capacitor in the power supply has to discharge before the dc voltage is zero.

6.3 THE *pn*-JUNCTION DIODE

Reasons for discussing the principles of diode operation.   The diode finds many uses as an electronic component. We have seen how diodes are used in power supplies to convert ac to dc, but this is only one of the many uses for diodes. Diodes are used extensively in analog electronic devices such as radios and audio systems, and are even more important in digital systems such as computers and digital watches.

The diode plays an important role in the study of electronics because it is a simple electronic device and hence presents a good starting place in exploring the bag of electronic tricks. The diode is also important in the teaching of electronics because it offers our first opportunity to investigate the physical basis of semiconductor electronics. The development of electronics based upon semiconductor processes has produced essentially all the electronic equipment we require and enjoy in our society. This section describes many of these semiconductor processes.

The main question.   What is a physical diode, and how does it work? The answer to that question will require the discussion of a sizable number of physical processes that occur in solids. Your author has a genuine problem in helping you understand how a diode works. It is commonly asserted, and reasonably so, that we should explain the unknown in terms of the known. But your author must explain the unknown (how a diode works) in terms of other unknowns (semiconductor processes such as holes, drift currents, uncovered charges, and depletion regions). Or, put another way, we are required to consider many physical processes before we can understand how a simple *pn*-junction diode operates. Whereas understanding of the diode itself might not merit the effort, our investment is worthwhile because modern electronics is founded on the manipulation of charges in semiconductor materials through such physical processes.

6.3.1 Semiconductor Processes and the *pn* Junction

Crystalline nature of solids. Matter is considered to exist in four states: solid, liquid, gas, and plasma. These may be understood in terms of the interaction between individual atoms or molecules comprising the matter in question. In a solid, the atoms remain in a fixed position relative to each other. Often the atoms exist in a regular crystalline order, held together by shared electrons in covalent bonds. Figure 6.16 shows a two-dimensional representation of such a lattice. The open circles represent the nuclei, the dots represent the valence electrons, and the curved lines represent the bonding effect of the electrons. The electrons are negatively charged and the nuclei are positively charged. Of course, there are many more electrons associated with each nucleus, but we have not represented the inner shells because these do not enter into the bonding process or the semiconductor processes we are investigating. We have represented the case where there are four electrons in the outer shell, which is typical of a semiconductor such as silicon. Because the individual atoms are electrically neutral, all the charge would be neutralized by the inner shells of electrons except for a positive charge of four times the electronic charge, which is thus the effective charge of the nucleus. This charge is indicated by the +4 in Fig. 6.16.

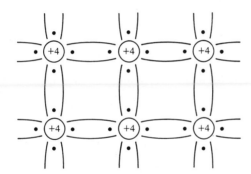

Figure 6.16 Insulator.

Insulators, conductors, and semiconductors. At zero absolute temperature, solids are either electrical conductors or insulators. The material represented by Fig. 6.16 would be an insulator because all the electrons are involved in the bonding. If such a material were placed in an electrical circuit, no current would flow because no electrons are free to move. In a conductor, each atom has one excess electron that is not involved in the bonding. Such excess electrons are known as conduction electrons and they move freely in response to electrical forces.

Thermal energy has a strong effect on the conduction properties of solids. Thermal energy is distributed throughout the electrons and nuclei of the materials and this energy is stored, among other ways, in the physical movement, or vibration, of the electrons and nuclei. The vibrations may cause some of the bonding electrons to break loose from their bonding positions to become conduction electrons.

Materials that have no excess electrons and retain all their electrons in bonds at normal temperatures remain good insulators. Carbon in a diamond crystalline

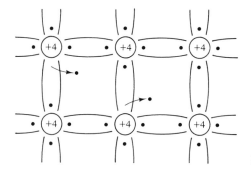

Figure 6.17 Semiconductor.

order and many plastics and ceramics furnish examples of good insulators at normal temperatures. In other materials that are good insulators at absolute zero, however, the electrons are not tightly held in the bonds, and in these materials some of the electrons escape their bonds at normal temperatures to become conduction electrons. We have represented this condition in Fig. 6.17 with two electrons out of their bonding positions. Such a material is known as a semiconductor because it becomes a conductor at normal temperatures due to the electrons that have become conduction electrons.

Holes and hole movement. When an electron leaves its bonding position, it leaves behind a vacant position, which is called a *hole*. We may think of the hole as having a positive charge because the nucleus adjacent to it now has charge not neutralized by the bonding electrons. Holes are also free to move under the influence of electrical forces because other bound electrons may move into the vacant location. For example, if in Fig. 6.17 there were an electrical force tending to move electrons from left to right, the conduction electrons would move in response to such a force. But the electron next to the vacant position will also tend to move to the right and may leave one bonding position for a vacant position toward the right. Thus we can envision the process represented in Fig. 6.18. First the electron breaks its bond and becomes a conduction electron, creating a hole. Because there is an electrical force tending to move electrons toward the right, there will also be a tendency for electrons remaining in bonds to move left to right; hence the transitions we have marked (2) through (5) are favored. The hole moves toward the left; in effect a positive charge moves right to left because there

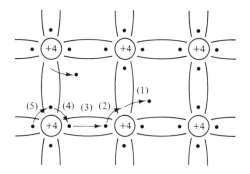

Figure 6.18 The orderly movement of bound electrons produces hole movement.

Introduction to Electronics   Chap. 6

is excess positive charge associated with the hole. If a semiconductor were placed in a circuit where the voltage created a current, the current would be carried by both holes and electrons, which are said to be *carriers*. Note that the currents carried by the movement of holes and electrons are additive; holes moving toward the left carry a positive current toward the left; and electrons moving to the right carry a negative current toward the right, which is a positive current toward the left. In a typical semiconductor, the holes move about one-third as fast as the electrons and hence carry about one-fourth the total current.

Importance of thermal energy.   We have stated that hole–electron pairs are created because of thermal energy, and we have shown how electrons changing bond positions due to thermal vibrations act like mobile positive charges. But we have described these processes as if they were orderly and sedate, which they are not. To get a feeling for what is happening in a semiconductor, you have to understand that the thermal processes are rapid and violent.

To hint at the time scale of these processes, we shall compute a representative velocity of a conduction electron using ideal gas law theory. From statistical thermodynamics we learn that, in a thermal system in equilibrium, the energy will distribute throughout the available energy-storage modes of the system, and the energy per mode will be proportional to the temperature. The constant of proportionality is one-half of Boltzmannn's constant, $k = 1.38 \times 10^{23}$ J/K. Consider an "electron gas" consisting of electrons moving around as conduction electrons in a semiconductor material. They can store kinetic energy in three directions of motion; hence their total kinetic energy will be

$$\tfrac{1}{2}m_e \times \text{time average of } (u_x^2 + u_y^2 + u_z^2) = 3 \times \tfrac{1}{2}kT \tag{6.5}$$

$$u = \sqrt{\frac{3kT}{m_e}} = 117 \text{ km/s}$$

where u is the Pythagorean sum of the x, y, and z components of velocity and would be the rms speed of the electron in three-dimensional space. When we set T to be 300 K and the mass of the electron, m_e, to its value of 9.11×10^{-31} kg, we calculate the rms velocity to be 117 km/s. Of course, this calculation is not highly accurate because the electrons are not an ideal gas, but the conclusion offered is valid—those electrons are moving rapidly. When you consider the small physical scale of the lattice structure, you must conclude that the time processes occur quite rapidly. For pure silicon at 300 K, there are approximately 1.5×10^{16} conduction electrons/m^3 and there are approximately 10^{22} hole–electron pairs/m^3 created and eliminated through recombination per second through thermal action.

Hence emerges a picture of violent, random thermal motion of conduction electrons in a semiconductor. Many hole–electron pairs are created and many recombinations occur during short periods of time.

Doping.   We can increase the concentration of carriers by adding small amounts of *impurities* to the pure (called *intrinsic*) semiconductor. Consider that we have intrinsic silicon, which has four bonding electrons per atom. If we add a small amount of an element that has five electrons in its valence shell, such as phosphorus, the impurity nuclei will bond into the lattice with one electron left over. This electron will be a conduction electron, as shown in Fig. 6.19, and there will also be one additional positive charge fixed into the lattice of nuclei because of the additional charge of the nucleus of the impurity atom. This process of adding impurities is called *doping*; Figure 6.19 shows *n*-type doping, so called because of the additional negative carriers. This would make an *n*-type semiconductor because it has an increased concentration of electrons. The impurity in this case would be called a *donor* atom because it donates an additional conduction electron to the semiconductor.

Similarly, if we add an impurity with three electrons in its valence shell, we would create a hole for each impurity atom, as shown in Fig. 6.20. Here we have shown the hole and we have indicated that the nucleus is deficient one positive charge with a $+3$. If the hole were filled by a conduction electron, that region would in effect have a negative charge built into the lattice structure of the semiconductor. Such an impurity is called an *acceptor* atom because it accepts an electron from the conduction electrons in the semiconductor and an extra hole is created. A semiconductor that is doped with acceptor atoms is called a *p*-type semiconductor because it has an excess of holes, which act like mobile positive charges.

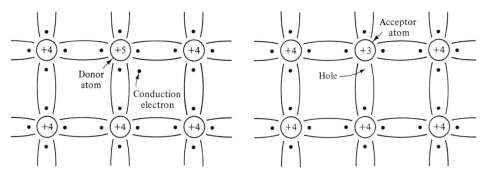

Figure 6.19 An *n*-type semiconductor has extra conduction electrons.

Figure 6.20 A *p*-type semiconductor has extra holes.

The electrons donated by the donor atoms and the holes created by the acceptor atoms do not remain near their associated nuclei but participate in the random thermal processes of all carriers. By doping the semiconductor, we have the ability to increase the carrier concentration (either holes or electrons) and to control what type of carriers will dominate the conduction processes in the semiconductor.

Drift and diffusion currents. We have already described what happens when a semiconductor is placed in an electric circuit. A current would flow via movement of both holes and electrons. If the semiconductor were *p*-type, most of the current would be carried by the holes. If the semiconductor were *n*-type, most of the current would be carried by the electrons. This type of current is called a *drift current* because the carriers drift in a certain direction as dictated by an external voltage. We may think of this type of current as being caused by an orderly process.

However, if we had an excess of carriers in a certain region, they would tend to distribute uniformly throughout the material because of their random thermal movement. Such a flow is known as a *diffusion flow*. A diffusion flow occurs when a bottle of a smelly chemical is opened in a room filled with calm air. People near the open bottle would smell it first, then those farther away. Eventually, everyone in the room would smell the chemical because the molecules would be distributed uniformly throughout the room by their thermal motion. In a similar way, carriers move away from regions of concentration due to their thermal motion. The resulting *diffusion current* is proportional to the rate of change of carrier concentration with respect to distance. Diffusion is a disorderly process because it is driven by thermal motion.

Summary. An intrinsic semiconductor will have many, but equal numbers, of holes and electrons because thermal energy causes electrons to leave their bond positions, become conduction electrons, and leave behind a hole. We can create *p*-type or *n*-type semiconductors by doping the pure material with either acceptor or donor impurity atoms. The additional holes or electrons participate in the violent, random thermal motion of the carriers in the material. The carriers form currents either by the orderly process of a drift current or by the disorderly process of diffusion. There are also charges bound into the lattice structure associated with the acceptor or donor nuclei. The acceptor atoms act as stationary negative charges because their nuclei are deficient one positive charge relative to the other nuclei in the vicinity. The donor atoms act as stationary positive charges because their nuclei have one additional positive charge relative to the surrounding nuclei. Semiconductors are neutral in total charge because the excess mobile carriers neutralize the bound charges of the impurity atoms. All of these processes occur in a *pn* junction, to which we now turn.

6.3.2 The *pn* Junction

Structure. A *pn* junction is formed at the boundary between regions of *p*-type semiconductor and *n*-type semiconductor, as shown by Fig. 6.21. When such a junction is formed, the sequence of events diagrammed in Fig. 6.22 occurs rapidly.

Figure 6.21 *pn* junction.

Diffusion currents
↓
Recombination
↓
Depletion region
↓
Uncovered charge
↓
Drift current
↓
Dynamic equilibrium

Figure 6.22 Processes involved in *pn*-junction formation.

- Electrons in the n-type material will diffuse toward the p-type material and holes in the p-type material will diffuse toward the n-type material. These strong diffusion currents will occur because the concentrations are unequal and the carriers are in violent thermal motion.

- Recombinations will occur immediately as the electrons that have diffused into the p-type material find holes to fill and holes that have diffused into the n-type material are filled by conduction electrons.

- Thus there will be a region on both sides of the junction that has a deficiency of carriers due to recombinations. This region is called a *depletion region*. We might anticipate a continual pouring of carriers into this depletion region were it not for another process that occurs.

- On the p-type material side of the junction in the depletion region, the acceptor atoms bound into the lattice structure are now *uncovered*, that is, because the electrons from the n side have recombined with many of the holes, the deficiency of charges in the nuclei of the acceptor atoms acts as an excess of negative charges fixed in this region. Similarly, the excess positive charges of the donor nuclei will act as positive charges bound into the lattice structure, now uncovered because the electrons that formerly neutralized them have recombined with holes from the p side. Thus we have uncovered charges bound into the lattice structure in the depletion region, as shown by Fig. 6.23. This charge distribution will act similar to a capacitor, as shown in Fig. 6.24. Here we have shown a capacitor with a charge placed on it by a battery. The bottom plate of the capacitor will have positive charges and the top will have negative charges. If an electron were in the region between the plates, it would move downward, attracted by the positive charges below and repelled by the negative charges above. Similarly, a positive charge between the plates would move upward. This would be an orderly process and would create a drift current in this region if there were electrons or positive charges between the plates. Similarly, in Fig. 6.23 excess carriers in the depletion region experience forces from the uncovered, bound charges.

- These forces would tend to create a drift current upward as electrons are sent back toward the n-type material and holes are sent back toward the

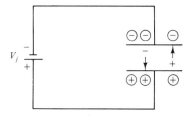

Figure 6.23 Uncovered charges.

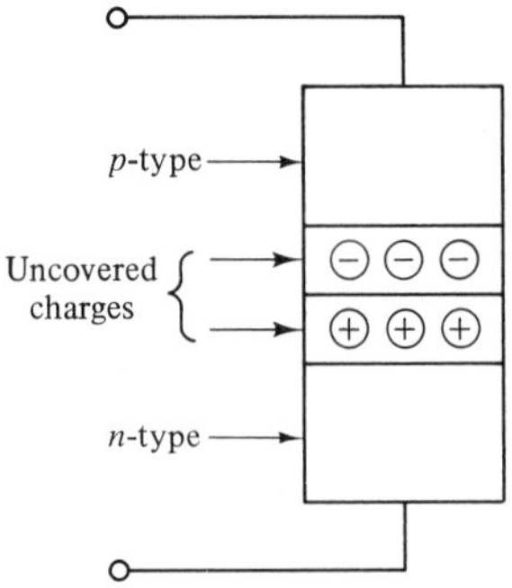

Figure 6.24 The uncovered charges act like a charged capacitor.

p-type material. Thus we can envision the holes and electrons moving across the junction due to diffusion but moving the other way by the drift current caused by the uncovered charges.

- These two processes rapidly reach dynamic equilibrium.

Another way to view this dynamic equilibrium is to consider that the uncovered charges constitute an internal battery–capacitor effect that automatically adjusts itself to stop the diffusion currents. Hence we have a dynamic equilibrium between a disorderly, random process (the diffusion process) and an orderly process (the battery–capacitor effect). Due to the uncovered charges, the *pn* junction diode will have an internal voltage of approximately 0.7 V across its junction, but no external voltage across the entire diode is produced.† No current will flow unless the equilibrium is disturbed by an external force.

Forward-bias characteristic. If we add an external battery to the diode, as shown by Fig. 6.25, and if the polarity of the external battery is such as to oppose the internal battery–capacitor effect, the equilibrium is disturbed and current will flow continually through the diode. In this case the battery will cause the holes in the *p*-type material and the electrons in the *n*-type material to move toward the junction. Because the drift process tending to prevent flow of carriers through the depletion region has been partly neutralized by the external battery, current is carried across the junction by the diffusion currents. The current increases dramatically with increases in the external voltage because it is driven by the energetic thermal motion of the carriers. If the external voltage exceeds the voltage of the internal battery–capacitor effect, large currents will flow. Because this voltage is only about 0.7 V for a silicon diode, the forward-bias region of the diode approximates that of an ideal diode. A germanium diode requires about 0.2 V to cause substantial conduction and hence is more nearly ideal than a silicon diode.

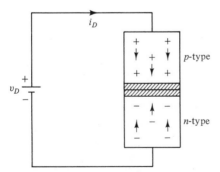

Figure 6.25 Forward-biased *pn* junction.

Reverse-bias characteristic. If the polarity of the external voltage reinforces the influence of the internal battery–capacitor effect, the holes in the *p*-type semiconductor and the electrons in the *n*-type semiconductor will tend to

---

† This voltage cannot be measured with an ordinary meter because the meter connections form other junctions that produce voltages to cancel the interval voltage, V_j.

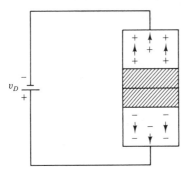

Figure 6.26 Reverse-biased *pn* junction.

move away from the junction, as shown in Fig. 6.26. The external voltage merely widens the depletion region and strengthens the restraining effect of the internal battery–capacitor effect. For this polarity of the external voltage, very little current will flow through the diode and an ideal diode is well approximated. The reverse-biased *pn* junction behaves as a capacitor due to the charge separation associated with the uncovered charges. In the next section we will give the equation of a *pn*-junction diode and see how well an ideal diode approximates a semiconductor *pn* junction.

6.3.3 Physical Properties of Real Diodes

***pn*-Junction equation.** The voltage–current characteristic of the *pn*-junction diode, as described in the previous sections, is well described by

$$i_D = I_0 \left[\exp\left(\frac{qv_D}{\eta kT}\right) - 1 \right] \tag{6.6}$$

where I_0 = a constant called the *reverse saturation current,* depends on the semiconductor materials, manner of junction formation, and junction size

$q = |e|$, the magnitude of the electronic charge, 1.60×10^{-19} C

k = Boltzmann's constant, 1.38×10^{-23} J/K

T = absolute temperature, K

η = a constant between 1 and 2, called the *ideality factor,* which depends on junction materials and method of formation

Figure 6.27 shows a plot of Eq. (6.6) for the parameters $\eta = 1.5$ and $I_0 = 10^{-9}$ A. These would be typical of a silicon diode at room temperature. If you wish to plot the *pn*-junction equation, it is useful to have v_D as a function of i_D.

$$v_D = \eta V_T \ln\left(\frac{i_D}{I_0} + 1\right) \tag{6.7}$$

where $V_T = kT/q$ is called the *voltage equivalent of temperature* and has the magnitude of about 26 mV at $T = 300$ K.

If you compare the characteristics of a *pn*-junction diode [Fig. 6.27 or Eq. (6.6)] with those of an ideal diode (Fig. 6.2), you might experience some disappointment, for the properties of a real diode appear quite nonideal. That impression is created by our plotting the *pn*-junction characteristic on an expanded volt-

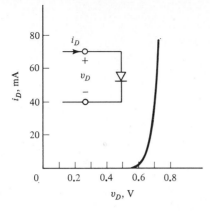

Figure 6.27 Typical current–voltage characteristic for a silicon diode.

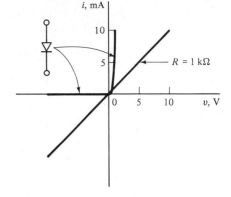

Figure 6.28 Diode characteristic compared with a 1-kΩ resistor.

age scale. Figure 6.28 shows the diode characteristics on the same scale as a 1-kΩ resistor.

The properties of the *pn*-junction diode may be summarized in this fashion: Essentially, zero current flows in the reverse-bias region. Negligible current flows in the forward-bias region until a small threshold voltage is reached, after which the magnitude of the current rises rapidly. Once the current begins to rise, the voltage remains fairly constant. The threshold voltage for Fig. 6.27 is about 0.7 V for a silicon diode. The *pn*-junction equation [Eq. (6.6)] gives the precise characteristic, but the concept of a threshold voltage is used to simplify analysis for many diode applications. For the rectifier circuits presented earlier in this chapter, we may assume that the diodes will have a voltage drop of about 0.7 V when they are ON and thus the rectifier output voltages and currents will be reduced accordingly. When characterized by a threshold voltage, the voltage drop of the diode is easy to take into account in design and analysis of diode circuits.

Power limits and heat transfer. Temperature strongly affects the semiconductor processes upon which diode operation depends. In Eq. (6.6) temperature appears explicitly in the exponential term, but the reverse saturation current, I_0, is also affected by temperature. For this reason, the electronics designer must prevent the diode from getting too hot. Heat is produced by the electrical power given to the diode. The power into a device is

$$p = vi$$

The ideal diode requires no power because in the forward bias region the voltage is zero and in the reverse bias region the current is zero; hence the product of voltage and current is always zero. A real diode inherently receives little power because it also has small voltage when the current is high (the forward bias region) and low current when the voltage is high (the reverse bias region). Even so, the small amount of power that is given to the diode is important because the thermal energy is generated in the junction region, which is physically small. Thus a small power can cause a significant rise in the junction temperature and affect diode

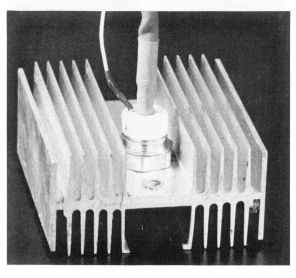

Figure 6.29 Power semiconductor mounted on a heat sink.

performance. For this reason diodes in power supplies are designed to have good heat conduction between the junction and the outer case of the diode, and the diode is mounted in such a way as to enhance heat transfer to the ambiance. Often *heat sinks* and even fans are used to improve cooling of the diode. Occasionally, water cooling is used on large power supplies. Figure 6.29 shows a power semi-conductor mounted on a heat sink.

Breakdown.   A physical diode is also limited by the amount of voltage it can withstand in the reverse-bias region. Conduction is prevented in the reverse-bias region by the reinforcement of the effect of the uncovered charges in the depletion region, as suggested in Fig. 6.26. When too much voltage appears across the depletion region, however, the forces on the bound electrons in that region become so great that they can be torn from their locations. In this condition several processes are involved, and a rapid buildup of current occurs, as shown in Fig. 6.30. The rapid buildup of current at the reverse-breakdown voltage, $-V_B$, increases the power into the junction region, and the diode can be quickly destroyed by the resulting heat.

Breakdown must be avoided in power supply operation. Designers have available diode types with breakdown voltages (peak inverse voltage, or PIV) in excess of -1000 V. In the unfiltered half-wave rectifier shown in Fig. 6.4, the maximum diode voltage is $-V_p$, which occurs when the diode is OFF and the source at its maximum negative value. For successful operation in such a power supply, the PIV of the diode must exceed this voltage.

In the filtered power supply in Fig. 6.12, the maximum voltage across the diode is approximately $-2V_p$, as shown in Fig. 6.31. This maximum occurs because the capacitor holds the load voltage at approximately $+V_p$ while the voltage source swings negative to $-V_p$. To operate successfully in such a power supply, the diode must be able to withstand $-2V_p$.

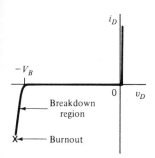

Figure 6.30 When reverse-bias voltage exceeds the breakdown voltage, the current increases rapidly.

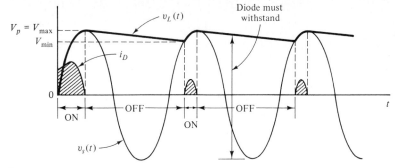

Figure 6.31 To operate successfully in a rectifier, a diode must never have its breakdown voltage exceeded.

6.3.4 Check Your Understanding

1. A intrinsic semiconductor has an excess of holes, conduction electrons, or neither?
2. In a *pn* junction, the *p* stands for semiconductor material that has an excess of holes or electrons?
3. Which of the following are nonlinear devices: resistor, *pn*-junction diode, capacitor, short circuit, ideal diode?
4. In the ON state, current crosses a *pn* junction under the influence of a drift or diffusion process?

   Answers: 1. Neither; 2. holes; 3. *pn*-junction diode and ideal diode; 4. diffusion.

4 BIPOLAR JUNCTION TRANSISTOR OPERATION

6.4.1 Importance of the Transistor

From your earlier experience with circuit theory, you will recall that one can do a fair amount of work solving a two- or three-loop ac circuit. That being true, did you ever wonder how electrical engineers can design circuits containing literally thousands of loops and nodes? For example, if you ever examine the circuit diagram for a relatively simple piece of electronics, say, a TV set or an FM radio, you would see hundreds of resistors and capacitors, not to mention diodes and transistors and other components. How do electrical engineers understand such complicated circuits?

   The design of such complicated circuits is simplified by the unidirectional properties of transistors. The transistor is a three-terminal device connected normally as shown in Fig. 6.32. The transistor has an input and an output, suggesting that the cause–effect relationship goes from left to right. The input affects the output but the output has little effect on the input. This one-directional causality

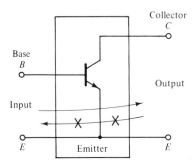

Figure 6.32 *npn* transistor symbol.

we have indicated by marking an arrow from left to right while crossing out the arrow from right to left. We might say that the transistor is a "one-way street" to the electrical signal. This one-directional property of the transistor isolates the output from the input and allows electrical engineers to build complex circuits. This complexity can be mastered because circuits can be designed (or analyzed) one part at a time, unlike the two- or three-loop circuits that we tackled in circuit theory, which must be analyzed all at once.

In addition to this isolating property, transistors have other important properties. Transistors can give signal gain, thus allowing small signals (such as a voltage induced in a car radio antenna) to be amplified by stages until large enough to power a radio speaker. Transistors can also be used to couple circuits of greatly differing impedance levels, allowing more efficient transfer of signals between them. Transistors are used in digital circuits as electrically controlled switches. Finally, transistors offer a variety of nonlinear effects that are used in communication circuits for manipulating signals in the frequency domain. In this section, however, we limit our attention to the isolating, amplifying, and switching properties of the transistor.

There are many types of transistors, all of which accomplish the purposes stated above. There are *npn* and *pnp* bipolar junction transistors (BJTs) and there are *p*-channel and *n*-channel field-effect transistors (FETs). FETs can be either junction field-effect transistors (JFETs) or they can be metal-oxide semiconductor field-effect transistors (MOSFETs). In this section we will deal with the *npn* bipolar junction transistor, which is symbolized in Fig. 6.32. In the next section we deal with field-effect transistors.

6.4.2 Bipolar Junction Transistor Characteristics

As shown in Fig. 6.32, the transistor is a three-terminal device that is connected with one terminal in common between the input and output circuits. The parts of the transistor are the emitter (E), the base (B), and the collector (C), each with appropriately labeled terminals (Fig. 6.32). Our goal in this section is to describe the input and output characteristics of a typical *npn* transistor.

Transistor input characteristic. Figure 6.33 shows a simple picture of the physical structure of an *npn* transistor in the common-emitter configuration. The structure is that of a sandwich, with the *p* material like a very thin piece of bologna between two thick pieces of *n*-material bread. The external connections

Introduction to Electronics   Chap. 6

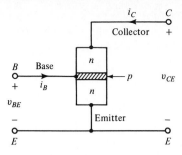

Figure 6.33 *npn* transistor structure.

are made through wires that are bonded to the three regions of the transistor. Depletion regions will form at the two *pn* junctions and, in the absence of external applied voltages, the orderly and disorderly processes will rapidly come to equilibrium at both junctions. First we consider the effect of placing a voltage at the base–emitter (*B–E*) junction, the input part of the transistor.

Ignoring for the present the effect of the collector–base junction, we note that the base–emitter junction forms a *pn*-junction diode. We conclude that the input *i–v* characteristic would be like that of a diode, as shown in Fig. 6.34. The base current will be very small until sufficient voltage exists across the junction to turn it ON, about 0.7 V for a silicon transistor. Once the junction is turned ON, the current increases rapidly, with the base–emitter voltage remaining constant at about 0.7 V. We can therefore model the base–emitter characteristic as either an open circuit (for $v_{BE} < 0.7$) or else a constant voltage of 0.7 V once the input voltage tries to go above that value. This model is shown in Fig. 6.35. In justifying this model, we have ignored the state of the output circuit (v_{CE} and i_C); but as we stressed above, the output has negligible effect on the input.

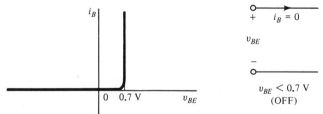

Figure 6.34 Transistor input characteristic.

Figure 6.35 *npn* transistor input circuit model.

Output Characteristics

Effect of doping. Like the input characteristic, the output characteristics depend on whether the collector–base *pn* junction is forward biased or reverse biased. We will assume that the input current to the base has been fixed at i_B, which requires approximately 0.7 V at the base region, considering the reference node to be the emitter. (Later in this discussion we will present a simple circuit to establish base current.) Now we will increase the collector–emitter voltage, beginning at zero volts, and observe the collector current.

First we consider the holes and electrons in the base region. When we discussed the *pn* junction earlier in describing diode operation, we implied that there is roughly the same density of holes in the *p* region as electrons in the *n* region. In this implied situation, forward biasing the *pn* junction will cause roughly as many holes to diffuse into the *n*-type material as electrons to diffuse into the *p*-type material. But if, say, the *n*-type material were more heavily doped than the *p*-type material, the electron density would greatly exceed the hole density. For this case a forward bias would produce many more electrons diffusing into the *p*-type material than holes diffusing into the *n*-type material. The diode would still work; however, the current would be carried across the junction largely by the electrons and most of the recombinations would occur in the *p*-type material.

When a transistor is made, the emitter is doped more heavily than the base; hence for an *npn* transistor the conditions are those described above—excess electrons diffuse into the base region. For zero volts between collector and emitter, the base–collector junction is also forward biased. Hence electrons also tend to diffuse into the base from the collector region. Although there is a buildup of excess electrons in the base region, the only current that flows is that permitted by recombination of electrons in the base: this current is the i_B we assumed in the base–emitter bias circuit.

Saturation region. If we now increase the collector–emitter voltage, we reinforce the orderly process forcing a drift current of electrons from the base to the collector. The result is a rapid buildup of collector current as excess electrons are permitted to pass into the collector region, where they flow through the collector–emitter circuit. This region of rapid increase in collector current is the *saturation region* marked in Fig. 6.36.

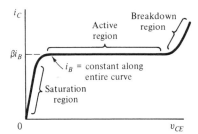

Figure 6.36 Output current–voltage characteristic with base current held constant.

Active region. By the time the collector base voltage is about 0.2 to 0.4 V, all the excess electrons in the base region are being drawn into the collector region and the curve levels off, the *active region* in Fig. 6.36. In the active region the collector current is controlled by the number of electrons injected into the base region by the emitter. This is controlled by the base–emitter voltage, which is controlled in turn by the amount of base current the base circuit allows to flow. Thus in the active region it is customary to consider the collector current as controlled by the base current.

Let us review the principal processes that occur in the active region. The external bias of about 0.7 V that is applied to the base–emitter junction diminishes the orderly process that might hold back the elecrons in the heavily doped emitter region. These electrons diffuse into the base region, where a few, say 1%, combine with holes and create the base current. The remaining 99% diffuse into the collector–base depletion region, where the orderly process of the uncovered bound charges forces them into the collector region. Thus in the active region the collector current is controlled by the base current.

Current gain of the transistor.   If we call α (alpha) the fraction of electrons that diffuse across the narrow base region, the fraction of electrons that recombine with holes in the base region to create the base current is $1 - \alpha$. The ratios of the base, collector, and emitter currents are thus

$$I_C = \alpha I_E \quad \text{and} \quad I_B = (1 - \alpha)I_E, \qquad \alpha < 1 \tag{6.8}$$

The value of α is fairly constant throughout the active region for a given transistor and characterizes the current gain of the device. Usually, the current gain is described in terms of the β (beta) of the transistor, defined as

$$\beta = \frac{I_C}{I_B} = \frac{\alpha}{1 - \alpha} \tag{6.9}$$

In terms of β, the ratios of the currents become

$$I_C = \beta I_B, \qquad I_E = \frac{\beta + 1}{\beta} I_C = (\beta + 1)I_B \tag{6.10}$$

We stress that Eqs. (6.10) are valid only in the active region.

Breakdown region.   Figure 6.36 also shows a breakdown region, where the current increases rapidly with increasing collector–emitter voltage. In this region, the power into the transistor becomes excessive and thermal failure often occurs.

Normally, the output characteristics are shown for many values of base current, as in Fig. 6.37. Here we have given typical characteristics for a transistor with a β of approximately 100 ($\alpha \approx 0.99$).

Figure 6.37 Typical *npn* transistor output characteristics.

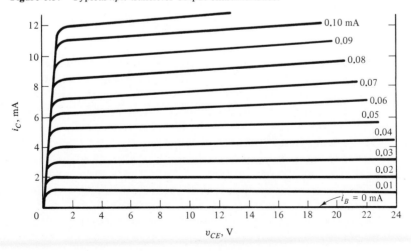

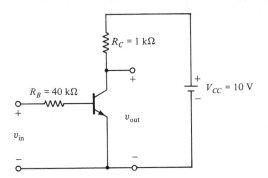

Figure 6.38 Transistor amplifier–switch circuit.

6.4.3 Transistor Amplifier–Switch Circuit Analysis

Problem statement.   Figure 6.38 shows the circuit we will analyze in this section. Here is an *npn* transistor in the common-emitter connection, meaning that the emitter of the transistor provides the common terminal between the input and output circuits. The circuit has an input circuit voltage, v_{in}, and an output circuit with its output voltage, v_{out}. A dc voltage source (V_{CC}) in the output circuit supplies the energy required by the circuit, and two resistors control the currents and voltages applied to the transistor. The transistor input characteristics are those of a silicon *pn* junction (Fig. 6.34) and the output characteristics are shown in Fig. 6.37. Our goal is to determine how the output voltage depends on the input voltage.

Input-circuit analysis.   We begin at the input circuit, shown in Fig. 6.39. We wish to determine the base current, i_B, because this current controls the state of the output circuit. Let us start with a negative value of v_{in} and increase this voltage to positive values. Comparing the input circuit with the model of the input characteristics in Fig. 6.35, we note that no current will flow until the input voltage becomes at least $+0.7$ V because the *pn* junction will be OFF. This is known as *cutoff*, for no current flows in the base or in the collector. When the input voltage exceeds $+0.7$ V, the base–emitter junction will turn ON and the second model circuit in Fig. 6.35 applies. Since in this case v_{BE} is constant at 0.7 V, KVL requires the base current to be

$$i_B = \frac{v_{in} - 0.7}{R_B}, \qquad v_{in} > 0.7 \text{ V} \tag{6.11}$$

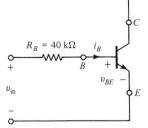

Figure 6.39 Input circuit.

Introduction to Electronics   Chap. 6

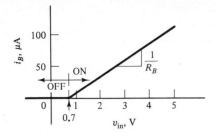

 — caption follows to the right:

Figure 6.40 Once the base-emitter junction is ON, the base current is limited by the base resistor, R_B.

This current is graphed in Fig. 6.40. This completes our analysis of the input circuit.

Output-circuit load-line analysis. The output part of the circuit is controlled by the input circuit. In the active region the base current controls the collector current, which in turn determines v_{out}. The concept of a *load line* offers an excellent way to understand the interaction of the transistor with the rest of the output circuit.

In Fig. 6.41, we have shown the output circuit broken at the transistor collector and emitter connections. If we stand at that break and look to the left, we see the i–v output characteristics of the transistor, which are given in Fig. 6.37. If we look to the right, we see the i–v characteristics of the external circuit, V_{CC} in series with R_C. Using Kirchhoff's voltage law and Ohm's law, we find the external circuit i–v characteristic to be

$$-V_{CC} + iR_C + v = 0 \Rightarrow i = \frac{V_{CC} - v}{R_C} \tag{6.12}$$

If we plot current versus voltage for Eq. (6.12), as in Fig. 6.42, we observe a straight line with a voltage intercept of V_{CC}, a current intercept of V_{CC}/R_C, and a slope of $-1/R_C$. This line, known as the load line, represents the output circuit external to the transistor.

Figure 6.43 repeats Fig. 6.37 with the load line drawn for our specific values of V_{CC} (10 V) and R_C (1 kΩ). The mental break in the output circuit was made for the purpose of examining independently the characteristics of the two parts

Figure 6.41 The current–voltage characteristics of the left and right halves of the output circuit may be considered separately.

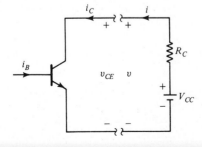

Figure 6.42 The right half of the circuit is characterized by a load line.

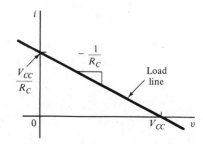

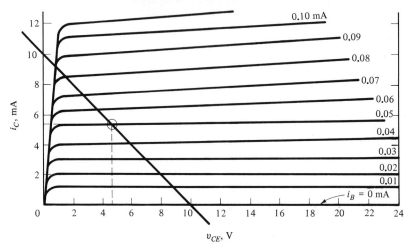

Figure 6.43 Transistor output characteristic with load line.

of the circuit. When we mentally reconnect the circuit, the two voltages must be the same and the two currents must be the same. These requirements determine the current and voltage in the output circuit.

Input–output characteristic. Consider now that the base current is 50 μA. According to the transistor output characteristics, the transistor must operate on the line corresponding to that base current. On the other hand, the external circuit must operate on the load line. Both requirements are satisfied at the intersection of the two lines. For 50-μA base current, the intersection occurs at a collector current of 5.3 mA and a collector–emitter voltage of 4.7 V, as indicated in Fig. 6.43. This result demonstrates the method for finding the output voltage from the input voltage. For each input voltage, we can determine the base current [Eq. (6.11) and Fig. 6.40], and for each value of base current we can determine the output voltage by noting the intersection with the load line. We can thus plot the input–output characteristic of the amplifier–switch. Such a plot is shown in Fig. 6.44.

Several features of Fig. 6.44 merit close attention. When the input voltage is less than $+0.7$ V, no base current flows and hence no collector current flows. The transistor is cut off. This cutoff condition fixes the output voltage at V_{CC},

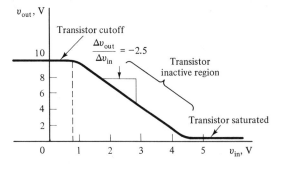

Figure 6.44 Transistor amplifier-switch input-output characteristic.

10 V in this case. Because no current flows in R_C, the full power supply voltage must appear across the transistor. This is like turning off a valve in a water pipe; the full pressure must be supported by the valve in the absence of flow. As the input voltage increases beyond $+0.7$ V, the base current begins to flow and the transistor moves out of cutoff into the active region. In the active region, the transistor acts as an amplifier to small changes in the input voltage. The incremental gain of the amplifier is the slope of the input–output characteristic in the active region. The amplifier in Fig. 6.38 has an incremental voltage gain of

$$A_v = \frac{\Delta v_{\text{out}}}{\Delta v_{\text{in}}} = \frac{-\Delta i_C R_C}{\Delta i_B R_B} = -\beta \frac{R_C}{R_B} = \frac{-100 \times 1 \text{ k}\Omega}{40 \text{ k}\Omega} = -2.5$$

in this case. The minus sign of the gain signifies the inversion of the incremental changes. That is, a small positive change in the input voltage would produce a larger negative change in the output voltage.

As the input voltage continues to increase, the base current eventually will saturate the transistor. This is like opening fully a valve in a water pipe; the valve relinquishes control of the flow rate to the capacity of the supply. With the transistor saturated, the output voltage will remain small, 0.2 to 0.4 V, and the collector current will remain at approximately $V_{CC}/R_C = 10$ mA, even though the base current will continue to increase as v_{in} continues to increase. Thus in the saturation region the output voltage will remain small for increasing values of the input voltage.

6.4.4 Transistor Applications

Operation as a switch. An *ideal* switch is either an open circuit (OFF) or a short circuit (ON). A *real* switch, like a wall switch for the lights, will have a large resistance when OFF and a small resistance when ON. To function properly as a switch, the device must have a large OFF resistance *compared with its load resistance* such that almost all the voltage appears across the switch and very little voltage appears across the load. Likewise in the ON state, the switch must have a small resistance *compared with its load resistance* so that almost all the voltage appears across the load with very little voltage across the switch.

Thus the transistor can function as an electronic switch in the circuit shown in Fig. 6.38 if its OFF resistance is much larger than its load resistor (R_C) and its ON resistance is much smaller than its load resistor.

If the input voltage is less than 0.7 V, the transistor is cutoff, that is, the electronic switch is OFF. If R_C represented a light bulb, no current would flow through the bulb and it would not glow. If the input voltage exceeded about 4.0 V, the transistor would be saturated and our electronic switch would be ON. If R_C were a light bulb, it would glow. Thus we can use the transistor as a voltage-controlled switch to turn the bulb on and off.

The switching action of the transistor is one of its most valuable properties. This is true because digital circuits—computers, calculators, digital watches, and digital instrumentation—utilize transistors in switching operation. Some of the important applications of transistors in digital circuits are explored in Chapters 7 and 9.

Operation as a large-signal amplifier.   An amplifier is called a large-signal amplifier when the signal levels require the full transistor characteristics, from near cutoff to near saturation. Such amplifiers can furnish moderate amounts of power to transducers such as loudspeakers or control motors. Our simple amplifier circuit is limited to an output voltage of 10 V, peak to peak, but this limitation can be overcome by increasing the power supply voltage or the circuit complexity. As we will see in Chapter 8, feedback techniques can be used to reduce distortion and generally improve the characteristics of large-signal amplifiers.

Small-signal amplifiers.   Most amplifiers are small-signal amplifiers. Consider, for example, a radio that receives a signal of 10 mV and produces an output voltage of 10 V. Such a radio would require a voltage gain of 1000. This gain would be accomplished in "stages," each transistor amplifier stage taking as its input the output of the previous stage. If each stage had a voltage gain of $\sqrt{10}$, six stages of amplification would provide the necessary gain. Of these, all but the last stage would be small-signal amplifiers.

A small-signal amplifier must have a means (bias circuit) for placing the transistor in its amplifying region, a means for introducing the input signal, and a means for supplying its output signal to the next stage. The circuit shown in Fig. 6.45 is a simple circuit that accomplishes these functions. This is our basic amplifier with added features to provide bias and coupling at input and output. We have symbolized the power supply connection with the terminal marked $+V_{CC}$, which appears as an open circuit but actually is connected through a dc power supply to ground. The two resistors R_1 and R_2 comprise a voltage divider to supply dc current to the transistor base to bias the transistor to its amplifying region. At the input we have a Thévenin equivalent circuit of the signal source; this might be the previous stage of the amplifier or it might be the origin of the signal such as a microphone or a radio antenna. The load, R_L, represents the input impedance of the next stage of the amplifier. We begin our analysis of the small-signal amplifier by examining the role of the input and output coupling capacitors.

Coupling capacitors.   Large capacitors block dc current but pass ac current. The infinite impedance of the capacitor at dc will allow the dc state of each

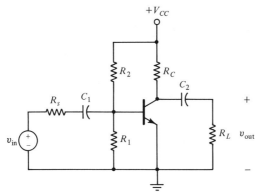

Figure 6.45   Small-signal amplifier.

amplifier stage to be independent of the adjoining stages. If the impedance of the capacitors is small relative to the resistors in the circuit, the time-varying signals will pass through the capacitors undiminished. Thus the stages are isolated for dc but coupled for ac signals by the coupling capacitors at input and output, C_1 and C_2.

Bias voltage divider. A dc current is supplied to the transistor base by the voltage divider, R_1 and R_2. Because the coupling capacitors act as blocks to the dc current, the equivalent circuit at dc is as shown in Fig. 6.46a. We have replaced the $+V_{CC}$ symbol with two voltages sources. We draw the circuit this way for two reasons: to emphasize that the power supply acts independently on the bias and collector circuits, and to help you recognize the voltage divider. The circuit in Fig. 6.46a can be reduced to the familiar circuit in Fig. 6.38 by converting the voltage divider to a Thévenin equivalent circuit, as shown in Fig. 6.46b. We have used the symbol V_{BB} for the open-circuit voltage at the base of the transistor. This would play the role of v_{in} in Fig. 6.38 and Eq. (6.11).

We continue this development for the specific circuit shown in Fig. 6.47. The base bias voltage is

$$V_{BB} = 10 \times \frac{5.6 \text{ k}\Omega}{50 \text{ k}\Omega + 5.6 \text{ k}\Omega} = 1.0 \text{ V}$$

and the value of R_B is

$$R_B = 5.6 \text{ k}\Omega \parallel 50 \text{ k}\Omega = 5.0 \text{ k}\Omega$$

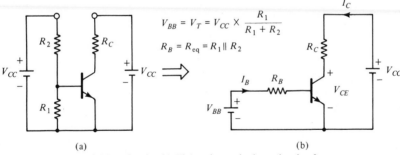

Figure 6.46 (a) DC bias circuit; (b) Thévenin equivalent circuit of input portion of bias circuit.

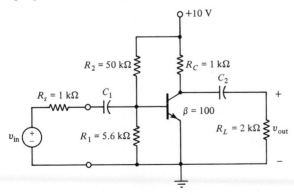

Figure 6.47 Small-signal amplifier. The capacitors are assumed to have negligible impedance to the signal.

Assuming a silicon transistor requiring $V_{BE} = 0.7$ V to turn ON the base–emitter junction, we can use the equivalent circuit in Fig. 6.35. Thus the dc base current is

$$I_B = \frac{V_{BB} - 0.7}{R_B} = \frac{1.0 - 0.7}{5.0 \text{ k}\Omega} = 60 \text{ }\mu\text{A}$$

Assuming the transistor to be in the active region, we calculate the collector dc current to be

$$I_C = \beta I_B = 100 \times 60 \text{ }\mu\text{A} = 6.0 \text{ mA}$$

To confirm that the transistor is in the active region (not saturated), we must determine the collector–emitter voltage, V_{CE}. We could use a load line, but KVL around the collector–emitter loop in Fig. 6.46b will do as well.

$$-V_{CC} + I_C R_C + V_{CE} = 0 \Rightarrow V_{CE} = 10 - (6.0 \text{ mA})(1 \text{ k}\Omega) = 4.0 \text{ V}$$

Thus the transistor is operating near the middle of its active region. If there were no input signal, the transistor would have a steady voltage of $+4.0$ V across it, but the voltage across the 2-kΩ load resistor would remain zero because the dc voltage would be blocked by the output coupling capacitor, C_2.

Small-signal equivalent circuit. We now consider the effect of an input signal. In general, all voltages and currents in the circuit will have two components, a dc component and a time-varying component, which we will call the signal component. For example, the base current would be

$$i_B(t) = I_B + i_b(t)$$

where $I_B = 60 \text{ }\mu\text{A}$ and $i_b(t)$ is the signal component. What is the equivalent circuit that the signals "see," in the sense that Fig. 6.46a is the circuit seen by the dc voltages and currents?

Beginning at the input and moving toward the output, we will now justify the circuit shown in Fig. 6.48 as the appropriate small-signal equivalent circuit. The input source and its resistance, R_s, are now coupled to the amplifier with a short circuit representing the negligible impedance of C_1 to the signal. The base bias resistor R_1 appears as expected, but R_2 now connects to the *signal ground*

Figure 6.48 Small-signal equivalent circuit.

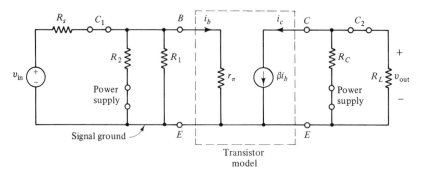

Introduction to Electronics   Chap. 6

because of the low impedance of the power supply to the signal. That is, the power supply circuit, not shown but represented by the V_{CC} symbol, would be connected to the circuit ground by a filter capacitor, as shown for example in Fig. 6.12. This filter capacitor not only filters the power supply output but also provides for the signal a low-impedance path to ground.

The input to the transistor is represented by a resistor, r_π, and the output by a dependent current source. Representing the base–emitter by a constant 0.7 V, as shown in Fig. 6.35, is adequate for the dc solution but is not accurate for the signal: r_π is a relatively low resistance in the range 300 to 1000 Ω that may be determined from theory (see Problem P6.20), from measured transistor input characteristics, or from published specifications of the transistor. The dependent current source, $i_c = \beta i_b$, represents the current gain of the transistor.

In the output circuit, the collector bias resistor, R_C, is shown connected to the signal ground for the reason given above for R_2. Finally, the collector is connected to the load resistor, R_L, through the output coupling capacitor, which acts as a short circuit to the signal.

Small-signal voltage gain. The small-signal equivalent circuit in Fig. 6.48 may be analyzed for the voltage gain, $A_v = v_{out}/v_{in}$, by the standard methods of circuit theory. Our analysis begins at the input with the calculation of the base signal current. From the base current we can calculate the collector signal current, then the output voltage.

The base current is easily calculated if the input source is converted to a Norton circuit, as shown in Fig. 6.49. We thus have a four-way current divider. The base current is

$$i_b = \frac{v_{in}}{R_s} \times \frac{R_s \parallel R_2 \parallel R_1 \parallel r_\pi}{r_\pi} \tag{6.13}$$

For the values in Fig. 6.47 and $r_\pi = 740\ \Omega$, Eq. (6.13) leads to $i_b = v_{in}/1.89\ \text{k}\Omega$.

The collector current is βi_b and the output voltage is thus

$$v_{out} = -\beta i_b(R_c \parallel R_L) = -100\, \frac{v_{in}}{1.89\ \text{k}\Omega}\, (1\ \text{k}\Omega \parallel 2\ \text{k}\Omega)$$

Thus the voltage gain of the amplifier is $A_v = v_{out}/v_{in} = -35.3$. As stated above, the minus sign represents inversion of the signal.

Summary. The calculation of the small-signal gain of the amplifier requires analysis of a dc equivalent circuit and a small-signal equivalent circuit. In the

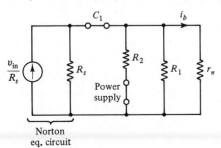

Figure 6.49 Analysis of the input circuit is simplified through use of a source transformation.

latter, the transistor is represented by its input resistance and a dependent current source. Because of the unilateral property of the transistor, the analysis proceeds from input to output.

6.4.5 Check Your Understanding

1. For a transistor, the conditions in the output (C–E) circuit have little influence on the input (B–E) circuit. (T/F?)
2. What type of semiconductor material is the base region for an *npn* transistor?
3. The input (B–E) voltage–current characteristic of a transistor is similar to that of a resistor, an ideal diode, a *pn*-junction diode, or an open circuit?
4. The load line intersects the current axis at the current that would flow if the transistor were replaced by a short circuit. (T/F?)
5. In the saturation region, the transistor acts as a switch that is open or closed?
6. To be used as an amplifier, the transistor must be biased into its cutoff, active, breakdown, or saturation region?
7. If the voltage gain of a transistor amplifier is negative, it means that the signal is diminished by the amplifier. (T/F?)
8. To increase the gain of a transistor amplifier, we should (a) increase the power supply voltage, (b) increase the collector resistor, (c) increase the base resistor, (d) get a transistor with a larger β. (More than one answer may be correct.)

Answers. 1. True; 2. *p*-type; 3. a *pn*-junction diode; 4. true; 5. closed; 6. active; 7. false; 8. b and d.

6.5 FIELD-EFFECT TRANSISTORS

The bipolar junction transistor (BJT) was invented in 1947 and developed to usable form in the late 1950s. The field-effect transistor (FET) was also invented in the late 1940s but was made practical by manufacturing developments in the 1970s. Today virtually all digital electronic systems such as watches and computers use integrated circuits of FETs operating as electronic switches.

Like the BJT, the FET comes in two polarity types: the *n*-channel corresponds to the *npn* and the *p*-channel corresponds to the *pnp*. We will limit our study to *n*-channel devices. Unlike the BJT, FETs come in two varieties: the junction FET (JFET) and the metal oxide semiconductor FET (MOSFET). We will consider the JFET in detail and then deal briefly with the MOSFET because its properties are similar.

6.5.1 Junction Field-effect Transistors

The physical construction of an *n*-channel JFET is shown in Fig. 6.50a. The transistor has three terminals, gate (G), drain (D), and source (S). The input (controlling) signal is the voltage applied between gate and source, v_{GS}, and the output (controlled) signal is the current from drain to source, i_D. The gate is connected to *p*-type semiconductor, and the drain and source are connected to the channel, which is *n*-type material. For proper operation, the *pn* junction be-

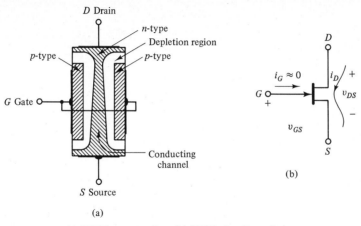

Figure 6.50 (a) JFET construction; (b) JFET circuit symbol.

tween the gate and channel regions must be reverse biased so that a depletion region is formed, as shown. The depletion region acts as a nonconductor, but the remainder of the channel acts as a conductor and thus forms a "resistor" with peculiar properties. The width of the depletion region and thus the width of the conducting channel depend on the gate voltage. The drain is operated at a positive voltage relative to the source, and hence the gate–drain end of the *pn* junction is more strongly reverse biased than the gate–source end. For this reason, we have shown the conducting channel to be more narrow near the drain. The circuit symbol for the JFET is shown in Fig. 6.50b. Notice that the arrowhead points from *p* to *n*, as for a diode.

Input characteristics. Because the *pn* junction between gate and channel is reverse biased, very little current flows into the gate. Thus we show i_G to be approximately zero in Fig. 6.50b, and the drain current flows through the channel and out the source connection. The gate controls the current in the channel through an electric field that affects the depletion region. Because very little current flows into the gate, the input impedance to the device is extremely high, up to 10^{12} Ω, and very little energy is required to control the device.

Output characteristics. Like the BJT, the FET has several regions of operation, which we may examine by fixing the input (v_{GS}) voltage and observing the output current as we vary the output voltage (v_{DS}). Figure 6.51a shows the circuit for the experiment, and Fig. 6.51b shows the results. Consider first the top curve, where $v_{GS} = 0$ V. For small values of drain–source voltage, the current increases as for a resistor. This is the *ohmic region*. However, as v_{GS} increases, the current begins to level off because the channel narrows at the gate–drain end of the device. At the negative of the *pinchoff* voltage $(-V_P)$, the conducting channel reaches a minimum size and the current becomes constant, independent of further increases in v_{GS}. This is the *saturation region*. (*Warning:* This "saturation region" is totally different from the saturation region for the BJT. Using the same term for both is unfortunate and confusing, but customary.) In the sat-

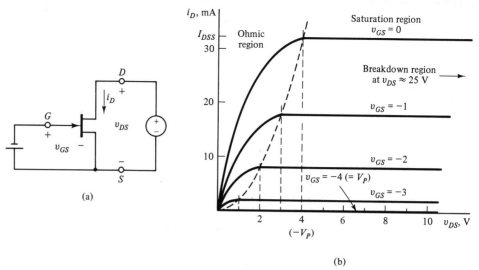

Figure 6.51 (a) Circuit for determining the output characteristic; (b) output characteristics for $V_P = -4$ V and $I_{DSS} = 32$ mA.

uration region, the drain current is controlled totally by the input signal, and thus the saturation region for the FET corresponds to the active region for the BJT. Off the graph, to the right, is a breakdown region, where the current increases rapidly.

For smaller values of v_{GS} (negative values), currents fall below the curve corresponding to $v_{GS} = 0$, and the device changes from the ohmic region to the saturation region at lower values of v_{DS}. For $v_{GS} < V_P$ the FET is *cut off*, meaning that no current flows, regardless of v_{DS}.

In the ohmic region, the drain current is given by

$$i_D = \frac{2I_{DSS}}{V_P^2}\left[(v_{GS} - V_P)v_{DS} - \frac{v_{DS}^2}{2}\right],\qquad (6.14)$$

$$0 < v_{DS} < (v_{GS} - V_P), \quad v_{GS} > V_P$$

where I_{DSS} is the saturation current for $v_{GS} = 0$. Equation (6.14) describes a parabola passing through the origin and tangent to the point of pinchoff. After pinchoff, the current is constant at the value

$$i_D = \frac{I_{DSS}}{V_P^2}(v_{GS} - V_P)^2, \qquad v_{GS} > V_P, \quad v_{DS} > v_{GS} - V_P \qquad (6.15)$$

Summary. The JFET has four regions of operation. In the cutoff region, no current flows. In the ohmic region, the device behaves like a nonlinear resistor, with the resistance controlled mainly by the gate–source voltage. In the saturation region, the device behaves as a current source, with the current controlled by the gate–source voltage. Finally, in the breakdown region, the drain current increases rapidly. With the exception of the breakdown region, which is to be avoided, all the regions are useful.

Determining the region of operation. The region of operation for a JFET can be determined by the simple rules given in Table 6.1. We consider that the transistor has two ends: a gate–source end and a gate–drain end. The ends can be ON or OFF, depending on the voltage between the gate and the source or drain. For example, in the device shown in Fig. 6.51b, the pinchoff voltage is -4 V. Therefore, if the gate–source voltage is greater than -4 V, the gate–source end of the transistor will be ON, and if the gate–source voltage is less than (more negative than) -4 V, the gate–source end of the transistor will be OFF. Likewise, if the gate–drain voltage is greater than -4 V, the gate–drain end of the transistor will be ON; otherwise, it is OFF. With these definitions of ON and OFF, the transistor has the four states shown in Table 6.1.

We have already discussed rows 1, 3, and 4 in Table 6.1. Row 2 deals with a condition not previously discussed. For the gate–source end to be OFF and the gate–drain end to be ON, the drain voltage must be negative relative to the source voltage; and hence source and drain exchange roles. In this case, the current will flow from source to drain (if we keep the same labels). This condition does not occur in the normal operation of the transistor.

TABLE 6.1

Gate–Source End	Gate–Drain End	Condition
OFF	OFF	Cutoff region
OFF	ON	Reverse saturation
ON	OFF	Saturation region
ON	ON	Ohmic region

Example. An n-channel JFET has a pinchoff voltage of -3 V. The gate–source voltage is -1 V and the drain–source voltage is $+2$ V. What is the region of operation? *Answer:* The gate–source voltage is greater than the pinchoff voltage so the gate–source end is ON. The gate–drain voltage may be determined from KVL and the law of subscripts:

$$v_{GD} = v_{GS} + v_{SD} = v_{GS} - v_{DS} = -1 - (+2) = -3 \text{ V} \qquad (6.16)$$

This is equal to the pinchoff voltage, and hence the gate–drain end of the transistor is borderline between ON and OFF. Hence the transistor is on the boundary between the ohmic region (row 4) and saturation (row 3).

JFET applications: variable resistance. In the ohmic region, the JFET acts as a resistor between drain and source whose resistance is controlled by v_{GS}. For small signals, the resistance can be determined from Eq. (6.14):

$$r_{DS} \approx \frac{1}{\partial i_D / \partial v_{DS}} = \frac{V_P^2}{2I_{DSS}(v_{GS} - V_P)}, \qquad v_{DS} \ll v_{GS} - V_P \qquad (6.17)$$

Hence the resistance is smallest for large v_{GS} and becomes infinite when $v_{GS} = V_P$. For example, for the transistor characteristics shown in Fig. 6.51a, the resistance with $v_{GS} = 0$ is about 63 Ω.

JFET applications: small-signal amplifier.   Like the bipolar junction transistor, the JFET may be used as a small-signal amplifier. In this application the JFET is biased into its saturation region, and small changes in v_{GS} produce changes in the drain current that, passed through a resistor in the drain circuit, produce an amplified signal. To represent the JFET for this application, we require a small-signal model for the transistor (Fig. 6.52). The voltage and current variables are signal components and must remain much smaller than the dc bias voltages and current, respectively.

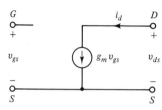

Figure 6.52   Small-signal model for the JFET. The voltages and current are signal components of the total voltages and current. The gain-related parameter g_m is the mutual conductance.

The gate circuit is shown as an open circuit because the gate draws no current. The drain–source circuit is shown as a voltage-controlled current source. The mutual conductance g_m may be derived from Eq. (6.15):

$$g_m = \frac{\partial i_D}{\partial v_{GS}} = \frac{2 I_{DSS}}{V_P^2} (v_{GS} - V_P) \big|_{v_{GS} = V_{GS}} \tag{6.18}$$

where V_{GS} is the dc bias voltage between gate and source. The use of Eq. (6.18) and the small-signal equivalent circuit will be illustrated after we discuss use of the JFET as a switch.

JFET applications: switch.   Figure 6.53a shows a circuit that uses a JFET as a voltage-controlled switch. Figure 6.53b shows the load line. With v_{GS} less than -4 V, the transistor is cut off and the output voltage is $+10$ V. With v_{GS}

Figure 6.53   (a) In this circuit, the JFET can operate as a switch; (b) by changing the input voltage from -4 V to 0 V, the JFET can be changed from OFF to ON and the output voltage changed from 10 V to near zero.

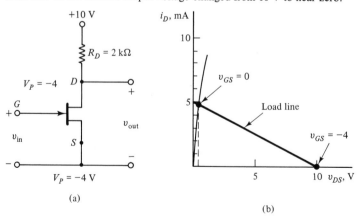

= 0 V, the transistor is in the ohmic region and, as shown, has a resistance of about 63 Ω. In series with 2 kΩ, this drops the output voltage to about 0.3 V. Thus the JFET acts as a voltage-controlled switch.

JFET amplifier. We will determine the small-signal gain of the amplifier shown in Fig. 6.54. The transistor has the output characteristics shown in Fig. 6.51b. Our first task is to determine the dc conditions in the circuit because the mutual conductance g_m depends on the dc conditions. To establish the bias condition, we must solve simultaneously for the dc gate–source voltage and the drain current. The source is not at ground voltage but is biased to a positive voltage by the current passing through the 250-Ω resistor. The gate is grounded through the 1-MΩ resistor, but since no current flows in this resistor the gate is at ground voltage. Consequently, the gate–source voltage is

$$v_{GS} = v_G - v_S = 0 - 0.250i_D \tag{6.19}$$

where current is expressed in milliamperes (mA) and the reference node is the power ground. We assume the JFET is saturated and hence we obtain another equation relating the gate–source voltage and the drain current from Eq. (6.15).

$$i_D = \frac{I_{DSS}}{V_P^2}(v_{GS} - V_P)^2 = \frac{32}{(-4)^2}[v_{GS} - (-4)]^2 \tag{6.20}$$

When we eliminate v_{GS} between Eqs. (6.19) and (6.20), we obtain a quadratic equation for the drain current. The two solutions are i_D = 8.0 and 32.0 mA. The second value is unrealistic for several reasons (see Problem P6.28). The I_D = 8.0 mA can be accepted tentatively, but we must verify our assumption that the transistor is in the saturation region. With a drain current of 8 mA, the source voltage is +2 V and hence v_{GS} = −2 V. The gate–source end of the transistor is ON, as required for saturation. The drain voltage can be obtained from KVL as +18 − 8 mA × 1.5 kΩ = +6 V. Hence the gate–drain voltage is −6 V and the gate–drain end of the transistor is OFF. Therefore, the transistor is operating in row 3 of Table 6.1, in the saturation region as assumed.

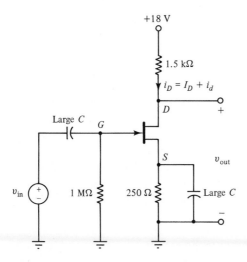

Figure 6.54 A JFET amplifier. The capacitors block dc but act as short circuits for the signal.

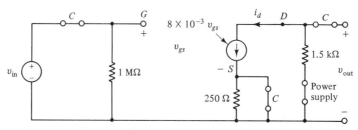

Figure 6.55 Small-signal equivalent circuit. The transistor has been replaced by its small-signal model, and the capacitors and the power supply have been replaced by short circuits.

Figure 6.55 shows the small-signal equivalent circuit for the amplifier in Fig. 6.54. Because the capacitors have a small impedance at the signal frequencies, we have replaced them with short circuits. The power supply also has a small impedance to the signals, and hence we have connected one end of the 1.5-kΩ resistor to the signal ground. The transistor had been replaced by its equivalent circuit from Fig. 6.52. The mutual conductance of 8×10^{-3} mhos has been determined from Eq. (6.18) for the dc operating conditions established above.

The analysis of the circuit proceeds from input to output. The gate–source signal voltage is equal to the input voltage because the source is connected to the signal ground. Thus the signal current in the drain–source circuit is $8 \times 10^{-3} v_{in}$. This current flows upward through the 1.5-kΩ resistor, and hence the output voltage is

$$v_{out} = -8 \times 10^{-3} v_{in} \times 1.5 \times 10^3 = -12.0 \, v_{in} \qquad (6.21)$$

Thus the voltage gain of the amplifier is -12.0, as before the minus sign indicates inversion of the signal.

Summary. We have in this section investigated the physical construction, input and output characteristics, and equivalent circuits appropriate to the various regions of operation of the n-channel JFET. We have investigated its applications as a switch, a voltage-controlled resistor, or a small-signal amplifier. In the next section we introduce briefly a class of devices very similar to the JFETs.

6.5.2 Metal Oxide Semiconductor Field-effect Transistors (MOSFETs)

Physical construction and operating principles. Figure 6.56a shows the physical structure of an n-channel depletion-mode MOSFET. The substrate of p material is maintained at a voltage equal to or less than any voltage anticipated at either source or drain so that a depletion region forms at the pn junction and isolates the channel from the substrate. Normally, the drain is operated positive relative to the source, and hence the substrate can be connected to the source to accomplish this purpose. Note that the arrow in the substrate connection points from p to n.

The gate is separated from the channel by an insulating layer and hence no current flows into the gate. If the gate has a large negative voltage relative to the

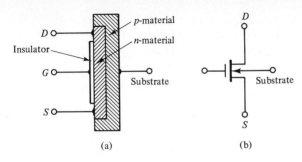

Figure 6.56 (a) Physical construction of an *n*-channel depletion-mode MOSFET. The substrate connection is kept at a dc voltage that is negative relative to all anticipated voltages at the source and drain. (b) The gate is shown insulated from the channel.

channel, the carrier electrons are driven from the channel, it becomes a nonconductor, and the channel is cut off. Like the JFET, there exists a critical voltage to cut off the channel. This is called the threshold voltage, V_T, and has the same effect as the pinchoff voltage for the JFET. Indeed, the depletion-mode MOSFET has similar characteristics to the JFET, except that the gate–source voltage can be positive. It has regions of cutoff, saturation, and ohmic operation as described by Table 6.1. For example, if a depletion-mode MOSFET has a threshold voltage of -5 V, the gate–source voltage is -2 V, and the drain–source voltage is $+2$ V, then both ends of the device are ON (voltage is greater than the threshold) and hence operation falls in the ohmic region (row 4 of Table 6.1).

The depletion-mode MOSFET differs from the JFET in its physical construction and operating principles, but otherwise it operates similarly. Indeed, the equations describing operation are identical to Eqs. (6.14) and (6.15) except for a change in notation and the permitting of positive gate–source voltage. It can operate as a voltage-controlled resistor, small-signal amplifier, or switch.

Enhancement-mode MOSFET.

Figure 6.57a shows the physical structure of an enhancement-mode MOSFET, which differs from the depletion-mode device by having no *n*-type channel. For proper operation, the substrate must be connected to a voltage more negative than any voltage anticipated at either source or drain.

Both the drain–substrate and source–substrate *pn* junctions are reverse biased, and no channel exists until external voltage is applied to the gate. Thus the device is cut off if the gate–source voltage is less than a positive threshold voltage. When the gate–source voltage exceeds the threshold voltage (V_T), the resulting electric field attracts electrons and repels holes in the *p*-type substrate, and a channel forms. The circuit symbol in Fig. 6.57b shows a broken line for a

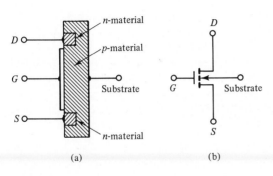

Figure 6.57 (a) Physical structure of an enhancement-mode MOSFET. No channel exists until the gate–source voltage is large enough to create a channel out of the *p*-type substrate. (b) The circuit symbol for an enhancement-mode MOSFET. The broken line for the "channel" indicates that a channel does not exist naturally but must be created by the electric field between the gate and source.

"channel" to indicate that a channel does not exist naturally. Note that the arrow on the substrate connection points from p to n, where n refers to the n channel formed by enhancement. The depletion- and enhancement-mode devices have identical characteristics except V_T is negative for the former and positive for the latter. Equations describing operation in the ohmic and saturation regions of both types of MOSFETs are identical to those of the JFET except for a change in notation. In the ohmic region, the drain current is given by

$$i_D = \frac{K}{(V_T)^2}\left[(v_{GS} - V_T)v_{DS} - \frac{v_{DS}^2}{2}\right], \qquad v_{GS} > V_T, \quad v_{DS} < v_{GS} - V_T \quad (6.22)$$

where K is a constant, and in the saturation region we have

$$i_D = \frac{K}{2(V_T)^2}(v_{GS} - V_T)^2 \qquad\qquad (6.23)$$

For $K = 16\,\text{mA}$ and $V_T = +4\,\text{V}$ (enhancement mode), we have the characteristics shown in Fig. 6.58.

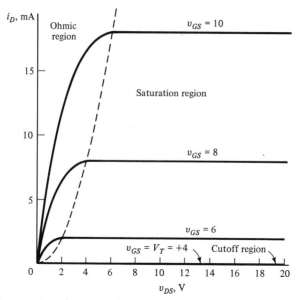

Figure 6.58 Device characteristics for an enhancement-mode n-channel MOSFET.

Cutoff, ohmic, and saturation regions. In the cutoff region, the drain–source circuit acts as an open circuit or a switch that is open. In the ohmic region, the enhancement-mode MOSFET acts as a voltage-controlled resistor, and in the saturation region it acts as a voltage-controlled current source. The region of operation may be determined from Table 6.1. For example, consider the top curve given in Fig. 6.58. With $v_{GS} = +10$ V, the source end of the transistor is ON because the gate–source voltage exceeds the threshold voltage of $+4$ V. If the gate–drain voltage is less than $+4$ V, then the gate–drain end of the transistor will be OFF and the transistor will be in the saturation region. This occurs for $v_{DS} > 10 - 4 = 6$ V. Hence the transistor will be in the saturation region for drain–source voltages greater than 6 V and in the ohmic region for voltages less than 6 V.

Circuit models for operation in the ohmic and saturation regions. In the ohmic region, the transistor may be modeled as a voltage-controlled resistance. We may determine the transistor resistance in the ohmic region from Eq. (6.22) in the manner used to derive Eq. (6.17). For small-signal operation in the saturation region, the circuit model shown in Fig. 6.52 is valid. We may determine the mutual conductance for the model from Eq. (6.23) in the same manner as Eq. (6.18) was derived.

Summary. The JFET and both types of MOSFETs have similar characteristics. The main difference between them is that the enhancement-mode MOSFET has a positive threshold voltage and the JFET and depletion-mode MOSFETs have a negative threshold voltage (called the pinchoff voltage for the JFET). Operation in the ohmic or saturation region may in all cases be determined from Table 6.1. These FETs can function as switches, voltage-controlled resistors, or small-signal amplifiers.

Circuit designers also have available p-channel JFETs and MOSFETs. These devices have identical characteristics except that all polarities of voltages and currents are reversed.

6.5.3 Check Your Understanding

1. Why is the channel shown as a dashed line for the enhancement-mode MOSFET, yet shown as a solid line for the depletion-mode device?
2. For a JFET with $V_P = -5$ V and $v_{GS} = -3$ V, what range of drain–source voltage corresponds to operation in the ohmic region? (Assume $v_{DS} > 0$ V).
3. For the device described in the previous problem and $I_{DSS} = 20$ mA, determine the resistance of the device for small values of drain–source voltage.
4. An n-channel MOSFET has a threshold voltage of -3 V. Is this a depletion- or enhancement-mode MOSFET?
5. For the device described in the previous problem, assume the gate–source voltage is -1.5 V. What range of drain–source voltages corresponds to operation in the saturation region (positive voltages only)?

Answers. 1. Because the channel exists naturally for the depletion-mode device but must be created by the gate voltage for the enhancement-mode device; 2. $0 < v_{DS} < +2$ V; 3. 313 Ω; 4. depletion mode; 5. $v_{DS} > 1.5$ V.

Section 6.2: Rectifiers and Power Supplies

P6.1. Figure P6.1 shows a sinusoidal voltage source, a diode, and a load resistor, R.

(a) A dc voltmeter measures 10 V when connected between A and G. What would an ac voltmeter measure between B and G? Assume an ideal diode.

(b) If a very large capacitor were connected between A and G, what then would the dc meter read? (Assume that $RC \gg T$, where T is the period of the sinusoid.)

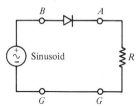

Figure P6.1

P6.2. Using the unfiltered half-wave rectifier circuit shown in Fig. 6.4, design a power supply to produce 35 mA of dc current into a 500-Ω load. Give the circuit and the value that an ac voltmeter would indicate for the input voltage.

P6.3. (a) Derive the equations for the dc voltage and ripple voltage for the full-wave rectifier shown in Fig. 6.6a with a capacitor across the load to reduce the ripple. Assume that the time constant of the RC part of the circuit is large compared with the period of the input sinusoid. Have f continue to represent the frequency of the input.

(b) Using your results from part (a), design a 60-Hz filtered power supply to provide 50 V dc, 50 mA dc to a load, with a ripple voltage of 0.5 V. You must specify the rms value of the input voltage and the value of the capacitor to be used. Assume ideal diodes. *Hint:* The dc voltage and current define an equivalent resistance for the load.

P6.4. Draw the circuit of a filtered half-wave rectifier that will deliver 20 V dc and 27 mA dc to a resistive load. Specify the value of all components, including the equivalent load resistor. Assume an ideal diode. The peak-to-peak ripple should not exceed 0.5 V. Also specify the rms input voltage (60 Hz). See hint in the previous problem.

P6.5. (a) Draw the circuit of a half-wave rectifier, with a generator, an ideal diode, and a 10-kΩ load.

(b) Consider now that the generator puts out the square wave shown in Fig. P6.5. What would be the dc current in the load?

(c) Now add a 10-μF capacitor across the load. What would in this case be the current in the load?

Answers. (b) 5.0 mA; (c) 9.76 mA

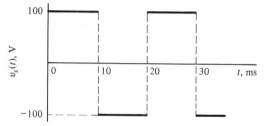

Figure P6.5

P6.6. A simple rectifier circuit has the unsymmetric square waveform for an input as shown in Fig. P6.6. What is the average load voltage, with the polarity markings shown? Assume an ideal diode.

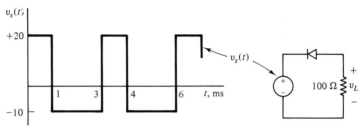

Figure P6.6

P6.7. For the circuit shown in Fig. P6.7, assume ideal diodes.
  (a) Is the circuit a half- or full-wave rectifier?
  (b) Is the circuit a bridge or center-tapped rectifier?
  (c) What is the maximum load voltage?
  (d) What is the minimum load voltage?
  (e) What is the dc load voltage?

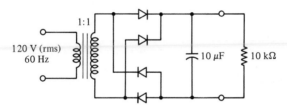

Figure P6.7

P6.8. In Fig. P6.8, the load requires 10 W of dc power. The capacitor has a value of 600 μF. Assuming an ideal diode, find:
  (a) DC load voltage
  (b) DC load current
  (c) Ripple voltage at the load
  (d) Maximum reverse voltage across the diode

Answers. (a) 168.9 V; (b) 58.9 mA; (c) 1.64 V; (d) 339 V

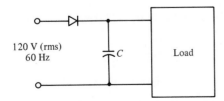

Figure P6.8

P6.9. For the power-supply circuit shown in Fig. P6.9, an ac voltmeter measures 60 V between A and B. Find:
  (a) Maximum load voltage
  (b) Minimum load voltage
  (c) DC load current
  (d) DC voltage across the load if C were removed.

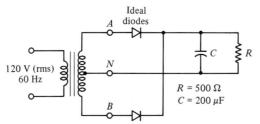

Figure P6.9

P6.10. The power-supply circuit shown in Fig. P6.10 converts 120-V (rms), 60-Hz voltage to 17 V dc, with 15 mA of dc current. Find:
  (a) Turns ratio of the transformer
  (b) Equivalent load resistance
  (c) Peak-to-peak ripple voltage

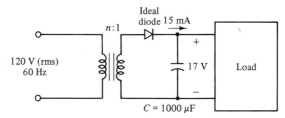

Figure P6.10

Section 6.3: The *pn*-Junction Diode

P6.11. **(a)** A *pn*-junction diode has a characteristic well described by Eq. (6.7). Plot the i–v characteristic in the region $0 < i_D < 100$ mA, assuming $I_0 = 10^{-10}$ A and an ideality factor of 1.4. Plot on a linear scale in both i_D and v_D.
  (b) If the diode in part (a) were placed in series with a 10-V source and a 250-Ω resistor, with the battery polarity such as to forward bias the diode, what current would flow? You may use trial and error or an analytic technique.
  (c) What current would flow in part (b) if the battery were reversed so as to reverse bias the diode?

P6.12. Figure P6.12 shows a piece of silicon that is doped with donor and acceptor atoms to form a *pn*-junction diode. Assume an ideality factor of 1.4.
  (a) Where does there exist (1) more carrier electrons than holes; (2) more holes than carrier electrons; and (3) relatively few carriers?
  (b) If you measured $i_D = 10$ mA of current for $v_D = 0.75$ V, what would you expect for $v_D = 0.80$ V? Assume a temperature of 290 K.
  (c) What current would you expect for $v_D = -0.3$ V.

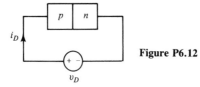

Figure P6.12

P6.13. (a) The usual symbol for a *pn*-junction diode is shown in Fig. P6.13 together with a corresponding piece of silicon. Indicate which half is *p*-type material and which *n*-type material. Which contains the donor and which the acceptor atoms?

(b) Indicate with an arrow the direction in which holes cross the junction due to diffusion (thermal motion) when the diode is ON.

(c) With the diode reverse biased, the current is determined to be 10^{-11} A. How much voltage would it take to turn ON the diode to a current of 1 mA? Assume an ideality factor of 1.7.

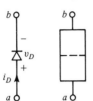

Figure P6.13

Section 6.4: Bipolar Junction Transistor Operation

P6.14. For the transistor output characteristics shown in Fig. P6.14, determine the approximate value of β in the active region. What value of α does this imply? If the collector current is saturated at a value of 15 mA, what is the collector–emitter voltage that results, and what is the base current required to saturate the transistor at this value of collector current?

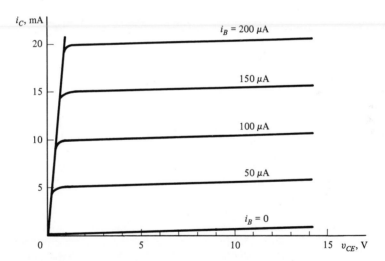

Figure P6.14

P6.15. For the transistor amplifier shown in Fig. 6.38, change $V_{CC} = 12$ V and $R_B = 15$ kΩ.

(a) Draw the new load line on the transistor characteristics in Fig. 6.43.

(b) Find the input and output voltages for the following base currents: 0, 20, 40, 60, 80, and 100 μA.

(c) From part (b), plot v_{out} versus v_{in}.

(d) Find the incremental gain from the slope, $A_v = \Delta v_{out}/\Delta v_{in}$.

P6.16. A simple transistor circuit is shown in Fig. P6.16. The transistor output characteristics are shown in Fig. P6.14.
    (a) Draw the load line on the characteristics. Find the required current into the base to give a collector–emitter voltage of 4 V.
    (b) What value of base voltage, V_{BB}, is required to give this amount of base current? Assume 0.7 V between base and emitter.
    (c) What value of V_{BB} is required to saturate the transistor for this circuit? What is the collector current at saturation?

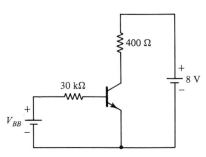

Figure P6.16

P6.17. A transistor amplifier–switch circuit in Fig. P6.17a has a load line as shown on the transistor characteristics in Fig. P6.17b. Assume that $v_{BE} = 0.7$ V when ON.
    (a) What are V_{CC}, R_C, and β for the transistor?
    (b) What value of v_{in} is required to saturate the transistor?
    (c) What is the output voltage with the transistor saturated?

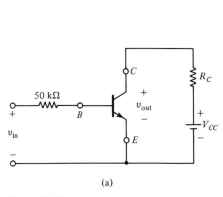

(a)

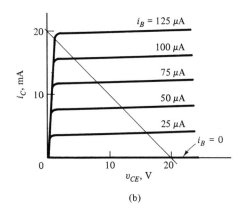

(b)

Figure P6.17

P6.18. A transistor circuit is shown in Fig. P6.18. Transistor characteristics are those shown in Fig. P6.14. Assume 0.7 V for the base–emitter junction when ON.
    (a) Draw the load line. What value of R_B will give a dc collector current of 8 mA?
    (b) Find the voltage between collector and emitter for this value of R_B.

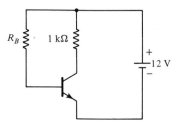

Figure P6.18

P6.19. For the transistor amplifier shown in Fig. P6.19:

(a) What is the collector–emitter voltage (v_{CE}) if the transistor is cut off?

(b) What is the collector current (i_C) if v_{CE} = 10 V?

(c) What is the minimum base current required to saturate the transistor?

(d) What is the minimum value of v_{in} to place the transistor in the active region?

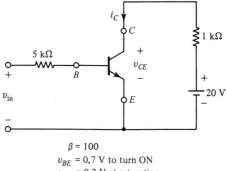

β = 100

v_{BE} = 0.7 V to turn ON

v_{CE} = 0.3 V at saturation

Figure P6.19

P6.20. The input characteristic of a transistor is that of a *pn* junction, Eq. (6.7). For a transistor, we would associate the diode current with the base current, i_B, and the diode voltage with the base–emitter voltage, v_{BE}. The appropriate input resistance for the transistor in a small-signal equivalent circuit would be the slope of the input characteristic at the dc base current level: $r_\pi = dv_{BE}/di_B$ at $i_B = I_B$. Using Eq. (6.7) and ignoring the +1 term in the ln term, derive an expression for the input resistance to the transistor base.

Answers. $r_\pi = \eta 26$ mV/I_B at 300 K. Also, evaluate with the result r_π for the circuit in Fig. 6.47 and confirm that a value of 740 Ω is reasonable.

P6.21. Figure P6.21a shows an amplifier–switch circuit and Fig. P6.21b shows the output characteristics of the transistor. The input voltage consists of a dc and an ac source in series. Assume 0.7 V for the base–emitter junction when ON.

(a) Draw the load line.

(b) Find V_{BB} to give an output voltage (with V_p = 0) of 8 V.

(c) With the value of V_{BB} found in part (b) what is the largest value of V_p that can be amplified without serious distortion?

Answers.   (b) 2.78 V;   (c) $\approx$2 V

Figure P6.21

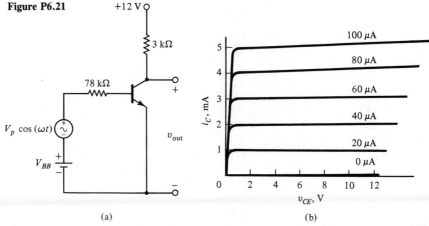

(a)

(b)

P6.22. Figure P6.22a shows a transistor amplifier–switch circuit and Fig. P6.22b the transistor output characteristics.
(a) What is β for the transistor?
(b) Draw the load line.
(c) *Estimate* the base current required to saturate.
(d) Find the collector current for $v_{CE} = 6$ V.
(e) What is the voltage across R_c if the transistor is cut off?

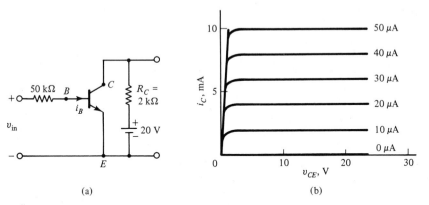

(a)

(b)

Figure P6.22

P6.23. For the transistor in Fig. P6.23, $\beta = 100$, $V_{CE(sat)} = 0.3$ V and $V_{BE(ON)} = 0.7$ V. Find:
(a) Values of v_{CE} and i_C at which the load line will cross the axes
(b) Collector current and base current at saturation
(c) Range (from ? to ?) of input voltages to have the transistor in the active region

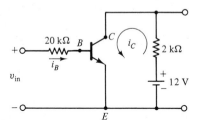

Figure P6.23

P6.24. Figure P6.24 shows a transistor amplifier, biased to operate as a small-signal amplifier.
(a) Draw the dc bias circuit and solve for the base and collector dc currents and the dc voltage from collector to emitter.
(b) Draw the small-signal equivalent circuit and solve for the voltage gain of the amplifier, v_{out}/v_s. Consider the capacitors as short circuits at the signal frequency. Let $r_\pi = 450$ Ω.
(c) Find the input impedance of the amplifier as seen by the input generator. This does not include the 1-kΩ source resistance.

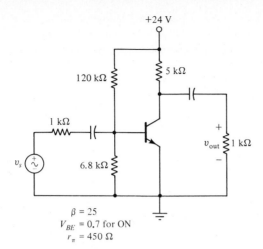

+24 V

120 kΩ

5 kΩ

1 kΩ

v_s

6.8 kΩ

$+$

v_{out} 1 kΩ

$-$

$\beta = 25$
$V_{BE} = 0.7$ for ON
$r_{\pi} = 450\ \Omega$

Figure P6.24

Section 6.5: Field–effect Transistors

P6.25. For a JFET with $V_P = -2$ V and $v_{DS} = 4$ V, what range of gate–source voltage corresponds to operation in the ohmic region? (Assume $v_{GS} < 0$ V).

P6.26. For the JFET amplifier circuit in Fig. 6.54 with characteristics shown in Fig. 6.51b:
   (a) Redesign the circuit to have $i_D = 6$ mA by changing the 250-Ω resistor to a different value. Find the corresponding value of v_{DS}.
   (b) Determine the small-signal gain of the amplifier at the new operating point.

P6.27. For the JFET amplifier circuit in Fig. 6.54, the designer wishes to increase the gain of the amplifier by changing the 1.5-kΩ resistor to a larger value.
   (a) How large a resistor value can be used and still have the transistor in the saturation region?
   (b) What is the small-signal gain with this limiting resistance?

   Answers. (a) 1.75 kΩ; (b) −14.0

P6.28. In solving for the dc conditions for the JFET amplifier circuit in Fig. 6.54, we referred to a quadratic equation derived from Eqs. (6.19) and (6.20).
   (a) Derive the quadratic equation and show that the solution is $I_D = 8.0$ mA and 32.0 mA.
   (b) Show that 32.0 mA is not a realistic solution because it violates the assumed device characteristics, KVL, or both.

P6.29. An n-channel MOSFET has a threshold voltage of -3 V, and the drain–source voltage is $+4$ V. What range of gate–source voltages corresponds to operation in the ohmic region?

P6.30. Show that Eqs. (6.14) and (6.15) match each other in both current and slope at the transition between the ohmic and saturation regions.

P6.31. The MOSFET in Fig. P6.31 has the characteristics described by Eqs. (6.22) and (6.23), with $K = 16$ mA and $V_T = 4$ V, and displayed in Fig. 6.58. Determine the input–output characteristic, v_{out} versus v_{in} for the range $0 < v_{in} < 10$ V. *Hint:* A load line on the characteristics would be a good place to start.

igure P6.31

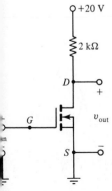

+20 V

2 kΩ

D

$+$

G

v_{out}

S

P6.32. The MOSFET in the small-signal amplifier in Fig. P6.32 has the characteristics described by Eqs. (6.22) and (6.23), with $K = 16$ mA and $V_T = 4$ V, and displayed in Fig. 6.58. The capacitors may be treated as dc open circuits and signal short circuits.

(a) Determine the operating point for the dc drain current and drain–source voltage.

(b) Calculate the mutual conductance for the transistor, g_m, and draw the small-signal equivalent circuit for the amplifier, replacing the transistor with its small-signal equivalent circuit.

(c) Determine the small-signal gain of the amplifier, $v_{\text{out}}/v_{\text{in}}$.

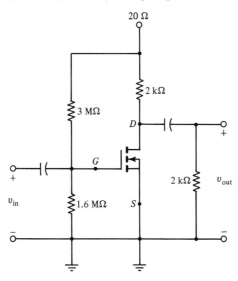

Figure P6.32

General Problems for Chapter 6

P6.33. A three-phase system with a neutral is used in a rectifier configuration as shown in Fig. P6.33. The time-domain line-to-neutral voltages are shown.

(a) If a voltmeter measures 240 V between A and B, what is the peak value of the voltage, V_p?

(b) Sketch the current in the load, $i_L(t)$.

(c) Find the dc current through the load.

Figure P6.33

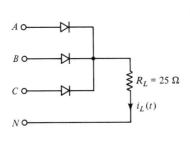

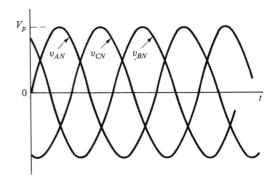

P6.34. A transistor circuit is shown in Fig. P6.34. The β for the transistor is 100. The base–emitter voltage is 0.7 V when ON and the collector–emitter voltage is 0.2 V when the transistor is saturated.

(a) What range of input voltages corresponds to the transistor's being cut off? What is the collector current of the transistor when cut off?

(b) What range of input voltages corresponds to the transistor's being saturated? What is the collector current of the transistor when saturated?

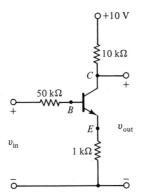

Figure P6.34

P6.35. The circuit shown in Fig. P6.35 will operate as an ac/dc voltmeter, depending on the switch setting. The meter movement responds to dc, 0 to 100 μA full scale.

(a) Explain how the circuit works on dc. Mention what would happen if the input polarity were reversed.

(b) Explain how the circuit works on ac. What would happen in this case if the input leads were reversed?

(c) Determine R_1 so that the dc meter indicates full scale with 100 V dc input.

(d) Determine R_2 for the meter to read full scale with 100 V ac (rms) input. Note that the meter movement registers full scale with 100 μA of dc current through it. Assume ideal diodes and that the capacitor is large.

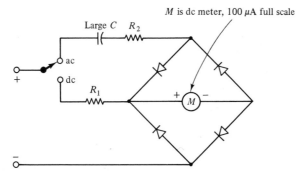

Figure P6.35

P6.36. A calculator "charger," shown in Fig. P6.36, is actually a simple transformer. The remainder of the power supply is located in the calculator, as shown. The battery pack has a voltage ranging from 3.0 V (discharged) to 3.6 V (charged). The current limit for the diode is 50 mA and its ON voltage is 0.7 V. No current should flow into the fully charged battery. Determine the turns ratio required for the "charger" and the minimum value of R.

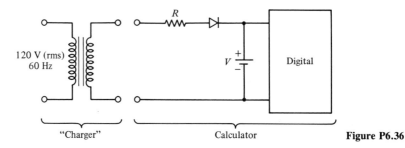

Figure P6.36

P6.37. For Fig. P6.37, find the dc current from a to b if:
  (a) The load is a short circuit.
  (b) The load is a 5-V battery, with + connected to a.
  (c) The load is an ideal diode, with the allowed direction from a to b.

Answers. (a) 0 A; (b) −0.1 A; (c) 45 mA

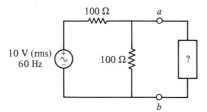

Figure P6.37

P6.38. A transistor circuit is shown in Fig. P6.38a. The transistor characteristics are shown in Fig. P6.38b. Determine R_2 to make the collector–emitter voltage $+5$ V, as shown on the circuit.

Figure P6.38

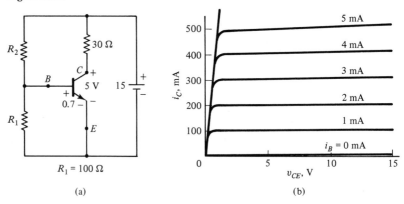

(a)

(b)

Introduction to Electronics   Chap. 6

P6.39. The transistor circuit shown in Fig. P6.39 has a constant voltage of $+5$ V at the base. The voltage at the emitter is 4.3 V because the ON voltage of the base–emitter junction is 0.7 V. The β for the transistor is 75.

(a) Is the transistor in the active, cutoff, or saturation region?

(b) What is the base current if a and b have no load?

(c) If a is shorted to b, what is the current that would flow in the short circuit?

(d) Give a Thévenin or Norton circuit with a and b as the output terminals. *Note:* Although the transistor is a nonlinear device, here it operates in a linear region of its characteristics. Hence we may use an equivalent circuit to represent circuit output characteristics in that linear region.

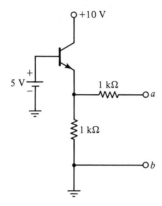

Figure P6.39

7 Digital Electronics

7.1 THE DIGITAL IDEA

7.1.1 What Is a Digital Signal?

A historical example. "Listen my children and you shall hear/Of the midnight ride of Paul Revere. . . ." According to Longfellow's poem, Paul Revere was sent riding through the New England countryside by a signal from the bell tower of the Old North Church in Boston. "One if by land and two if by sea." That is, one light was to be displayed if the British forces were advancing toward Concord by the road from Boston, and two lights were to be displayed if they were crossing the Mystic River to take an indirect route.

The message received by the patriot was coded in digital form. We would say today that two "bits" of information were conveyed by the code.† The first light signaled that the British were advancing by either means; no light indicated that they were not yet advancing. The second light indicated by what route they were coming. Because only two routes of advance were anticipated, this second bit of information could be interpreted as indicating one of the two routes.

Information can be communicated in digital form if a message is capable of being defined by a series of yes/no statements. There can be only two states of each variable used in conveying the information. Reducing information to a series

---

† Strictly speaking, two bits could indicate four possible messages and would require distinguishable lights, say, one red and one white.

of yes/no statements might appear to be a severe limitation, but the method is in fact quite powerful. Numbers can be represented in base 2, and the alphabet by a digital code. Indeed, any situation with a finite number of outcomes can be reduced to a digital code. Specifically, n digital bits can represent 2^n states or possible outcomes. Digital communication takes a well-defined code known to the parties at both ends, as in our historical example.

Analysis of the Revere communication code. To fix further the idea of digital information, we shall define two digital variables that describe the Paul Revere communication system. Let B describe whether the British are coming, and L describe the route by which they are coming, provided that they are coming. The variables, B and L, are unusual mathematical variables because each can have only two values. We may call those two values by any names we wish: yes/no, true/false, one/zero, high/low, black/white. When this type of mathematics was used primarily for analysis of philosophical arguments through symbolic logic, the values of the variables were called true or false, according to the validity of the logical propositions being represented. Recently, the names one/zero have come to be preferred by engineers and programmers dealing with digital codes. These names have the obvious advantage of fitting with the binary (base 2) number system for representation of numerical information. We will use 1/0 or one/zero as the two possible states of the digital variables B and L. Thus the definitions of B and L are

$$B = 1 \qquad \text{if the British are coming}$$
$$B = 0 \qquad \text{if the British are not coming}$$
$$L = 1 \qquad \text{if coming by sea}$$
$$L = 0 \qquad \text{if coming by land}$$

The first variable (B) uniquely determines if he should ride. When the first light appears, he mounts his horse. But he cannot leave until the second light (L) appears (or fails to appear), for the second light reveals the route of the British and hence defines in part the message to be announced.

Representing digital information electrically. In digital electronics, digital variables are represented by logic levels. At any given time, a voltage is expected to have one value or another, or more precisely to lie within one region or another. In a typical system, a voltage between 0 and 0.8 V would be considered a digital zero, a voltage above 2 V would be considered a digital one, and anything between 0.8 and 2 V would be forbidden; that is, if the voltage fell within this range, you would know that the digital equipment needs repair. These definitions are shown in Fig. 7.1.

As an illustration of a digital circuit, we will analyze the BJT amplifier–switch we studied in Chapter 6 as a NOT circuit. The output of a NOT circuit is the digital complement, or the opposite, of the input. First we represent the definition of the NOT circuit with the *truth table* shown in Fig. 7.2: A represents the input, which may be either 1 or 0; B represents the output, which may also be 1

Voltage

Digital one (1)

Forbidden

Digital zero (0)

Figure 7.1 Ranges of voltage represent digital ones and zeros.

Figure 7.2 NOT binary function.

A	B
0	1
1	0

$B = \text{NOT } A = A'$

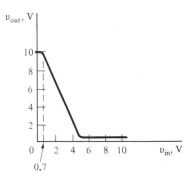

Figure 7.3 Amplifier-switch input-output characteristic.

or 0 but depends on the input. The NOT, or logical complement, operation is indicated algebraically by the equation under the truth table.

We will now define logic levels for the amplifier–switch we studied in Chapter 6 such that it performs the NOT function. The input–output characteristic of the circuit is repeated in Fig. 7.3. Clearly, we wish 10 V to be in the region for a 1 and 0.7 V to be in the region for a 0. That is, if the input were 10 V (digital 1), the output should be less than 0.7 V (digital 0), and vice versa. Hence we might consider making the region for a digital 0 to be from 0 to 1 V, and the region for a digital 1 to be, say, from 8 to 10 V. This will work but leaves insufficient range for a working digital system. That is, it is desirable to broaden the range of values in the regions for 1 and 0 to allow for variations in transistors or power-supply voltage, noise that might get mixed with the signal, and the like. In the present case, we can by trial and error determine that the region 0 to 1.5 V as a digital 0 works well with 5 to 10 V as a digital 1; with these definitions the circuit operates as a NOT circuit, that is, it performs the logical complement. These logic levels are shown in Fig. 7.4.

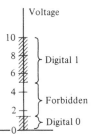

Figure 7.4 Digital definitions for amplifier-switch.

7.1.2 Digital Representation of Information

Elevator door controller. Having explained the nature of digital signals, how digital information is represented electrically, and how transistor circuits have the possibility of performing digital operations, we turn to a more complete example showing how to represent a situation in digital form. Our purpose is to introduce the AND and OR digital functions and to illustrate further the language and mathematics of the digital approach.

The door on a typical elevator has a timer that closes the door automatically if no one enters the elevator. For safety, it also has an "electric eye" to prevent the door from closing on a passenger. Let us represent a command to the door-closing mechanism with the binary variable D ($D = 1$ if the door is to close). The state of the door (D) will be controlled by three binary variables: T represents the state of the timer ($T = 1$ means that the timer is running, time has not yet elapsed); B represents someone's pushing a button for another floor ($B = 1$ means that a button has been pushed); and S represents the state of the safety device ($S = 1$ means that someone is in the door). We see that D is the dependent variable and is a function of three independent variables (T, B, and S). Keep in mind that these are all binary variables and hence can be only 1 or 0.

Digital Electronics   Chap. 7

$$D = f(T, B, S)$$

T	B	S	D
0	0	0	1
0	0	1	0
0	1	0	1
0	1	1	0
1	0	0	0
1	0	1	0
1	1	0	1
1	1	1	0

Figure 7.5 Truth table for elevator door controller.

One useful method for describing a binary function is a truth table, as shown in Fig. 7.5. Here we have enumerated all possible combinations of the independent variables and shown the appropriate value of the dependent variable. In general, when there are n independent variables, each having two possible states, there will be 2^n possible combinations (2^3 in this case), which may be enumerated to define the function. The truth table offers a systematic form for displaying such an enumeration. The name *truth table* originated historically from the application of this type of representation to the systematic investigation of logical arguments. Because of this association, digital circuits are often called *logic circuits*.

Truth-table representation. Figure 7.5 presents the truth table for the elevator door function. The 1's and 0's in the first three columns resulted from a systematic counting of the eight possibilities, or *states*. We call it "counting" because the pattern we have used constitutes counting in the base 2 number system. The pattern is clear: we alternated the 1's and 0's fastest for S, slower for B, and slowest for T, thus covering all possible combinations. These represent the values of our independent variables. For filling out the 1's and 0's in the last column, we looked first at the S column, which represents the safety switch. We do not want the door to close when the safety switch indicates that someone stands in the door ($S = 1$), so we put 0 in the D column ($D = 0$ means do not close the door) for every 1 in the S column. This accounts for four of the eight states. The other four states depend on the button and the timer. If $S = 0$ (nothing blocking the door), the door should close if either the button is pushed ($B = 1$) or the timer expires ($T = 0$). We examine the remaining four states and put a 1 in column D if there is a 1 in the B column or a 0 in the T column. Using this rule, we find three combinations leading to the closing of the door. The first, all 0's, represents the timer running out to close the door. The second represents a button being pushed and the timer running out simultaneously, and the third represents a button being pushed before the timer runs out.

NOT function. The truth-table method is a brute-force way for describing a binary function. The same information can be represented algebraically through the AND, OR, and NOT binary functions. Consider first the NOT function, the logical complement, described in the truth table of Fig. 7.2. The NOT function is involved in this problem because NOT S allows the door to close, and NOT T prompts the closing of the door by the timer. The NOT function is represented algebraically by a prime added to the variable or expression to be NOTed: S' means NOT S. We need the NOT when a 0 is to trigger the OR or AND combinations because these trigger on a 1.

Figure 7.6 OR function. $C = A$ OR B

A	B	C
0	0	0
0	1	1
1	0	1
1	1	1

OR function. The OR binary function is defined in Fig. 7.6. The dependent variable $C = A$ OR B, is 1 when either A or B (or both) is (are) 1. This is thus the *inclusive* OR because it includes the case where both A and B are 1. The OR function is involved in our elevator problem in describing the combined effect of the timer and the button. We wish the door to get the signal to close when the timer elapses ($T = 0$) OR the button is pushed ($B = 1$) or both. The way to

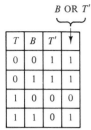

B OR *T'*

T	*B*	*T'*	↓
0	0	1	1
0	1	1	1
1	0	0	0
1	1	0	1

Figure 7.7 Binary function *T'* OR *B*.

C = *A* AND *B*

A	*B*	*C*
0	0	0
0	1	0
1	0	0
1	1	1

Figure 7.8 AND function.

express this algebraically is *T'* OR *B*. The truth table for this function is shown in Fig. 7.7. In constructing the truth table in Fig. 7.7, we added a NOT *T* column. Then we put a 1 in the last column wherever there is a 1 in either of the previous two columns because these are the variables we are ORing.

AND function. Next we need to account for the safety switch. The AND function is required because we must express the simultaneous occurrence of an impulse to close the door and the lack of an obstacle in the door. The truth table for the AND function appears in Fig. 7.8. Here we get a 1 only when both *A* and *B* are 1. To complete the truth table for our door-closing variable (*D*), we must AND *S'* with the last column in Fig. 7.7 to cover all the possibilities. Thus we may state the door-closing function as

$$D = (T' \text{ OR } B) \text{ AND } (S') \tag{7.1}$$

When interpreted as a binary or logical function, Eq. (7.1) states algebraically the same information as the truth table in Fig. 7.5, which we worked out by considering all possible combinations. We have now introduced a method of representing information in digital form and we have defined some basic logic relationships. We turn next to describing how electrical circuits can perform logical operations such as AND and OR.

7.1.3 Check Your Understanding

1. A truth table having five independent and three dependent binary variables would have how many rows?
2. How many bits of information (binary variables) are required to specify one of the 50 states in the United States?
3. If a signal in a digital system is found to be between the regions for a 1 and a 0, you should assume 1, assume 0, or repair the circuit. (Which?)

Answers. 1. 32; 2. 6 bits; 3. repair the circuit.

7.2 THE ELECTRONICS OF DIGITAL SIGNALS

7.2.1 NOT Circuit

Circuit improvements. In Section 7.1.1 we showed that the transistor amplifier-switch circuit performs the digital NOT function, provided that we define a digital 1 as any voltage between 5 and 10 V and a digital 0 as any voltage between 0 and 1.5 V. We propose here three modifications of the circuit to improve its performance as a NOT circuit. Specifically, we will add two diodes to the base circuit, lower the resistance in the base circuit to 10 kΩ, and lower the power-supply voltage to 5 V, all shown in Fig. 7.9.

Analysis of the modified circuit. Let us investigate the input–output characteristic of the modified circuit to appreciate the improvements brought by these changes. In our investigation we will assume the diodes and the base–emitter

Digital Electronics   Chap. 7

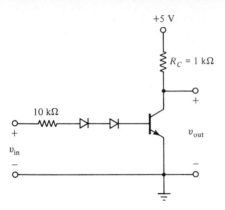

Figure 7.9 Improved NOT circuit.

pn junction are OFF when the voltage across each is less than 0.7 V. After they are turned ON and current begins to flow, their voltage will remain at 0.7 V each. This is the simple model of the *pn* junction shown in Fig. 6.35.

We will now derive the input–output characteristic of the circuit in Fig. 7.9. Let us increase the input voltage, beginning at zero volts. No current will flow into the base of the transistor until both diodes and the base–emitter junction turn ON. This requires about 3×0.7 or 2.1 V at the input. Once the voltage at the input rises to 2.1 V, therefore, the three *pn* junctions will turn ON, and the transistor will move out of cutoff. Once this occurs, Kirchhoff's voltage law in the base–emitter loop takes the form

$$i_B = \frac{v_{in} - 2.1}{R_B} = \frac{v_{in} - 2.1}{10 \ k\Omega} \tag{7.2}$$

The base current will control the flow of current in the collector–emitter loop, which is the output part of the circuit. While the transistor is cut off, no collector current will flow and the output voltage will remain at +5 V. As the base current begins to flow, however, the transistor will move into the active region and the collector (output) voltage will begin to fall. This moves the operating point from cutoff toward saturation as shown on the load line in Fig. 7.10. Because we changed the power-supply voltage, the load line now goes from +5 V on the voltage axis (V_{CC}) to 5/1 kΩ = 5 mA on the current axis. Because the transistor beta (β) is about 100, a base current of about 5 mA/100 = 50 μA will be required

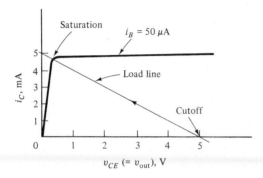

Figure 7.10 Load line for improved circuit.

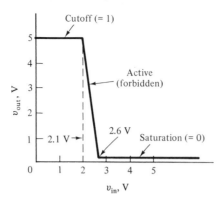

Figure 7.11 Input–output characteristic of improved NOT circuit.

for saturation; hence we have put the 50-μA characteristic on Fig. 7.10. Base currents between 0 and 50 μA put the transistor in the active region. Equation (7.2) requires, therefore, that input voltages between 2.1 and 2.6 V correspond to the active region. After the input voltage exceeds 2.6 V, the transistor will be saturated; that is, the collector voltage (v_{out}) will have fallen to about 0.4 V and further increases in base current cause little change in the output. Thus the input–output characteristic of the modified amplifier–switch is shown in Fig. 7.11.

Benefits of changing the circuit. If you compare the characteristic in Fig. 7.11 with that in Fig. 7.3, you will note several differences. The slope is greater (higher gain) in the active region, a result of decreasing the resistance in the base circuit. We want higher gain so that the transistor passes through the active region (the forbidden region in digital operation) with a smaller range of input voltage and hence passes through with greater speed. Next we notice that the modified circuit remains in cutoff until the input exceeds 2.1 V, in contrast to 0.7 V for the unmodified circuit. This results from placing the two diodes in the base circuit. Their function is to increase the range for a digital 0 and to enhance symmetry between the allowed ranges of the digital 0 and the digital 1. Finally, we note that the output voltage for a 1 is decreased from +10 V to +5 V. We lowered the value of V_{CC} for two reasons, to save diodes and to save power. We added two diodes to raise the threshold of the active region to roughly half of the +5 V. If we had to add enough diodes to raise the active region up to one-half of 10 V, we would have had to use five or six diodes; thus we save diodes by lowering the power-supply voltage. We also save power by lowering the voltage since the saturation current is lowered correspondingly. The power used by the circuit when saturated is approximately V_{CC} times the saturation current; hence we use one-fourth the power with the smaller V_{CC}.

Logic levels for the modified NOT circuit. Appropriate digital levels for the modified amplifier–switch are shown in Fig. 7.12. The range 0 to 1.0 V defines a digital 0, 4.0 to 5.0 V defines a digital 1, and the range from 1.0 to 4.0 V is forbidden. We desire these broad, symmetrical regions for several reasons. Digital equipment is reliable and inexpensive because we do not have to be careful about the exact voltage levels, as long as they lie in the ranges for a 0 or 1. Furthermore, broad regions for the logic levels immunize the digital circuits to noise and false

Figure 7.12 Logic levels for improved NOT circuit.

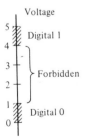

Digital Electronics   Chap. 7

signals as far as possible. We will use these voltage ranges as our logic levels for the remainder of this section; that is, when we write 1 (meaning a digital 1, a symbol and not a number) we mean a voltage between 4.0 and 5.0 V, and when we write 0 we mean a voltage between 0 and 1.0 V.

7.2.2 Simple gates

NOR gate. Digital gates are circuits that pass or block signals moving through a logic circuit. We will now examine the NOR gate, a circuit that combines the OR function with the NOT function. A simple NOR circuit is shown in Fig. 7.13a. The inputs are A and B, and the output is C. We will justify the truth table shown in Fig. 7.13b. Because there are two inputs, there will be 2^n ($n = 2$) or 4 possible input combinations for A and B, which we have listed systematically in the truth table. The first of these (00) corresponds to having voltages below 1.0 V at both inputs. Although the four pn junctions (three diodes and the base–emitter junction) are slightly forward biased, insufficient voltage is present to turn them ON and in particular the transistor base–emitter junction will not turn ON; hence the transistor will remain OFF. This cutoff condition causes the output to be $+5$ V, a digital 1; hence we place 1 in the C column, first row. The next row has a digital 1 at B and a digital 0 at A. The voltage at B will exceed 4 V, which will turn ON diodes D_2, D_3, and the base–emitter junction, leading to saturation and a digital 0 at the output. Notice that the voltage at P is at least $4 - 0.7 = 3.3$ V and the voltage at A is at most 1.0 V; hence D_1 is OFF. This diode, acting as an open circuit in its OFF condition, prevents the signal at B from coupling back into the source of A. The last two rows in the truth table follow from similar considerations: clearly, if either (or both) of the inputs is (are) a 1, the transistor will turn ON, current will flow in the saturated transistor, and the output voltage will drop into the range for a digital 0. The diodes perform the OR operation and the transistor gives the NOT. Incidentally, we are not limited to two inputs; we can have three, four, or more inputs coming into the OR part of the circuit.

Figure 7.13 (a) NOR circuit; (b) truth table for NOR function.

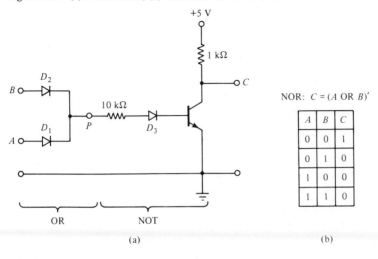

NOR: $C = (A \text{ OR } B)'$

A	B	C
0	0	1
0	1	0
1	0	0
1	1	0

(a)

(b)

The output C is isolated by the transistor from affecting the inputs. The transistor firms up the output of the diode OR gate and isolates the input circuit from the output circuit. In the process of offering these benefits, the transistor inverts the digital signal and we thus pick up the NOT operation. If we require an OR circuit, we would put C into a NOT circuit; this would NOT the NOR to give the OR operation.

NAND gate. The NAND circuit shown in Fig. 7.14 combines the AND and the NOT operations. Ignore for the moment the two inputs with their diodes and think of the circuit as a NOT circuit with the input (to the 10-kΩ resistor) connected to the $+5$-V power supply. Without action at the A and B inputs, the circuit would be a NOT circuit with the input locked at a digital 1 and the output locked therefore at 0. Now let us consider the effect of the inputs. The first row of the truth table has both A and B at low voltage, below 1 V. This will cause both D_1 and D_2 to turn ON and current will flow through the 10-kΩ resistor, through the diodes, and to ground through whatever is controlling A and B. The voltage at P will be at most $1 + 0.7 = 1.7$ V, not enough to turn ON diodes D_3 and D_4 and the base–emitter junction. Hence the transistor will be cut off and the output voltage will be $+5$ V, a digital 1. The next state (01) leads to similar operation, the only difference being that the voltage at B is now at least 4 V and hence D_2 is OFF. But the voltage at P remains no higher than 1.7 V and the output remains at 1. Only if both A and B are a digital 1 does the current through the 10-kΩ resistor return to the transistor base, turn ON the base–emitter junction, saturate the transistor, and drop the output voltage to a digital 0. The input diodes perform the logical AND function and the transistor the NOT function, giving the entire circuit a NAND function. If we wish the AND function, we can invert C with a NOT circuit. Note that if we add other inputs in parallel with A and B, all would have to be at a digital 1 to give a digital 0 at the output.

Figure 7.14 (a) NAND circuit; (b) truth table for NAND function.

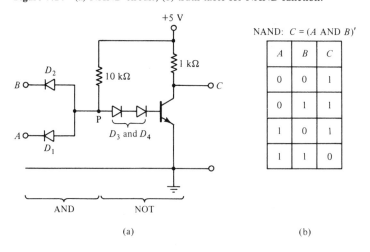

(a)

(b)

7.2.3 MOSFET Gates

A number of features make MOSFETs attractive as logic gates: they are small and easily fabricated in integrated circuits, they have a high input impedance, and they require very little power to hold and change logic states. In Section 6.5.2, we showed how a MOSFET can operate as a switch when used in series with a load resistor. The circuits used in practice replace the load resistor with another FET.

Using an *n*-channel depletion-mode MOSFET for a load. Figure 7.15a shows a NOT gate that uses an enhancement-mode MOSFET as a switch, or *driver*, transistor and a depletion-mode MOSFET as a *load* transistor. When the input voltage is less than the threshold voltage for the driver transistor, it will be OFF and no current will flow through the transistor. The load transistor will be in the ohmic region because the output voltage will be very near V_{DD}.

When the input voltage to the driver transistor is much greater than its threshold voltage, it allows current to flow, the output voltage drops, and the driver transistor enters the ohmic region. The load transistor is in its saturation region and thus acts as a current source having a high impedance. Since a large current would require high power, the load transistor is fabricated to have a relatively small current when in saturation with $v_{GS} = 0$.

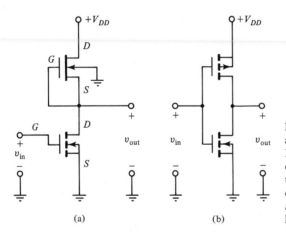

Figure 7.15 (a) A NOT gate using an enhancement-mode *n*-channel FET for a driver transistor and a depletion-mode FET for a load transistor; (b) a NOT gate using two enhancement-mode MOSFETs, an *n*-channel driver and a *p*-channel load transistor.

Using a *p*-channel enhancement-mode MOSFET for a load. The NOT gate shown in Fig. 7.15b uses a *p*-channel enhancement-mode MOSFET for a load transistor. Here a low input voltage will cause the driver transistor to be cut off and the load transistor to be in the ohmic region; hence the output voltage will be nearly V_{DD}. Likewise, an input voltage near V_{DD} will put the driver transistor into the ohmic region and the load transistor into cutoff; hence the output voltage will be small. Note that in both states one of the transistors is cut off and no dc current flows through the inverter. Current flows in the gate only during transitions, and hence this circuit uses very little power.

NMOS and CMOS logic families.   The type of circuit shown in Figure 7.15a is called NMOS because the logic circuits are constructed entirely of *n*-channel MOS transistors. The type of circuit in Fig. 7.15b is called CMOS because it uses *p*-channel and *n*-channel, or complementary, transistors. Although we have shown only the NOT gate, families of logic elements including NAND and NOR gates are constructed of these transistors without use of resistors or capacitors.

The NOR, NAND, and NOT circuits are the basic building blocks from which digital systems are constructed. We will consider how large systems are formed after we have refined our mathematical language.

7.2.4 Check Your Understanding

1. The transistors in a digital watch spend the least time in the cutoff, saturation, or active region?
2. A transistor, among other things, produces the complement (or NOT) of a binary signal. (T/F?)
3. In the NAND circuit in Fig. 7.14, the "AND" part is done by the diodes or the transistor?
4. If the collector voltage is used to represent a digital signal, the output of a transistor in saturation would be 1 or 0?
5. The diodes in the base circuit of Fig. 7.9 are there to increase the gain (T/F?)
6. In a three-input NAND gate, how many input states correspond to 1 at the output?
7. Which type of NOT gate would use less power, the BJT or the CMOS?

Answers.   1. Active;   2. true;   3. by the diodes;   4. 0;   5. false;   6. seven; 7. CMOS.

7.3 THE MATHEMATICS OF DIGITAL ELECTRONICS

7.3.1 Need for a Mathematical Language

We showed earlier how to express a digital function with a truth table. We termed this a brute-force method, for a truth table lists all possible states of the independent variables (the inputs) and lists the corresponding values of the dependent variables (the outputs). This way of expressing a digital function has several limitations. The truth table offers little guidance about how to accomplish the required logical operations with digital circuits to produce the desired output. That is, the truth table yields few hints about how to "realize" the required logical function. In our example about the elevator door, for example, we were able to translate the problem description into a logical function by common sense rather than through examination of the truth table. The mathematics we will develop in this section not only suggests a realization of the logic function but also permits manipulation of the function into different forms, thus offering alternative realizations. Also, we will demonstrate ways to simplify logical expressions, thus permitting simpler realizations.

Origins of Boolean algebra. The mathematics of two-valued variables is called Boolean algebra after George Boole, an English mathematician who first investigated this type of mathematics. Boole was interested in symbolic logic, the formal examination of logical arguments to establish their soundness or expose their fallacies. As we have already remarked, this early application of the mathematics has influenced the terminology of digital electronics.

A warning. Whereas the ideas, theorems, and applications of Boolean mathematics are relatively simple, the nomenclature and language can be confusing. One difficulty is that this mathematics uses some of the same symbols as ordinary algebra, but with different meanings. Thus $A + B = C$ is a meaningful equation in both systems but has totally different meaning and is read differently as a binary expression. For example, the equation $1 + 1 = 1$ is correct in Boolean mathematics, but is incorrect in ordinary algebra. Another difficulty is that common English words such as "and" are given technical meanings. Until you become accustomed to this new usage, many of the statements about Boolean variables sound like double talk.

7.3.2 Common Boolean Theorems

Boolean variables. We have already stated that a Boolean (or digital, or binary) variable has two values, which we call 1 and 0. The values are *defined* to satisfy the definitions of OR, AND, and NOT in Fig. 7.16. We note that some of these definitions are invalid if 1 and 0 are considered numbers, but are valid definitions for 1 and 0 as symbolic representations of the two possible states of a binary variable. We note that the $+$ sign is used for OR, the "$\cdot$" symbol for AND, and the prime for NOT or logical complement.

OR	AND	NOT
$1 + 1 = 1$	$1 \cdot 1 = 1$	$1' = 0$
$1 + 0 = 1$	$1 \cdot 0 = 0$	$0' = 1$
$0 + 1 = 1$	$0 \cdot 1 = 0$	
$0 + 0 = 0$	$0 \cdot 0 = 0$	

Figure 7.16 Basic definitions of OR, AND, and NOT functions.

One-variable theorems. The theorems involving one variable, here A, are shown in Fig. 7.17. All of these may be verified by testing the validity of the expression for both states of A and comparing with the definitions in Fig. 7.16. For example, the bottom expression under AND is verified by the truth table shown in Fig. 7.18.

Two or three variables. Some useful theorems and properties involving two or three binary variables are shown in Fig. 7.19. Again we see that many of these are deceptively similar to the familiar properties of algebra. Also note that we show the expressions involving AND operations both with and without the

OR	AND	NOT
$1 + A = 1$ $0 + A = A$	$1 \cdot A = A$ $0 \cdot A = 0$	$(A')' = A$
$A + A = A$ $A + A' = 1$	$A \cdot A = A$ $A \cdot A' = 0$	

Figure 7.17 Boolean theorems for one variable.

A	$A \cdot A'$	DEF
1	$1 \cdot 1'$	$1 \cdot 0 = 0$
0	$0 \cdot 0'$	$0 \cdot 1 = 0$

Figure 7.18 Proof of a theorem with a truth table.

Commutation: $A + B = B + A;\ A \cdot B = B \cdot A;\ AB = BA$

Association: $A + (B + C) = (A + B) + C$
$A \cdot (B \cdot C) = (A \cdot B) \cdot C;\ A(BC) = (AB)C$

Absorption: $A + (A \cdot B) = A;\ A \cdot (A + B) = A;\ A(A + B) = A$

Distribution: $A \cdot (B + C) = (A \cdot B + A \cdot C);\ A(B + C) = AB + AC$
$A + (B \cdot C) = (A + B) \cdot (A + C);\ A + BC = (A + B)(A + C)$

DeMorgan's Theorems: $(A + B)' = A' \cdot B';\ (A + B)' = A'B'$
$(A \cdot B)' = A' + B';\ (AB)' = A' + B'$

Figure 7.19 Some useful theorems and properties involving two or three binary variables.

"·" symbol. Writing the AND without any symbol is common, even though the confusion with ordinary multiplication is compounded.

All the relationships in Fig. 7.19 may be verified by examining all possible states and then applying the earlier properties and definitions. Because the absorption rules appear the strangest, we select the first of these for verification in Fig. 7.20. Note that the last column is identical to the column for A, showing that the theorem is valid. Examination of the truth table shows how B gets "absorbed" by A: if $A = 1$, B does not matter; and if $A = 0$, B does not matter. Hence B is absorbed. The proof of the second absorption rule we will reserve for a homework problem at the end of the chapter.

De Morgan's theorems. De Morgan's theorems reveal how to distribute the NOT over variables that are ANDed or ORed. Figure 7.21 gives a proof for the first theorem. Notice that the fourth and seventh columns are identical, thus proving the theorem. Proof of the second of the theorems we will save for a

Figure 7.20 Truth-table proof of an absorption theorem.

A	B	AB	$A + AB$
0	0	0	0
0	1	0	0
1	0	0	1
1	1	1	1

Figure 7.21 Truth-table proof of one of De Morgan's theorems.

A	B	$A + B$	$(A + B)'$	A'	B'	$A' \cdot B'$
0	0	0	1	1	1	1
0	1	1	0	1	0	0
1	0	1	0	0	1	0
1	1	1	0	0	0	0

homework problem. The form of De Morgan's theorems is that on the left side of the identities, the NOT covers two variables that are either ANDed or ORed, and on the right side the NOT has been distributed to the individual variables. Both rules are summarized by the following statement: A NOT can be distributed in a logical expression involving two variables provided that ANDs are changed to ORs, and vice versa. This rule can be applied to expressions involving more than two variables, provided care is taken with the grouping of variables.

The importance of De Morgan's theorems in digital electronics issues from the inversion of the output signal relative to the input(s) by the transistor. We saw before in examining NOR and NAND gates that the diodes at the input perform the OR or AND logic and the transistor firms up the decision, isolates the input from the output, and inverts the signal in the process, thus adding the NOT unavoidably to the logical operation. This "NOT" makes De Morgan's theorem useful and important in digital electronics.

We will explore the usefulness of De Morgan's theorems in digital electronics after we have introduced symbols for the logic gates. For now, we will illustrate the use of the various theorems in simplifying logical expressions. In Eq. (7.3) we show that a logical expression that appears to depend on two input variables is in fact constant and remains in the 0 state, regardless of the values of A and B.

$$[(A{\cdot}B)' + A]' = (A' + B' + A)' = (1 + B')' = 1' = 0 \qquad (7.3)$$

In Eq. (7.4) we simplify an expression and discover that the second term adds nothing new to the first. Notice that we used the absorption rule to go from the second form to the third.

$$(A{\cdot}B)' + (A' + B)' = A' + B' + A''{\cdot}B' = A' + B' = (A{\cdot}B)' \qquad (7.4)$$

EXCLUSIVE OR and the equality function.   The OR operation is, as we pointed out, an *inclusive* OR, meaning that it includes the case where both variables are 1. It is also useful to define an EXCLUSIVE OR function, that is, a logical function that takes the value 1 if either variable is 1 but takes the value 0 if both variables are either 1 or 0. The truth table for such a function is given in Fig. 7.22. The third column contains the definition of the EXCLUSIVE OR; note that the symbol for this operation is a + inside a circle. The bottom row of the table shows the exclusion of the state where both inputs are 1. The EXCLUSIVE OR indicates inequality between the variables, for the output 1 of the EXCLUSIVE OR indicates that A and B are unequal. It follows that the complement of the EXCLUSIVE OR would indicate equality, as the last column of Fig. 7.22 shows.

A	B	$A \oplus B$	$(A \oplus B)'$
0	0	0	1
0	1	1	0
1	0	1	0
1	1	0	1

Figure 7.22   EXCLUSIVE OR function.

7.3.3 Check Your Understanding

1. Evaluate $D = B' + B(C + A')$ if $A = 1$, $B = 1$, and $C = 0$.
2. A digital system has three inputs, A, B, and C, and one output, $(AB)' + C$. How many of the possible input states correspond to a 1 at the output?
3. Under what condition is the Boolean expression $A + A = 1$ valid?
4. Simplify $[A'(1 + A') + B']'$.
5. As a Boolean equation, $C + C = C$ is valid. (T/F?)

Answers. 1. 0; 2. 7 states; 3. if $A = 1$; 4. AB; 5. true.

7.4 ASYNCHRONOUS DIGITAL SYSTEMS

7.4.1 Logic Symbols and Logic Families

Logic symbols. Digital systems consist of vast numbers of NAND, NOR, and NOT gates, plus memory and timing circuits that we will discuss later, all interconnected to perform some useful task such as count and display time, measure a voltage, or perform arithmetic operations. If we were to draw a circuit diagram for such a system, including all the resistors, diodes, transistors, and interconnections, we would face an overwhelming and unnecessary task. The task would be unnecessary because anyone who read the circuit diagram would in his or her mind group the components together into standard circuits and think in terms of the "system" functions of the individual gates. For this reason, we design and draw digital circuits with standard logic symbols, as shown in Fig. 7.23.† The small circle at the output indicates the inversion of the signal. Thus without the small circle, the triangle would represent an amplifier (or buffer) with a gain of unity, and the second symbol would indicate an OR gate. As stated above, however, the common circuits are those that invert. These logic symbols show only the input and output connections. When wired into a digital circuit, the gates would have power-supply (V_{CC}) and grounding connections as well. Figure 7.24 shows the connections for a quadruple, two-input NAND gate. Notice that the supply voltage is applied between pins 14 and 7, with 7 grounded.

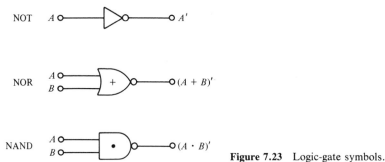

Figure 7.23 Logic-gate symbols.

† The + (OR) and · (AND) markings in the gates are optional. The shape of the gate symbol defines its function.

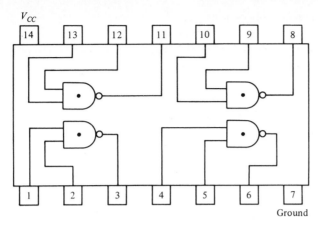

Figure 7.24 Quad-NAND chip.

Figure 7.25 Physical appearance of a logic chip.

Logic families. If you wished to construct a digital circuit, you would not assemble a pile of diodes, resistors, and transistors and proceed to wire them together, first into standard gate circuits, then into larger functions. You would purchase the gates already fabricated on an integrated circuit (IC), and packaged in a plastic capsule, as suggested by Fig. 7.25. This commercial IC chip includes four NAND gates packaged together. An important part of the design would be to select a particular *logic family,* depending on the nature and working environment of your eventual product. The logic families are composed of a large selection of compatible circuits that can be connected together to make digital systems. The logic families differ in the details of the circuits used to perform the logical operations. Here are some of the logic families.

1. Diode–transistor logic (DTL) circuits are similar to the circuits in Figs. 7.13 and 7.14. These circuits are now obsolete. We used this simple type of logic only to illustrate the principles of logic gates.

2. High-threshold logic (HTL) circuits are similar to DTL gates but include a special diode in place of the two series diodes in Figs. 7.13 and 7.14, and use a larger power supply voltage, V_{CC}. The special diode raises the threshold level for switching the transistor and hence separates the voltage regions for a 1 and 0 by a large margin, say 10 V. This logic family is useful where electrical noise is a problem, to prevent moderate noise signals, which might leak into the circuit, from affecting the circuit. If, for example, your circuit must operate adjacent to a large dc motor or an arc welding machine, you would use HTL circuits.

3. Transistor–transistor logic (TTL) circuits use special-purpose transistors in place of the diodes. These circuits are widely used because they switch rapidly, require modest power to operate, and are inexpensive. The circuits used in this logic family are considerably more complicated than the DTL circuits.

4. Complementary metal oxide semiconductor (CMOS) logic circuits use field-effect transistors (FETs). These circuits require very little power to operate

and are used where low power consumption is an important requirement, as in battery-operated calculators.

5. Emitter-coupled logic (ECL) gates switch very fast and are used in high-speed circuits such as high-frequency counters.

The design of logic circuits is highly sophisticated, and specialists in this area must become intimately familiar with all the possible products and logic families that are available at a given time.

7.4.2 Realization of Logic Functions

NOT function. Often when a NOT circuit is required, the designer will make one out of a NOR or a NAND circuit. Figure 7.26 shows the two ways to make a NOT (or inverter) out of a NAND circuit. These realizations are based on the first and third rows under AND in Fig. 7.17. The input to be fixed at a digital 1 in the lower realization would be attached to the V_{CC} power supply through a resistor. Similarly, if we required an input to be fixed at 0, this input would be grounded. Grounding an input can be used to realize the NOT function with a NOR gate, which we will leave for a homework problem at the end of the chapter.

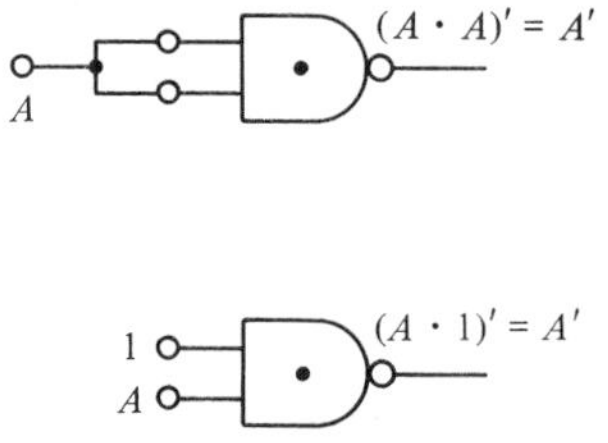

Figure 7.26 NOT from a NAND.

Realizing the elevator door function. In Section 7.1.2 we derived a logical expression for closing an elevator door ($D = 1$ means close) based on a timer ($T = 1$ means the timer is still running, do not close the door), a button switch ($B = 1$ means someone has pushed the button for another floor, close the door), and a safety device ($S = 1$ means that someone is blocking the door, do not close the door). The logical expression, recast into the notation we have developed, is given in Fig. 7.27. We first will realize the function with NAND and/or NOR gates. The realization in Fig. 7.27 utilizes six gates: one NOR, one NAND, and four NOTs, which were accomplished with NANDs. This realization is based on direct translation of the logical expression into logic-gate symbols.

We may accomplish a simpler realization by manipulating the expression for D into a more convenient form. Equation (7.5) shows such a manipulation.

$$D = [[(T' + B) \cdot S']']' = [(T' + B)' + S'']' = [(T' + B)' + S]' \quad (7.5)$$

In Eq. (7.5) the first form is the same expression for D as in Fig. 7.27, except that we NOTed it twice. We do this because we want the final result for D to be primed, that is, to be the NOT of something. This is required if we are to realize the final operation with a NOR or NAND gate. The second form results from using De Morgan's theorem to distribute one of the NOTs to the individual

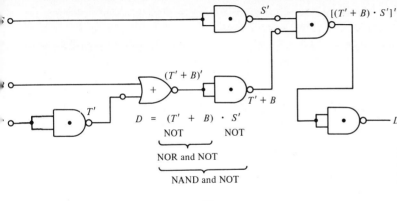

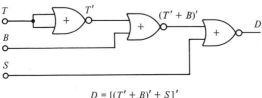

Figure 7.27 Straightforward realization of the elevatordoor function using NAND and NOR gates.

$$D = [(T' + B)' + S]'$$

Figure 7.28 Simple realization of the elevator door function using NOR gates.

terms. The third form is the same as the second, except that we have removed the double complement from S. This final form proves convenient for realization with NOR gates. Figure 7.28 shows the realization. Note that we made a NOT out of NOR similar to the way we realized a NOT with a NAND earlier. De Morgan's theorem thus leads to a simpler realization.

7.4.3 Binary arithmetic

Number bases. High-speed arithmetic is one of the spectacular achievements of digital electronics. Computers perform arithmetic with logic circuits through the representation of numbers as *binary numbers,* that is, in base 2 form. In this section we present the bare outlines of how binary arithmetic can be accomplished by digital circuits.

The development of efficient arithmetic methods was retarded for centuries by the lack of a convenient system and notation for the representation of numbers. The breakthrough came when the 10 Arabic† numerals were used to write numbers in base 10, that is, allowing the repetition of numerals, with the position representing powers of 10. For example, the number 806.1 means

$$806.1_{10} = 8 \times 10^2 + 0 \times 10^1 + 6 \times 10^0 + 1 \times 10^{-1}$$

The advantage of such a system is that the addition and multiplication tables assume manageable size; and the rules for applying these properties follow simple patterns.

The triumph of the base 10 number system was so successful that, until recently, few could write and perform arithmetic in some other base, say, base 5. This is no longer true because "modern math" in the secondary school system

---

† Actually, these symbols are thought to have been first used in ancient India and introduced into Western society through Islamic culture.

includes arithmetic in nondecimal number bases. You presumably required little explanation of the binary counting already used in Fig. 7.5. In that table, the first row represents zero, the second one, and the seventh, for example, represents six in binary form:

$$6_{10} = 110_2 = 1 \times 2^2 + 1 \times 2^1 + 0 \times 2^0$$

Thus it takes three binary digits, or *bits,* to represent six in base 2, or binary, form. In general, n bits can represent numbers 0 to $2^n - 1$ and, if we wish to consider signed numbers (plus or minus), we require another bit to represent the sign. An ordered grouping of binary information, a group of bits, is called a *word.* Thus a digital computer would represent a number as a word of, say, 32 bits, and hence be limited to numbers smaller than about 1 billion. We are speaking here of an integer format (or straight binary) for the numbers; we can, of course, use a floating (exponential) form to represent a wider range of numbers.

Binary-coded decimal (BCD).　Thus far we have presented pure decimal and pure binary representation of numbers. A hybrid system, binary-coded decimal (BCD), is frequently used in calculators and digital instrumentation. With BCD, the decimal format of the numbers is preserved, but each digit is represented in binary form. Because we must represent 10 numerals (0, 1, 2, . . . , 9), we require 4 bits of digital information for each numeral to be represented. Figure 7.29 offers an example, representing 906_{10} as a 12-bit BCD word. From Fig. 7.29 we deduce that n bits (where n is a multiple of 4) can represent numbers up to $10^{n/4} - 1$ in BCD form. This represents a reduction from what we can represent with pure binary, but facilitates input and output interactions between system and operator. Because we think in decimal, for example, we want 10 keys on our calculators for entering numbers and we also require outputs in decimal. The internal manipulation of the binary information is complicated somewhat by the BCD form, but this is the problem of calculator designers, who obviously are up to the challenge.

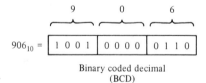

Binary coded decimal
(BCD)

Figure 7.29　BCD representation of decimal 906.

Hexadecimal system.　With the emergence of microcomputers, which work with words of 8 and 16 bits, the hexadecimal system has gained importance. Hexadecimal is base 16 and requires six "new" number symbols in addition to the 10 Arabic numerals. For convenience in using standard printers the letters A through F are used for 10 through 15, respectively. Thus in hexadecimal the number A8F represents

$$A8F_{16} = 10 \times 16^2 + 8 \times 16^1 + 15 \times 16^0 = 2703_{10}$$

The numbers zero through sixteen are represented in decimal, hexadecimal, and binary in Table 7.1.

TABLE 7.1 NUMBERS IN DECIMAL, HEXADECIMAL, AND BINARY

Decimal	Hexadecimal	Binary
0	0	0000
1	1	0001
2	2	0010
3	3	0011
4	4	0100
5	5	0101
6	6	0110
7	7	0111
8	8	1000
9	9	1001
10	A	1010
11	B	1011
12	C	1100
13	D	1101
14	E	1110
15	F	1111
16	10	10000

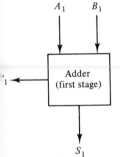

Figure 7.30 The first stage of the adder has two inputs and two outputs.

Figure 7.31 Truth-table representation of the adder inputs and outputs.

Inputs		Outputs	
A_1	B_1	C_1	S_1
0	0	0	0
0	1	0	1
1	0	0	1
1	1	1	0

7.4.4 Digital Arithmetic Circuits

BCD adder. Let us design a logic circuit to add two decimal digits represented in BCD form. The problem is symbolized by

$$A + B = S \Rightarrow A_4A_3A_2A_1 + B_4B_3B_2B_1 = S_4S_3S_2S_1 \tag{7.6}$$

where in the second form the A's and B's with the subscripts are binary variables representing the 4 bits required for the BCD representation† and the + represents addition.

First stage of the adder. Addition in both decimal and binary is shown below.

$$
\begin{array}{cc}
 & 1 \\
9_{10} & 1001_2 \\
5_{10} & 0101_2 \\
\hline
14_{10} & 1110_2
\end{array}
$$

We have shown the carry for the binary addition in the traditional manner to emphasize that we must consider carries in the design of our circuit. That is, when we add in the lowest-order bits, 1 and 1, we obtain 10_2, which is written as a 0 with a carry of 1. Thus the binary circuit that adds the lowest order bits must produce two binary outputs: the lowest-order bit of the sum and the carry to be added with the next-higher-order bits. Figure 7.30 indicates what the first stage of addition must accomplish. It must take two binary inputs, A_1 and B_1, and produce two outputs, the lowest-order bit of the sum (S_1) and the carry (C_1) to the next higher stage of addition. Examination of the truth table in Fig. 7.31 reveals

† In the development that follows, you will see that we need an additional carry bit to handle all possibilities.

that the carry bit is the AND of the inputs and the sum bit is the EXCLUSIVE OR of the two inputs,

$$C_1 = A_1B_1 \quad \text{and} \quad S_1 = A_1 \oplus B_1 \tag{7.7}$$

We thus need a realization for the EXCLUSIVE OR. We may express the EXCLUSIVE OR in terms of OR and AND functions through either of the following equivalent forms:

$$A_1 \oplus B_1 = A_1B_1' + A_1'B_1 = (A_1 + B_1)(A_1B_1)'$$

The first form expresses the two ways the function can be 1, namely, either (A_1 is 1 AND B_1 is 0) OR (A_1 is 0 AND B_1 is 1). The second form uses the ordinary OR and then removes the case where both A_1 and B_1 are 1 by ANDing with $(A_1B_1)'$. We can manipulate either expression into the form suitable for realization with NAND and NOR gates, but in this case the second expression should be chosen because it involves the AND of the two inputs, which we need for the carry bit. Using the technique illustrated on page 284, we express the EXCLUSIVE OR in the form

$$A_1 \oplus B_1 = [[(A_1 + B_1)(A_1B_1)']']' = [(A_1 + B_1)' + (A_1B_1)'']' \tag{7.8}$$

We will use only NAND and NOR gates in our realization and there is hence no benefit in canceling the double NOT on the A_1B_1 term. Figure 7.32 shows the realization of the lowest bit adder and represents in detail what is indicated by Fig. 7.30.

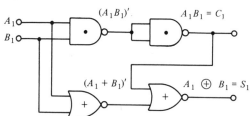

Figure 7.32 Realization for the first stage of the adder.

Higher stages of the adder. The second and higher stages of the adder must consider the carry from the next-lower stage. Thus these stages of addition must function with three inputs and two outputs, as indicated by Fig. 7.33 for the second stage. Figure 7.34 shows the truth table relating the two outputs to the three inputs. We will not work out the complete realization for the higher-order stages of the BCD adder, beyond developing Boolean expressions for the outputs.

Examination of the truth table shows there to be four ways to achieve a 1 in the output bit, S_2. These may be expressed as the ORs of a group of ANDs. This form is called a *sum of products*. In this form, the second output bit of the sum would be

$$S_2 = A_2'B_2'C_1 + A_2'B_2C_1' + A_2B_2'C_1' + A_2B_2C_1 \tag{7.9}$$

Similarly, the carry bit from the second stage can be expressed

$$C_2 = A_2'B_2C_1 + A_2B_2'C_1 + A_2B_2C_1' + A_2B_2C_1 \tag{7.10}$$

The theorems of Boolean algebra permit simplification of Eq. (7.10). Notice the

Digital Electronics   Chap. 7

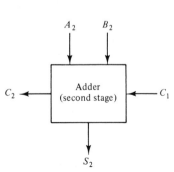

Inputs			Outputs	
A_2	B_2	C_1	C_2	S_2
0	0	0	0	0
0	0	1	0	1
0	1	0	0	1
0	1	1	1	0
1	0	0	0	1
1	0	1	1	0
1	1	0	1	0
1	1	1	1	1

Figure 7.33 The second stage of the adder has three inputs and two outputs.

Figure 7.34 Truth-table representation of the second-stage inputs and outputs.

last two terms. They can be combined as shown in Eq. (7.11). Thus we can replace two terms by a simpler term.

$$A_2B_2C_1' + A_2B_2C_1 = A_2B_2(C_1' + C_1) = A_2B_2(1) = A_2B_2 \qquad (7.11)$$

Notice also that we could have used the same trick by combining the last term with either of the first two terms. Too bad we do not have more terms like that last term in Eq. (7.10) to use in simplifying the first two terms. Fortunately, the theorems of Boolean algebra allow us to add the desired terms. Notice the third property of the OR in Fig. 7.17: $A = A + A + A + \cdots$, where we have expanded the expression to as many A's as we require. Using this property, we can insert two additional $A_2B_2C_1$ terms in Eq. (7.10), then combine them with each of the first three terms in the manner shown in Eq. (7.11), and hence reduce Eq. (7.10) to

$$C_2 = B_2C_1 + A_2C_1 + A_2B_2 \qquad (7.12)$$

Given this success in simplifying Eq. (7.10), you might try to simplify Eq. (7.9) in a similar way, but your effort would be unsuccessful because Eq. (7.9) is already in its simplest form. Clearly, we could complete the realization of the BCD adder with NAND or NOR gates by manipulating these expressions for the sum and carry bits of the higher-order stages of the adder into suitable forms, as we did in Eq. (7.8).

7.4.5 Karnaugh Maps

In the previous section we used the theorems of Boolean algebra to reduce Eq. (7.10) to Eq. (7.12). It is unclear, however, when such simplifications can be applied to Boolean expressions. The *Karnaugh map,* which furnishes a technique for deriving a minimal Boolean expression, is a geometric representation of the truth table, arranged such that patterns of 1's on the map are linked to algebraic relationships.

Making a map. The procedure for making a Karnaugh is simple: make a rectangular grid having 4, 8, 16, and so on, bins. Figure 7.35 shows such a map for S_2 as a function of A_2, B_2, and C_1. The map has eight bins because there are eight states to be represented. Each bin corresponds to one row of the truth table and is marked at top and side with its coordinates. For example, the bin in the lower-right corner of the map corresponds to $A_2 = 1$, $B_2 = 0$, and $C_1 = 1$, as marked at the top and side.

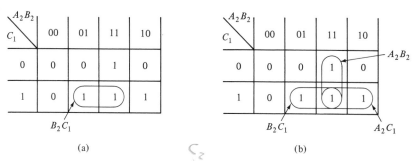

(a)   (b)

Figure 7.35 (a) Karnaugh map showing S_2 as a function of A_2, B_2, and C_1; (b) the same Karnaugh map with all three patterns identified.

The ordering of the bins is arbitrary, provided (1) all combinations are represented only once and (2) each bin differs from adjoining bins by a change of only one bit. For this reason, we cannot count across the top in the usual way (00, 01, 10, 11) because 2 bits change between the second and third states. This second rule for arranging the coordinates ensures that spatial proximity is linked to algebraic closeness.

The values of S_2 from the truth table go into the bins; hence the four 1's under S_2 in Fig. 7.34 correspond to the four 1's in Fig. 7.35, each in the bin representing its row in the truth table. We now look for square and rectangular patterns of 1's. We have circled one such pattern in Fig. 7.35a and marked it with B_2C_1 because $B_2 = 1$ AND $C_1 = 1$ uniquely identifies those adjacent bins. In other words, A_2 drops out because it is both 0 and 1 in that rectangle. The reason why we do not circle the three adjacent 1's is that this property of variables dropping out occurs only in patterns of two, four, eight, and so on, adjacent bins.

Figure 7.35b shows the same Karnaugh map with three patterns circled and identified. We now see that the 1's in the map may come from B_2C_1 OR A_2B_2 OR A_2C_1, which leads to Eq. (7.12).

Further properties of Karnaugh maps. Other properties of the Karnaugh map are:

- There is no "edge" to the map. Bits on the far left are adjacent to bits on the far right. Figure 7.36a shows the same information as Fig. 7.35b rearranged in a different order. The pattern for A_2C_1 now rolls over to the left.

Digital Electronics   Chap. 7

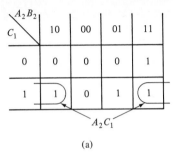
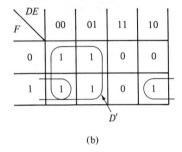

(a)   (b)

Figure 7.36 (a) The same Karnaugh map as Fig. 7.35, with a different ordering of the columns. The A_2C_1 pattern now rolls over the edge of the map. (b) The identified pattern corresponds to $D' = 1$.

- Patterns corresponding to independent variables that are 0 are identified as complements. For example, the four 1's in Fig. 7.36b correspond to $D = 0$; hence we may identify them as $D' = 1$. The 1's in Fig. 7.36b may thus be identified as D' OR $E'F$.
- Don't care's† are marked with ×'s in the Karnaugh map and may be considered either as 0's or 1's, whichever gives the largest patterns of 1's. Since these states never occur in practice, we may safely assign them either value to give the simplest algebraic expression.

We leave for a homework problem the Karnaugh map for C_2, which has no rectangular groupings of 1's. Thus no simplification of the expression for C_2 is possible.

7.4.6 Check Your Understanding

1. What is 1011_2 in hexadecimal (base 16)?
2. What is C_{16} in decimal?
3. How many bits are in a hexadecimal number?
4. Why would it be incorrect to label a Karnaugh map with 01, 11, 00, 10 across the top?
5. If the state $A_2B_2C_1 = 010$ were impossible (a don't care) in the Karnaugh map in Fig. 7.35, what would be the simplest expression for S_2?

   Answers. 1. B_{16}; 2. 12_{10}; 3. 4 bits; 4. because more than one bit changes between 11 and 00 and also between 10 and 01; 5. $S_2 = B_2 + A_2C_1$.

7.5 SEQUENTIAL DIGITAL SYSTEMS

The logic circuits in Section 7.4.4 are called *combinational* circuits because the output responds immediately to the inputs and there is no memory. When a logic circuit has memory, the system is called *sequential* because its output depends on the inputs *plus* its history. In this section we show how memory is developed

† See note in Problem P7.4.

in logic circuits and how memory elements increase greatly the applications of logic circuits.

7.5.1 Bistable Circuit

The basic memory circuit is the bistable circuit, which we examine in this section. In Fig. 7.37 we show two amplifier–switch circuits in cascade, the output of the first (T_1) providing the input to the second (T_2). This is called a two-stage amplifier, for amplification takes place in two distinct stages. Each of these stages is identical to the original amplifier–switch we analyzed in Section 6.4.3 to obtain the input–output characteristic shown in Fig. 6.44. Placing the two stages of amplification in cascade requires that we consider the output of the first stage as the input of the second stage. In Fig. 7.38, we have shown the overall input–output characteristic of the amplifier in Fig. 7.37. We derived this characteristic from Fig. 6.44 by increasing the input voltage starting with zero volts. With zero volts input, the first transistor (T_1) will be cut off and the second transistor (T_2) will be saturated. As the input voltage rises, the first transistor will leave cutoff as the input voltage rises above 0.7 V, but the second transistor will remain saturated until the output of the first stage drops to about 4 V. This output of 4 V requires an input of about 3 V (see Fig. 6.44); hence the input must increase to approximately 3 V before T_2 comes out of saturation and the output voltage of the entire two-stage amplifier begins to rise. In this region, both transistors are in the active region and the output rises rapidly. The second transistor will reach cutoff when the output of the first transistor falls below 0.7 V, which will occur when the input of T_1 exceeds above 4 V. Thus both transistors are in the active region for input voltages between 3 and 4 V.

What will happen if we connect the output of the two-stage amplifier to its input? Figure 7.39a shows the circuit redrawn with this connection and with T_1 turned around to emphasize the symmetry of the resulting circuit. We have also added inputs, which we will discuss presently. Mathematically, this connection requires $v_{in} = v_{out}$, which defines a straight line passing through the origin and

Figure 7.37 Two-stage amplifier.

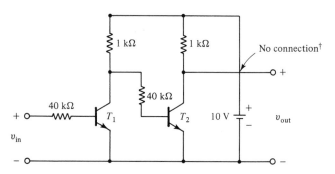

Figure 7.38 Input–output characteristic of two-stage amplifier.

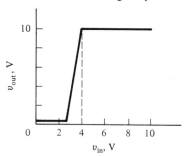

† When wires cross in a circuit diagram, no electrical connection is implied unless a dot is placed at the intersection.

Digital Electronics   Chap. 7

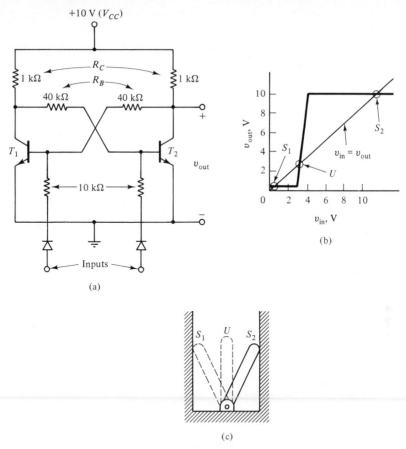

Figure 7.39   (a) Two-stage amplifier with output connected to input; (b) the circuit has two stable and one unstable operating point; (c) mechanical analog.

having a slope of unity. In Fig. 7.39b we have drawn this line on the amplifier characteristic, which also has to be satisfied. This straight line is not a load line, but the same reasoning that we followed in thinking about load lines applies here: to satisfy both characteristics, the solution must lie at their intersection(s). Consequently, we have marked and labeled the three intersections on the graph. The two intersections labeled S_1 and S_2 are stable solutions, but the intersection labeled U is unstable. Figure 7.39c suggests a mechanical analog: the lever will have stable equilibria when resting against either wall but with a frictionless pivot the balanced position will be unstable and will not occur in practice. The stable position marked S_1 occurs with transistor T_2 saturated and transistor T_1 cut off, and the stable position marked S_2 has transistor T_2 cut off and transistor T_1 saturated. The circuit will remain in one of these stable states forever unless an external signal forces it to the other stable state, just as the lever in Fig. 7.39c will lean against one wall unless an external force moves it to the other wall. By applying sufficient positive voltage to the input of the transistor that is cut off, we can switch the state of the circuit. The diodes are placed in the inputs to ensure that the state of the circuit will not affect the input drivers.

The circuit shown in Fig. 7.39a is called a *bistable* (or *latch*) *circuit* and it provides electronic memory. When you depress a button on your calculator, the signal sets latch circuits in the calculator to retain the keyed information after you release the button. The information thus retained is then available for processing after all numerical information is entered.

7.5.2 Flip-Flops

R-S flip-flop. We can realize the latch function with standard logic gates. Figure 7.40 shows a latch constructed from two NOR gates. The output of each NOR provides one of the inputs for the other NOR. The other inputs are labeled S (for SET) and R (for RESET). The outputs are labeled Q and Q' because the latch provides complementary outputs. This circuit, called an *R-S flip-flop* (FF), is similar in its operation to the bistable in Fig. 7.39a.

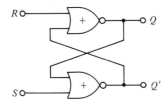

Figure 7.40 Latch from NOR gates.

We can make a brute-force analysis of the circuit in Fig. 7.40 by pretending that the inputs are independent of the outputs, as shown in Fig. 7.41a. Thus the inputs to the NOR gates are R, S, Q_{in}, and Q'_{in}. With these "inputs" we can determine the ouputs, Q_{out} and Q'_{out}, from the characteristics of the NOR gates and identify states that are self-consistent. The truth table in Fig. 7.41b gives the results. We require 16 rows for four input variables, with two output variables, Q_{out} and Q'_{out}. We fill in the 0's and 1's in the first four columns to cover all possible input states by the usual method of binary counting. The columns under Q_{out} and Q'_{out} are filled in by using the outputs of the NOR gates, Fig. 7.13b. For example, in the second row $R = 0$ and $Q'_{in} = 1$, so $Q_{out} = 0$, and $S = 0$ and $Q_{in} = 0$, so $Q'_{out} = 1$.

After filling in all 0's and 1's, we compare the input Q's with the output Q's, and where they are different we know that this state is impossible. Impossible states are labeled "inconsistent." We identify five consistent states, which therefore are stable states for the circuit. Two of these states we have identified as the MEMORY states; one state we have labled SET, one RESET, and one FORBIDDEN. Thus we reduce the truth table in Fig. 7.41b to that in Fig. 7.42a.

The interpretations in Figs. 7.41b and 7.42a make sense in the following context: information comes to the FF in pulses of 1's that come to S or R. If we get a pulse at S, the FF output (Q) is set to 1. If we get a pulse at R, the output is reset to 0. If we get no pulse at either input, the output state remains in (remembers) its present state. If it gets simultaneous pulses at S and R, both outputs go to 0 (which does not hurt anything) but then go to an indeterminate state at the end of the input pulses. This is undesirable because it destroys information; hence this state is called FORBIDDEN.

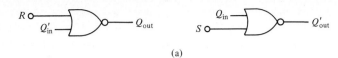

(a)

R	S	Q_{in}	Q'_{in}	Q_{out}	Q'_{out}	Interpretation
0	0	0	0	1	1	Inconsistent
0	0	0	1	0	1	Consistent, MEMORY with $Q_{out} = 0$
0	0	1	0	1	0	Consistent, MEMORY with $Q_{out} = 1$
0	0	1	1	0	0	Inconsistent
0	1	0	0	1	0	Inconsistent
0	1	0	1	0	0	Inconsistent
0	1	1	0	1	0	Consistent, SET
0	1	1	1	0	0	Inconsistent
1	0	0	0	0	1	Inconsistent
1	0	0	1	0	1	Consistent, RESET
1	0	1	0	0	0	Inconsistent
1	0	1	1	0	0	Inconsistent
1	1	0	0	0	0	Consistent but FORBIDDEN
1	1	0	1	0	0	Inconsistent
1	1	1	0	0	0	Inconsistent
1	1	1	1	0	0	Inconsistent

(b)

Figure 7.41 (a) R-S flip-flop with inputs independent of outputs; (b) truth table for R-S flip-flop with identification and interpretation of the possible states.

Figure 7.42 (a) Truth table for R-S flip-flop made with NOR gates; (b) an R-S flip-flop made with NAND gates. The properties of this flip-flop are the same as shown in (a) except the outputs go to $Q = 1$ and $Q' = 1$ in the FORBIDDEN state.

S	R	Q	Q'	Interpretation
0	0	1 / 0	0 / 1	MEMORY
0	1	0	1	RESETS $Q \rightarrow 0$
1	0	1	0	SET $Q \rightarrow 1$
1	1	0	0	FORBIDDEN $Q = Q'$

(a)

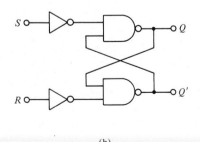

(b)

The *RS*-FF made from NOT and NAND gates shown in Fig. 7.42b has the same properties as the *RS*-FF we have analyzed except that both its outputs go to 1's when 1's appear at both inputs. This type of *RS*-FF forms the basis for the gated FF.

Gated and clocked flip-flops. The *R-S* flip-flop requires a number of refinements to achieve its full potential for memory and digital signal processing. One problem is that the *R-S* flip-flop responds to its input signals at *R* and *S* immediately and at all times. Timing problems can occur when logic signals that are supposed to arrive at the same time actually arrive at slightly different times due to separate delays. Such timing problems can create short, unwanted pulses called *glitches*.

The gated flip-flop in Fig. 7.43a will respond to the *R* or *S* inputs only when a gating signal arrives at the *G*(gate) input. Note that here we have built the flip-flop out of NAND gates. In this form, the forbidden state at the inputs to the cross-coupled NANDs is 00, which corresponds to 11 at the *R* and *S* inputs, as before. This flip-flop also has Preset (Pr) and Clear (Cr) inputs that set the latch independent of the input gates. These are active when in the 0 state, as indicated by the circle at their inputs on the logic symbol (in Fig. 7.43b). The truth table in Fig. 7.43c now lists the output state (Q_{n+1}) *after* the gating pulse as a function of the *R* and *S* inputs and the state (Q_n) *prior* to the gating pulse. Specifically, with $RS = 00$, the gating signal produces no change in the output state $(Q_{n+1} = Q_n)$. Notice that by using an inverter on the gate input, we could have the gating occur at $G = 0$. This would be indicated by a circle at the gating input (G) of the flip-flop symbol in Fig. 7.43b.

Figure 7.43 (a) Gated flip-flop; (b) logic symbol; (c) truth table.

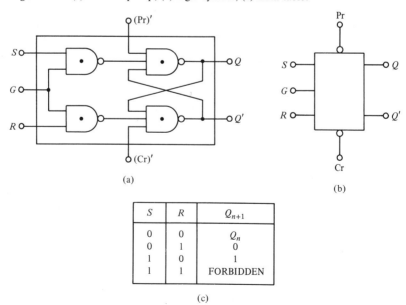

S	R	Q_{n+1}
0	0	Q_n
0	1	0
1	0	1
1	1	FORBIDDEN

(c)

The R and S inputs are thus active when the signal at the gate input is 1. Normally, such timing, or synchronizing, signals are distributed throughout a digital system by clock pulses, as shown in Fig. 7.44. The symmetrical clock signal provides two times each period when switching may be accomplished [i.e., when Ck $\Rightarrow$ 1 and when (Ck)$'$ $\Rightarrow$ 1].

The gating time of the inputs can be further reduced by differentiating the clock signal in an *RC* circuit as shown in Fig. 7.45 and applying the result to the input gate. This *edge triggering* is accomplished by building a small coupling capacitor into the input of the integrated circuit. The circuit can be designed to trigger at the leading or trailing edge of the clock. The symbol for an edge-triggered flip-flop is shown in Fig. 7.46 for both leading and trailing edge triggering. The distinguishing mark for edge triggering is a triangle at the clock input. Triggering at the edges of the waveform limits the time during which the inputs are active and thus serves to eliminate glitches. By using circuits that trigger at either the leading or trailing edges, the designer can pass signals in a circuit at two times in each clock cycle.

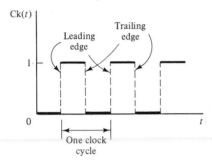

Figure 7.44 Clock signal.

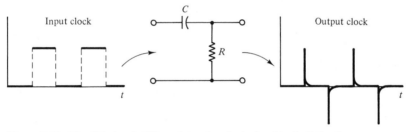

Figure 7.45 The *RC* circuit differentiates the clock signal to limit the time during which state changes can occur.

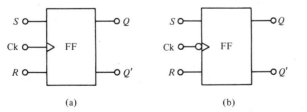

Figure 7.46 Logic symbols for edge-triggered flip-flops: (a) leading edge triggering; (b) trailing edge triggering.

The *J-K* flip-flop. Another problem with the basic *R-S* flip-flop is the forbidden state at the input. This can be eliminated by ANDing the inputs with the output of the flip-flop, thus blocking one of the inputs, as shown in Fig. 7.47.

Sec. 7.5   Sequential Digital Systems

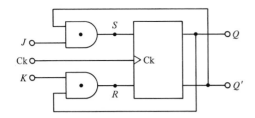

Figure 7.47 *J-K* flip-flop.

J	K	Q_{n+1}
0	0	Q_n
0	1	0
1	0	1
1	1	Q'_n

Figure 7.48 Truth table for *J-K* flip-flop. Q_n represents the state before the clock pulse and Q_{n+1} the state after the clock pulse.

The added gates here have the effect of inhibiting the 1 input to the gate whose output is 1. Therefore, the input (*J* or *K*) that is passed will always change the state of the output. The truth table for the *J-K* flip-flop (Fig. 7.48) is the same as the truth table in Fig. 7.43, except that we indicate a change of output state ($Q_{n+1} = Q'_n$) for the hitherto forbidden input state. The *J-K* flip-flop thus gives us, in addition to a latched memory of the input, the capacity to *toggle*† when both inputs are 1. This toggle feature reveals why we must use edge triggering for this flip-flop; for if the clock pulse were extended in time, the state would oscillate back and forth and the eventual output would be indeterminate. The toggle mode of the *J-K* flip-flop is useful in counters and frequency dividers.

D- and T-type flip-flops. The *J-K* (or *R-S*) flip-flop can be converted to a *D*-type flip-flop (*D* for delay) by connecting an inverter between the inputs as shown in Fig. 7.49. This eliminates the forbidden state for the *R-S* flip-flop and has the effect of delaying the output by one clock cycle, as shown by Fig. 7.50. Although the input signal and the clock appear here to change at the same time, the clock transition occurs *before* the input transition marked with the arrow on the leading edge by a prescribed time. Thus the output is assured to have the value that was present at the input during the *previous* clock cycle (i.e., the output is delayed one clock cycle).

Tying the *J* and *K* inputs together produces a *T*-type flip-flop (*T* for toggle). The *T*-type flip-flop toggles with the clock pulse when $T = 1$ and does not toggle when $T = 0$. This is useful for counters and divide-by-2 applications. The logic symbol and truth table for the *T*-type flip-flop are shown in Fig. 7.51.

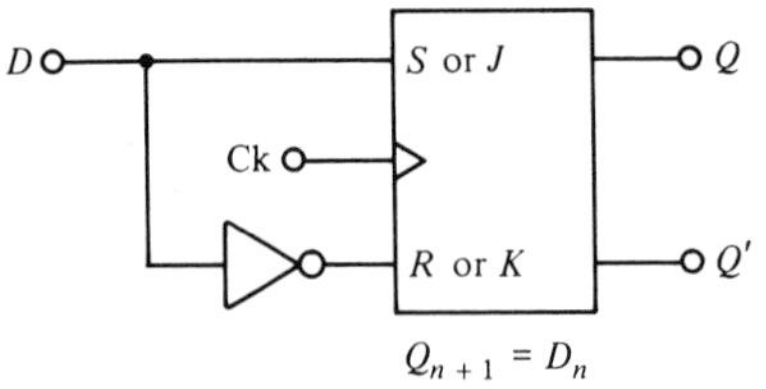

Figure 7.50 The output is delayed one clock cycle.

Figure 7.49 *D*-type flip-flop.

$$Q_{n+1} = D_n$$

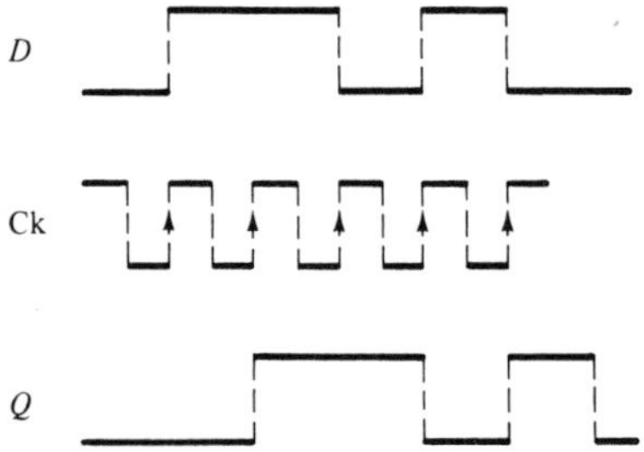

† A lever-actuated switch, like the ordinary light on–off switch, is called a toggle switch. Thus to toggle means to switch from one state to another.

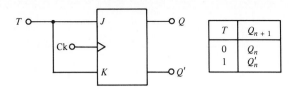

T	Q_{n+1}
0	Q_n
1	Q_n'

Figure 7.51 T-type flip-flop with truth table.

Summary. In this section we have presented the R-S flip-flop as the basic memory element in logic circuits. A variety of refinements were given, leading to several types of flip-flops. In all cases, we have shown only one way to realize the characteristic of the different flip-flops. In the various logic families, these circuits could be realized through many variations, depending on the properties of the specific family.

In the next section we show some common applications of flip-flops. Our purpose is to indicate broadly how these components can be used in digital systems.

7.5.3 Flip-Flop Applications

Frequency dividers and counters. The clock frequency can be halved with a J-K flip-flop by connecting the J and K inputs to 1 and letting the clock toggle the output.

Figure 7.52 illustrates the counting process. We begin with a cleared counter, that is, with all output Q's at zero. As shown, every trailing edge of the clock pulse changes the state of FF0; hence the output frequency of FF0 is half the clock frequency. Since this output is used as a clock for FF1, the frequency is again divided by 2, and so on down the counter. We have drawn two dashed lines to verify that the repeated divide-by-2 operation counts the input clock pulses. For example, after the end of the seventh input pulse, the state of the Q's is 0111 ($= 7_{10}$), considering Q_0 as the lowest-order bit. The reader can verify the count after 12 pulses.

Division by a factor that is not an exact power of 2 can be accomplished by using an AND gate to detect the appropriate state. Figure 7.53 shows a divide-by-3 counter. If the two flip-flops start off cleared (both Q's = 0), then at the end of the third clock cycle both Q's would be 1, which would produce a 1 at the AND output. This 1 clears the flip-flops and acts as output. Note that the chain counts input pulses in binary even if the pulses are unevenly spaced.

Figure 7.54 shows how to convert a binary counter to a decade counter. The AND gate detects a count of ten, clears the counter, and provides an input to the next stage. The $Q_3Q_2Q_1Q_0$ output from the stage could be transferred to a BCD-to-decimal display while the counter is counting another sample of the input.

The types of counters that we have presented above are called *ripple counters* because the flip-flop transitions move in sequence from left to right through the counter. By contrast, a synchronous counter uses a simultaneous clock pulse at all flip-flops, and controls the counting operation with external gates. Also possible are up–down counters that increase or decrease the count depending on an input signal.

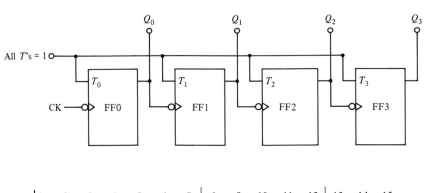

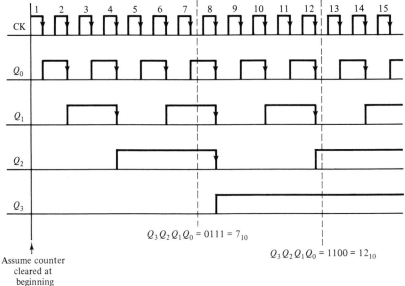

$$Q_3 Q_2 Q_1 Q_0 = 0111 = 7_{10}$$

$$Q_3 Q_2 Q_1 Q_0 = 1100 = 12_{10}$$

Assume counter
cleared at
beginning

Figure 7.52 Basic counter operation.

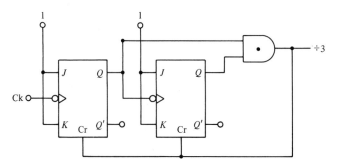

Figure 7.53 Divide-by-3 circuit.

The digital designer has available frequency dividers, counters, and many other useful system functions on LSI (large-scale integration) chips. Such LSI circuits manage the internal connections and provide the external inputs and outputs.

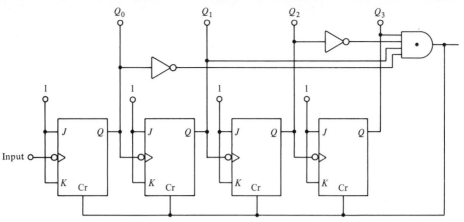

Figure 7.54 One stage of a decade counter.

Registers. A register is a series of flip-flops arranged for organized storage or processing of binary information. Before describing several types of registers, we must introduce some concepts that relate to the use of registers in computers.

Information is represented in a computer by groups of 1's and 0's called *words*. Thus a word in a 8-bit microprocessor might be 01101101, or 6D in hexadecimal. An 8-bit word is also called a *byte* and is a convenient unit for digital information. One byte can represent two BCD digits or an ASCII (American Standard Code for Information Interchange) code alphanumeric symbol. Current microprocessors work with words of 4, 8, 16, or 32 bits, whereas larger computers work with words of 32 or more bits. A register in a computer with 8-bit words would require eight flip-flops to store or process simultaneously the 8 bits of information.

Words of information are moved around in a computer or other digital system on a *bus*. As on a city bus line where passengers can enter or leave at a variety of points along the way, so on a computer bus the words can originate at any of the several registers or arrive at any of several destination registers. The bus itself consists of the required number of wires† connecting all potential source registers with all potential destination registers.

Figure 7.55 shows a bus of four wires connected to a destination register of four *D*-type flip-flops. At the leading edge of the LOAD signal, the information on the bus will be stored in the register. The information does not come down the bus and get off at the register (that is pushing the analogy too far); rather, the information appears simultaneously all along the bus and may be loaded simultaneously into several registers. If the register is part of a computer memory, the register will have an address consisting of one or more words, and this address is decoded to give the LOAD signal at the appropriate location in memory.

Whereas many registers can be loaded simultaneously from the bus, only one register can put information on the bus at one time. We need a way to connect

---

† Actually, conducting paths is more appropriate, for no wires are used for internal information transfer in a microprocessor.

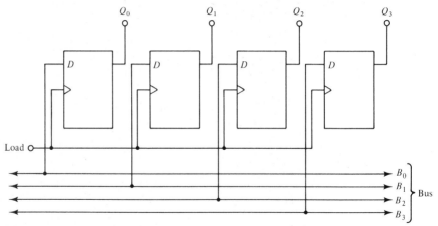

Figure 7.55 Loading a register from a bus.

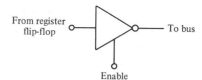

Figure 7.56 Three-stage gate. When ENABLE is 0, the output is disconnected from the bus.

the outputs of all source registers to the bus, such that only one register can transfer its output word to the bus at any time. An ordinary gate will not accomplish this, for its output must be either 1 or 0 and hence connecting the outputs of all source registers would result in a tug of war. A *three-state gate,* shown in Fig. 7.56, is required. In the absence of an ENABLE signal, the output of the gate approximates an open circuit and the gate is disconnected from the bus. When the ENABLE signal is present, the gate is connected to the bus and the output is 1 or 0 according to the input. Thus we can connect all source registers to the bus with three-state gates and ENABLE one register at a time to transfer data to the bus.

Shift register. The data-storage register described in the preceding section transfers information with parallel input and parallel output. Sometimes digital information must be sent over one wire, as when a telephone circuit is used. In this case bits are sent in time sequence, or serial form. When digital information must be received in serial form, a shift register may be used to accept the serial information and convert it to parallel form.

The D-type flip-flops in Fig. 7.57 will act as a shift register. Recall that the D flip-flop input immediately before the clock pulse shifts to its output after the clock pulse. The input to D_3 will appear at Q_3 after one clock pulse, at Q_2 after two, at Q_1 after three, and at Q_0 after four clock pulses. Thus the 4-bit sequence (word) input at D_3 will fill the register $Q_3Q_2Q_1Q_0$ after four clock pulses. The word can then be transferred to a parallel bus with three-state gates.

The shift register takes a serial input and produces parallel output when operated in the manner of Fig. 7.57. The same register can accept a parallel word

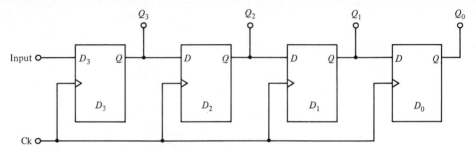

Figure 7.57 Shift register.

at the Preset inputs to the flip-flops. The word can then be driven out Q_0 by the clock to produce serial output.

The shift register in Fig. 7.57 shifts its bit pattern one unit to the right with each clock pulse. A more versatile shift register results when the flip-flop inputs are connected with gates to their neighbors on left or right. Such a register can shift its bit pattern to the right or left, depending on which set of gates are EN-ABLED. This facility is useful in arithmetic operations. Multiplication by 10 in the decimal number system can be accomplished by shifting the decimal point one digit to the right. Similarly, multiplication by 2 in binary can be accomplished by moving the binary point one place to the right. This can be effected in hardware by shifting all bits one place to the left in a shift register, and clearly multiplication by 2^n requires n left shifts. A combination of shifting and adding intermediate results is required for multiplication by numbers that are not exact powers of 2. In like manner, division is accomplished by shifting bits to the right in a shift register.

Summary. In this section we have shown how flip-flops are used to store digital information. Groups of flip-flops called registers can receive or deliver words of information in parallel with a bus, or information can be stored and delivered in serial form. The shift register is useful in binary arithmetic operations. These are the principal components of computers, to which we now turn.

7.5.4 Check Your Understanding

1. In the R-S flip-flop made with NOR gates in Fig. 7.40, the forbidden state is $R \cdot S = 1$ or 0?
2. Tying J and K together in a J-K flip-flop makes what type of flip-flop?
3. An edge-triggered flip-flop has a narrower gate time than a flip-flop that does not have edge triggering. (T/F?)
4. If $Q = 1$ and $T = 0$ in the T-type flip-flop in Fig. 7.51, what will Q' be after the clock pulse (Ck)?
5. How many flip-flops are required to store a byte of information?
6. Which of the following have a forbidden state: RS-FF, JK-FF, D-type FF, T-type FF?

Answers. 1. $R \cdot S = 1$; 2. T, or toggle FF; 3. true; 4. $Q' = 0$ after the clock pulse; 5. 8 FFs; 6. RS-FF.

7.6.1 Introduction

Electrical engineers have produced some passing fads, but computers are here to stay. These versatile devices increasingly influence modern business and pleasure; seers predict an even broader place for computers in the future. The computer epitomizes many of the themes of this book. In Chapter 1 we stressed the speed with which electrical phenomena occur: computer magic arises, for the most part, out of the speed with which computers operate. We stressed in Chapters 1 and 6 that microscopic matter is electrical in nature: from manipulation of the electrical properties of matter come the tiny transistor switches and electrical connections that physically constitute computer circuits. The digital idea finds its fullest expression in computers.

Computers have changed drastically since their development four decades ago. The original computers were big and expensive; only large institutions could justify their purchase for demanding computational tasks. Costs were in hundreds of thousands, if not millions, of dollars. Large (mainframe) computers still command an important place in the computer market. Then came the minicomputer, about the size of a suitcase. The electronics of minicomputers utilize LSI digital circuits. These computers made possible the automation of process control and the processing of data in real time; costs were in tens of thousands of dollars.

Now we have microcomputers. The low cost of the basic computer chip is revealed by its use in toys that retail for less than $25. Small personal computers including memory, keyboard, and elementary software sell for hundreds of dollars (you provide a TV set for a display). The computer on a chip finds more and more applications: smart typewriters, cash registers, appliances, and sewing machines; in the automobile, in electronic instruments, at the video arcade, in the nursery. Perhaps the personal computer itself best demonstrates the potential of microprocessor technology.

Computers are complicated systems, not easily explained. Aside from the fundamental principles of the computer itself, a myriad of related topics could be discussed: interfacing the computer to peripheral devices such as keyboards, CRTs, and printers; software development, including programming languages, editors, assemblers, and compilers; information representation matters such as fixed- and floating-point representation of numbers, data structures, codes for representing text; and potential applications such as numerical calculation, word processing, accounting systems, real-time process control, and time sharing. To add to the complexity, the world of computers has developed its own esoteric and colorful language.

We have listed above topics that, for the most part, we are not going to discuss in this section. Our brief introduction to this subject will address two basic questions: What is a computer, and how does it work?

7.6.2 What Is a Computer?

The four basic elements of a computer are memory, an arithmetic-logic circuit, a control circuit, and input–output. These elements are present as well in a hand-

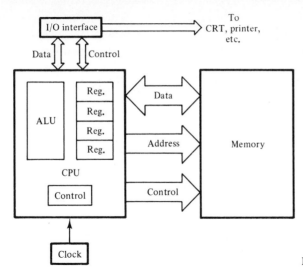

Figure 7.58 Elements of a computer.

held calculator. In that case you supply the program by pushing the various buttons that sequence the calculations. When the calculator has the capability of storing a "program" of keystrokes, it is a computer, albeit a limited, slow, and highly specialized computer.

Central processing unit (CPU).   Figure 7.58 shows the system configuration of a typical computer. The central processing unit (CPU) contains an arithmetic-logic unit (ALU), control circuits, and several registers for storage and general manipulation of words of data. (We will hereafter speak as if our computer were a 8-bit microprocessor.) The CPU is the brains of the computer and, at our level of understanding, the most complex and mysterious. Although we have touched on hardware implementation of binary arithmetic and data storage, the complexity of the CPU places it far beyond the level of this introduction.

The CPU has an instruction register that is sequentially loaded with words from the program in memory. These instructions, mere strings of 1's and 0's, are decoded and executed by the CPU: words are brought from memory, placed in designated registers, processed according to the instruction, then stored in memory at a specified location or perhaps transferred to an output device. Within the CPU are a number of registers to hold data and addresses of special places in memory relating to the program or to blocks of data. A program-counter register is incremented after each instruction is "fetched" from memory into the instruction register so that the CPU knows where to get the next instruction. The CPU communicates with memory by means of a two-way (8-bit) data bus, an address bus (16 bits for a 64K† memory) controls the location in memory to furnish or receive data or program words, and a control bus coordinates timing and order.

Memory.   Computer memory may be of several types. A random-access memory (RAM) is an array of memory registers with which data may be ex-

† In computer talk, K means 2^{10}, or 1024. Thus a 16-bit address can specify one location out of 2^{16}, or 64 K, or 65,536 words.

changed. The memory access is random because any memory location is equally accessible for reading or writing of data or program instructions. Normally, the program would not write into the portion of the memory containing program instructions, but the entire program must initially be written (loaded) into memory. Most semiconductor memories are "volatile" because they lose the stored information when the computer is turned off. For this reason, and generally to store large amounts of information, magnetic disks or tapes are used for storage of digital information.

Also important are read-only memories (ROMs), which contain information (usually programs) that can be read but not modified by the computer. Programmable ROMs (PROMs) allow the user to store information by burning microscopic fuses in the ROM. An erasable PROM (EPROM) can be reprogrammed after its information has been erased by ultraviolet light.

Input and output (I/O). Although self-contained for their internal calculations and data manipulations, computers must interact with the outside world. Often computer–person interaction is provided through terminals with keyboard and display and through printers and plotters. Such devices translate between people-oriented symbols, such as alphanumeric text or graphical representation of information, and computer-oriented representation, bits (1's and 0's) of information.

Computers interact with external systems such as robots, electronic instruments, and manufacturing processes. Such interactions often involve digital-to-analog (D/A) and analog-to-digital (A/D) conversion. The computer communicates with external devices over an external data bus, which is separate from the computer's internal data bus. Several protocols exist for announcing which component has information to transfer to the computer, for keeping two devices from "talking" at once, and for assuring that the target component "heard what was said."

7.6.3 Programming

Programming languages. Programming is the art of translating a problem into words of 1's and 0's that the computer CPU executes to solve the problem. Figure 7.59 summarizes the communication problem: we think in terms of language, mathematical notation, accounting conventions, and so on, and the computer "thinks" in 1's and 0's. Programming languages provide the bridge between human thought processes and the binary words that control computer operations. Computer languages that are deliberately close to human thought processes or notations are called high-level languages: FORTRAN, BASIC, LISP, ALGOL, C, FORTH, and Pascal are examples of high-level languages. Low-level languages are called *assembly languages* and these are oriented toward the instructions that the specific computer can execute. The binary words are loaded into computer memory and executed as machine or object code. One line of code in a high-level language can produce many lines of machine code, whereas one line of assembly code, being close to the computer operations, will produce one or at most a few lines of machine code. For the microprocessor, the assembly language is also called the *op code* for the machine.

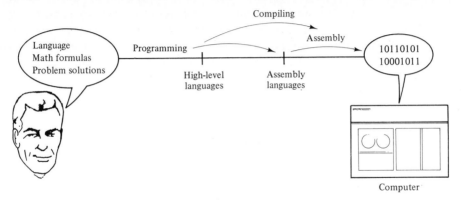

Figure 7.59 Communicating with a computer.

Compilers, assemblers, loaders, and interpreters. A compiler is a program that accepts as input a program in a high-level language (usually as a string of ASCII characters) and produces as an output either an assembly-language program or a machine-code program. The compiler is thus a computer program that performs language translation from one computer language to another language closer to what the computer uses. An assembler performs the same function, using the assembly code as input (source code) and producing output (object code) in machine code.

Compilers and assemblers are large programs that function on a specific computer. Such programs would thus be developed and furnished by the computer manufacturer (often on a ROM chip) or perhaps by an institution providing computer services. The object code that is produced usually is stored on disk or magnetic tape or in memory at a different location from where it will eventually reside. After the compiler or assembler does its work, it often would be erased from memory to make room for the program machine code and whatever data the program requires. A small program called a *loader* will load the machine code into memory for execution. When the program consists of several separate parts (main program, subroutines, functions), the loader will load these into consecutive memory and, in effect, tell each where all the others are so that they can communicate. When one begins with a completely empty memory, a small program called a *bootstrap* can be loaded into memory by means of a keyboard or front panel switches. The bootstrap program can load the loader, which loads the program and signals the beginning of execution.

Some high-level languages, such as BASIC, are interpreted instead of compiled. This means that the program is converted into machine language and executed line by line. This is inefficient in computer time but efficient for finding certain program errors because mistakes can be corrected as they occur without having to reexecute the program up to the point where the error has occurred. This is especially appropriate when the programmer is working on a small computer.

In the past, programs were usually punched on cards or perhaps paper tapes and read into the computer with a card (or paper-tape) reader. Modern systems use editors that allow typing of alphanumeric text directly into the computer and offer the programmer easy modification through a variety of editing commands. Indeed the storage, editing, and manipulation of text has opened new computer applications for the production of printed matter. Such "text processing" computer systems offer many advantages over writing or typing by hand and revising a number of times to produce final copy.

PROBLEMS

Section 7.1: The Digital Idea

P7.1. A room has a three-way switch system, meaning that changing either switch changes the state of the light. Let S_A represent the switch at door A and S_B represent the switch at door B. Let $S_A = 0$ AND $S_B = 0$ be a state with the light OFF ($L = 0$).
   (a) Develop a truth table relating the switch states to the digital variable L representing the light.
   (b) Give a Boolean expression for L as a function of S_A and S_B.

P7.2. An outside floodlight has an automatic device turning it OFF during daylight hours ($D = 1$ during daylight hours). It also has a switch ($S = 1$ for the switch ON, controlling power to the automatic device), and $B = 1$ means the floodlight bulb is functional (not burned out).
   (a) Let $F = 1$ indicate that there is light from the floodlight. Give a truth table for F as a function of D, S, and B.
   (b) Write a Boolean expression for F.

P7.3. A student is allowed to take a course ($C = 1$) if he or she pays the registration fee ($R = 1$) and either has the prerequisites ($P = 1$) or has the instructor's approval ($A = 1$).
   (a) Give a truth table for C as a function of R, P, and A.
   (b) Write a Boolean expression for F.

   Answer. $C = R$ AND (P OR A)

P7.4. A burglary alarm system sounds an alarm ($A = 1$) if the detectors detect activity ($D = 1$). The alarm system has a key to disable it during working hours ($K = 1$ disables the alarm) and also has a test button ($T = 1$ sounds the alarm if the system is not disabled).
   (a) Give a truth table relating the output A to the inputs D, K, and T. Use $\times$ for don't-care states, that is, states that will never happen in practice.
   (b) Give a Boolean expression for A. Count "don't cares" as 1.
   Comment: In this situation, some states are impossible. We have excluded the case where $T = 1$ and $K = 1$. Such states are called *don't-care* states and are indicated by an $\times$ under A in the truth table rather than a 1 or 0. The logical function is still valid; certain combinations of the independent variables never occur in practice. The designer therefore doesn't care about what the system does in these states.

P7.5. Susie is sure to accept a date ($D = 1$) provided that she likes the guy ($L = 1$) and has no test the next day ($T = 0$). But if it's Henry ($H = 1$), she will go even if she has a test.

(a) Make a truth table relating the dependent variable, D, to the independent variables: L, T, and H.

(b) Express the Susie dating function, $D(L, T, H)$, in terms of NOTs, ANDs, and ORs. Count don't cares as 1.

See comment on don't care states in Problem P7.4.

P7.6. When in the 1980 NBA playoffs the Philadelphia 76ers basketball team led the Boston Celtics three games to one, the 76ers could have won the best-of-seven series by winning any of the fifth, sixth, or seventh games. Let G_5, G_6, and G_7 be digital variables to describe the outcome of those games, with $G_5 = 1$ if the 76ers win the fifth game and $G_5 = 0$ if Boston wins, and so forth for G_6 and G_7. Complete the truth table even if all games are not played. The dependent variable is $P = 1$ if the 76ers win the series and $P = 0$ if the Celtics win. Make a truth table showing the relationship between G_5, G_6, and G_7 as input (independent) variables and P as output (dependent) variable.

Section 7.2: Electronics of Digital Signals

P7.7. Figure P7.7 shows the input–output characteristic for a transistor circuit. The β of the transistor is 100.

(a) Give a circuit with this characteristic. Let one of your resistors be 1 kΩ.

(b) Define logic levels for 0 and 1 such that the circuit exhibits the NOT function.

(c) Modify the circuit such that it exhibits NOR behavior.

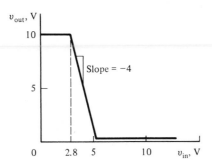

Figure P7.7

P7.8. The circuit in Fig. P7.8 operates as a digital inverter. Assume a voltage of 0.7 V for a pn junction that is ON, and a saturation voltage of 0.3 V for the transistor. The β of the transistor is 50.

(a) Find the minimum value of v_{in} to saturate the transistor. What is the collector current in saturation?

(b) Find the maximum of v_{in} to leave the transistor in cutoff. What is the output (collector) voltage when the transistor is cut off?

Answers. (a) 0.7 V plus a little, 7.7 mA; (b) 0.7 V minus a little, 8 V

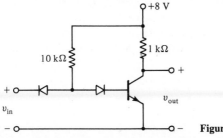

Figure P7.8

P7.9. The circuit in Fig. P7.9 uses switches to create logic inputs to a circuit. $S_A = 1$ indicates closure, and so on. Give the truth table for the circuit operation, using the usual convention for the output C. For the transistor, $\beta = 100$.

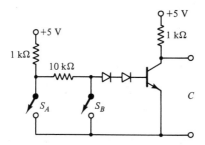

Figure P7.9

P7.10. For the NOT gate in Fig. 7.9 to operate properly, the transistor must be saturated when the input is a digital 1. Assume that the transistor characteristics are those in Fig. 6.37 and that the minimum voltage input for a 1 is 4.0 V. Assume that the pn junctions have a voltage of 0.7 V when ON.

    (a) What is the largest value of R_B (10 kΩ in Fig. 7.9 but now a variable) that allows the transistor to remain saturated for an input 1?

    (b) Keeping R_B at 10 kΩ, what is the smallest value of the transistor β that will keep the transistor saturated? (Now the transistor characteristics are different from those in Fig. 6.37.) Assume that $V_{CE(\text{sat})} = 0.5$ V.

Section 7.3: The Mathematics of Digital Electronics

P7.11. Make a truth table with two inputs, A and B, and the outputs OR, NOR, AND, NAND, EXCLUSIVE OR, and the equality function.

P7.12. With a truth table verify the second of the absorption rules in Fig. 7.19.

P7.13. Use the theorems of Section 7.3.2 to simplify the following digital functions:

    (a) $[A' + (B + A)B']'$

    (b) $A + [B(1 + A')]'A$

    (c) $(ABC)' + A'B'(C + A)$

P7.14. Simplify each of the following Boolean expressions by applying the theorems in Section 7.3.2.

    (a) $ABC' + (ABC')'$

    (b) $A + B'C + D'(A + B'C)$

    (c) $AB'(C + D) + (C + D)'$

    (d) $(AB' + C' + DE')(AB' + C')$

Section 7.4: Asynchronous Digital Systems

P7.15. Show two ways for realizing a NOT function with a NOR gate.

P7.16. Give a realization of the Susie dating function of Problem P7.5 with NOR and NAND functions. Do this directly with the function as you derived it and then use De Morgan's theorem to place the function into a more favorable form for a simpler realization.

P7.17. Show a way to make a three-input OR gate out of two two-input OR gates.

P7.18. The radio in a car should be ON ($R = 1$) if the ignition switch is either ON ($I = 1$) or in the accessory position ($A = 1$) and the radio on/off switch is also ON ($S = 1$). Watch for don't cares.

   (a) Write a Boolean expression for R as a function of I, A, and S.

   (b) Give a realization with NOT, NOR, and NAND gates.

P7.19. The safety system in Fig. P7.19 has three inputs. If two or more of these are 1 at the same time, an alarm should sound.

   (a) Give a truth table for A.

   (b) Give a Boolean function for A.

   (c) Simplify the results of part (b) if possible.

   (d) Use De Morgan's theorem to put the Boolean expression in a form for NAND gate synthesis.

   (e) Give a logic circuit to perform the alarm function using only 2-input and 3-input NAND gates.

    Answers.   **(b)** $A = S_1S_3 + S_1S_2 + S_2S_3 + S_1S_2S_3$ **(c)** $A = S_1S_3 + S_1S_2 + S_2S_3$ **(d)** $A'' = [(S_1S_3)' \cdot (S_1S_2)' \cdot (S_2S_3)']'$

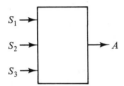

Figure P7.19

P7.20. A student will pass ($P = 1$) if he has a good average ($A = 1$) and gets in all the required work ($W = 1$). But if he fails the final exam ($F = 1$), he will not pass.

   (a) Make a truth table for P as a function of A, W, and F.

   (b) Give a Boolean expression for P.

   (c) Give a realization using only two input NOR gates.

P7.21. Give a truth table for the logic circuit shown in Fig. P7.21. *Hint:* Consider the second input to the NOR gate to be independent of B, and then eliminate those states in which this input is not B'.

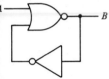

Figure P7.21

P7.22. Make a Karnaugh map for C_2 in Figure 7.34 and show that no rectangular patterns of 1's are present. This shows that no simplification of Eq. (7.10) is possible.

P7.23. On the Karnaugh map in Fig. P7.23, mark the regions corresponding to

    (a) $AC' = 1$ and $B' = 1$

    (b) $B'C' = 1$ and $A = 1$

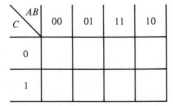

Figure P7.23

P7.24. Make a Karnaugh map of the elevator-door function described in Fig. 7.5, and from the map determine the logic function for D.

P7.25. The seven-segment numerical display shown in Fig. P7.25 can display the integers 0 through 9, depending on which segments are illuminated. The driver accepts a *BCD* input and gives output 1's to the segments that should be lit and 0's to the segments that should not be lit.

    (a) Construct the truth table for segments *a* through *g*. Note that inputs corresponding to 10_{10} through 15_{10} do not occur and result in DON'T CARE'S in the truth table.

    (b) Develop a Karnaugh map for the *a* segment.

    (c) From the map, determine the logic function for *a*.

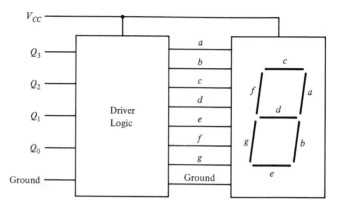

Figure P7.25

P7.26. Give the truth table for the logic circuit in Fig. P7.26.

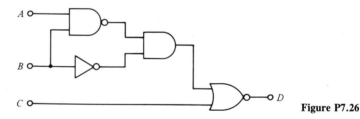

Figure P7.26

Section 7.5: Sequential Digital Systems

P7.27. For the bistable circuit in Fig. 7.39a to operate as described, sufficient current must flow into the base of the ON transistor to permit saturation (i.e., $i_B \geq \beta i_C$). For the resistor values shown, this imposes a minimum value of β.

    (a) What is the minimum value of β for the circuit in Fig. 7.39a to function as a bistable? Assume 0.7 V for the ON base–emitter voltage and 0.3 V for $V_{CE(\text{sat})}$.

    (b) Find the formula for the minimum value of β in terms of the power-supply voltage (V_{CC}), base resistor (R_B), and collector resistor (R_C).

    (c) For the circuit in Fig. 7.39a, and for a β of 60, what voltage at the input to the OFF transistor is required to switch the circuit? Assume 0.7 V to turn ON the diode and transistor *pn* junctions. *Note:* The bistable will switch if the current into the ON transistor is dropped below the value required to saturate the transistor. This implies a certain voltage at the collector of the OFF (but coming ON) transistor. This implies in turn a certain collector current, and so on.

Digital Electronics   Chap. 7

P7.28. Design an R-S flip-flop circuit with NAND gates rather than the NOR gates shown in Fig. 7.40. Determine the inputs (1 or 0) required to SET and RESET the flip-flop.

P7.29. Consider a 1-MHz square-wave clock signal with voltage levels of 0 and 5 V, as shown in Fig. P7.29. Find the RC time constant of the circuit in Fig. 7.45 to differentiate this signal to produce a gating time of 30 ns. Assume that the gate is open when the voltage is above 2.0 V.

Answer. 32.7 ns

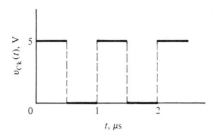

Figure P7.29

P7.30. In a digital circuit the edge-triggered flip-flops would use an input circuit similar to that shown in Fig. P7.30. Assume the gates treat a logical zero as any voltage in the range from 0 to 0.8 V and a logical one as any voltage in the range 2.3 and 5.0 V. In this case the AND gates need to be open for 1 μs to pass the J and K signals. The capacitor is part of the integrated circuit and has a value of 30×10^{-12} farad (30 pF). A 4-V pulse is used for the clock. Find the value of R to enable the edge-triggering feature to operate correctly.

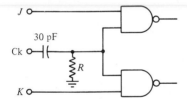

Figure P7.30

P7.31. Construct a truth table for the J-K flip-flop with J, K, Q_n, and Q'_n as inputs, R and S as intermediate outputs, and Q_{n+1} and Q'_{n+1} as outputs. Use the truth table to confirm the truth table in Fig. 7.48. You must eliminate all rows that contradict the definitions for the gates or the R-S flip-flop.

P7.32. The logic circuit in Fig. P7.32 uses a J-K flip-flop. There is one input, J, and the output is Q. Find the output, Q_{n+1}, after the clock pulse in terms of the input, J, and the output before the clock pulse, Q_n.

Answer. $Q_{n+1} = JQ'_n$

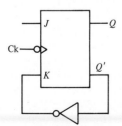

Figure P7.32

P7.33. Using *T*-type flip-flops and an AND gate, design a circuit to divide the clock frequency by 5.

P7.34. Repeat Problem P7.33, except use a NOR gate rather than an AND gate.

P7.35. The *R-S* flip-flop shown in Fig. P7.35 is made with NOR gates, as in Fig. 7.40. At $t = 0$, $Q = 0$. Give Q for $t > 0$ on the axis provided. If Q is indeterminant, mark ×'s on the time axis.

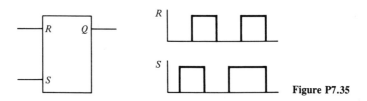

Figure P7.35

P7.36. Figure P7.36a shows the inputs and the initial state of the output of the *T*-flip-flop in Fig. P7.36b. Give the output function $Q(t)$ if $Q = 1$ at $t = 0$.

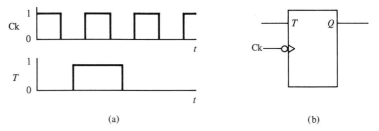

(a) (b)

Figure P7.36

P7.37. In the circuit shown in Fig. P7.37, Q_0 and Q_1 are both 0 at the beginning. Give Q_0 and Q_1 as functions of time under a sketch of the clock pulses.

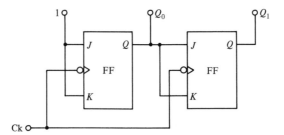

Figure P7.37

P7.38. Two *J-K* flip-flops are connected as shown in Fig. P7.38. Give Q_0 and Q_1 under a sketch of the clock pulses. $Q_0 = Q_1 = 0$ at $t = 0$

Digital Electronics   Chap. 7

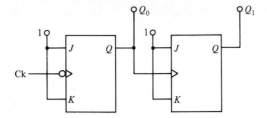

Figure P7.38

P7.39. Figure P7.39 shows a chain of T-type flip-flops. The initial state is $Q_0 = 1$, $Q_1 = 0$, and $Q_2 = 1$.
(a) What will be the state of the outputs after one clock pulse?
(b) What will the circuit do generally?

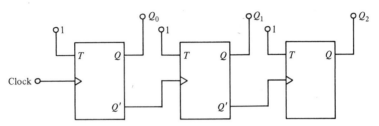

Figure P7.39

P7.40. Figure P7.40 shows a clocked circuit with two binary inputs and one binary output. What should go in the box such that the output after the clock pulse is equal to the binary product of the two inputs, with inputs and output considered base 2 numbers? That is, $C_{n+1} = A_n \times B_n$. You may use standard gates and flip-flops.

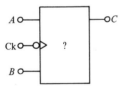

Figure P7.40

P7.41. Figure P7.41 shows an ordinary RS-FF, except that an RC circuit is used in one of the connections. The time constant of the RC circuit is 0.5 second.
(a) If $R = 0$ and $S = 0$ for a long time, what is Q?
(b) Now R receives a 1-μs pulse. What happens?
(c) Then 1 s later, S receives a 1-μs pulse. Explain what happens.

Answer. (a) 0; (b) nothing; (c) Q goes to 1 for about 0.5 s, then reverts back to 0

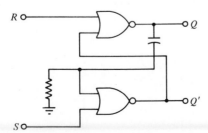

Figure P7.41

P7.42. The circuits that perform multiplication in a calculator must handle the signs of the component numbers according to the usual rules of algebra. However, if the product is exactly zero, the sign should be set to positive so that a negative zero is not displayed. In Fig. P7.42, we display the portion of the circuit that determines the sign to be displayed. Consider the multiplication of N_1 and N_2. Let $S_1 = 1$ if the sign of N_1 is $+$, $S_2 = 1$ if the sign of N_2 is $+$, $Z = 1$ if the product is exactly zero, and $S_R = 1$ if the sign of the product is to be displayed as a positive number (including a positive zero).

(a) Make a truth table for the sign of the result, S_R, as a function of S_1, S_2, and Z.

(b) Give a Boolean expression for $S_R(S_1, S_2, Z)$. You may use $\oplus$ for the EXCLUSIVE OR if you wish. You may use a Karnaugh map.

(c) Give a realization using two-input gates.

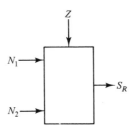

Figure P7.42

P7.43. Figure P7.43 shows four T-type flip-flops clocked for synchronous counting. Using AND gates, design logic such that at the clock leading edge, FF_1 changes states when $Q_0 = 1$, FF_2 changes states when $Q_0Q_1 = 1$, and FF_3 changes states when $Q_0Q_1Q_2 = 1$. What is the function of this circuit?

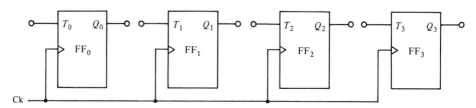

Figure P7.43

8 Analog Electronics

Contrast between analog and digital electronics. We have already explored how transistors and diodes are used as switching devices to process information that is represented in digital form. Digital electronics uses transistors as electrically controlled switches: transistors are either cut off or saturated. The active region is used only in transition between states.

By contrast, analog electronics depends on the active region of transistors and other types of amplifying devices. The Greek roots of "analog" mean "in due ratio," signifying in this usage that information is encoded into an electrical signal that is proportional to the quantity being represented.

In Fig. 8.1 our information originates physically in a musical instrument. The radiated sound is best understood as sound waves. These produce motion in the diaphragm of a microphone, which in turn produces an electrical signal. The variations in the electrical signal are a proportional representation of the sound waves. The electrical signal is amplified electronically, with an increase in signal power occurring at the expense of the input ac power to the amplifier. The amplifier output drives a recording head and produces a wavy groove on a disk. If the system is good, every acoustic variation of the air will be recorded on the disk and, when the record is played back through a similar system and the signal reradiated as sound energy by a loudspeaker, the resulting sound should faithfully reproduce the original music.

Systems based on analog principles form an important class of electronic devices. Radio and TV broadcasting are examples of analog systems, as are many

317

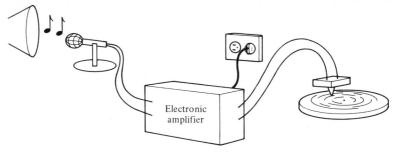

Figure 8.1 Analog system.

electrical instruments used in monitoring deflection (strain gages, for example), motion (tachometers), and temperature (thermocouples). Many electrical instruments—voltmeters, ohmmeters, ammeters, and oscilloscopes—utilize analog techniques, at least in part.

Contents of this chapter. Analog techniques employ the frequency-domain viewpoint extensively. We begin by expanding our concept of the frequency domain to include periodic, nonperiodic, and random signals. We will see that most analog signals and processes can be represented in the frequency domain. We will introduce the concept of a *spectrum,* which is the representation of a signal as the simultaneous existence of many frequencies. Bandwidth (the width of a spectrum) in the frequency domain will be related to information rate in the time domain.

This expanded concept of the frequency domain also helps us distinguish the effects of linear and nonlinear analog devices. In this chapter, linear circuits are shown to be capable of "filtering" out unwanted frequency components. Chapter 10 shows that new frequencies can be created by nonlinear devices such as diodes and transistors.

Next we study the concept of *feedback*, a technique by which gain in analog systems is exchanged for other desirable qualities such as linearity or wider bandwidth. Without feedback, analog systems such as audio amplifiers would at best offer poor performance. Understanding of the benefits of feedback provides the foundation for appreciating the many uses of operational amplifiers in analog electronics.

Operational amplifiers (op amps, for short) provide basic building blocks for analog circuits in the same way that NOR and NAND gates are basic building blocks for digital circuits. We conclude by describing some of the more common applications of op amps.

8.2 FREQUENCY-DOMAIN REPRESENTATION OF SIGNALS

8.2.1 Frequency-Domain Concepts

We introduced the frequency domain in Chapter 4 for the analysis of ac circuits. There we showed that sinusoidal sources can be represented by complex numbers through

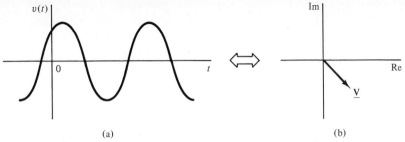

(a)   (b)

Figure 8.2 Sinusoid in (a) the time domain and (b) the frequency domain.

$$v(t) = V_p \cos(\omega t + \theta) = \text{Re}[\underline{\mathbf{V}} e^{j\omega t}] \qquad (8.1)$$

where the phasor voltage, $\underline{\mathbf{V}} = V_p\underline{/\theta}$, is a complex number representing the sinusoidal function in the frequency domain. When only one frequency is present, the frequency is not stated explicitly. The time-domain and frequency-domain representations for a sinusoidal signal appear in Fig. 8.2.

8.2.2 Frequency-Domain Representation of Periodic Functions

Example. Our frequency-domain representation of signals is enriched when we consider periodic signals that are nonsinusoidal. As an example, we will consider the function

$$v(t) = \frac{V}{2} + \frac{2V}{\pi}\left(\cos \omega_0 t - \frac{1}{3}\cos 3\omega_0 t\right)$$

$$= \text{Re}\left[\frac{V}{2} + \frac{2V}{\pi}e^{j\omega_0 t} + \frac{2V}{3\pi}\underline{/180°}\,e^{j3\omega_0 t}\right] \qquad (8.2)$$

Leaving aside for the moment the significance of this mathematical function, let us consider the time-domain representation in Fig. 8.3. In the time domain, we see nothing unusual except that we now have a more complicated function than a simple sinusoidal function. The dc component of the voltage is easily identified in Eq. (8.2) as $V/2$. The time-varying part of the function consists of two sinusoidal terms. Because the frequency of the second term is an integral multiple of the

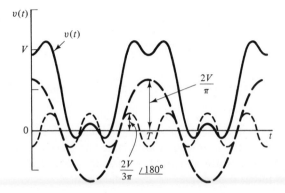

Figure 8.3 Time function.

frequency of the first term, the frequencies are said to be *harmonically related*. The term "harmonic" (from the Greek, meaning "to fit together") is used because in music the tones that share harmonics fit together to form pleasant-sounding chords. Because the two sinusoids are harmonically related, they form a stable pattern and thus repeat with the period of the lower frequency, $T = 2\pi/\omega_0$. To summarize, we would describe this function in the time domain as having three harmonic components: a dc component (zero frequency), a fundamental (or first harmonic), and a third harmonic. The dc component is described by its magnitude, and the two sinusoid components are described by their amplitudes and phases.

Spectrum. Because we now have three frequencies involved, we no longer can describe the signal in the frequency domain by a point in the complex plane. Frequency becomes important in a new way and cannot be relegated to the memory or the margin of the page. A common way to handle this new complexity is shown in Fig. 8.4. We have made frequency (ω) the independent variable and shown the three harmonics as lines at their respective frequencies. The magnitude (for the dc component) and phasors (for the ac components) have been placed beside the various frequency components, drawn roughly to scale.

Figure 8.4 is called a *voltage spectrum*. The voltage spectrum in Fig. 8.4 is the frequency-domain representation of the time-domain voltage in Fig. 8.3. This expanded concept of the frequency domain enriches greatly the concept presented in Chapter 4. Figures 8.3 and 8.4 portray the complementary nature of the two domains: time and frequency are shown as complementary variables.

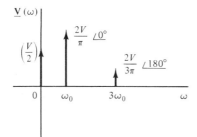

Figure 8.4 Spectrum consists of three frequencies.

Fourier series. In 1822, a French artillery officer immortalized his name among electrical engineers yet unborn by proving that any periodic function can be expressed as a series of harmonically related sinusoids. Fourier's theorem expressed mathematically what musicians had long known by ear, that a steady tone consists of many harmonically related frequencies. The mathematical expression of Fourier's theorem is

$$f(t) = C_0 + C_1 \cos(\omega_0 t + \theta_1) + C_2 \cos(2\omega_0 t + \theta_2) + \cdots \qquad (8.3)$$
$$+ C_n \cos(n\omega_0 t + \theta_n) + \cdots$$

where $\omega_0 = 2\pi/T$, T = the period of $f(t)$, C_n is the amplitude of the nth harmonic, and θ_n its phase. Fourier's theorem shows that any periodic function can be expressed as a spectrum consisting of dc, a fundamental frequency, and all frequencies that are integral multiples of the fundamental.

Fourier's theorem gives a mathematical explanation for many musical phenomena. Musicians know that the overtones of musical instruments are important. The piano sounds different from the harpsichord, for example, because the harpsichord strings are excited in a way that creates more sound energy at the higher harmonics than piano strings produce. In the language of the frequency domain, the various musical instruments sound different because they produce different acoustic spectra.

The musical scale is based on the harmonic relationship between the frequencies of the notes. The octave is a factor of 2 in frequency. Two tones are said to be consonant, to sound good together, when they are harmonically related, that is, when their harmonic structures are compatible. For example, in the pure scale the frequency of G would be $\frac{3}{2}$ that of C. It follows that the third, sixth, ninth, . . . harmonics of C coincide with the second, fourth, sixth, . . . of G. This shared harmonic structure defines all major fifth intervals in the pure musical scale, and other consonant intervals share harmonics similarly.

Our digression on music was to show that frequency-domain concepts are commonplace in certain areas of experience. For our purposes Fourier's theorem, as expressed by Eq. (8.3), shows that all periodic functions can be described by harmonic spectra. We will bypass the mathematical procedures for computing the amplitudes and phases of the harmonics of a periodic function; our primary purpose is the exploration of the expanded concept of the frequency domain.

Two spectra. We will give two examples of Fourier series. The first is the Fourier series representation of the periodic series of pulses shown in Fig. 8.5a. The Fourier series for the square wave in Fig. 8.5a is

$$v(t) = \frac{V}{2} + \frac{2V}{\pi} \cos(\omega_0 t) - \frac{2V}{3\pi} \cos(3\omega_0 t) + \frac{2V}{5\pi} \cos(5\omega_0 t) + \cdots \quad (8.4)$$

We have shown only the first four components of the frequency-domain representation in Fig. 8.5b. Comparison of Eqs. (8.2) and (8.4) shows that the first three frequency components of the square wave are those plotted in Fig. 8.3. Indeed, you should see some resemblance between the series of pulses and the sum of the dc, the fundamental, and the third harmonic in Fig. 8.3. If the fifth harmonic were added, it would steepen the slope at the sides and flatten the top

Figure 8.5 (a) Periodic time function; (b) voltage spectrum.

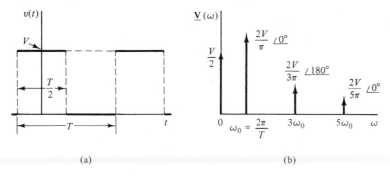

(a)   (b)

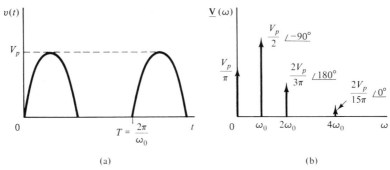

(a)   (b)

Figure 8.6 (a) Half-wave rectified sinusoid; (b) voltage spectrum.

and bottom. With the addition of each higher harmonic the series would approximate more closely the periodic pulses.

A second example of the Fourier series of a periodic function is shown in Fig. 8.6a, the half-wave rectified sinusoid. The Fourier series is

$$v(t) = V_p \left(\frac{1}{\pi} + \frac{1}{2} \sin \omega_0 t - \frac{2}{3\pi} \cos 2\omega_0 t + \frac{2}{15\pi} \cos 4\omega_0 t - \cdots \right) \quad (8.5)$$

The spectrum (Fig. 8.6b) consists mainly in the even harmonics, unlike that in Fig. 8.5b, which contains odd harmonics only. These are but two common spectra, which we introduce as examples to be explored in this and later sections.

Voltage spectra and power spectra.   The spectra shown in Figs. 8.5b and 8.6b are called *voltage spectra* because they correspond to the amplitudes and phases of the harmonics in their respective signals. In many applications the power in a signal carries great importance; hence we extend the concept of a spectrum to include power and energy concepts.

Power is defined as energy flow per unit time. Figure 8.7 shows a sinusoidal voltage source connected to a resistive load. The power into the resistor, as given on the figure, depends only on the amplitude and the resistance; frequency and phase do not matter. The factor of 2 in the denominator results from time averaging [this result was worked out in Eq. (5.7)] and can be absorbed into the voltage term if we use the rms value of the voltage instead of the peak.

The presence of the resistance in the power relationship in Fig. 8.7 poses to us a mild dilemma in developing the concept of the power spectrum of a signal. One possibility is to state explicitly the impedance level of an assumed load (say, 50 Ω) and to compute the power that each harmonic would deliver to such a load. With this approach, watts would be the units of the power spectrum. The second possibility leaves the resistance term unstated. In this case the "power" would be $(V_p)^2/2$, it being understood that to get actual power one has to divide by an appropriate resistance. In this second approach, which is more general, the units of "power" are (volts)^2. One would be wise to state explicitly these units to stress that the power spectrum is of this second variety. Because of its generality, this second approach is frequently preferred and will be used here.

We will therefore write the power spectrum of a signal in units of (volts)^2.

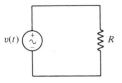

$v(t) = V_p \cos(\omega t + \theta)$ volts

$P_R = V_p^2/2R$ watts

Figure 8.7 Average power in a resistor.

TABLE 8.1 VOLTAGE AND POWER SPECTRA

Harmonic	Frequency	General Voltage Spectrum (V)	General Power Spectrum (V^2)	Series of Pulses Voltage Spectrum (V)	Series of Pulses Power Spectrum (V^2)
0	dc	C_0	C_0^2	$V/2$	$V^2/4$
1	1000	$C_1\underline{/\theta_1}$	$C_1^2/2$	$2V/\pi\underline{/0°}$	$2V^2/\pi^2$
2	2000	$C_2\underline{/\theta_2}$	$C_2^2/2$	0	0
3	3000	$C_3\underline{/\theta_3}$	$C_3^2/2$	$2V/3\pi\underline{/180°}$	$2V^2/9\pi^2$
4	4000	$C_4\underline{/\theta_4}$	$C_4^2/2$	0	0
5	5000	$C_5\underline{/\theta_5}$	$C_5^2/2$	$2V/5\pi\underline{/0°}$	$2V^2/25\pi^2$
.	.	.	.	.	.
.	.	.	.	.	.
.	.	.	.	.	.

Table 8.1 states the power spectrum of an arbitrary signal, plus the specific application to the case of the series of pulses in Fig. 8.5. Note that the dc term in the fourth column is not divided by the factor of 2 because the time average of the square of the dc term is equal to the dc value squared. The power spectrum cannot be written in the form of a Fourier series such as Eq. (8.3), but it can be graphed in the manner of Fig. 8.5b. Indeed, a graphical presentation of the last column in Table 8.1 would look like the voltage spectrum plot in Fig. 8.5b, except that the powers of the higher harmonics diminish at a greater rate due to the squaring.

Accounting for all the power. The legitimacy of the concept of a power spectrum must rest on a proper accounting for all the power in the frequency domain. We will now confirm equivalence of power between the two domains for the series of pulses in Fig. 8.5a. The time-average power is

$$P_{avg} = \frac{2}{T} \int_0^{T/2} v^2(t)\, dt = \frac{2}{T} \int_0^{T/4} V^2\, dt = \frac{2V^2}{T} \times \frac{T}{4} = \frac{V^2}{2} \quad \text{(volts)}^2 \quad (8.6)$$

The power in the frequency domain would be the sum of the harmonics in the last column of Table 8.1.

$$P = \frac{V^2}{4} + \frac{2V^2}{\pi^2} + \frac{2V^2}{9\pi^2} + \frac{2V^2}{25\pi^2}$$
$$= V^2 \left[\frac{1}{4} + \frac{2}{\pi^2} \left(1 + \frac{1}{9} + \frac{1}{25} + \cdots \right) \right] \quad \text{(volts)}^2 \quad (8.7)$$

The infinite sum in Eq. (8.7) converges rapidly to the value $\pi^2/8$, as you can confirm from math tables or by summing terms on your calculator. We see that the sum of the power-spectrum harmonics equals the power in the time domain. We note that half the power comes from the dc component and half from the ac harmonics. The equal division of power between dc and ac components is not a general result, but the accounting of the power between the time and frequency domains will always come out correct. We will leave for a homework problem verification of this equality for the spectrum of the half-wave rectified sine wave.

Summary. We have shown in this section that a periodic signal can be represented by an infinite series of sinusoidal components at frequencies harmonically related to the fundamental frequency of the signal. The voltage spectrum of a periodic signal consists, therefore, of a series of harmonics, each being described by an amplitude and phase. The power spectrum has no phase information and may be described in either watts or (volts)^2. The power spectrum accounts for the total power as the sum of the powers in the individual harmonics.

8.2.3 Spectra of Nonperiodic Signals

An interesting question. Granted this success in representing periodic signals in the frequency domain through Fourier series, we are emboldened to ask: How far can we carry this line of development? This is an important question, for in electronic systems we must deal with both periodic and nonperiodic signals. Examples of periodic signals are: ac voltages, pulses sent out by radar transmitters, timing signals in a TV signal, clocking signals in a computer, and steady musical tones. Examples of nonperiodic signals are: music, speech, information pulses in a computer, and radar pulse returns from a maneuvering target or a diffuse target, such as a thunderstorm. Thus we must press for the extension of frequency-domain concepts to include a wider class of signals.

Spectrum of a single pulse. How, then, might we explore the possibility of having a spectrum for a nonperiodic signal, such as a pulse that happens only once? One approach, which works in this case, begins with a series of pulses and then increases the period of the series, keeping the width of the individual pulses constant. As we let the period of the repeating pulses get larger and larger, in the limit we would have a single pulse. This limiting process in the time domain will affect the spectrum, and the resulting spectrum would correspond to the spectrum of a single pulse. Figure 8.8 suggests such a series of pulses by showing two pulses. The individual pulse widths are τ, which will be kept constant, and the period is T, which will be allowed to increase so as to isolate a single pulse. The time origin is the center of one of the pulses; hence the signal has even symmetry about the origin. This means that the Fourier series for this series of pulses will contain harmonics in the form of Eq. (8.3), but every harmonic will have a phase

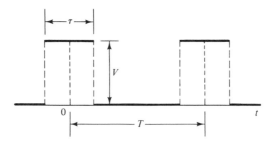

Figure 8.8 Series of pulses.

angle (θ) of 0 or 180°, because these phases alone possess even symmetry. Because the phase of 180° is equivalent to a negative sign in front of the amplitude, the Fourier series for this signal must be of the form

$$v(t) = A_0 + A_1 \cos \omega_0 t + A_2 \cos 2\omega_0 t + \cdots + A_n \cos n\omega_0 t + \cdots \quad (8.8)$$

where $\omega_0 = 2\pi/T$ and the A's are amplitudes of the harmonics, which may be negative. The amplitudes can be computed by the standard methods of Fourier series, with the result

$$A_n = \frac{2V\tau}{T} \frac{\sin(n\omega_0\tau/2)}{n\omega_0\tau/2} = \frac{2V\tau}{T} \frac{\sin(n\pi\tau/T)}{n\pi\tau/T} \text{ V} \quad (8.9)$$

Recall that the spectrum consists of components at $\omega = 0, \omega_0, 2\omega_0, \ldots, n\omega_0,$ $\ldots$, with harmonic amplitudes, except for $n = 0$, given by Eq. (8.9). We reserve for a homework problem the proof that Eq. (8.9) yields the results presented in Eq. (8.4) when $\tau = T/2$. As an example, we have plotted in Fig. 8.9 the spectrum for the case where $T = 10\tau$. We see in Fig. 8.9 that there are now many harmonics that are significant, unlike the spectrum in Fig. 8.5b, where the amplitudes fall off rapidly after the first few harmonics. The negative amplitudes indicate a phase of 180° from the tenth to the twentieth harmonics. Thus we note that the spectrum consists of a large number of harmonics spaced apart $\Delta f = 1/T$, where T is the period of the time function.

Although for graphing purposes we have normalized the harmonic amplitudes to unity relative to the dc component, Eq. (8.9) reveals that the individual harmonics become small as T is made large. There are two reasons for this decrease. As the period is increased (with τ kept constant), the pulses come less and less frequently and hence the power in the signal decreases accordingly. Specifically, the average power must decrease as $1/T$ due to the spreading of the pulses in time. Compounded with this effect, the power in any individual harmonic must decrease even faster because the power in the series of pulses distributes between an increasingly larger number of harmonics. As Eq. (8.9) shows, the

Figure 8.9 (a) Time-domain and (b) frequency-domain representations for a series of pulses with period T and width $\tau = T/10$.

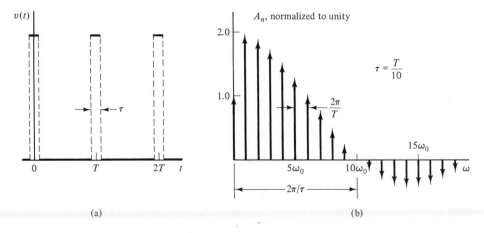

(a)   (b)

power in, say, the first harmonic ($n = 1$) decreases as $(1/T)^2$ because power is proportional to the square of the amplitude.

Isolating a single pulse.   With the aid of Fig. 8.9 and a measure of imagination, we can now anticipate what will happen to the spectrum as we allow the period, T, to approach infinity. As $T \rightarrow \infty$, the frequencies of the individual harmonics will crowd closer and closer together and, in the limit, constitute a continuous spectrum. Individual harmonics must disappear because there is no longer any specific period; hence frequency becomes a continuous variable. We conclude that a single pulse in the time domain contains all frequencies in the frequency domain. The shape of the spectrum, however, will not change as T becomes large. The period, T, affects the spacing of the harmonics, but the shape of the spectrum depends only on τ. Thus the shape of the spectrum will retain the form in Eq. (8.9), which approaches

$$A(\omega) = E_0 \frac{\sin(\omega\tau/2)}{\omega\tau/2} \quad \text{V/Hz} \tag{8.10}$$

where the continuous variable ω replaces the discrete variable $n\omega_0$ and E_0 is a constant. This continuous spectrum is pictured in Fig. 8.10b, where we have called the spectrum $S_V(\omega)$ because this is a voltage spectrum. We wish also to investigate the power spectrum, as we did in Table 8.1, but here we must be careful. As implied earlier, the single pulse carries no average power because the energy in the pulse is spread over all time, whereas power implies the continuous flow of energy as a time process. Consequently, we may meaningfully discuss the *energy* in the pulse but not the power. Our voltage spectrum in Eq. (8.10), when squared, becomes an energy spectrum, not a power spectrum. The energy in the pulse distributes between the infinite number of frequencies making up the spectrum. Each individual frequency carries an infinitesimal quantity of energy, but taken together account for the total energy of the pulse.

As in the case of periodic power spectra, we have the option of including

Figure 8.10   (a) Time-domain and (b) frequency-domain representations for a single pulse.

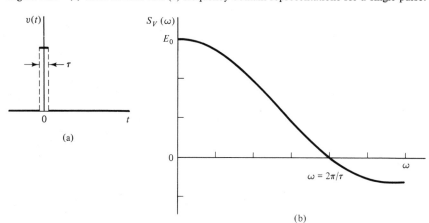

Analog Electronics   Chap. 8

resistance explicitly or omitting it to leave the definition of an energy spectrum more general. The units of the spectrum given in Eq. (8.10) are volts per hertz. The energy spectrum has the units joule-ohms per hertz, which requires division by a resistance and multiplication by a bandwidth in hertz to have the units for energy.

Spectrum amplitude.   Let us apply this interpretation of the square of the voltage spectrum in computing the energy in the pulse in the frequency domain. We can equate this energy to the energy in the time domain to evaluate the constant E_0 in the spectrum in Eq. (8.10). To have a physical picture, and to get the units to work out to the proper units of energy, let us consider that the pulse has an amplitude of V and is applied to a resistor of value R, as pictured in Fig. 8.11. The total energy in the pulse is the integral of the power, Eq. (1.13). The required calculation is

$$W_R = \int_{-\infty}^{+\infty} vi\, dt = \int_{-\tau/2}^{+\tau/2} \frac{v^2}{R}\, dt = \frac{V^2}{R} \int_{-\tau/2}^{\tau/2} dt = \frac{V^2 \tau}{R} \quad \text{J} \qquad (8.11)$$

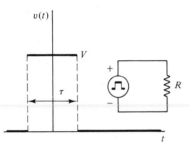

Figure 8.11   Computation of the energy in a single pulse.

The frequency-domain calculation follows the same reasoning as we used in Eq. (8.7), except that here we replace the sum of the contributions of each individual harmonic with the integral of the continuous spectrum. We shall use the rms value of each contributing sinusoid; hence the amplitudes must be divided by $\sqrt{2}$. We no longer must treat the dc term separately because it carries an infinitesimal amount of energy. Equation (8.12) summarizes the required computation.

$$W_R = \frac{1}{R} \int_0^{\infty} \left[\frac{E_0}{\sqrt{2}} \frac{\sin(\omega\tau/2)}{\omega\tau/2} \right]^2 d\left(\frac{\omega}{2\pi}\right) = \frac{E_0^2}{2R}\left(\frac{1}{\pi\tau}\right) \int_0^{\infty} \frac{\sin^2 x}{x^2}\, dx = \frac{E_0^2}{4R\tau} \quad \text{J} \qquad (8.12)$$

You will note that we are integrating with respect to $\omega/2\pi$, which is frequency in hertz. The second form of the integral results from a change in variables from ω to $x = \omega\tau/2$. The definite integral has the value $\pi/2$, leading to the final result. We now equate the two energy calculations, with the result

$$\frac{E_0^2}{4R\tau} = \frac{V^2\tau}{R} \Rightarrow E_0 = 2V\tau \quad V/\sqrt{H_z} \qquad (8.13)$$

Bandwidth and pulse width.   The spectrum of the single pulse in Eq. (8.10) illustrates an important general relationship between the time and frequency

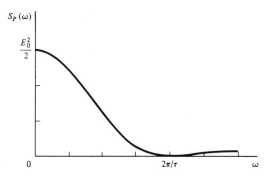

$S_P(\omega)$

$\dfrac{E_0^2}{2}$

0        $2\pi/\tau$ ω

Figure 8.12 Power spectrum of a single pulse.

domains. In Fig. 8.12, we show the energy spectrum of the pulse, $S_P(\omega)$,† which is the square of the voltage spectrum in Fig. 8.10. Clearly, most of the energy is contained in the frequency bandwidth below the first null at $2\pi/\tau$. Integration of the energy spectrum reveals that 90% of the total energy of the single pulse lies in the bandwidth between 0 and the first null in the spectrum. Thus, if we were to pass this spectrum through an ideal lowpass filter that eliminated all frequencies above the first null of the spectrum, most of the energy would get through and the pulse would appear in the time domain without serious change. Similarly, the representation of the periodic pulses (Fig. 8.5a) by the dc, first, and third harmonics contains 95% of the energy in the entire series of pulses, as you can verify from Eq. (8.7). Figure 8.3 verifies that the shape of the pulses is relatively unchanged by the elimination of the higher harmonics. Later in this chapter we will investigate filtering in the frequency domain; here we merely wish to establish that there is an effective bandwidth in the frequency domain associated with a pulse.

We may determine this effective bandwidth by setting $\omega\tau$ equal to 2π, the angle in radians where the spectrum first goes to zero. Solving this equation and converting to frequency in hertz, we obtain

$$\omega\tau = 2\pi \Rightarrow BW = \frac{2\pi}{2\pi\tau} = \frac{1}{\tau} \quad Hz \tag{8.14}$$

where BW is the significant bandwidth in hertz. Important consequences follow from this approximate relationship between pulse duration in the time domain and bandwidth in the frequency domain. For one, we must provide adequate bandwidth if pulses are to pass through communication or computing systems without serious distortion. If, for example, we have a digital system that utilizes pulses of 1-μs duration, we must provide at least 1×10^6 Hz or 1 MHz of bandwidth to send those pulses to another location for recording or processing.

This reciprocal relationship between pulse length and bandwidth applies also to periodic pulses. The spectrum of the periodic pulses shown in Fig. 8.5b has the same general shape, and hence the same bandwidth, as that of the single pulse. Furthermore, this relationship between time duration and bandwidth does not depend on the exact shape of the pulses. The pulses may be rounded on top or

---

† The subscript "P" indicates a power spectrum. In this context, this means that the voltage spectrum has been squared, not that the units are watts.

triangular in shape: the same approximate bandwidth would be needed to preserve the identity of the pulses in passing through a communication system.

Example from mechanics.   Let us imagine that we have a large bell and we strike this bell with a metal hammer. We know, of course, that we will "ring" the bell, meaning that the mechanical resonances of the bell will be excited to produce a ringing sound. The force between the hammer and the bell would be of short duration, and this short pulse of force will excite the bell structure. Because the energy spectrum of the short pulse is broad in bandwidth, the higher resonances of the bell will be excited. Pick up now a rubber hammer and again strike the bell. We would expect the bell to sound less harsh, more mellow. We can understand this change in sound in terms of the energy spectrum of the force from the rubber hammer. The rubber hammer will remain in contact with the bell much longer because the rubber hammer is more compliant than the metal hammer. The result of this longer pulse of force will be a smaller bandwidth of excitation; consequently, the higher-frequency resonances of the bell would not be greatly excited. These higher resonances, which give the bell a harsh sound, are eliminated; hence a mellow sound results with a rubber hammer.

In Section 8.2.5 we will explore the relationship between bandwidth and information rate in a communication system. There we will show another consequence of the relationship between bandwidth in the frequency domain and duration in the time domain. This relationship also underlies an important result in the next section on random signals.

8.2.4 Spectra of Random Signals

Importance of random signals.   A random signal is any signal that is unpredictable to us, either because it originates from chance events or because its complexity defies simple analysis. Examples of random signals are broadcast signals over radio and TV channels, digital signals passing from computer to computer, the output of a microphone placed before a symphony orchestra, and so on.

Random signals can be carriers of information. We say that a signal contains information when we know something new because we received the signal. This implies that a signal containing the information is unpredictable to us, that is, is random in the sense given above. The examples of random signals given above illustrate the information-carrying capacity of random signals. An exception might be the music, which evokes an esthetic as well as an intellectual response.

Characterization of random signals.   If we are to design electronic circuits to monitor, record, transmit, receive, and process random signals, we must develop appropriate concepts to describe, or characterize, such signals. Probability and statistics are the branches of mathematics that deal with random phenomena. These types of mathematics do not allow us to predict random events but they do allow us to describe the structure of the randomness. Consequently, the application of probability theory to communication systems has yielded many important results.

It is not our aim here to explore communication theory. We will limit our characterization of random signals to that required for the discussion of simple communications systems, such as an AM radio. Our major goal is to show that random signals can be characterized by a power spectrum in the frequency domain. First we discuss the power in a random signal; then we discuss the time structure and show, in consequence, how the power in the signal is distributed in the frequency domain.

Power structure of random signals. We have sketched a random signal in Fig. 8.13. This represents, let us say, an electrical signal monitoring the temperature at a certain point in a chemical plant. We note that the average temperature would be indicated by the average value of the voltage, v_{dc}. We can denote this time average as

$$v_{dc} = \langle v(t) \rangle \qquad (8.15)$$

where the angle brackets indicate time average. This average could be approximated by averaging over a long period of time

$$v_{dc} = \langle v(t) \rangle = \frac{1}{T} \int_0^T v(t) \, dt \qquad (8.16)$$

Here T would have to be large enough to include many random fluctuations, say, a period of hours for the function in Fig. 8.13. We acknowledge the possibility that at some future time the plant might shut down and cool off, and hence the average temperature might change. But for normal operation we assume there to be a meaningful average temperature as represented by the dc voltage in Eq. (8.16).

We may define the "ac" component of the random signal as the remainder after the average is subtracted:

$$v_{ac}(t) = v(t) - v_{dc} = v(t) - \langle v(t) \rangle \qquad (8.17)$$

By "ac" we mean everything but the dc, and not "alternating current." Thus "ac" means "fluctuating" in this context. Note that, by definition, the time average of the ac component is zero:

$$\langle v_{ac}(t) \rangle = \langle v(t) - v_{dc} \rangle = \langle v(t) \rangle - v_{dc} = v_{dc} - v_{dc} = 0 \qquad (8.18)$$

Because averaging is a linear process, we have distributed the averaging operation to the individual terms in Eq. (8.18).

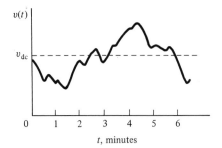

Figure 8.13   Random signal.

The time-average power of the random signal holds special interest to us. Leaving out the resistance for generality, we may define the power, P, to be

$$P = \langle v^2(t) \rangle = \langle [v_{ac}(t) + v_{dc}]^2 \rangle$$
$$= \langle v_{ac}^2(t) + 2v_{ac}(t)v_{dc} + v_{dc}^2 \rangle \quad \text{(volts)}^2$$

Again we may distribute the time average.

$$P = \langle v_{ac}^2(t) \rangle + 2\langle v_{ac}(t) \rangle v_{dc} + v_{dc}^2$$
$$= \langle v_{ac}^2(t) \rangle + v_{dc}^2 \quad \text{(volts)}^2 \qquad (8.19)$$

The middle term drops out because the time average of the ac component vanishes, as shown in Eq. (8.18). Equation (8.19) indicates that the total power of the signal may be considered as the sum of the powers in the dc and ac components, acting independently.

$$P = P_{ac} + P_{dc}$$

These two types of power produce distinct components in the power spectrum of the random signal.

Time structure and the power spectrum. The time structure of a random function is characterized by a *correlation time*. This is the time, on the average, during which the fluctuations in the signal become independent of the past. In Fig. 8.13, for example, the correlation time is about 1 min. This correlation time suggests how far into the future we might be able to predict the signal. If we know the voltage in Fig. 8.13 at some instant of time, we could predict its value 1 or 2 s later with confidence. But our uncertainty would grow as the prediction time moved into the future, and we can say little if anything about the value of the fluctuations, say, some 3 minutes hence. Thus correlation time characterizes the time structure of a random signal in a crude way.

Probability theory gives a precise definition of the correlation time and furnishes numerical methods for estimating it for a random signal, given that we analyze a sufficiently long sample of the signal. Investigation of such mathematical techniques, however, would lead us far beyond our goal of developing the ideas we need to understand basic communication systems.

Given that we know the correlation time for a signal, we can approximate the signal with a train of pulses, as shown in Fig. 8.14. We recognize that the approximation is poor for details of the fluctuation of the random signal. We introduce it merely as a crude representation of the time structure of the signal.

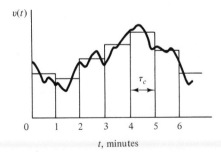

Figure 8.14 Random signal approximated by pulses.

Let us now consider the spectrum of the random signal. A spectrum is a representation in the frequency domain of a time-domain signal, that is, a representation as a sum of sinusoids. At first glance it would appear to be impossible to represent an unpredictable random signal with stable, predictable sinusoids. This is true in part, for a random signal cannot be represented with a stable sum of sinusoids. The signal contains sinusoids, but the phases of these sinusoids never stabilize but rather vary randomly with time. In effect, an amplitude spectrum exists for random signals but phase is meaningless. A power spectrum exists for a random signal but a voltage spectrum cannot be meaningfully defined.

The exact shape of the power spectrum of a random signal depends on the details of the nature of the fluctuations, but the meaningful bandwidth depends on the correlation time. If we were to consider the power spectrum of the pulses approximating the random signal in Fig. 8.14, all of them would have a significant bandwidth of $1/\tau_c$ hertz, where τ_c denotes the correlation time. Thus we would require approximately $\frac{1}{60}$ Hz of bandwidth to handle the random signal in Fig. 8.13, but if the correlation time were 1 μs, we would require about 1 MHz of bandwidth.

Thus the power spectrum $[S_P(f)]$ of a random signal will have one component due to the dc component and one component due to the ac component, as shown in Fig. 8.15. The dc component is represented by a discrete line at zero frequency, but the ac component is represented by a continuous spectrum. The units of this continuous spectrum are (volts)^2/hertz or (volts)^2-second. As shown, the area of the continuous power spectrum represents the total power in the random signal fluctuations and must give the same power as the time-domain representation,

$$P_{\text{ac}} = \langle v_{\text{ac}}^2(t) \rangle = \int_0^\infty S_{\text{ac}}(f)\, df \tag{8.20}$$

where $S_{\text{ac}}(f)$ represents the power spectrum of the fluctuations in the random signal. On Fig. 8.15 we have also shown the relationship between significant bandwidth and correlation time, as asserted above.

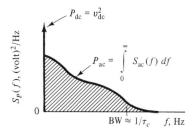

Figure 8.15 Power spectrum of a random signal.

8.2.5 Bandwidth and Information Rate

Information. Information is a word having meaning in both technical and ordinary language. We might describe a book as containing information, and we might understand the strange gestures of a baseball coach as offering information to his base runners. In these instances, we think of information as offering knowledge—matters to be observed and acted upon.

The technical definition of information can cover these possibilities but deals primarily with the machine transmission or storage of symbols. With this theory, we may study the exchange of information between two computers or we may calculate the information stored in a genetic code. This mathematical theory of information is of recent origin, springing from the work of Claude Shannon (1916–). This theory is based on probability concepts and deals with the probability of deciphering a known code that has been transmitted or stored in the presence of random disturbance. Information theory generally supports and enhances the commonsense understanding of information but also allows nonsense to be considered valid information. According to the mathematical theory, for example, the received message, "3 = 5," is valid information if that is what the transmitter sent. In other words, the theory measures and describes objective information (messages sent and received), not subjective information (the meaning, validity, and importance of such messages).

Information rate and bandwidth. Our purpose here is to show an important relationship between the information rate of a communication system and the bandwidth required by that communication system in the frequency domain. For our purposes, it is unnecessary to distinguish between the common understanding of information and that of the mathematical theory because this relationship is valid in both senses. Our examples will be based on the commonsense understanding of information. The conclusion we reach, however, can be supported with the mathematical theory. We consider one analog and one digital example.

First consider spoken language. What is the band of audio frequencies necessary to communicate, say, over a telephone line? Experiments have established that the essential bandwidth of audio frequencies lies between about 800 and 2400 Hz. That is, if someone spoke to you via a telephone system that passed that band of frequencies, you could certainly understand what they said. Although this would be the minimum bandwidth for communication to occur, the phone company allocates the band of frequencies between about 400 and 3300 Hz. The additional bandwidth is not required for intelligibility but rather for recognition of the speaker. For satisfactory telephone service we require not only that we can hear and understand the message but also that we be able to recognize the speaker's voice. This recognition factor represents additional information content to the phone message that is very important subjectively even if it does not add greatly to the content of the message.

An AM radio signal uses audio frequencies between 100 and 5000 Hz. This additional bandwidth is required for satisfactory transmission of music. Of course, FM radios sound better because their audio bandwidth is from 50 to 15,000 Hz. The large bandwidth enhances the quality of the music, a subjective measure of its information content.

A TV signal contains audio (voice) and video (picture). The audio bandwidth is similar to that of an FM radio, but the video requires much more bandwidth because the information rate is much higher. Of the 6-MHz bandwidth required for transmission of the entire TV signal, over 99% of the bandwidth is required for the video and synchronization signals, the remainder being used for the audio.

Thus we see a clear relationship between bandwidth and information rate in analog systems.

Consider now a digital communication system. You have probably done some interactive computing using a remote terminal connected to a computer over telephone lines. You have no doubt noticed that some computer terminals write a line of text slowly, such that you can read the words as they appear, whereas some write faster than one can read. The difference lies in the information rate between the computer and the terminal. Information rate in this case is described by the *baud rate*, where a *baud* is the unit for a bit per second. Typical baud rates are 110, 300, 1200, and 3000 baud. The lower two rates are appropriate for mechanical printers, which must print the symbols on paper, and the faster rates for CRT terminals.

Let us consider the bandwidth required for a data rate of 1200 baud. Each bit (pulse) will require $\frac{1}{1200}$ s to transmit, so the individual pulses will look as shown in Fig. 8.16. The pulse itself would occupy half the time space and hence the pulse width would be about $\frac{1}{2400}$ s. Using the relation given in Eq. (8.14), we would expect that a bandwidth of about 2400 Hz would be required to send such a pulse without significant distortion; and therefore we would expect that an ordinary phone line would be adequate for connecting the computer with the remote terminal. This is true for rates up to 1200 baud, but the 3000-baud system requires a special phone line. Therefore, for both the analog and digital systems, the higher information rates require larger bandwidth. One achievement of the mathematical theory of information was to place this relationship between bandwidth and information rate on a sound theoretical basis.

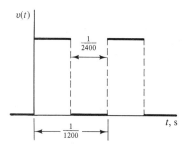

Figure 8.16 Digital pulse train, 1200 pulses/s.

8.2.6 Check Your Understanding

1. If the period of a periodic signal is 12 ms, what is the frequency in hertz of the third harmonic?
2. The most important harmonic output frequency of a power supply is the zeroth, first, second, or third harmonic of the input frequency. (Which?)
3. A random signal has a continuous spectrum except for one discrete frequency. What is that frequency?
4. If a pulse has a width of 10 ms, what is the bandwidth in hertz required to carry the pulse without significant distortion?
5. If a computer terminal writes 30 characters/second, and each character requires 8 bits, what is the baud rate and the required bandwidth of the channel supplying data to the terminal?

6. The voltage spectrum or power spectrum gives amplitude *and* phase information? (Which?)

7. A random signal has a correlation time of 2 ms. What is the required bandwidth for a communication channel to carry this signal?

8. Which type(s) of signals has(have) a continuous spectrum: sinusoidal, periodic, a single pulse, a random signal?

     Answers. 1. 250 Hz; 2. the dc is the zeroth harmonic; 3. zero hertz; 4. 100 Hz; 5. 240 baud, 480 Hz; 6. voltage spectrum; 7. 500 Hz; 8. single pulse and random signal.

.3 FILTERS

8.3.1 Simple *RC* Filters

    RC **low-pass filter.** Figure 8.17 shows a simple low-pass filter, so named because it will pass low frequencies and filter out, or eliminate, high frequencies. We will now derive the *filter function* of this circuit and show its effect on the spectra of signals. The analysis is based on the techniques we developed earlier in dealing with ac circuits. Specifically, we will use the principle of superposition and the methodology of ac circuits. Consider, for example, that the input to this circuit is a periodic series of pulses of height V volts as shown in Fig. 8.5a. To determine the output in the time domain, we must solve a differential equation. The DE is not impossible, but this method of solution generates little insight into the behavior of the circuit. We will approach the problem in the frequency domain.

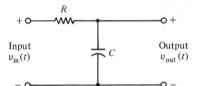

Figure 8.17 Low-pass filter.

    Filter response in the frequency domain. In Fig. 8.18 we show the impedance of the resistor and capacitor and derive the filter function, $\underline{F}(\omega)$, by treating the filter as a voltage divider. The characteristics of the low-pass filter can be understood best through considering the effect of frequency on the impedance of the capacitor. Recall that the impedance of a capacitor has a magnitude of $1/\omega C$. At dc the capacitor will act as an open circuit; hence all the input voltage will appear at the output, independent of R. At very low ac frequencies, the impedance of the capacitor will still be very high compared to R. Specifically, as long as $1/\omega C \gg R$, the output voltage will be approximately equal to the input voltage, and thus the filter gain will be near unity.

    At very high frequencies, the impedance of the capacitor will approach that of a short circuit and hence little voltage will appear across the capacitor. As long as $R \gg 1/\omega C$, the current will be approximately $\underline{V}_{in}/R$ and the magnitude of the voltage across the capacitor, which is the output voltage, will be $|\underline{V}_{in}|/\omega RC$ and hence will approach very small values as ω increases. Thus the gain of the filter

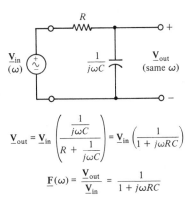

$$\mathbf{V}_{out} = \mathbf{V}_{in} \left(\dfrac{\dfrac{1}{j\omega C}}{R + \dfrac{1}{j\omega C}} \right) = \mathbf{V}_{in} \left(\dfrac{1}{1 + j\omega RC} \right)$$

$$\mathbf{F}(\omega) = \dfrac{\mathbf{V}_{out}}{\mathbf{V}_{in}} = \dfrac{1}{1 + j\omega RC}$$

Figure 8.18 Low-pass-filter function derivation.

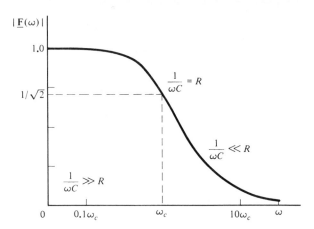

Figure 8.19 Low-pass-filter amplitude function.

becomes very small at high frequencies. This means that a high-frequency signal would be greatly reduced, whereas low-frequency components would not be reduced by the filter; that is, it will "pass" only components of the signal at low frequencies. The filter characteristics are shown in Fig. 8.19. We have used a log scale to show a large range of frequencies.

The region of transition between these two types of behavior is centered on the frequency where the impedance of the capacitor is equal to that of the resistance:

$$R = \dfrac{1}{\omega_c C} \Rightarrow \omega_c = \dfrac{1}{RC} \tag{8.21}$$

where ω_c is called the *cutoff frequency* and characterizes the region of frequency where filtering begins. Notice that the filter function can be expressed in terms of the cutoff frequency:

$$\mathbf{F}(\omega) = \dfrac{1}{1 + j\omega RC} = \dfrac{1}{1 + j(\omega/\omega_c)} = \dfrac{1}{1 + j(f/f_c)} \tag{8.22}$$

At $\omega = \omega_c$ (or $f = f_c$), the filter function has the value $1/(1 + j1)$, which has a magnitude of $1/\sqrt{2} = 0.707$. (This has nothing to do with the rms value of a sine wave.) The significance of a voltage gain of 0.707 lies in the relationship between input and output power. Power is proportional to the square of the voltage; hence at ω_c the output power is reduced by a factor of 2 from what it would have been in the absence of the filter. For this reason, the cutoff frequency (expressed in either rad/s or hertz) is often called the *half-power frequency*.

Phase effects. Figure 8.19 shows only the magnitude of the filter function, $|\mathbf{F}(\omega)|$. The filter function is a complex function with real and imaginary parts. Thus the filter affects both the magnitude and the phase of sinusoids passing through it. For example, at the cutoff frequency the filter function has the value $1/(1 + j1) = 0.707\underline{/-45°}$. This phase shift in an unavoidable by-product of the filtering process that can cause problems in some applications. In audio systems few problems occur due to phase shift because the ear is largely insensitive to phase, but in video systems the phase shift produced by filtering operations can degrade the appearance of the picture.

Frequency-domain analysis. We return to the task of passing a series of pulses through the low-pass filter. We can express the periodic pulses as a sum of sinusoids, each having a different frequency. Each of these represents an ac source and hence leads to a straight-forward ac circuit problem, which we can solve using phasors and impedance. Specifically, if we have an input sinusoid of frequency ω and phasor magnitude, $\underline{V}_{in}$, the output phasor can be determined by considering the circuit as a voltage divider or as a filter. Before exploring the implications of this result, however, let us review the approach that we are following. The basic idea is indicated by Fig. 8.20. This is similar to Fig. 4.4, except that now we are willing to let the input phasor be considered a spectrum of frequencies. Specifically, we must consider the effect of the filter on each input frequency. We know that the circuit, being a linear circuit, will create no new frequencies; each harmonic in the input will produce a harmonic in the output. Put another way, the frequencies in the output will be the same harmonics as exist in the input. The filter will affect each of these frequencies differently because the impedance of the capacitor varies with frequency. This frequency dependence of the circuit is symbolized by the filter function, $\underline{F}(\omega)$. For this circuit the filter function is derived in Figure 8.18 as

$$\underline{F}(\omega) = \frac{1}{1 + j\omega RC} \tag{8.23}$$

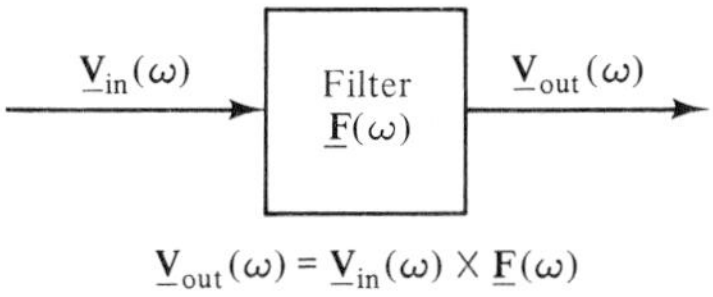

$$\underline{V}_{out}(\omega) = \underline{V}_{in}(\omega) \times \underline{F}(\omega)$$

Figure 8.20 Filtering a spectrum.

Passing the signal through the filter. We now can pass the series of pulses in Fig. 8.5a through the filter in Fig. 8.17 by treating separately each frequency component of the signal. Our approach is suggested by Fig. 8.21, where we have indicated that the input pulses can be represented as a series of voltage sources representing the dc component, the fundamental, the third harmonic, the fifth harmonic, and so on. (Recall that the even harmonics are missing from this Fourier series.) Our analysis is based on superposition, adding up at the output the effect of each input sinusoidal source, considered separately. This exchanges the solution of a DE in the time domain for a host of ac circuit solutions in the frequency domain. Fourier analysis allows us to determine the amplitude and phase of the harmonics of the input signal. This is symbolized as

$$v_{in}(t) = \text{Re}\{V_{in(dc)} + \underline{V}_{in(1)}e^{j\omega_0 t} + \underline{V}_{in(2)}e^{j2\omega_0 t} + \cdots + \underline{V}_{in(n)}e^{jn\omega_0 t} + \cdots\} \tag{8.24}$$

where $\underline{V}_{in(n)}$ represents the amplitude and phase of the nth harmonic of the input signal. We must next evaluate the filter function at each harmonic frequency to

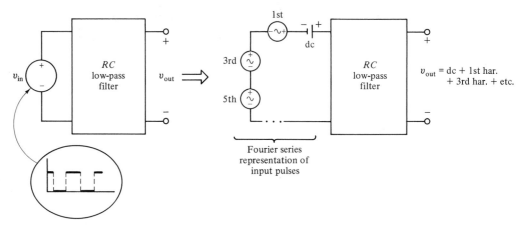

Figure 8.21 Filtering a periodic signal in the frequency domain.

see how each is affected in amplitude and phase by the filter. This is symbolized by

$$V_{\text{out(dc)}} = \mathbf{\underline{F}}(0) \times V_{\text{in(dc)}}$$

$$\mathbf{\underline{V}}_{\text{out(1)}} = \mathbf{\underline{F}}(\omega_0) \times \mathbf{\underline{V}}_{\text{in(1)}} \qquad (8.25)$$

$$\mathbf{\underline{V}}_{\text{out(n)}} = \mathbf{\underline{F}}(n\omega_0) \times \mathbf{\underline{V}}_{\text{in(n)}}$$

where $\mathbf{\underline{V}}_{\text{out(n)}}$ represents the amplitude and phase of the nth output harmonic and $\mathbf{\underline{F}}(n\omega_0)$ is the filter function evaluated at the frequency of the nth harmonic. Finally, we sum the frequency components of the output and transform back to the time domain, as shown in

$$v_{\text{out}}(t) = \text{Re}\{V_{\text{out(dc)}} + \mathbf{\underline{V}}_{\text{out(1)}}e^{j\omega_0 t} + \mathbf{\underline{V}}_{\text{out(2)}}e^{j2\omega_0 t}$$

$$+ \cdots + \mathbf{\underline{V}}_{\text{out(n)}}e^{jn\omega_0 t} + \cdots\} \quad (8.26)$$

Numerical example. We will work through the details for one case. We let the input signal consist of pulses of 1-V amplitude, $\frac{1}{2}$-ms duration, and a 1-ms repetition period. This produces a fundamental frequency of 1 kHz, a third harmonic at 3 kHz, and so on. For our filter we will use a resistance of 2 kΩ and a capacitor of 0.1 μF. This corresponds to a cutoff frequency of 5000 rad/s, or about 800 Hz. Thus the fundamental of the input signal is approximately equal to the cutoff frequency of the filter and all the higher harmonics well exceed the cutoff frequency. We would therefore anticipate that the higher harmonics would be reduced substantially by the filter compared with the fundamental. The details are given in Table 8.2. The Fourier series of the output voltage is the sum of the harmonics in the last column of Table 8.2, as expressed in

$$v_{\text{out}}(t) = 0.5 + 0.397 \cos(\omega_0 t - 51.6°) + 0.0544 \cos(3\omega_0 t - 255.1°)$$

$$+ 0.0200 \cos(5\omega_0 t - 81.0°) + 0.0103 \cos(7\omega_0 t - 264°) + \cdots \qquad (8.27)$$

This voltage is plotted in Fig. 8.22. The shape of the pulses is radically altered by the filter. The dc component and the fundamental are evident, but the sharp

TABLE 8.2 CALCULATION OF OUTPUT HARMONICS

Harmonic Number, n	Frequency, $\omega_n = 2\pi(nf_0)$	Input Phasor, $\underline{V}_{in(n)}$	Filter Function, $\underline{F}(\omega_n) = \dfrac{1}{1 + j\omega_n RC}$	Output Phasor, $\underline{V}_{out(n)}$
0	dc	0.5	1	0.5
1	$2\pi(1000)$	$0.637\underline{/0°}$	$\dfrac{1}{1 + j1.26} = 0.623\underline{/-51.6°}$	$0.397\underline{/-51.6°}$
3	$2\pi(3000)$	$0.212\underline{/-180°}$	$\dfrac{1}{1 + j3.77} = 0.256\underline{/-75.1°}$	$0.0544\underline{/-255.1°}$
5	$2\pi(5000)$	$0.127\underline{/0°}$	$\dfrac{1}{1 + j6.28} = 0.157\underline{/-81.0°}$	$0.0200\underline{/-81.0°}$
7	$2\pi(7000)$	$0.091\underline{/-180°}$	$\dfrac{1}{1 + j8.80} = 0.113\underline{/-83.5°}$	$0.0103\underline{/-264°}$

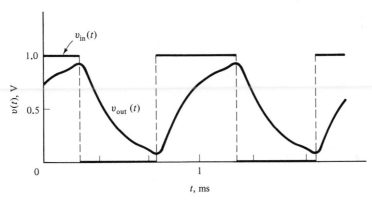

Figure 8.22 Filter input and output in the time domain.

corners of the input pulses are gone. This results from filtering out the higher harmonics, which originate in the steep slope and sharp corners of the input pulses.

8.3.2 Bode Plot

Decibel scale. The gain of amplifiers and the loss of filters are frequently specified in *decibels* (dB). This unit refers to a logarithmic measure of the ratio of output power to input power, as defined in

$$\text{gain (dB)} = 10 \log \frac{P_{out}}{P_{in}} \qquad (8.28)$$

Because the power gain is proportional to the square of the voltage gain, the voltage gain is defined to be

$$\text{gain (dB)} = 20 \log \left| \frac{\underline{V}_{out}}{\underline{V}_{in}} \right| \tag{8.29}$$

For example, a voltage gain of 50 would be 20 log (50) = 34.0 dB. We note in comparing Eqs. (8.28) and (8.29) that power ratios require a factor of 10 in the dB calculation and voltage ratios a factor of 20.

A loss can also be described by Eq. (8.28) but the "dB gain" is negative for a loss: for example, the gain of a lowpass filter [Eq. (8.22)] at $f = f_c$ was shown to be $1/\sqrt{2}$; Eq. (8.29) yields

$$G_{dB} = 20 \log | \underline{F}(\omega_c) | = 20 \log \frac{1}{\sqrt{2}} = -3.0 \text{ dB}$$

Thus we could say that the gain of the filter at its half-power frequency is negative 3.0 dB or, equivalently, that its loss is 3.0 dB.

The dB scale is also useful for expressing the power in a signal. In this context, the power is expressed as a logarithmic ratio between the signal power and an assumed power level, usually 1 W or 1 mW. Thus dBw, dB relative to 1 W, is defined as

$$\text{dBw} = 10 \log \frac{P}{1 \text{ watt}}$$

where P is a power and dBw is that same power relative to 1 W, expressed in dB. For example, a satellite transmitter with an output power of 1000 W would have an output power of 30 dBw. This power could also be called 60 dBm, where dBm means dB relative to a milliwatt.

Why dB is useful. There are three reasons why electrical engineers prefer the dB scale for describing gains, losses, and signal levels. Two of these reasons can be demonstrated by an example. Figure 8.23 suggests a communication satellite in stationary orbit around the earth. The satellite transmitter produces 1000 W and operates at a frequency of 5 GHz [λ (wavelength) = 6 cm]. We will calculate the power collected by an earth antenna. The formula for received power is derived in Chapter 10.

$$P_{rec} = \frac{P_t G_t G_r}{(4\pi D/\lambda)^2} \tag{8.30}$$

where P_{rec} = received power, to be calculated

P_t = transmitted power; assume 1000 W (30.0 dBw)

G_t = transmitting antenna gain; assume 400 ($G_{t(dB)}$ = 26.0 dB)

G_r = receiving antenna gain; assume 5000 ($G_{r(dB)}$ = 37.0 dB)

D/λ = distance between antennas/wavelength of the signal; assume 22,000 miles/6 cm

With these values, the term in the denominator, which is called the *space loss* (*SL*), has the value 5.50×10^{19} or 197.4 dB.

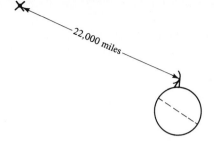

Figure 8.23 Satellite communication system.

If we evaluate Eq. (8.30) in the normal fashion, we get

$$P_{\text{rec}} = \frac{(1000)(400)(5000)}{5.50 \times 10^{19}} = 3.64 \times 10^{-11} \text{ W}$$

To use dB in the calculation of received power, we take the common log of Eq. (8.30) and multiply by 10 (it is a power equation), with the result

$$P_{\text{rec(dB)}} = P_{t\text{(dB)}} + G_{t\text{(dB)}} + G_{r\text{(dB)}} - SL_{\text{(dB)}}$$

$$= 30.0 \text{ dBw} + 26.0 \text{ dB} + 37.0 \text{ dB} - 197.4 \text{ dB} = -104.4 \text{ dBw}$$

The minus sign on the space-loss term occurs because this term is in the denominator. The answers are equivalent: $10 \log (3.64 \times 10^{-11}) = -104.4$ dBw.

This example illustrates two virtues of dB. Because of the compressive nature of the logarithmic function, the numbers involved in the dB calculation are quite moderate compared to the other calculation. This feature of dB measure is particularly useful when plotting quantities that vary greatly in magnitude, as will be illustrated below. The second virtue of dB measure is that the calculation is accomplished by addition and subtraction, rather than by multiplication and division. This feature results from the familiar technique for multiplying numbers by adding their logarithms. In these days of calculators this seems no great benefit, but the custom of using dB became universal among electrical engineers in an earlier day and will no doubt continue in the future.

Bode plot.   A third virtue of dB measure involves a special way of plotting the characteristic of a filter, amplifier, or spectrum. The Bode plot (named for H. W. Bode, 1905–) is a log power versus log frequency plot. For example, Fig. 8.24 shows the Bode plot of the function

$$\underline{A}(f) = \frac{100}{1 + j(f/10)}$$

which combines an amplifier (voltage gain of $100 = 40$ dB) with a low-pass filter ($f_c = 10$ Hz). In Fig. 8.24 the vertical axis is proportional to log power expressed in dB. The horizontal axis expresses frequency on a log scale, although normally the frequency (not the logarithm of the frequency) is marked on the graph, as in Fig. 8.24. This is a log plot because equal spacings on the graph are given to equal ratios of frequency (1:10, 10:100, and so on).

Sec. 8.3   Filters

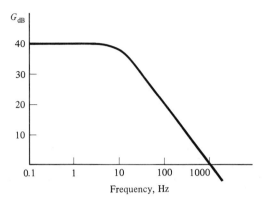

Figure 8.24 Bode plot.

The Bode plot is useful in two ways. As mentioned earlier, the use of a log scale allows the representation of a wide range of both power and frequency. In this case, the vertical scale of 0 to 40 dB represents a range of power gain from 1 to 10^4 (or a voltage gain from 1 to 100). Similarly, in the horizontal scale, we have represented four decades (factors of 10) in frequency. These wide ranges are possible because of the compressive nature of the logarithmic function.

The other advantage of the Bode plot is that a filter characteristic can often be well represented by straight lines. The straight lines are evident in Fig. 8.24: the characteristic is quite flat for frequencies well below $f_c = 10$, and slopes downward with constant slope for frequencies well above this value. Only in the vicinity of f_c does the exact Bode plot depart significantly from the straight lines. In the following section we examine in more detail the relationship between the (exact) Bode plot and the straight-line approximation, which is called the *asymptotic Bode plot*.

Bode plot for a low-pass filter. As shown in Fig. 8.18 and Eq. (8.22), the low-pass filter in Fig. 8.25 has the characteristic

$$\mathbf{F}(f) = \frac{1}{1 + j(f/f_c)}$$

The Bode plot represents the magnitude of the voltage gain and hence requires the absolute value

$$G_{dB}(f) = 20 \log \left| \frac{1}{1 + j(f/f_c)} \right| = 20 \log \frac{1}{\sqrt{1 + (f/f_c)^2}}$$

$$= -10 \log \left[1 + \left(\frac{f}{f_c} \right)^2 \right]$$

(8.31)

We may generate the exact Bode plot using normalized frequency f/f_c by numerical calculation, as shown in Fig. 8.26, where we emphasize the tendency of the exact Bode plot to approach a straight line above and below the cutoff frequency, $f/f_c = 1$ on the normalized frequency scale. At the cutoff frequency, the filter gain is -3.0 dB, as noted earlier.

We may derive the asymptotic Bode plot by taking the limiting forms of Eq.

Analog Electronics   Chap. 8

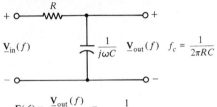

$$\underline{F}(f) = \frac{\underline{V}_{out}(f)}{\underline{V}_{in}(f)} = \frac{1}{1 + j(f/f_c)}$$

Figure 8.25 Low-pass filter.

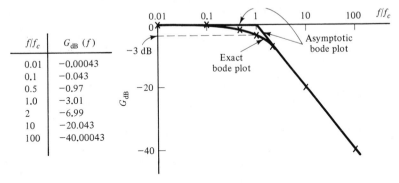

f/f_c	$G_{dB}(f)$
0.01	−0.00043
0.1	−0.043
0.5	−0.97
1.0	−3.01
2	−6.99
10	−20.043
100	−40.00043

Figure 8.26 Bode plot for the low-pass filter.

(8.31). For frequencies well below the cutoff frequency, $f \ll f_c$, the gain is approximately

$$G_{dB} \approx -10\log(1 + small) \approx 10\log(1) = 0, \qquad f \ll f_c \qquad (8.32)$$

This accounts for the flat section. For frequencies well above the cutoff frequency, the gain approaches

$$G_{dB} \to -10\log\left(\frac{f}{f_c}\right)^2 = -20\log f + 20\log f_c, \qquad f \gg f_c \qquad (8.33)$$

Recalling that our horizontal variable is $\log f$, we see that Eq. (8.33) is the form of a straight line, $y = mx + b$, where $m = -20$, x is $\log f$, and b is $20\log f_c$. Thus the high-frequency asymptote will be a straight line in this coordinate system. When $\log f = \log f_c$, the y of the straight line has zero value and hence the line passes through the horizontal axis at $f = f_c$. The slope is -20: usually we say that the slope is negative 20 dB/decade because dB of gain and decades of frequency are the units for a Bode plot.

The low- and high-frequency asymptotes of the exact Bode plot combine to form the asymptotic Bode plot. As shown in Fig. 8.26, the exact Bode plot is well represented by this approximation, the largest error being 3.0 dB at the cutoff frequency. For most filter applications, we may use the asymptotic Bode plot to give an adequate picture of the filter chartacteristics.

Bode plot for a high-pass filter. Figure 8.27 shows a filter that will block low frequencies and pass high frequencies. The filter characteristic can be determined by considering the circuit as a voltage divider in the frequency domain

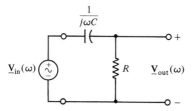

Figure 8.27 High-pass filter.

$$\underline{F}(\omega) = \frac{\underline{V}_{out}(\omega)}{\underline{V}_{in}(\omega)} = \frac{R}{R + 1/j\omega C} = \frac{j\omega RC}{1 + j\omega RC} = \frac{j(f/f_c)}{1 + j(f/f_c)} \quad (8.34)$$

where $f_c = 1/2\pi RC$ is the cutoff frequency. At low frequencies, $f \ll f_c$, the capacitor has a high impedance and allows little current; hence little voltage will develop across the resistor. Indeed, at zero frequency (dc), the capacitor acts as an open circuit and no voltage appears at the output. At high frequencies, $f \gg f_c$, the capacitor impedance will be low, and essentially all the voltage appears across the resistor. Consequently, the filter passes high frequencies and blocks low frequencies. The transition between these two regimes occurs when the impedance of the capacitor is comparable to that of the resistor, and the cutoff frequency occurs where they are equal, as given by Eq. (8.21).

The asymptotic Bode plot for the filter function given in Eq. (8.34) may be derived by the same method as we used for the low-pass filter. The Bode plot is derived from the absolute value of the filter function,

$$G_{dB} = 20 \log \left| \frac{j(f/f_c)}{1 + j(f/f_c)} \right| = 20 \log \frac{f/f_c}{\sqrt{1 + (f/f_c)^2}} \quad (8.35)$$

At frequencies well below the cutoff frequency, the frequency term in the denominator can be neglected and the asymptote becomes

$$G_{dB} \rightarrow 20 \log \frac{f}{f_c} = 20 \log f - 20 \log f_c$$

Plotted against log f, this is a straight line having a slope of $+20$ dB/decade and passing through the horizontal axis at $f = f_c$, as shown in Fig. 8.28. For high frequencies, the filter function approaches unity, or 0 dB; this asymptote is also shown in Fig. 8.28. The exact Bode plot for the filter combines these two asymptotes and makes a smooth transition between them near the cutoff frequency. At

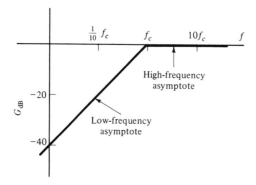

Figure 8.28 Asymptotic Bode plot for a high-pass filter.

Analog Electronics   Chap. 8

the cutoff frequency, the exact Bode plot lies 3 dB below the intersection of the low- and high-frequency asymptotes.

Combining filter characteristics. Figure 8.29 shows a combination of a low-pass filter, an amplifier, and a high-pass filter. We will examine their combined characteristic. The overall input–output characteristics proves to be the product of the individual characteristics, as shown by

$$\underline{F}(f) = \frac{\underline{V}_4}{\underline{V}_1} = \frac{\underline{V}_2}{\underline{V}_1} \times \frac{\underline{V}_3}{\underline{V}_2} \times \frac{\underline{V}_4}{\underline{V}_3} = \underline{F}_1(f) \times \underline{F}_2(f) \times \underline{F}_3(f) \qquad (8.36)$$

By expressing the total gain in dB, we convert the product to a sum

$$G_{dB} = 20 \log |\underline{F}_1 \times \underline{F}_2 \times \underline{F}_3| = 20 \log |\underline{F}_1| + 20 \log |\underline{F}_2| + 20 \log |\underline{F}_3|$$

$$= G_{1(dB)}(f) + G_{2(dB)}(f) + G_{3(dB)}(f)$$

Figure 8.30 shows the effect of adding up the individual Bode plots. The amplifier has a constant gain of 20 dB and raises the sum of the filter Bode plots by that amount. The high-pass filter blocks the low frequencies, and the low-pass filter blocks the high frequencies. The result is a bandpass filter; the "band" being passed in this case contains the frequencies between 100 and 1000 Hz. Thus we combine filter effects by adding the Bode plots.

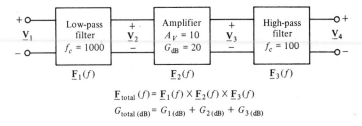

Figure 8.29 Two filters and an amplifier in cascade.

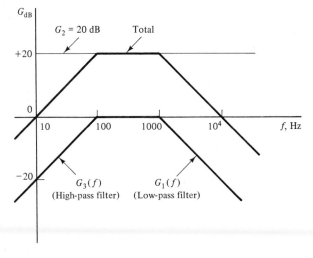

Figure 8.30 Adding Bode plots.

Narrowband *RLC* filters. As shown above, we can produce a filter to pass the frequencies in a certain band by combining high-pass and low-pass *RC* filters. This approach works well only when the passband of frequencies is rather broad. However, most communication systems, such as radios, require filters that pass a narrow band of frequencies. This is normally accomplished with an *RLC* filter such as appears in Fig. 8.31. This particular *RLC* filter produces a parallel resonance between the inductor and capacitor at about 1125 kHz, which is near the middle of the AM radio band. The resistance shown in parallel with the inductor is not present in the physical circuit but is placed in the circuit to represent the losses of the inductor. For frequencies far away from the resonant frequency, the impedance of either the inductor or the capacitor becomes small and the filter response drops. At the resonant frequency, the inductance and capacitance resonate and the output is maximum.

Figure 8.32 shows the passband characteristics of this filter in linear and dB scales. It might be noted that this filter has an inherent loss, even at its maximum output, -20.8 dB in this case. Thus such a filter must be used in conjunction with an amplifier to compensate for this loss. This is no disadvantage in most communication circuits, for the radio already must provide considerable amplification—this merely requires a bit more. We might mention in closing this section that the filter characteristic of the *RLC* filter in Fig. 4.39 is similar to that in Fig. 8.31. The parallel form of the filter is accomplished more easily in electronic circuits with realistic inductors and capacitors, and hence this is the form of narrowband filter commonly found in radios. In Chapter 10, after we have dis-

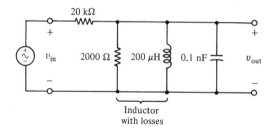

Figure 8.31 *RLC* filter.

Figure 8.32 Narrowband filter characteristics: (a) linear plot; (b) dB plot.

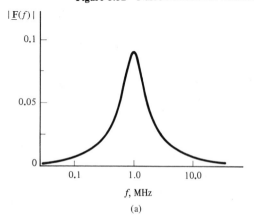

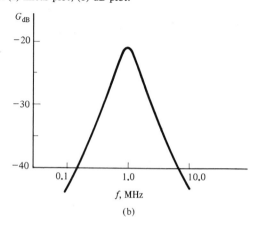

(a)

(b)

Analog Electronics   Chap. 8

cussed the properties of nonlinear circuits, we will discuss the role of such filters in typical radio circuits.

8.3.3 Check Your Understanding

1. A filter will normally change the phase as well as the amplitude spectrum of the input signal. (T/F?)
2. An RC high-pass filter will pass a dc signal. (T/F?)
3. If one decade is a factor of 10, what ratio corresponds to one-fourth of a decade?
4. If the input to an amplifier is 5 mW and the output is 0.1 W, what is its gain in dB if input and output impedance levels are the same?
5. If an amplifier reduces the signal voltage by a factor of 4, what is its gain in dB?

Answers. 1. True; 2. false; 3. 1.778; 4. 13.0 dB; 5. −12.0 dB.

8.4 FEEDBACK CONCEPT

8.4.1 Example

Amplifier with feedback. Figure 8.33 shows an amplifier that has been modified from straight amplification by having a sample of the output brought back to interact with the input. A main amplifier is represented by the top box: its input (v_i) and its output (v_{out}) are related by the main amplifier gain A, as indicated in the box. We assume A to be real and positive. In the main amplifier, the signal goes from left to right.

A feedback circuit, represented by the bottom box, consists in this example of two resistors arranged as a voltage divider. In the feedback circuit, the signal goes from right to left, that is, from the output of the main amplifier back to its input. The input to the feedback circuit is v_{out} and its output is v_f.

A voltage source, v_{in}, is located at the input to the entire "amplifier with feedback," that is, the entire system, feedback and all. The output of the feedback

Figure 8.33 Amplifier employing feedback.

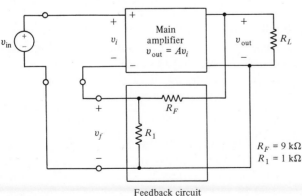

Feedback circuit

circuit, v_f, is part of the input loop, which also contains the voltage source and the input to the main amplifier.

Analysis of amplifier gain. Our goal in this section is to calculate the gain of the amplifier with feedback, A_f. The gain with feedback is the ratio of the output to the source voltage

$$A_f = \frac{v_{\text{out}}}{v_{\text{in}}} \tag{8.37}$$

We can write three equations describing the system. These are:

1. Definition of the main amplifier:

$$v_{\text{out}} = Av_i \tag{8.38}$$

2. Voltage divider for the feedback circuit:

$$v_f = \frac{R_1}{R_1 + R_F} v_{\text{out}} \tag{8.39}$$

3. KVL around the input loop:

$$-v_{\text{in}} + v_i + v_f = 0 \tag{8.40}$$

Equations (8.38) through (8.40) give us three equations relating four voltages. We can eliminate two voltages between the equations, reducing to one equation in two voltages. We require the ratio of v_{out} to v_{in} for computing the gain in Eq. (8.37), so we eliminate v_f and v_i. Some straightforward algebra results in

$$A_f = \frac{A(R_1 + R_F)}{R_1 + R_F + R_1 A} = \frac{A}{1 + [R_1/(R_1 + R_F)]A} \tag{8.41}$$

Equation (8.41) is representative of a more general relationship that we will develop presently. We conclude from Eq. (8.41) that the gain with feedback, A_f, is smaller than the gain without feedback, A. Having lost gain, what have we achieved? The answer is that we have improved the reliability, linearity, sensitivity, and bandwidth of the amplifier. We will verify these benefits later in this section. First we investigate how feedback works.

Signal levels. We now examine the signal levels for a typical amplifier with feedback. The gain of our main amplifier will be $A = 200$ and the feedback circuit will consist of $R_F = 9$ kΩ in series with $R_1 = 1$ kΩ. According to Eq. (8.41), the gain of the amplifier with feedback would be

$$A_f = \frac{200}{1 + [1 \text{ k}\Omega/(1 \text{ k}\Omega + 9 \text{ k}\Omega)](200)} = 9.52 \tag{8.42}$$

We assume that the output voltage is 10 V and calculate the signal levels throughout the circuit. With 10 V at its output, the input voltage to the main amplifier, v_i, must be $10/200 = 0.05$ V. The input voltage to the entire amplifier would be $v_{\text{in}} = 10/9.52 = 1.050$ V. The output of the feedback circuit, v_f,

$$v_f = \frac{R_1}{R_1 + R_F} v_{\text{out}} = \frac{1\ k\Omega}{1\ k\Omega + 9\ k\Omega} (10) = 1.000\ \text{V}$$

Hence we confirm that KVL is satisfied in the input loop, as it must be if our derivation is valid.

The calculation performed in the preceding paragraph moves from the output back to the input. The signal would go the other way, so let us think of the operation of the amplifier in time sequence. We apply 1.050 V at the input at some instant of time. Electronic circuits respond quickly but not instantaneously; hence the output signal at this initial instant is zero. Thus the feedback voltage, v_f, will at first be zero and the entire 1.050-V input would appear at the main amplifier input, v_i. This signal is amplified by the main amplifier and the output voltage increases toward 200×1.050; but as the output increases, so does the feedback signal. Because the feedback signal subtracts from the 1.050-V input, the input voltage to the main amplifier diminishes as the output voltage rises. Thus we have competing effects: the higher the output voltage rises, the more voltage feeds back and the more the input voltage to the main amplifier is reduced. In the end, the feedback signal subtracts 1.000 V from the input voltage of 1.050 V and the remaining 0.050 V is amplified by the main amplifier to yield an output voltage of 10 V.

8.4.2 System Model

System notation. A system representation of the feedback amplifier is shown in Fig. 8.34. This representation is characterized by the various blocks and circles connected with lines representing signal flow. Although our signals are voltages, customarily denoted with v's, we have used a neutral symbol x to denote the signals on the system diagram in Fig. 8.34. Use of a neutral symbol is important because the *signal*, not the physical variable that represents the signal, is of primary importance in the system viewpoint. At some places in the system the signal might be represented by a voltage, at others a current, at yet others a pressure or the physical displacement of a mechanical component. The system representation embraces such hybrid systems.

System components. We have shown four system components in Fig. 8.34. The main amplifier and the feedback circuit are obvious carryovers from

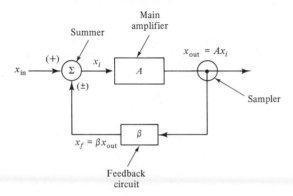

Feedback
circuit

Figure 8.34 System representation of feedback.

the original formulation of the amplifier. Note that the direction of signal flow is indicated with arrowheads. At the output we have shown a "sampling" connection; this represents the parallel connection to the output of the main amplifier. In other feedback arrangements, the feedback network might be placed in series with the load and main amplifier output; with this arrangement the feedback network would sample the output current. Both possibilities are covered by the sampling symbol.

At the input we have drawn a circle containing a summation symbol to indicate the interaction of signals at the input of the system. In our previous example this circle represents KVL in Eq. (8.40), which can be put into the form

$$v_i = +v_{\text{in}} - v_f \qquad (8.43)$$

where the $+$ and $-$ signs on the summer in Fig. 8.34 are associated with the input and the feedback signals, respectively. We may think of the summer as having a gain of $+1$ for v_{in} and a gain of -1 for v_f. The summer thus works as a "differencer" in this particular case, but the symbol covers both possibilities. In other contexts the summer at the input might be called a "comparator" because it compares the input with a sample of the output, furnishing the difference as input to the main amplifier.

Analysis of the system. The equations of the system are:

1. Main amplifier:

$$x_{\text{out}} = Ax_i$$

2. Feedback circuit:

$$x_f = \beta x_{\text{out}}$$

3. Summer:

$$x_i = x_{\text{in}} \pm x_f$$

In our equation for the summer, we have considered that the gain for the input signal is $+1$ but allowed for either $+1$ or -1 for the feedback signal. The main amplifier gain, A, may be positive or negative. As before, we can eliminate two of the variables and solve for the ratio of the output and input signals to obtain the gain with feedback.

$$A_f = \frac{x_{\text{out}}}{x_{\text{in}}} = \frac{A}{1 - (\pm 1)(\beta)(A)} \qquad (8.44)$$

Equation (8.44) reduces to our earlier result [Eq. (8.41)] when the minus sign is used for the summer and $R_1/(R_1 + R_F)$, the gain of the voltage-divider circuit, is used for β.

Loop gain. Another form for Eq. (8.44) is

$$A_f = \frac{A}{1 - L} \qquad (8.45)$$

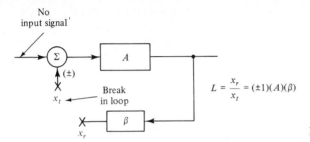

$$L = \frac{x_r}{x_t} = (\pm 1)(A)(\beta)$$

Figure 8.35 Derivation of loop gain.

where $L = (\pm 1)(\beta)(A)$ is called the *loop gain*. The loop gain is an important concept in feedback theory and is defined as the product of the gains of all the system components around the feedback loop. The basic idea of the loop gain is suggested in Fig. 8.35. To calculate (or measure in the laboratory) the loop gain, break the feedback loop at some convenient point, insert a test signal, x_t, and calculate (or measure) the return signal, x_r. The loop gain is the ratio $L = x_r/x_t$, and is thus the gain around the loop. The loop may be broken, both in analysis and in the laboratory, only at a point where the function of the various system components is unimpaired. Care must be taken to terminate the break with the same impedance as the circuit saw before the break. For example, it would be inconvenient in the circuit in Fig. 8.33 to open the loop between R_F and R_1.

Importance of the loop gain. The sign of the loop gain indicates the nature of the feedback. Negative loop gain indicates negative feedback. With negative feedback, the feedback signal subtracts from the input signal and the gain is reduced. Positive loop gain indicates positive feedback and is rarely used. Positive feedback increases the amplifier gain; indeed, for $L = +1$ the gain becomes infinite. This condition is used in the design of electronic oscillators, which can be considered amplifiers with output but no input.

The magnitude of the loop gain indicates the importance of feedback properties in influencing the system characteristics. When the magnitude of the loop gain is much larger than unity, the system properties are controlled largely by feedback considerations. We will illustrate the importance of loop gain in the next section.

8.4.3 Benefits of Negative Feedback

Improved static stability. By *static stability* we mean insensitivity of performance to changes in the system parameters. For instance, in our earlier example the gain of the main amplifier might decrease due to the deterioration of a circuit component or replacement of a transistor. Let us say the gain of the main amplifier decreases from 200 to 100. Without feedback, this change might jeopardize the usefulness of the amplifier. With feedback, however, the gain becomes

$$A_f' = \frac{100}{1 - (-1)(0.1)(100)} = 9.09$$

Thus a 50% decrease in the gain of the main amplifier results in a 5% decrease in the overall gain of the system. We can see from Eq. (8.45) why this happens. When the loop gain is much greater than unity, the gain with feedback is approximately

$$A_f = \frac{A}{1 - L} \approx \frac{A}{-L} = \frac{A}{-(-1)(\beta)(A)} = \frac{1}{\beta} \qquad (8.46)$$

We see that, as loop gain increases, the gain approaches a value dependent on β. Thus the reduction of the main amplifier gain has little effect as long as the loop gain remains much larger than unity. The value of β could depend on resistors only, as in Fig. 8.33, and thus be stable over long periods of time. In our example the reciprocal of β is 10, and we see that the gain of the system, both before and after the gain decrease, falls close to this value. Often in a practical system, the main amplifier merely provides sufficient gain to keep the loop gain much larger than unity, for in this case the β of the feedback circuit determines the overall gain with feedback.

We have used the word "static" to distinguish this type of stability from dynamic stability, the tendency of the system to vibrate under the influence of external stimuli. With a bridge, for example, the static stability might be good, meaning that the bridge footings are sound and the bridge members are sufficiently stiff to hold the bridge solidly in place. But under the influence of traffic or wind, the bridge might shake and even collapse, as did the Tacoma Narrows bridge in 1940, and hence exhibit poor dynamic stability.

Although negative feedback improves static stability, feedback can cause dynamic instability in a system. Time delays, or the frequency-domain equivalent (phase shift), can change negative feedback to positive feedback and hence cause dynamic instability or oscillations. This aspect of feedback theory will be discussed in Chapter 11.

Improved linearity.   Semiconductor devices such as transistors can be moderately linear in their active regions if signal variations are kept small, but large signals are often required. For example, the output amplifier in an audio system must produce a large signal to drive the speakers. Without feedback, such an amplifier would produce substantial distortion, but feedback techniques can be used to improve the linearity of the amplifier and thus reduce the distortion.

Equation (8.46) suggests how this is accomplished. The main amplifier introduces distortion into the system. If the loop gain is high, the system gain is determined by the β of the feedback circuit, which depends on two resistors in this case. The resistors are linear components; hence the amplifier will produce minimal distortion if the loop gain is high. Thus a large loop gain is always used when good linearity is required for large-signal amplification.

Disturbance reduction.   Feedback can reduce unavoidable disturbances that enter a system. As an example, we will consider the thermostatically controlled oven shown in Fig. 8.36. The input voltage to the system is compared with a voltage from a temperature sensor in the oven. The difference (v_d) is amplified and applied to a heater element (v_H). Because of the delay between the signal to

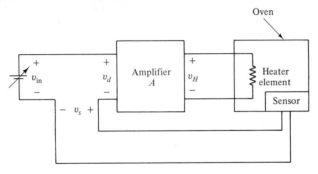

Figure 8.36 Oven controller.

the heater and the corresponding response from the sensor, the dynamic stability of this system is problematic. Chapter 11 considers the dynamic stability of this system.

A simplified system diagram is shown in Fig. 8.37. The heater element has been represented as a box having a gain A_H, which is defined as the change in box temperature divided by the change in the heater voltage. Although the relationship between heat (power) and voltage is nonlinear, we are for simplicity treating small changes as linear effects. The oven temperature is influenced both by the internal heater and by the ambient temperature, whose changes represent a disturbance in the system. This effect is represented with a summer. The sensor is also represented by a gain factor, A_S, which would represent changes in sensor output voltage divided by changes in oven temperature.

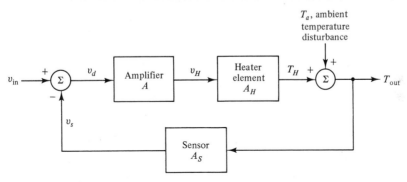

Figure 8.37 System diagram for oven controller.

The loop gain of the system is

$$L = -AA_H A_S$$

Thus the relationship between oven temperature and input voltage would be

$$T_{out} = v_{in} \times \frac{AA_H}{1 - L}$$

The disturbance signal, T_a, represents a second input to the system. We redraw the system diagram in Fig. 8.38 to clarify this role. Considering the gain of the

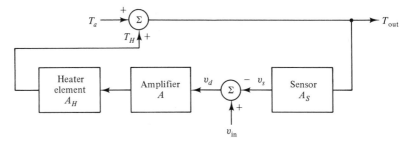

Figure 8.38 System diagram for controller with T_a as input.

summer to be $+1$, we can determine a relationship between ambient and oven temperature:

$$T_o = T_a \frac{1}{1 - L}$$

We want a large loop gain to reduce the effect of changes in the ambient temperature. For example, with a loop gain of -100, a $10°$ change in ambient temperature would cause a $0.1°$ change in oven temperature. Thus the effect of ambient temperature can be reduced to low levels, provided the system can be stabilized dynamically. This system is analyzed in greater detail in Chapter 11.

Improved response time.   Negative feedback can improve the response time of a system. Consider, for example, an electrical relay, as pictured in Fig. 8.39. A relay is basically an electromagnet. When current passes through the coil, a magnetic force is produced at the gap and the lever closes against a spring. Such relays are used typically to activate switches and to operate locks and valves. We will consider a relay having the following properties: current required to activate $= 20$ mA; maximum current $= 30$ mA; coil resistance $(R_c) = 200$ Ω; and coil inductance $(L_c) = 2$ H.

A suitable circuit for energizing the relay is shown in Fig. 8.40. When the relay is to be activated, the switch is closed. Current, initially at $i_0 = 0$, increases exponentially toward a final value of $i_\infty = V_s/R_c$. As current builds up, the relay will close when the current exceeds 20 mA. Let us assume that we wish the relay to close as quickly as possible. In that case, we require the voltage source to be as large as possible, which in this case is limited by the maximum current to be 30 mA $\times$ 200 Ω = 6 V. With that value of voltage, the time required for the relay to close is easily shown to be $\tau \ln 3$, where $\tau = L_c/R_c$ is the time constant, 2H/200 Ω = 10 ms. Hence in this case the relay will close about 11.0 ms after the switch is closed.

Throughout this period when the relay current is increasing, the diode remains OFF; but when the switch is opened, the inductor will react to keep the current going. This will cause the diode to conduct and the current will flow for a time through coil and diode, even with the switch open. The diode is thus placed across the relay coil to de-energize the inductance. Without the diode, a spark would appear at the switch contacts, as discussed on page 101.

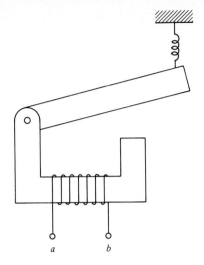

Figure 8.39   Relay.

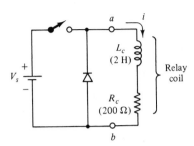

Figure 8.40   Controlling circuit.

Let us suppose that the 11-ms response time is too slow. We can use a feedback technique in the relay circuit to improve the relay speed. As shown in Fig. 8.41, we have added an amplifier and a resistor to the relay circuit. The resistor, R_s, produces a feedback signal that is proportional to the current in the relay coil. After the switch is closed, the equations of the system are: for the amplifier

$$v_{\text{out}} = A(V_s - v_f) = A(V_s - iR_s) \tag{8.47}$$

and for the coil and feedback resistor

$$v_{\text{out}} = L_c \frac{di}{dt} + (R_c + R_s)i \tag{8.48}$$

When we combine Eqs. (8.47) and (8.48), we obtain a differential equation describing the feedback system:

$$L_c \frac{di}{dt} + [R_c + (1 + A)R_s]i = AV_s$$

Figure 8.41   Feedback circuit to improve response time.

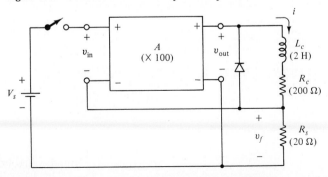

355

which can be put into the standard form

$$\frac{di}{dt} + \frac{i}{\tau_f} = \frac{AV_s}{L_c}$$

where τ_f would be $L_c/[R_c + (1 + A)R_s]$ and would be shorter than the original time constant. With the numerical values we have shown in Fig. 8.41, the time constant drops from 10 ms to 0.91 ms and the response time would be reduced accordingly, from 11.0 ms to 1.0 ms. Other aspects of this example we will leave for a homework problem.

Improvement of frequency response.   As might be anticipated from the preceding example, and from the correspondence between speed of response in the time domain and bandwidth in the frequency domain, feedback can often improve the frequency response of a system. Let us reconsider the amplifier example in Fig. 8.33, except that we now introduce a bandwidth limitation to the main amplifier. We have shown the circuit in Fig. 8.42 (frequency-domain version), with the main amplifier now having a low-frequency gain of A_0 and a frequency cutoff of ω_c. The main amplifier frequency response has the form of a low-pass filter, the loss of the high frequencies resulting from a limitation within the amplifier. This limitation of the amplifier can be alleviated through negative feedback.

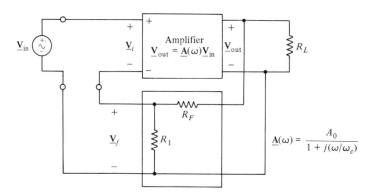

Figure 8.42   Improving bandwidth with feedback.

The equations of the system [Eqs. (8.38) to (8.40)] transform into the frequency domain changed only in notation, and hence the amplifier characteristic with feedback becomes the frequency-domain version of Eq. (8.41):

$$\underline{A}_f(\omega) = \frac{\underline{A}(\omega)}{1 + [R_1/(R_1 + R_F)]\underline{A}(\omega)} = \frac{A_0/[1 + j(\omega/\omega_c)]}{1 + [\beta]A_0/[1 + j(\omega/\omega_c)]}$$

where $\beta = R_1/(R_1 + R_F)$. This can be put into the form

$$\underline{A}_f(\omega) = \frac{A_0/(1 + \beta A_0)}{1 + j(\omega/\omega_{cf})} \tag{8.49}$$

where $\omega_{cf} = (1 + \beta A_0)\omega_c$ is the cutoff frequency with feedback. Recognizing that the loop gain (L) is $-\beta A_0$, we see from Eq. (8.49) that the amplifier gain at low frequncies is reduced by the factor $1 - L$ and the cutoff frequency is increased by the same factor.

Bode plots for the amplifier with and without feedback are shown in Fig. 8.43. This application furnishes a good example of our claim that negative feedback trades gain for some other useful property. In this case gain is exchanged for bandwidth. The inherent bandwidth limitation of the amplifier has been overcome through the use of negative feedback.

We have shown that negative feedback increases the bandwidth of a first-order system. Higher-order systems cannot always be improved by these means.

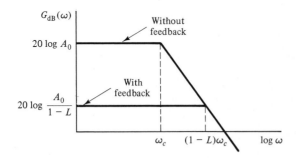

Figure 8.43 Feedback reduces gain but increases bandwidth.

Impedance control. Feedback techniques can be used to change the effective impedance of an electronic circuit. This is useful for effecting maximum power transfer and for reducing loading effects. However, the system approach we have taken in this section precludes consideration of impedance levels. Impedance control is extremely important in electronic circuits, and feedback techniques are frequently used for this purpose. The details, however, rapidly become complicated and we shall not explore this subject in detail.

8.4.4 Check Your Understanding

1. Sometimes the "summer" in a feedback system represents taking the difference, not the sum, of two signals. (T/F?)
2. For negative feedback, the loop gain must be positive or negative?
3. The loop gain is calculated or measured with the input source replaced by a short circuit. (T/F?)
4. If the loop gain in a feedback system is -0.2, feedback effects will not be very important in defining system characteristics. (T/F?)
5. Negative feedback, compounded with phase shift in the loop, can cause a tendency for dynamic instability. (T/F?)

Answers. 1. True; 2. negative; 3. only if the amplifier has voltage as the input signal; 4. true; 5. true.

8.5.1 Introduction

Importance of op amps.   An operational amplifier is a high-gain electronic amplifier, controlled by negative feedback, that accomplishes many functions or "operations" in analog circuits. Such amplifiers were originally developed to accomplish operations such as integration and summation for solving differential equations with analog computers. Applications of op amps have increased until, at the present time, most analog electronic circuits are based on op-amp techniques. If, for example, you required an amplifier with a gain of -10, rarely would you design a circuit of the type in Fig. 6.45; convenience, reliability, and cost considerations would dictate the use of an op amp. Thus op amps form the basic building blocks of analog electronic circuits much as NOR and NAND gates provide the basic building blocks of digital circuits.

Op-amp model and typical properties.   The typical op amp is a sophisticated transistor amplifier utilizing a dozen or more transistors, several diodes, many resistors, and perhaps a few capacitors. Such amplifiers are mass produced on semiconductor chips and sell for less than \$1 each. These parts are reliable and rugged, approaching the ideal in their electronic properties.

Figure 8.44a shows the symbol and the op-amp properties that interact with external signals. The two input voltages, v_+ and v_-, are subtracted and amplified with a large voltage gain, A, typically 10^5 to 10^6. The input resistance, R_i, is large, typically exceeding 10 MΩ; the output resistance, R_{out}, is small, 10 to 100 Ω. The amplifier is often supplied with dc power from positive ($+V_{CC}$) and negative ($-V_{CC}$) power supplies. For this case, the output voltage lies between the power-supply voltages, as shown by Fig. 8.44b. Sometimes one power connection is grounded (for example, "$-V_{CC}$" $= 0$). In this case the output lies between 0 V and V_{CC}. The power connections are seldom drawn on circuit diagrams; it is assumed that the op amp is connected to the appropriate power source.

Realistic op amps.   The model in Fig. 8.44a approaches the ideal voltage amplifier: high input impedance, low output impedance, and high gain. Real op amps have these properties but also have undesirable features that limit their performance and influence circuit design. The circuit given in Fig. 8.45 introduces a number of the limitations of realistic op amps. These are:

- A dc offset voltage, V_{os}, is shown to indicate that the output of the op amp is nonzero with zero input voltage, that is, with $v_+ = v_-$.
- Input dc bias currents, I_{B+} and I_{B-}, are shown. Normally, these bias currents are expressed in terms of an offset bias current, $I_{os} = I_{B+} - I_{B-}$, and an average bias current $I_B = (I_{B+} + I_{B-})/2$.
- A noise voltage $v_n(t)$ represents broadband noise originating in the op amp. The magnitude of this noise is proportional to the square root of the bandwidth of the op amp in the circuit application.

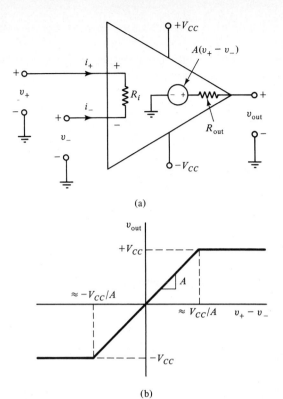

(a)

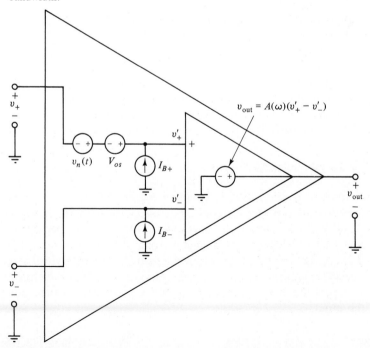

(b)

Figure 8.44 (a) Op-amp model; (b) input–output characteristic of an op amp. The sloped region is the amplifying region. The op amp saturates at the power-supply voltages.

Figure 8.45 An op amp has noise, offset voltage, bias currents, and limited bandwidth.

- Gain–bandwidth limitation is indicated by making the op-amp gain $A(\omega)$ a function of frequency. The gain–bandwidth product is limited by inherent capacitance in the circuit and the semiconductor devices in the op amp, and often is deliberately limited by the chip designer to control oscillations at high frequencies.

Reducing the effects of op-amp imperfections.   The effect of dc offset voltage and current (V_{os} and I_{os}) can be eliminated by introducing a dc voltage to null the output voltage with no input signal present. Application literature provided by the manufacturer gives information about canceling the effect of offset voltage and current.

The effect of the bias current (I_B) can be eliminated by providing equal dc impedance at the two inputs. In this way, the bias current produces equal voltages that cancel in the subtraction process.

The effect of input noise can be reduced by limiting the bandwidth of the system to that required by the signal. However, noise cannot be eliminated entirely and must be considered in the error analysis of the circuit employing the op amp.

The gain–bandwidth limitation of the op amp limits the product of the gain and bandwidth of the resulting amplifier employing the op amp. Generally, op amps with large gain–bandwidth products are more sophisticated and expensive than basic op amps.

Summary.   The high gain of the op amp is converted to other useful features through the use of strong negative feedback. All the benefits of negative feedback are utilized by op-amp circuits. To those listed earlier in this chapter, we would add three more: low expense, ease of design, and simple construction.

Contents of this section.   We next analyze two common op-amp applications, inverting and noninverting amplifiers. We derive the gain of these amplifiers by a method that may be applied simply and effectively to any op-amp circuit operating in its linear amplifying region. We then discuss op-amp circuits for adding and subtracting signals, for converting a current signal to a voltage signal, and for differentiating and integrating signals. We then consider nonlinear applications of op amps: comparators, a rectifier circuit, and log and antilog amplifiers. In the next chapter we consider additional applications of op amps.

8.5.2 Linear Op-amp Circuits

Inverting amplifier.   The inverting amplifier, shown in Fig. 8.46, uses an op amp plus three resistors. The positive (+) input to the op amp is grounded through R_2 (zero signal); the negative (−) input is connected to the input signal (via R_1) and to the feedback signal from the output (via R_F). Note that $R_2 = R_1 \parallel R_F$ so that the bias current produces no unbalance. Furthermore, throughout this section on linear op-amp circuits, we consider that the effect of offset voltage and current has been canceled to produce zero volts at the output for zero volts at the input.

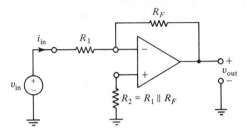

Figure 8.46 Inverting amplifier.

One potential source of confusion in the following discussion is that we must speak of two amplifiers simultaneously. The op amp is an amplifier that forms the amplifying element in a feedback amplifier that contains the op amp plus the associated resistors. To lessen confusion, we reserve the term "amplifier" to apply only to the overall, feedback amplifier. The op amp will never be called an "amplifier"; it will be called "the op amp." For example, if we refer to the input current to the amplifier, we are referring to the current through R_1, not the current into the op amp.

We could solve for the gain of the inverting amplifier in Fig. 8.46 either by solving the basic circuit laws (KVL and KCL) or by attempting to divide the circuit, in the style of Fig. 8.34, into summer, main amplifier, and feedback system blocks. We will, however, present another approach based on the assumption that the op-amp gain is very high, effectively infinite. In the following, we will give a general assumption, which may be applied to any op-amp circuit; then we will apply this assumption specifically to the present circuit. As a result, we will establish the gain and input resistance of the inverting amplifier.

1. We assume that the amplifier operates in its linear amplifying region, the sloping region in Fig. 8.44b. It follows that the output lies between the power-supply voltages. In other words, we assume that the negative feedback stabilizes the amplifier such that moderate input voltages produce moderate output voltages. If the power supplies are $+10$ V and -10 V, for example, the output would have to lie between these limits.

2. Therefore, the difference between input voltages *to the op amp* is very small, essentially zero, because this difference is the output voltage divided by the large voltage gain of the op amp:

$$v_+ - v_- \approx 0 \Rightarrow v_+ \approx v_-$$

For example, if $|v_{out}| < 10$ V and $A = 10^5$, then $|v_+ - v_-| < 10/10^5 = 100$ μV. Thus, normally v_+ and v_- are equal within 100 μV or less, for any op-amp circuit. For the inverting amplifier in Fig. 8.46, v_+ is grounded through R_2; therefore, $v_+ \approx 0$ and $v_- \approx 0$. Consequently, the current at the input to the amplifier would be

$$i_{in} = \frac{v_{in} - v_-}{R_1} \approx \frac{v_{in}}{R_1} \tag{8.50}$$

3. Because $v_+ \approx v_-$ and R_i is large, the current into the $+$ and $-$ op-amp inputs will be very small, essentially zero:

$$|i_+| = |i_-| = \frac{|v_- - v_+|}{R_i} \approx 0 \tag{8.51}$$

For example, for $R_{in} = 10\ M\Omega$, $|i_-| < 10^{-4}/10^7 = 10^{-11}\ A$.

For the inverting amplifier, Eq. (8.51) implies that the current at the input, i_{in}, flows through R_F, as shown in Fig. 8.47. This allows us to compute the output voltage. The voltage across R_F would be $i_{in}R_F$ and, because one end of R_F is connected to $v_- \approx 0$,

$$v_{out} = -i_{in}R_F = -\frac{v_{in}}{R_1} \times R_F$$

Thus the voltage gain would be

$$A_v = \frac{v_{out}}{v_{in}} = -\frac{R_F}{R_1} \tag{8.52}$$

The minus sign in the gain expression means that the output will be inverted relative to the input: a positive signal at the input will produce a negative signal at the output. Equation (8.52) shows the gain to depend on the ratio of R_F to R_1. This would imply that only the ratio matters, not the individual values of R_1 and R_F. This would be true if only the gain of the amplifier were important, but the input resistance to an amplifier is also important. The input resistance to the inverting amplifier would follow from Eq. (8.50):

$$R_{in} = \frac{v_{in}}{i_{in}} \approx R_1 \tag{8.53}$$

For a voltage amplifier, the input resistance is an important factor, for if R_{in} were too low the signal source (of v_{in}) could be loaded down by R_{in}. Thus in a design, R_1 must be sufficiently high to avoid this loading problem. Once R_1 is fixed, R_F may be selected to achieve the required gain. Thus the values of the individual resistors become important because they affect the input resistance to the amplifier.

Let us design an inverting amplifier to have a gain of -8. The input signal is to come from a voltage source having an output resistance of $100\ \Omega$. To reduce loading, the input resistor, R_1, must be much larger than $100\ \Omega$. For a 5% loading reduction, we would set $R_1 = 2000\ \Omega$. To achieve a gain of -8 (actually 95% of -8, considering loading), we require that $R_F = 8 \times 2000 = 16\ k\Omega$.

Feedback effects dominate the characteristics of the amplifier. When an input voltage is applied, the value of v_- will increase. This will cause v_{out} to

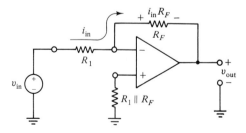

Figure 8.47   The input current flows through the feedback circuit.

increase rapidly in the negative direction. This negative voltage will increase to the value where the effect of v_{out} on the $-$ input via R_F cancels the effect of v_{in} through R_1. Put another way, the output will adjust itself to withdraw through R_F any current that v_{in} injects through R_1, since the input current to the op amp is extremely small. In this way the output depends only on R_F and R_1.

Noninverting amplifier.   For the noninverting amplifier shown in Fig. 8.48 the input is connected to the $+$ input through $R_2 = R_1 \parallel R_F$. The feedback from the output connects still to the $-$ op amp input, as required for negative feedback. To determine the gain, we apply the assumptions outlined above.

1. Because the input current to the op amp is very small, no signal voltage is created across R_2 and hence $v_+ = v_{\text{in}}$.
2. Because $v_+ \approx v_-$, it follows that

$$v_- \approx v_{\text{in}} \tag{8.54}$$

3. Because $i_- \approx 0$, R_F and R_1 carry the same current. Hence v_{out} is related to v_- through a voltage-divider relationship

$$v_- = v_{\text{out}} \frac{R_1}{R_1 + R_F} \tag{8.55}$$

Combining Eqs. (8.54) and (8.55), we establish the gain to be

$$v_{\text{in}} = v_{\text{out}} \frac{R_1}{R_1 + R_F} \Rightarrow A_v = + \left(1 + \frac{R_F}{R_1}\right) \tag{8.56}$$

The $+$ sign before the gain expression emphasizes that the output of the amplifier has the same polarity as the input: a positive input signal produces a positive output signal. Again we see that the ratio of R_F and R_1 determines the gain of the amplifier.

When a voltage is applied to the amplifier, the output voltage increases rapidly and will continue to rise until the voltage across R_1 reaches the input voltage. Thus negligible input current will flow into the amplifier, and the gain depends only on R_1 and R_F. The input resistance to the noninverting amplifier will be very high because the input current to the amplifier is also the input current to the op amp, i_+, which must be extremely small. Input resistance values exceeding 1000 MΩ are easily achieved with this circuit. This feature of high input

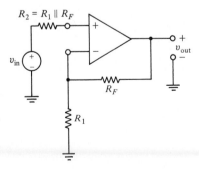

Figure 8.48 Noninverting amplifier.

resistance is an important virtue of the noninverting amplifier because loading of the input source is not an issue.

With the noninverting amplifier, the values of R_1 and R_F may be chosen from a broad range as long as the ratio gives the desired gain. However, if the impedance level is too low, comparable to the output resistance of the op amp, the op amp might be loaded by its own feedback network, which would be undesirable. At the other extreme, very large resistors can act as sources of significant noise in the amplifier, increasing the output noise above that inherent to the op amp. Resistor noise will be discussed in Chapter 10.

Buffer or voltage follower. An amplifier with a gain of $+1$ results from eliminating R_1 ($= \infty$) in the noninverting amplifier, as shown in Fig. 8.49. This "buffer" is used to control impedance levels in the circuit. The input impedance to the buffer is very high and its output impedance is low. The output voltage from a source with high output impedance can, via the buffer, supply signal to one or more loads that have a low impedance. The value of R_F is arbitrary and can be chosen equal to the source impedance, R_s, to eliminate entirely the need for R_2. Many times R_F is replaced by a short circuit when dealing only with ac signals.

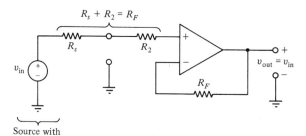

Source with high output impedance

Figure 8.49 The buffer or voltage-follower circuit.

Current-to-voltage converter. Many devices and circuits produce current signals, whereas most electronic circuits require voltage signals. An op-amp circuit for converting a current signal to a voltage signal is shown in Fig. 8.50. This is the inverting amplifier with $R_1 = 0$. It has zero input impedance and hence does not load the signal current source. The input current is converted to a voltage by flowing through the feedback resistor, as explained earlier.

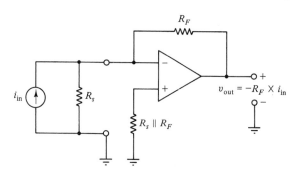

Figure 8.50 A current-to-voltage converter.

Analog Electronics   Chap. 8

Summing amplifier. The inverting amplifier can accept two or more inputs and produce a weighted sum. Figure 8.51 shows a summer with two inputs. We may understand the operation of the circuit by applying the same reasoning we used earlier to understand the inverting amplifier. Since $v_- \approx 0$, the sum of the currents through R_1 and R_2 is

$$i_{in} = \frac{v_1}{R_1} + \frac{v_2}{R_2} \tag{8.57}$$

The output voltage will adjust itself to draw this current through R_F, and hence the output voltage will be

$$v_{out} = -i_{in}R_F = -\left(v_1 \times \frac{R_F}{R_1} + v_2 \times \frac{R_F}{R_2}\right)$$

The output will thus be the sum of v_1 and v_2, weighted by the gain factors, R_F/R_1 and R_F/R_2, respectively. If the inversion produced by the summer is unwanted, the summer can be followed by an inverting amplifier with a gain of -1. Clearly, we could add other inputs in parallel with R_1 and R_2.

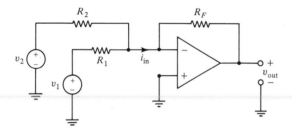

Figure 8.51 Summer circuit.

Differencing amplifier. The circuit in Fig. 8.52 produces an output proportional to the difference between the two inputs. The circuit is linear, so we may use superposition. The output due to v_2 is given by Eq. (8.52). The signal to the noninverting input is reduced by the voltage divider of R_1 in series with R_F. The output thus combines a voltage-divider relationship with the gain given by Eq. (8.56):

$$v_{out} = v_1 \times \frac{R_F}{R_1 + R_F} \times \left(1 + \frac{R_F}{R_1}\right) = v_1 \times \frac{R_F}{R_1}$$

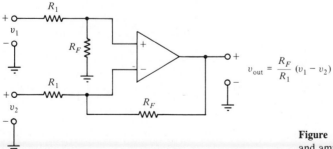

$$v_{out} = \frac{R_F}{R_1}(v_1 - v_2)$$

Figure 8.52 This circuit subtracts and amplifies the inputs.

Combining the effects from each input, we determine the output voltage to be

$$v_{\text{out}} = \frac{R_F}{R_1}(v_1 - v_2) \tag{8.58}$$

Thus the amplifier subtracts the inputs and amplifies their difference.

Two features of the subtractor circuit should be mentioned. Perfect subtraction requires a balance of the resistor values and thus may call for careful adjustment of the resulting circuit. Furthermore, the input impedances to the inputs are different, and therefore the balance could be affected if loading effects were significant. On the other hand, loading problems can be eliminated by reducing the resistors in the voltage divider by the ratio $R_1/(R_1 + R_F)$, but this would forfeit the impedance balance seen at the op-amp inputs. In practice, a compromise might be required.

Integrator. The op-amp circuit in Fig. 8.53 uses negative feedback through a capacitor to perform integration. We have charged the capacitor in the feedback path to an initial value of V_1 and then removed this prebias voltage at $t = 0$. Let us examine the initial state of the circuit before investigating what will happen after the switch is opened. Since v_+ is approximately zero, so will be v_-, and hence the output voltage is fixed at $-V_1$. The input current to the amplifier, v_{in}/R, will flow through the V_1 voltage source and into the output of the op amp. Thus the output voltage will remain at $-V_1$ until the switch is opened.

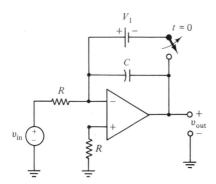

Figure 8.53 Integrator circuit.

After the switch is opened at $t = 0$, the input current will flow through the capacitor and hence the v_c will be

$$v_c(t) = v_c(0) + \frac{1}{C}\int_0^t \frac{v_{\text{in}}(t')}{R}\,dt'$$

Thus the output voltage of the circuit is

$$v_{\text{out}}(t) = -v_c(t) = -V_1 - \frac{1}{RC}\int_0^t v_{\text{in}}(t')\,dt', \qquad t > 0 \tag{8.59}$$

Except for the minus sign, the output is the integral of v_{in} scaled by $1/RC$.

Example. Design a noninverting circuit with an input impedance of 1 kΩ resistive and an output that is ten times the integral of the input voltage and has an initial value of +3 V. To have a noninverted output, we require an inverting ampifier and an integrator, which may be placed in either order. However, if the integrator were placed first, its "drift" due to unbalance would be amplified, so we use the circuit in Fig. 8.54. The input impedance requirement fixes R_1 as 1 kΩ, and the initial voltage requirement fixes V as +3 V. Two of the three components (C, R, and R_F) may be chosen arbitrarily and the third adjusted to give the correct overall gain. We will choose C as a convenient value of 10 μF and choose R_F as 20 kΩ to give the inverting amplifier a gain of -20. It follows from Eq. (8.59) that the integrator must have an RC value of 2, and hence the value of R must be 200 kΩ.

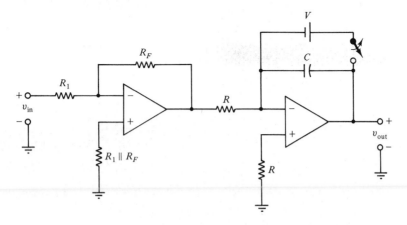

Figure 8.54 An inverting amplifier followed by an integrator.

Differentiator circuit. A circuit that differentiates the input signal uses a capacitor in the input circuit, as shown in Fig. 8.55.

Applying Eq. (3.6), we determine the input current to be

$$i_{in} = C \frac{d}{dt}(v_{in} - v_-) \approx C \frac{dv_{in}}{dt} \tag{8.60}$$

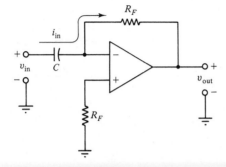

Figure 8.55 The output of this circuit is proportional to the derivative of the input signal.

This current will flow through the feedback resistor and produce an output

$$v_{\text{out}} = -R_F \times i_{\text{in}} \approx -R_F C \frac{dv_{\text{in}}}{dt} \tag{8.61}$$

The resistor used to ground the noninverting input is chosen to balance the dc impedance seen by the inverting input.

8.5.3 Nonlinear Op-amp Circuits

Many circuits employing op amps are nonlinear, either because nonlinear elements such as diodes are used in the circuit or because the op amp operates outside its linear amplifying region. In this section we discuss several nonlinear applications of op amps. Many more circuits are detailed in standard handbooks and application literature provided by manufacturers.

Comparators. Figure 8.56a shows a basic comparator, and Fig. 8.56b shows its input–output characteristic. Notice that no feedback is employed. When the input is less than V_{th}, the op amp is saturated at the value of the negative power supply, which is grounded in this case. As the input voltage is increased past V_{th}, the op amp passes rapidly through its linear region and saturates at the voltage of the positive power supply. Because there is no feedback, transition between the two saturation voltages requires an input voltage range of a few millivolts at most. This circuit therefore gives a digital output for an analog input, depending on the region of the input voltage. Comparators have many applications, including alarm circuits, control circuits, and analog-to-digital converters.

A noisy signal, however, causes problems, as shown in Fig. 8.57a. The transition region is so narrow that an erratic output is likely to occur, as shown in the bottom graph. The circuit shown in Fig. 8.57b remedies this problem through the use of positive feedback. Figure 8.57c shows the input–output characteristic of the modified circuit. As the input voltage increases, the output transition occurs when the input voltage reaches a threshold voltage, which in this case is established from the power supply with a voltage divider. Because the input voltage

Figure 8.56 (a) Basic comparator circuit; (b) the input–output characteristic.

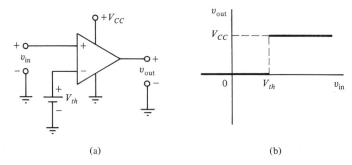

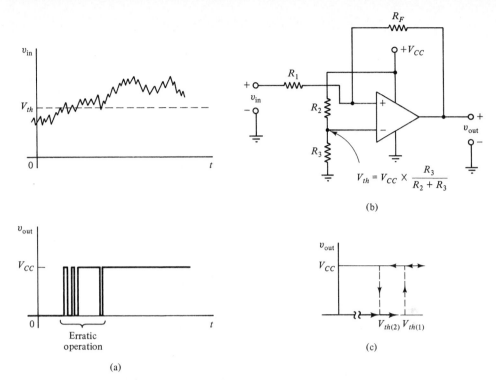

Figure 8.57 (a) A noisy signal causes erratic operation; (b) a comparator with hysteresis created by positive feedback; (c) the input–output characteristic of the comparator with hysteresis.

to the op amp is reduced slightly by the feedback resistor, the transition occurs at the value given in Eq. (8.62).

$$\frac{V_{th(1)} - V_{th}}{R_1} = \frac{V_{th} - 0}{R_F} \Rightarrow V_{th(1)} = V_{th}\left(1 + \frac{R_1}{R_F}\right) \tag{8.62}$$

As the op amp comes out of saturation in the lower state, the positive feedback effects a rapid transition to saturation in the upper state. As the input voltage is lowered, the input voltage must be reduced to a value given by Eq. (8.63).

$$\frac{V_{th(2)} - V_{th}}{R_1} = \frac{V_{th} - V_{CC}}{R_F} \Rightarrow V_{th(2)} = V_{th}\left(1 + \frac{R_1}{R_F}\right) - V_{CC} \times \frac{R_1}{R_F} \tag{8.63}$$

$$= V_{th(1)} - V_{CC} \times \frac{R_1}{R_F}$$

Note that the width of the threshold region is $V_{CC} \times R_1/R_F$.

Example. Design a comparator with a hysteresis region 30 mV wide, centered on 5 V. The positive and negative power supplies are 15 V and 0 V, respectively. The input impedance to the op amp should exceed 100 kΩ.

The specified threshold values are $V_{th(1)} = 5.015$ V and $V_{th(2)} = 4.985$ V. From the width of the threshold region, we find $R_1/R_F = 0.030/15 = 0.002$, and from Eq. (8.62) $V_{th} = 5.015/(1 + 0.002) = 5.005$ V. The input impedance is the sum of R_1 and R_F because the op amp draws no current. We choose $R_F = 1$ MΩ and hence $R_1 = 2000$ Ω. Here it is unnecessary to balance the input dc impedances at the op-amp inputs because the op amp is always saturated, and therefore bias currents have no effect. The voltage divider of R_2 and R_3 may be chosen at convenient values to give the required value of V_{th}.

Improved half-wave rectifier. The op amp in Fig. 8.58 drives a half-wave rectifier. When the input voltage is negative the output of the op amp will be negative and the diode will be OFF; hence the output will be zero. When the output is positive the diode will turn ON and the output will be identical to the input, because the circuit will perform as a voltage follower. Use of the op amp effectively reduces the diode turn-on voltage. If the input voltage is greater than $0.7/A$, where A is the voltage gain of the op amp, the output voltage will exceed 0.7 V and turn ON the diode. Hence the turn-on voltage is reduced from 0.7 to $0.7/A$.

This circuit would not be used in a power-supply circuit; rather, it would be used in a detector or other circuit processing small signals, where the turn-on voltage of the diode would be a problem.

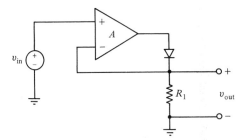

Figure 8.58 Improved half-wave rectifier.

Logarithmic amplifier. By placing a diode in the feedback path, as shown in Fig. 8.59, we create an amplifier whose output is proportional to the logarithm of the input. Following our standard line of reasoning, we see the input current to be v_{in}/R. The output voltage will assume a value to draw this current through the diode. The diode characteristic is well represented by the ideal pn-junction equation in Eq. (6.7); hence the output voltage will be

$$v_{out} = -\eta V_T \ln \left(\frac{v_{in}}{RI_0} + 1 \right) \approx -\eta V_T \ln v_{in} + \eta V_T \ln RI_0 \qquad (8.64)$$

The output thus contains a factor proportional to the logarithm of the input voltage. The circuit in Fig. 8.59 can be followed by another op-amp circuit that subtracts the constant term and adjusts the gain of the log amplifier to the required value. For example, the gain can be adjusted to produce an output in dB relative to 1 V. The circuit in its present form cannot deal with signals that go negative.

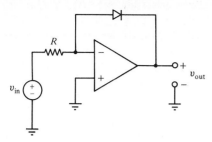

Figure 8.59 Logarithmic amplifier.

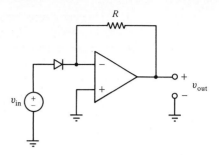

Figure 8.60 Antilog amplifier.

Antilog amplifier. With a diode in the input circuit as shown in Fig. 8.60, the output voltage can be made proportional to the antilog, or exponential function. We leave the derivation of the amplifier characteristics for a homework problem.

The existence of log and antilog amplifiers makes possible analog circuits for multiplication and powers of analog signals. For example, if we wished a circuit to produce the product of two analog signals, we could take the log of each, add the logs, and take the antilog. If one or both of the inputs could go both positive and negative, they would have to be put through circuits that produce the absolute value, with the signs handled by a separate logic circuit.

Excellent analog multipliers and other nonlinear circuits are commercially available. Such circuits operate according to the principles outlined above but contain more complicated circuits to compensate for temperature effects and generally improve performance. Our purpose here has been to show some representative nonlinear applications of op amps.

Summary. In this section we have introduced the op amp as an inexpensive and versatile circuit component for processing analog signals. We have described a method for determining circuit operation when the op amp is operating in its linear region. We have dealt with representative circuits for amplification, addition, subtraction, integration, and differentiation of signals. We have also described several nonlinear applications of op amps, including comparators, log, and antilog converters. In Chapter 9, we show additional applications of op amps, focusing mainly on the design of active filters.

8.5.4 Check Your Understanding

1. An ideal op amp has infinite gain, infinite input impedance, and infinite output impedance. (T/F?)
2. In a noninverting op-amp amplifier the feedback resistor is five times the value of the resistor between the inverting input and ground. If the output is -6 V, what is the voltage at the input?
3. For a voltage gain of $+5$ in an op-amp amplifier, the feedback comes from the output to which input terminal? Which input gets the input signal?

4. What operation is performed in an inverting amplifier if the feedback resistor is replaced by a capacitor? By a diode?

5. If an op amp has ± 8 V for plus and minus power-supply values and a gain of 80 dB, at what value of $v_+ - v_-$ does the op amp saturate?

6. The voltage-follower amplifier inverts the signal. (T/F?)

Answers. 1. False; 2. -1.0 V; 3. feedback to inverting $(-)$, input to noninverting $(+)$; 4. becomes an integrator; becomes a log amplifier; 5. ± 0.8 mV; 6. false.

PROBLEMS

Section 8.2: Frequency-Domain Representation of Signals

P8.1. In Austin, Texas, the maximum average daily temperature occurs in mid-July and is 84.5°F. The minimum daily average temperature is 49.1°F and occurs in mid-January. Based on this information, and assuming that the daily average temperature $T(t)$ is well represented by the dc and first harmonic terms of a Fourier series such as Eq. (8.3), find C_0, C_1, θ_1, and ω_0. Let t be the time in months, with $t = 0$ at January 1, $t = 1$ at February 1, and so on. From your result, estimate the expected average temperature for Christmas Day.

Answer. $T(\text{Christmas}) = 50.3°$

P8.2. Consider a square wave similar to the one in Fig. 8.5a, except that it goes from $+2$ to -2 V and has a pulse width of 2.5 ms. Find the frequency (in hertz), the amplitude, and the phase (in degrees) of the eleventh harmonic.

Answer. $C_{11} = 0.231$, $\theta = \pm 180°$, $f_{11} = 2200$ Hz

P8.3. What would be the Fourier series for the half-wave rectified sinusoid in Fig. 8.6a if the origin were drawn through the middle of the pulse? *Hint:* Change variables from $t \rightarrow t' + T/4$, where t' is the origin in the new time system.

P8.4. Show that the powers in the time and frequency domain are equal for the half-wave rectified sinusoid (Fig. 8.6a), whose Fourier series is indicated in Eq. (8.5). *Hints:* The time-domain power must be one-half that of an unrectified sinusoid; the nth harmonic amplitude is $2V_p/\pi(n^2 - 1)$, n even, with the signs alternating; sum up enough terms on your calculator to show approximate equality between the two domains. This series is difficult to sum mathematically.

P8.5. A Fourier series is

$$v(t) = 10 + 5\cos(200\pi t) + 3\cos(400\pi t + 90°) + \cdots$$

(a) Would this signal have a continuous or a discrete spectrum?
(b) What is the "power" in (volts)^2 in the three harmonics given above?
(c) If the voltage were increased by 5 V dc, how would the Fourier series change?
(d) What is the frequency of the third harmonic in hertz?

P8.6. Figure P8.6 shows a series of pulses in the time domain. Find:
(a) Frequency of the third harmonic
(b) Power in (volts)^2 in the dc component
(c) Amplitude of the third harmonic
(d) Total power in (volts)^2 in all ac harmonics

Answers. (a) 150 Hz; (b) 100 (volts)^2; (c) 4.24 V; (d) 100 (volts)^2

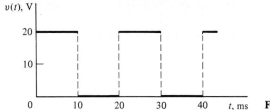

$v(t)$, V

20

10

0   10   20   30   40 t, ms   **Figure P8.6**

P8.7. At a certain location, the maximum elevation angle of the sun at summer solstice (June 23) is about 83° and at winter solstice (December 23) about 37°. The maximum elevation can be described by a Fourier series of only two terms. What will be the minimum length of the shadow of a building 306 feet in height on the Fourth of July?

P8.8. An unfiltered half-wave power supply produces 12.3 V dc from a 60-Hz sinusoid. What would be its output power in (volts)^2 in the range of frequencies between 100 and 320 Hz?

P8.9. A periodic waveform is shown in Fig. P8.9. This voltage can be represented by the following Fourier series:

$$v(t) = C_0 + C_1 \cos(\omega_0 t + \theta_1) + C_2 \cos(2\omega_0 + \theta_2) + C_3 \cos(3\omega_0 t + \theta_3) + \cdots$$

Find:

(a) C_0
(b) ω_0
(c) C_1
(d) θ_2
(e) C_3

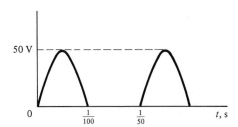

50 V

0    $\frac{1}{100}$ $\frac{1}{50}$ t, s   **Figure P8.9**

P8.10. Show that Eq. (8.9) for $\tau = T/2$ gives the same harmonic amplitudes and phases as those in Eq. (8.4).

P8.11. The nominal bandwidth of a standard telephone line is 4 kHz. What is the approximate duration of the shortest pulse that can be sent over such a telephone line? Given that the width between pulses should be equal to the pulse width, how many pulses per second can be sent over a phone line?

Answers.   0.25 ms, 2000 pulses/s or 2000 band

P8.12. For the half-wave rectified sinusoid in Fig. 8.6a, what fraction of the total power is carried by the dc component? *Hint:* For the ac power, just sum up the terms given. Note that the magnitudes diminish rapidly. See P8.4 for additional information.

Answer.   40.5%

P8.13. The "power" spectrum of a random signal is shown in Fig. P8.13.

    (a) What is the dc value of the signal?

    (b) What power would this signal give to a 5-Ω resistor?

    (c) What is the approximate correlation time for this signal?

Answers. (a) ± 3.16 V; (b) 4 W; (c) 0.1 s

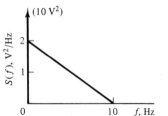

Figure P8.13

P8.14. The effective bandwidth (BW) of standard AM radio is 5000 Hz. If such a radio were tuned off a station, receiving only static, what would be the approximate correlation time for the static?

P8.15. Figure P8.15 shows the power spectrum of a random signal.

    (a) What is the average value of the signal?

    (b) What would be the approximate correlation time of the signal?

    (c) What would be the total power from this signal into a 50-Ω resistor?

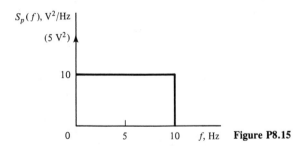

Figure P8.15

P8.16. A computer-based communication system has an information rate of 1000 characters/second, where each character is represented by a byte of digital information. What is the approximate bandwidth required by this system? State assumptions.

P8.17. A 60-Hz half-wave rectified sinusoid with a peak value of 10 V is filtered by a low-pass filter, as shown in Fig. P8.17. Calculate the dc and the peak value of the fundamental at the output.

Answers. 3.18 V, 0.133 V peak

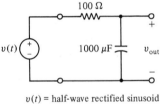

$v(t)$ = half-wave rectified sinusoid
$V_p = 10$ V
$f = 60$ Hz

Figure P8.17

P8.18. Make asymptotic Bode plots of the magnitude of the following filter functions:

(a) $\underline{F}(\omega) = \dfrac{j\omega}{10 + j2\omega}$

(b) $\underline{F}(f) = \dfrac{1 + jf/10}{1 + jf/100}$

P8.19. Show that, at a frequency one octave (factor of 2) below the cutoff, the loss of the low-pass filter in Fig. 8.17 is approximately 1 dB.

P8.20. The gain in dB of a filter is shown in the Bode plot in Fig. P8.20.
(a) Is this a high-pass, low-pass, or bandpass filter?
(b) What is the gain in dB at 500 Hz?
(c) Estimate the cutoff frequency of the filter.
(d) If the input voltage to the filter were $v_{in}(t) = 5 \cos(2000t)$ V, what would be the output in the form $v_{out}(t) = A \cos(2000t + \theta)$? In other words, find A and θ.

Answers. (a) high-pass; (b) -3 dB; (c) 500 Hz; (d) 2.685, $+57.5°$

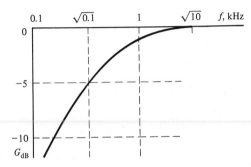

Figure P8.20

P8.21. The input to a filter is the time function $v_{in}(t) = 10 + 5 \sin(725t)$. The asymptotic Bode plot of the filter characteristic is shown in Fig. P8.21.
(a) Is the filter high pass or low pass?
(b) What is the frequency in hertz of the fundamental of the input?
(c) What is the input "power" in (volts)^2?
(d) What is the output "power" in (volts)^2?

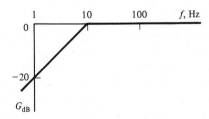

Figure P8.21

P8.22. For the square wave shown in Fig. P8.22, determine
  (a) DC component
  (b) Frequency of the third harmonic
  (c) RMS amplitude of the third harmonic
  (d) After passing through a low-pass filter with a cutoff frequency of 100 Hz, what would be the output peak amplitude of the fundamental?

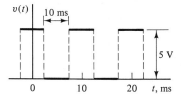

Figure P8.22

P8.23. The half-wave rectified sinusoid in Fig. P8.23a is passed through an R-L filter, as shown in Fig. P8.23b. What percent of the output "power" in (volts)^2 is in its second harmonic?

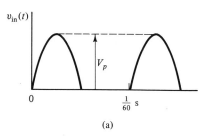

(a)   (b)

Figure P8.23

Section 8.4: Feedback Concept

P8.24. Work out the signal levels with $v_{out} = 10$ V for the amplifier in Fig. 8.33 with the reduced gain, $A = 75$. Make a table comparing with those calculated for $A = 200$ in Section 8.4.1.

P8.25. How large does the amplifier gain, A, have to be in the feedback amplifier in Fig. 8.33 before the gain with feedback is within 1% of $1/\beta$ with negative feedback? What is the loop gain for this value of A? Based on this result, guess what loop gain would be required for a 0.1% difference between the exact gain and $1/\beta$.

P8.26. A voltage amplifier has a gain of -1000. Ten percent of the output voltage is added to an input voltage to provide the input voltage to the amplifier.
  (a) What is the gain of the composite system?
  (b) If you wished to double this gain, how could this be accomplished?

P8.27. A feedback system employs negative voltage feedback. Assume that for a test the feedback path is opened at the point shown in Fig. 8.35, and 10 mV is fed into the input to the amplifier. Under these conditions the output voltage is observed to be 5 V and the output from the feedback loop is 1 V.
  (a) What is the loop gain?
  (b) What is the gain of the feedback amplifier with the loop closed?

P8.28. The feedback amplifier shown in Fig. P8.28 uses a current amplifier and has a single resistor, R_F, as a feedback circuit. In your analysis, assume that $R_F \gg R_L$, such that the output voltage is very nearly $v_{out} = i_{out} R_L$. Assume that the input impedance of the current amplifier is zero.

   (a) Derive the gain with feedback, $A_f = v_{out}/i_{in}$, by using KCL, Ohm's law, and the gain equation for the current amplifier.

   (b) For $A_i = 500$ and $R_L = 10\ \Omega$, find R_F for an overall gain of $-500\ \Omega$.

   (c) Put the results from part (a) in the form of Eq. (8.44) and draw the system diagram corresponding to this amplifier.

   (d) What is the loop gain of the amplifier?

   (e) If $A_i \to \infty$, what is the limiting form for A_f?

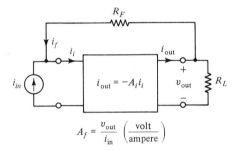

$$A_f = \frac{v_{out}}{i_{in}} \left(\frac{volt}{ampere} \right)$$

Figure P8.28

P8.29. Assume that the amplifier in Fig. 8.33 has a severe nonlinearity, as shown by the input–output characteristic in Fig. P8.29. Without feedback this would cause distortion in the amplifier output. Derive and sketch the overall (with feedback) characteristics of v_{out} versus v_{in} with feedback to demonstrate the benefits of feedback in reducing the degree of nonlinearity. *Hint:* The final results will consist of straight-line segments. Assume selected values at the output and work back to the input, as was done in Section 8.4.1.

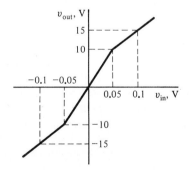

Figure P8.29

P8.30. For the relay circuit with feedback shown in Fig. 8.41:

   (a) What would be the maximum input voltage to keep the diode current less than 30 mA.

   (b) Assume the relay opens at a current of 14 mA. In the original circuit, how long after the switch opens will it take for the relay to open, assuming that the relay has been closed a long time? Assume an ideal diode.

   (c) Would the feedback circuit in Fig. 8.41 affect the time calculated in part (b)?

Section 8.5: Operational-amplifier Circuits

P8.31. Design an op-amp amplifier having a gain of $+10$ dB, a phase shift of $0°$, and an input resistance of 500 Ω.

P8.32. Design an amplifier having a gain of $+15$ dB, a phase shift of $180°$, and an input resistance of 500 Ω.

P8.33. For the amplifier shown in Fig. P8.33:
  (a) What is the voltage gain in dB?
  (b) What is the input resistance?

Answers. (a) 6.02 dB; (b) $3R$

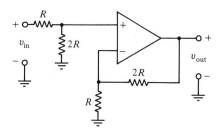

Figure P8.33

P8.34. Is the output voltage of the circuit in Fig. P8.34 $+2$ V, -2V, 0 V, or none of these? Explain your answer.

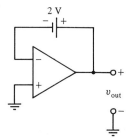

Figure P8.34

P8.35. Certain digital devices use a "20-mA current loop" to encode a digital signal. This means that a digital one is a current of 20 mA and a digital zero is 0 mA. Assume that the output impedance of the loop is 1 MΩ. Design an op-amp interface to convert this to a voltage signal with logic levels of 0 and $+5$ V.

P8.36. A transducer puts out a current signal in the range from 5 to 25 mA. For compatibility with a computer analog-to-digital input, this signal must be converted to a voltage signal in the range from -5 to $+5$ V. Design an op-amp circuit to accomplish this conversion.

P8.37. Design a comparator circuit that has output levels of 0 and $+5$ V (these are the power-supply values) and makes its transition at an input voltage of $+3$ V for increasing inputs and $+2$ V for decreasing inputs.

P8.38. Design an integrator with an input impedance of 5 kΩ that will provide an output

$$v_{out}(t) = 10 - 5 \int_0^t v_{in}(t')\, dt'$$

when the switch is opened at $t = 0$.

P8.39. The op amp in Fig. P8.39 may be considered ideal.

    (a) Write a DE for the output voltage in terms of the input voltage and the circuit components.

    (b) Solve for the output voltage if the input is a battery of V volts in series with a switch that is closed at $t = 0$. Assume that the op amp behaves in a linear manner.

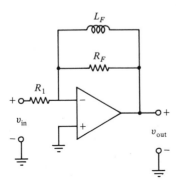

Figure P8.39

P8.40. For the op-amp circuit shown in Fig. P8.40, what is the voltage at points a and b relative to ground?

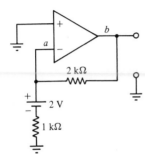

Figure P8.40

P8.41. The op-amp circuit shown in Fig. P8.41 has plus and minus power supplies at ± 10 V. The switch is closed at $t = 0$.

    (a) Find the value that the output would reach if the output transistors in the op amp did not saturate.

    (b) Find the time when the output saturates, assuming that this occurs at 10 V of output voltage.

Answers. (a) $+24$ V; (b) 5.39 ms

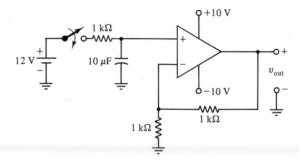

Figure P8.41

Chap. 8   Problems

P8.42. For the op-amp circuit shown in Fig. P8.42, find:
  (a) Input resistance seen by v_{in}.
  (b) Voltage gain v_2/v_{in}.
  (c) Voltage gain v_3/v_{in}.
  (d) If $v_{in} = 1$ V and the switch opened at $t = 0$, as shown, how long would it be until the magnitude of the output voltage were 1 V?

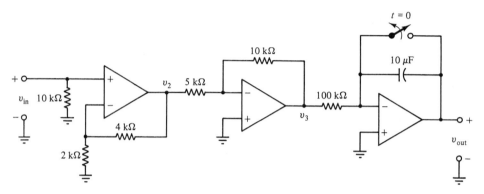

Figure P8.42

General Problems for Chapter 8

P8.43. The circuit in Fig. P8.43 is a comparator with hysteresis. The op amp has a voltage gain of 10^5 and its output saturates at ± 9 V.
  (a) Is the feedback positive or negative?
  (b) Determine the voltage that will cause the output to change from -9 V to $+9$ V for increasing inputs.
  (c) Determine the voltage that will cause the output to change from $+9$ V to -9 V for decreasing inputs.

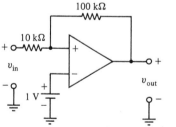

Figure P8.43

P8.44. The filter shown in Fig. P8.44 uses a resistor, a capacitor, and an ideal transformer with a turns ratio of 2:1. Make a Bode plot of the gain of this filter.

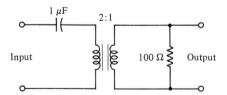

Figure P8.44

P8.45. The frequency characteristic of a filter circuit is described by the Bode plot shown in Fig. P8.45a. Find the output voltage of the circuit, $v_{out}(t)$, if a 10-V battery is connected for 2 ms, as indicated by Fig. P8.45b.

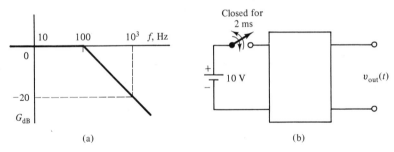

(a)　　　　　　　　　　(b)

Figure P8.45

P8.46. The filter shown in Fig. P8.46b has no input initially; then a series of pulses begins, as shown in Fig. P8.46a.
 (a) What will be the output voltage at the end of the first pulse, at $t = 50$ ms?
 (b) After a long period of time with the pulses continuing, what will be the amplitude of the dc and first-harmonic components in the output spectrum?

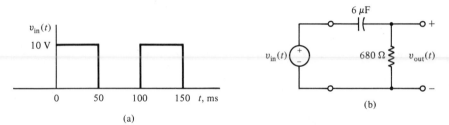

(a)　　　　　　　　　　(b)

Figure P8.46

P8.47. Figure P8.47 shows a standard noninverting op-amp amplifier circuit.
 (a) Represent the entire amplifier in a system diagram with main amplifier, summer, β network, and sampler. What are A, β, and the sign of the summer in this amplifier?
 (b) Find the loop gain for this amplifier.
 (c) Find the exact gain of the amplifier (no approximation) and compare with the approximate gain derived in the text.

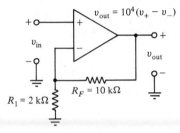

Figure P8.47

P8.48. Using the circuit in Fig. P8.48, design an amplifier that will produce an output of $20 \log (v_{in}/1 \text{ V})$, that is, the output is the input voltage in dB compared with 1 V. Assume that the diode characteristic is described by Eq. (6.7) for $\eta = 1.5$, $I_0 = 10^{-10}$ A, and $V_T = 0.026$ V. Specify R_F for the required gain and V_B to remove the constant term in Eq. (8.64). (*Note:* V_B could be obtained with a voltage divider connected to the negative power supply.)

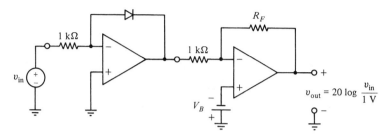

Figure P8.48

P8.49. The analysis of the passive low-pass filter in Section 8.3.1 ignores the loading at input and output. Consider now the circuit in Fig. P8.49, which contains a source resistance R_s and a load resistor R_L. The filter function $\underline{F}(\omega) = \underline{V}_{out}/\underline{V}_{in}$ will be identical in form to that in Eq. (8.22), except (1) the gain will no longer be unity in the region below the critical frequency, ω_c, and (2) the equation for ω_c is changed because it depends on the source and load resistances. Determine the revised expressions for ω_c and $\underline{F}(\omega)$.

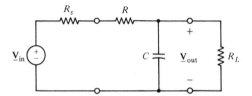

Figure P8.49

P8.50. The small-signal amplifier in Fig. 6.48 has an input impedance

$$Z_{in} = R_B \| r_\pi, \qquad \text{where } R_B = R_1 \| R_2$$

which often is unacceptably low for a voltage amplifier because r_π is small. A circuit in which feedback improves the input impedance is shown in Fig. P8.50a. The small-signal equivalent circuit for the amplifier is shown in Fig. P8.50b. This circuit uses feedback to control impedance. The feedback is proportional to the emitter current and is subtracted from the input voltage at the base of the base–emitter *pn* junction. Show that the input impedance to the transistor is raised from r_π to $r_\pi + (1 + \beta)R_E$, where β is the current gain of the transistor.

Analog Electronics   Chap. 8

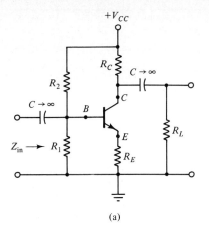

(a)

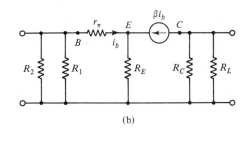

(b)

Figure P8.50

9 Data-Acquisition Systems

9.1 INTRODUCTION TO DATA-ACQUISITION SYSTEMS

9.1.1 General Considerations

Data gathering and reduction were once a simple but tedious task. Meters were read by eye and the results were recorded by hand. At best, a continuously recording "strip chart" left a wiggly line of data to be examined for trends, interpolated by eye, and analyzed by hand calculations.

Computers have made data gathering and reduction quick, automatic, and sophisticated. The computer can monitor many inputs, adjust the gain of analog channels to keep data within a prescribed range, filter incoming signals to improve data quality, process and record data, furnish displays, and produce control outputs.

Figure 9.1 shows a data-acquisition system that employs analog and digital processing. We now discuss the various subsystems to introduce the structure of this chapter.

Transducer. The transducer produces an electrical output indicative of some physical quantity such as pressure, temperature, or angular position. Tranducers, also called *sensors*, are discussed later in this introductory section.

Analog processing. Amplification and filtering are normally required to condition the signal for conversion to digital form. We show a digital control signal

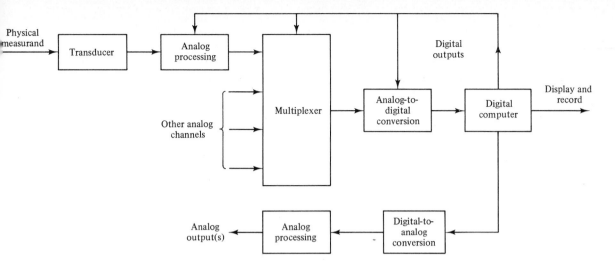

Figure 9.1 Data-acquisition system.

coming into this subsystem because often the gain of an analog channel is controlled by the digital system to keep the signal within a prescribed range. Analog processing is discussed in Section 9.2.

Multiplexing.   Typically, several analog channels are processed simultaneously through the *multiplexer*, which is a digitally controlled switch. The multiplexer accepts parallel inputs from several analog channels and provides one analog output at a time for conversion to digital form. Multiplexers are discussed in Section 9.3.

Analog-to-digital (A/D) conversion.   The A/D converts the information from analog to digital form. Often the time structure of the analog signal must be arrested with a *sample-and-hold* circuit while A/D conversion is taking place. Many types of A/D converters are available; we discuss two typical types in Section 9.3.

Digital computer.   The brains of the entire operation, and the immediate recipient of the information, is a digital computer. This might be a microprocessor dedicated to the data acquisition system or it might be a general-purpose computer that is structured to perform the required data-acquisition function simultaneously with other activities. For example, a desk-top personal computer can be adapted to accept analog and digital data inputs, and standard programs are available to supervise the data-gathering function. In this chapter, we will not investigate the processing and storage functions of the computer.

Digital-to-analog (D/A) conversion.   Often the computer must provide outputs in analog form. If, for example, the data monitor is part of a control system, the computer might furnish analog output signals as feedback to the controller of the process of which the physical measurand is a part. We will discuss D/A converters and basic control systems in Section 9.4.

Processing of analog outputs.   Analog outputs often require filtering and amplification for controlling process functions. However, no new topics are introduced, so we have no further discussion of analog outputs.

9.1.2 Transducers

A *transducer*, also called a *sensor*, converts a physical quantity to an electrical signal. Most transducers produce an analog output, but some produce digital outputs directly. Many types of transducers exist for most types of physical measurands; they differ in physical processes, noise, accuracy, linearity, ruggedness, output impedance, frequency response, and need for frequent calibration. After discussing bridge circuits and strain gages, we describe some common transducers used to measure position and angle, pressure, flow rate, and temperature.

Bridge circuits.   Many transducers employ a resistor whose resistance changes as a function of the physical measurand. The most common circuit used to convert this resistance change to a voltage change is the Wheatstone bridge circuit, shown in Fig. 9.2. The circuit consists of two voltage dividers, and the output voltage is the difference in the voltages created by the voltage dividers. Application of the voltage-divider relationship gives the output voltage:

$$v_{\text{out}} = V \left[\frac{R_0}{R_1 + R_0} - \frac{R_2}{R_3 + R_2} \right] \tag{9.1}$$

The bridge is said to be *balanced* when the output voltage is zero, which requires

$$\frac{R_0}{R_1} = \frac{R_2}{R_3} \tag{9.2}$$

If the bridge is approximately balanced, small changes in R_0, ΔR_0, produce corresponding small changes in output voltage, given by

$$\Delta v_{\text{out}} = V \frac{R_1 \, \Delta R_0}{(R_0 + R_1)^2} \tag{9.3}$$

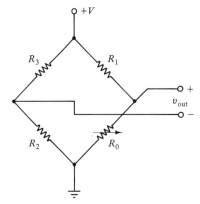

Figure 9.2   The Wheatstone bridge circuit converts a resistance change in R_o to a change in output voltage.

but large changes in R_0 are related to the changes in output voltage by the nonlinear relationship given by Eq. (9.1). Furthermore, the change in the resistance of the transducer is often nonlinear. Thus the measurement system may have to process this nonlinear information to determine the measurand. With a computer-based system, this conversion is often performed by means of a *look-up table* that stores the nonlinear function in the computer.

Strain gages. Another common measurement component is the strain gage, which can be used to indicate force, pressure, acceleration, or strain. A strain gage is a resistive element that is attached to a mechanical member to measure the elongation of the member. A bonded strain gage, shown in Fig. 9.3a, is a resistance that is bonded to the member with an adhesive. Elongation of the member changes the resistance of the gage by stretching the wire. The bonded strain gage is placed in a bridge circuit to produce an output voltage. Bonded strain gages are temperature sensitive, and normally a second strain gage that shares the thermal environment, but not the strain, is used in the same bridge to minimize thermal effects.

An unbonded strain gage is more sensitive than the bonded variety. As shown in Fig. 9.3b, the entire bridge circuit is placed in a fixture that applies the strain to all four resistors, and temperature effects are self-canceling. More sensitive yet is the semiconductor strain gage, in which a special semiconductor material is bonded to the mechanical member. However, the semiconductor strain gage must be used in a temperature-controlled environment because its resistance is influenced strongly by temperature.

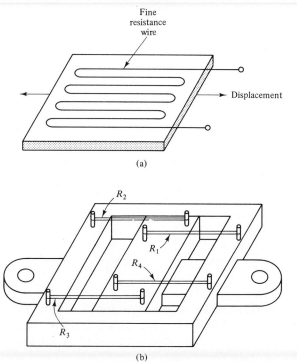

Figure 9.3 (a) Bonded strain gage; (b) unbonded strain gage.

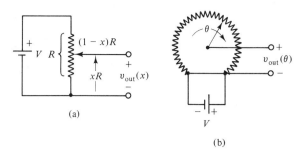

(a)

(b)

Figure 9.4 (a) Translational potentiometer; (b) rotational potentiometer.

Position transducers: potentiometers. Figure 9.4a shows a translational potentiometer, and Fig. 9.4b a rotational potentiometer. The potentiometer consists of a fixed resistor with a movable contact that responds to physical movement, thus changing the resistance ratio and hence the output voltage of the voltage divider. The resulting change in resistance could be used in a bridge circuit or could produce a voltage directly, as shown in Fig. 9.4. Potentiometers are available in which the output voltage is related to the physical movement through linear or logarithmic functions. Rotational potentiometers are available in single- or multiturn versions.

Position transducers: linear variable differential transformers. The linear-variable differential transformer uses variable coupling between the primary and two secondaries of a transformer to create ac voltages that depend on the position of a magnetic slug. The diodes and resistors shown in Fig. 9.5 rectify the ac voltages and produce a dc output voltage that, after suitable filtering, gives an indication of the position of the magnetic slug. With the slug in the center, equal voltages are created across the two resistors and the output voltage is zero. With the slug moved off center, the voltages are unequal and an output voltage is produced. The polarity of the resulting dc voltage indicates the direction of movement.

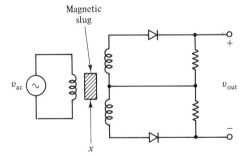

Figure 9.5 The linear-variable differential transformer.

Pressure transducers: diaphragm type. Pressure can be converted to force or displacement through the use of a diaphragm, bellows, or spiral tube. Therefore, pressure can be measured with a strain gage or some other means for monitoring force or displacement.

Pressure transducers: integrated-circuit pressure cells.   Pressure cells based on semiconductor sensors are manufactured complete with bridge circuit and amplifiers. In effect, the input to the cell is a pressure, and the output is an electrical signal.

Fluid flow transducers.   Fluid flow in a pipe can be monitored through the differential pressure across an orifice or screen mesh. Fluid flow can also be measured through a turbine placed in the stream. The rotation rate of the turbine can be measured digitally to determine flow rate.

Temperature transducers.   Many types of temperature sensors are available because almost all physical processes are affected by temperature. Sensors differ according to range and precision; obviously, we would use a different transducer to measure temperatures in the range of 5000°F than to monitor the temperature for an air-conditioning system. In the following, we mention two common types of temperature transducers that produce electrical outputs.

Temperature transducers: thermistors.   A *thermistor* is a resistor made of semiconductor material. The change of intrinsic carrier concentration with temperature produces a resistance that decreases strongly with increasing temperature. Typical temperature range for a thermistor is $-100°$ to $+300°C$, and the resistance change is nonlinear. For monitoring of small temperature changes, the thermistor may be placed in a bridge circuit, but for large temperature changes, other circuit arrangements may be required.

Temperature transducers: thermocouples.   A junction between two dissimilar metals produces a voltage that depends on the junction temperature. Thermocouples utilize this effect to monitor temperatures in a wide range, up to 2500°C. As shown in Fig. 9.6, two junctions are required, with a reference junction kept at constant temperature, usually 0°C. Many types of metals are used, depending on the temperature range required, and typical temperature coefficients are $+50 \mu V/°C$.

Summary.   In this section, we have discussed the more common transducers used to measure displacement, force, pressure, fluid flow, and tempera-

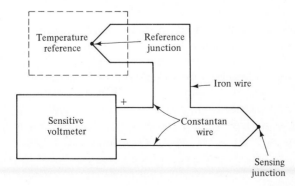

Figure 9.6 A thermocouple-based temperature measurement requires two junctions, one at a known temperature.

ture. The technology of measurement is advanced and diverse, and the best source of information is application literature from manufacturers.

9.1.3 Error Analysis

Need for error analysis.   Errors are introduced into the measurement system by the transducer, by the noise and nonlinearities of the analog electronics, by A/D conversion, and by digital signal processing. Error analysis combines the errors from the various components and processes to estimate the total error in the measurand. Some of the contributing errors are random errors describable by statistical methods, and some of the errors are estimates of uncertainty in calibration and system properties. Such errors may be combined only by making assumptions; and error analysis, even when undertaken with total objectivity, can be a source of controversy. This section presents basic methods of error analysis.

Exact and linearized error analysis.   The output of a measurement system depends on many factors, and in principle all can introduce errors. For example, the output voltage of the Wheatstone bridge in Eq. (9.1) gives the output voltage in terms of the power-supply voltage and four resistor values. We will assume that the errors due to R_1, R_2, and R_3 are eliminated by temperature compensation and calibration techniques, and consider only the effects of errors in V and R_0. Let the error in V be ΔV and the error in R_0 be ΔR_0. The interpretation of these errors depends on the method of analysis employed, as will be discussed below. These errors are not known quantities but rather uncertainties that we are willing to estimate. For example, ΔV would depend on output ripple, noise introduced into the wiring, the influence of variations in line voltage and temperature on the properties of the power supply, and whatever else might conceivably influence the voltage applied to the bridge. Likewise, ΔR_0 might be estimated from temperature and vibration effects and from calibration uncertainty. Usually the magnitude of each error is estimated, the sign assumed randomly plus or minus. Our purpose in this section is to show how such "known" errors may be combined to estimate an overall error. We will use the Wheatstone bridge in Fig. 9.2 to illustrate the process.

Equation (9.1) may be written

$$v_{\text{out}} \pm \Delta v_{\text{out}} = (V \pm \Delta V) \left[\frac{R_0 \pm \Delta R_0}{R_1 + R_0 \pm \Delta R_0} - \frac{R_2}{R_3 + R_2} \right] \tag{9.4}$$

where Δv_{out} is the uncertainty in the output voltage. Equation (9.4) may be used to determine Δv_{out} by either the *worst-case* or the *statistical* error analysis shown below. Of course, Eq. (9.4) is a simple expression, but usually large numbers of factors are involved in the error analysis, and the large number of variables and the nonlinear character of the expressions can lead to mathematical difficulties. For this reason, a linearized analysis is often performed. A linearized approximation to Eq. (9.4) is legitimate if the errors are small on a percent basis. We may linearize Eq. (9.4) by power-series expansion to the form

$$v_{out}(1 \pm \epsilon_v) = V(1 \pm \epsilon_V)\left[\frac{R_0(1 \pm \epsilon_R)}{R_1 + R_0(1 \pm \epsilon_R)} - \frac{R_2}{R_3 + R_2}\right]$$

$$= V\left[\frac{R_0}{R_1 + R_0} - \frac{R_2}{R_3 + R_2}\right] + \epsilon_V \times V\left[\frac{R_0}{R_1 + R_0} - \frac{R_2}{R_3 + R_2}\right]$$

$$+ \epsilon_R \times V\frac{R_0 R_1}{(R_0 + R_1)^2} + \text{higher-order terms in } \epsilon\text{'s}$$

$$(9.5)$$

where $\epsilon_v = \Delta v_{out}/v_{out}$, $\epsilon_V = \Delta V/V$, and $\epsilon_R = \Delta R_0/R_0$ are the normalized errors. The first term on the right side does not involve the errors and is the nominal output voltage. We may drop the higher-order terms and use Eq. (9.1) to cancel the zeroth-order terms on both sides of Eq. (9.5) to obtain a form involving only the errors:

$$\pm\epsilon_v = \pm\epsilon_V \frac{V}{v_{out}}\left[\frac{R_0}{R_1 + R_0} - \frac{R_2}{R_3 + R_2}\right] \pm \epsilon_R \frac{V}{v_{out}}\frac{R_0 R_1}{(R_1 + R_0)^2}$$

$$(9.6)$$

$$= \pm\epsilon_V \pm \epsilon_R \frac{R_0 R_1}{(R_0 + R_1)^2} \bigg/ \left[\left(\frac{R_0}{R_1 + R_0}\right) - \left(\frac{R_2}{R_3 + R_2}\right)\right]$$

Note that Eq. (9.6) would be valid with the ϵ's as normalized errors or as percent errors.

For example, let $V = 12$ V, $R_1 = R_2 = R_3 = 1000\ \Omega$ and $R_0 = 1100\ \Omega$. With these values, $v_{out} = 286$ mV, and Eq. (9.6) reduces to

$$\pm\epsilon_v = \pm\epsilon_V \pm 10.5\epsilon_R \qquad (9.7)$$

Thus the error in the supply voltage reflects directly into uncertainty in the output voltage, but the error in the sensor resistor is increased in importance by the near balance of the bridge.

Summary. We are analyzing the bridge circuit in Fig. 9.2 to illustrate the methods of error analysis. In this section we introduced linearization. In the next two sections we explain two common methods of error analysis.

Worst-case error analysis. *Worst-case* error analysis, as the name implies, assumes that the errors are at their maximum and add in the worst possible way. The errors in this case are interpreted to be the maximum errors in the various factors, and the signs are chosen such that the effects of all errors are cumulative.

Continuing the example from above, we consider that the maximum error in V is 5% and the maximum error in R_0 is 2%. Both errors tend to increase the output voltage, so in Eq. (9.7) for the maximum value of v_{out} we use the $+$ signs and for the minimum value of v_{out} we use the $-$ signs. Equation (9.4) gives a maximum output voltage of

$$v_{out} + \Delta v_{out} = 12(1 + 0.05)$$

$$\times \left[\frac{1100(1 + 0.02)}{1000 + 1100(1 + 0.02)} - \frac{1000}{1000 + 1000} \right] = 0.362 \text{ V} \quad (9.8)$$

and a minimum output voltage of

$$v_{out} - \Delta v_{out} = 12(1 - 0.05)$$

$$\times \left[\frac{1100(1 - 0.02)}{1000 + 1100(1 - 0.02)} - \frac{1000}{1000 + 1000} \right] = 0.214 \text{ V} \quad (9.9)$$

Equations (9.8) and (9.9) may be solved for v_{out} and Δv_{out} to yield

$$v_{out} = 288 \text{ mV} \quad \text{and} \quad \Delta v_{out} = 74 \text{ mV} \quad (9.10)$$

which may be expressed as

$$v_{out} = 288 \pm 74 \text{ mV} \quad \text{(worst case)} \quad (9.11)$$

Note that the average value given by Eq. (9.10) is very near to the nominal value of 286 mV calculated above, because the relative errors are small. Note that we put "worst-case" by our stated error to make explicit the method of error calculation.

Because the relative errors are small, we may use the linearized expression in Eq. (9.7) to estimate the worst-case error. In this case the nominal value would be 286 mV and the relative error would be

$$\pm \epsilon_v = \pm 5\% \pm 2\% \times 10.5 = \pm 26.0\% \quad (9.12)$$

for an error of $0.260 \times 286 = 74.2$ mV. Hence the linearized expression gives an output voltage of 286 ± 74 mV (worst case).

Worst-case analysis is pessimistic because it assumes that all errors are at their maximum values and that all the effects of errors accumulate. For this reason, worst-case error analysis is used primarily when life, property, or product acceptability is dependent on measurement accuracy.

Statistical error analysis. A *statistical* analysis of system errors considers that errors are random variables described by probability density functions. The underlying mathematical models are beyond the scope of our treatment, but are detailed in books on statistics and probability theory. Use of statistical techniques with the full nonlinear expressions are difficult at best, and usually a linearized analysis is performed.

With this model, the quoted error of the ϵ's are standard deviations and they combine in a Pythagorean addition. If we assume a standard deviation of 5% in V and 2% in R_0.

$$\epsilon_V = \sqrt{(5)^2 + (10.5 \times 2)^2} = 21.5\% \quad \text{(1 s.d.)} \quad (9.13)$$

Note the following:

- The weighting factors in Eq. (9.7) must be included.
- With Pythagorean addition, the total errors are strongly dominated by the

larger contributors. Small errors do not matter unless there are many of them.

- The "(1 s.d.)" means one standard deviation and indicates the statistical method for combining errors.

For our example, the statistical method yields an output voltage of 286 ±61.6 mV (1 s.d.). Normally, Gaussian statistics are assumed, and the results imply that the true output voltage lies within the error bounds with a probability of 68% and within twice the error bounds with a probability of 95%.

Statistical error analysis is optimistic because it assumes that the errors are not always near their limits and that error cancellations occur to some extent. Nevertheless, experience shows that this method is valid, and the simplicity of the analysis makes it popular.

9.1.4 Check Your Understanding

1. The voltage applied to a Wheatstone bridge will influence the output voltage unless the bridge is balanced. (T/F?)
2. Reversing both diodes in the linear-variable differential transformer will not change the output voltage. (T/F?)
3. A measurement system has three sources of error, which produce maximum errors of 1%, 1.8%, and 2.7%. What is the worst-case error possible in the output?
4. Repeat the previous problem if the errors are standard deviations. What is the standard deviation due to the combined errors?

   Answers. 1. True; 2. false: the output will change sign; 3. 5.5% (worst case); 4. 3.40% (1 standard deviation).

9.2 ANALOG SIGNAL PROCESSING

The purpose of the analog section of the data-acquisition system is to provide the A/D converter with analog information, as free from noise as possible and in the prescribed voltage range. This involves amplification, gain control, removal of unwanted dc voltage, filtering, and cabling the system to minimize interference. In this section, we first treat matters relating to amplification and gain control, then deal with analog filtering techniques, and end with a brief section on grounding and shielding of instrumentation cables.

9.2.1 Instrumentation Amplifiers

Grounds, single-ended, and floating signals.   In Chapter 5 we discussed grounding for safety. The *power ground* is the point of a circuit connected to earth by a large wire; the *signal ground* of an electronic circuit is a connected set of conductors distributed to many points of the circuit. Signal ground may or may not correspond to power ground. When the signal voltage appears between a wire carrying the signal and the signal ground, as shown in Fig. 9.7a, the signal line is said to be *single-ended*. When the signal voltage appears between two wires

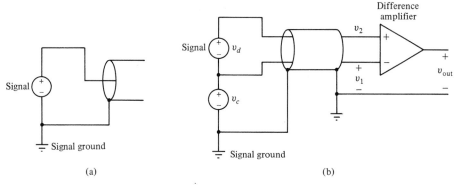

Figure 9.7 (a) Single-ended signal; (b) floating signal.

independent of signal ground, as shown in Fig. 9.7b, the signal is said to be *floating*. The voltage source v_c in Fig. 9.7b might be voltage introduced by the circuit of the transducer or might represent spurious voltage induced in both wires carrying v_d, which is the signal. The floating signal can be considered two single-ended signals, with the real signal being their difference in voltage.

With a single-ended signal, amplification would be accomplished with the standard op-amp amplifiers presented in Chapter 8. The noninverting amplifier would be favored because its high input impedance minimizes loading effects. The floating input requires a "difference amplifier," that is, an amplifier that ideally subtracts the signals between the two input conductors and amplifies their difference. We presented a circuit that accomplishes this in Fig. 8.52, but this circuit suffers from a low input impedance. After discussing difference- and common-mode signals we will present an amplifier suitable for floating inputs.

Difference- and common-mode signals. Signals of the type shown in Fig. 9.7b are described by a difference-mode signal and a common-mode signal. As shown in Fig. 9.8, the difference-mode voltage, v_{dm}, is the difference between the two input voltages and is equal to v_d in Fig. 9.7b. The common-mode signal is the average between v_1 and v_2 and is the voltage that the difference amplifier is required to reject, or at least minimize. The equations relating common- and difference-mode voltage to the single-ended voltages are

$$v_{dm} = v_2 - v_1 \qquad v_2 = v_{cm} + \frac{v_{dm}}{2}$$

$$\Leftrightarrow \qquad\qquad (9.14)$$

$$v_{cm} = \frac{v_2 + v_1}{2} \qquad v_1 = v_{cm} - \frac{v_{dm}}{2}$$

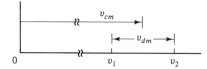

Figure 9.8 The difference-mode signal is the difference between two voltages, and the common-mode signal is their average.

The amplifier with a floating input ideally should have equal but opposite gains for both inputs, but in practice will have slightly different gains. Let the gain for the noninverting input be $+A_2$ and the gain for the inverting input be $-A_1$, where both A's are positive. The output will therefore be

$$v_{out} = +A_2 v_2 - A_1 v_1 = \left(\frac{A_1 + A_2}{2}\right) v_{dm} + (A_2 - A_1) v_{cm}$$

$$= A_{dm} v_{dm} + A_{cm} v_{cm}$$

(9.15)

where the difference-mode gain is the average of A_2 and A_1 and the common-mode gain is the difference between A_2 and A_1. The common-mode rejection ratio (CMRR) is the ratio

$$\text{CMRR} = \left| \frac{A_{dm}}{A_{cm}} \right|$$

(9.16)

A difference amplifier should have a large common-mode rejection ratio. A typical op amp would have a CMRR of 70 dB, which means that the gain for the difference-mode signal is $10^{+70/20} = 3162$ times the gain for the common-mode signal. In a given circuit application, the amplifier utilizing the op amp would have effects from the external circuit that would degrade the CMRR from that of the op amp alone. The common-mode voltage also has to lie within acceptable bounds for the op amp to operate satisfactorily.

Example. In the previous example, the Wheatstone bridge in Fig. 9.2 operates with $V = 12$ V, $R_1 = R_2 = R_3 = 1000 \ \Omega$ and $R_0 = 1100 \ \Omega$. With these values, the voltage across R_0 is $v_2 = 6.286$ V and the voltage across R_2 is $v_1 = 6.000$ V. We assume that this signal is amplified by a difference amplifier with a CMRR of 70 dB. What is the normalized error in the output due to the common-mode component?

The common-mode and difference-mode can be determined from Eqs. (9.14), with the result $v_{dm} = 0.286$ V and $v_{cm} = 6.143$ V. Assuming a difference-mode gain of A_{dm}, the common-mode gain would be smaller by 70 dB: $A_{cm} = A_{dm}/3162$. The normalized error in the output due to common-mode signal is therefore

$$\frac{6.143 \ A_{dm}/3162}{0.286 \ A_{dm}} = 6.80 \times 10^{-3} = 0.68\%$$

(9.17)

and thus is a minor factor. However, if the bridge were more nearly balanced, the common-mode error would be more important.

Instrumentation amplifier. The instrumentation amplifier shown in Fig. 9.9 has high (and equal) input impedances in both inputs and an output voltage that amplifies the difference between the input voltages. We recognize that the output op amp provides subtraction and amplification with a gain of R_F/R_1, as shown in Eq. 8.58. We may analyze the input stage by the methods of Chapter 8. The voltages at the inverting inputs to the top and bottom input op amps are v_2 and v_1, respectively. The current in R_2 is

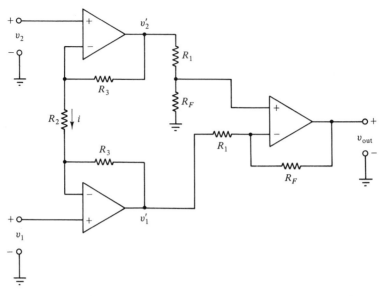

Figure 9.9 The instrumentation amplifier has a combined noninverting amplifier for high input impedance, followed by a difference amplifier.

$$i = \frac{v_2 - v_1}{R_2} \tag{9.18}$$

Since the op amp inputs have insignificant input current, the current through R_2 must also go through both R_3's, so the difference in output voltages will be

$$v_2' - v_1' = iR_2 + 2iR_3 = (v_2 - v_1)\left(1 + \frac{2R_3}{R_2}\right) \tag{9.19}$$

Hence the input stage, in addition to providing a high input impedance to the sources of v_2 and v_1, also gives a gain of $(1 + 2R_3/R_2)$. Therefore, the difference-mode gain of the instrumentation amplifier shown in Fig. 9.9 is

$$A_{dm} = \frac{R_F}{R_1}\left(1 + \frac{2R_3}{R_2}\right) \tag{9.20}$$

The common-mode gain of the circuit depends on the CMRR for the output amplifier and the degree to which the R_F/R_1 ratios are matched for the two inputs to the subtractor circuit.

Amplifier with bias removal. If the information appears as small changes in a dc voltage, we can use the amplifier shown in Fig. 9.10 to remove all or part of the dc voltage from the signal. This circuit consists of a voltage-follower input stage, for high input impedance, followed by a subtractor. The dc input to the subtractor is derived from a potentiometer. The resistance of the potentiometer, R_p, should be somewhat less than R to avoid loading effects.

Programmable attenuator. An *attenuator* is a device to reduce the signal level by a prescribed amount. An attenuator is required when the signal level is

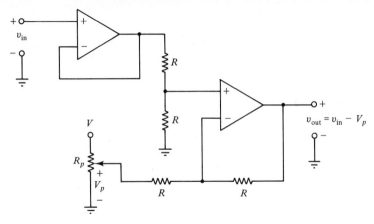

Figure 9.10 This amplifier can be used to remove a dc bias from a signal.

too large for the A/D converter or some other component in the system. The inverting amplifier shown in Fig. 9.11 has gains of 1, 0.5, 0.1, 0.05, or 0.01, depending on the switch settings. The truth table shows the usable settings, with a 1 indicating that the switch is closed. The switches would be relay switches or FET switches, controlled automatically by an overrange indication from the A/D converter.

Summary. In this section we have discussed a number of op-amp circuits that are useful in analog conditioning of signals. We defined difference-mode and common-mode signals and introduced the common-mode rejection ratio for an amplifier. In the next section we discuss filtering of analog signals.

Figure 9.11 This programmable attenuator reduces the signal level by factors up to 100, depending on the switch settings.

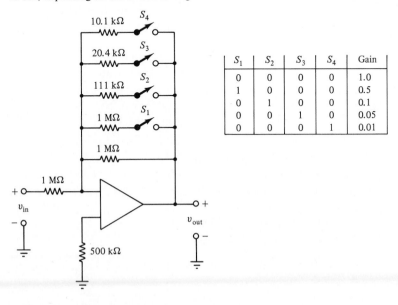

S_1	S_2	S_3	S_4	Gain
0	0	0	0	1.0
1	0	0	0	0.5
0	1	0	0	0.1
0	0	1	0	0.05
0	0	0	1	0.01

9.2.2 Analog Active Filters

The analog section of a data-acquisition system normally includes some filtering. The filters typically are *active filters* employing op amps with frequency-dependent feedback networks. In this section we first discuss sources of analog noise and then present representative active-filter circuits for low-pass, high-pass, and band-pass filtering.

Signal and noise spectra. The transducer output signal has a time structure that may be described in the frequency domain by a spectrum. As explained in Chapter 8, a periodic signal will have a harmonic spectrum, and a random signal will have a continuous spectrum. These spectra must be estimated as a basis for designing the band-pass of the analog section of the system. Noise from several sources will also be present in the system. The purpose of the analog filter is to pass the transducer signal and to eliminate as much of the noise as possible. Because the signal and noise share the same bandwidth, the filter cannot fully eliminate the noise.

The spectrum of the signal is determined by the time variations in the measurand and the construction of the transducer. The specifications of the manufacturer of the transducer should give some indication of time response. For example, a transducer with a 15-ms time constant would have a bandwidth of approximately $1/0.015 = 66.7$ Hz, and variations in the measurand at higher frequencies would be lost.

The principal components of the noise in the system are wideband noise, $1/f$ noise, and interference. Figure 9.12 shows the spectra of wideband and $1/f$ noise. The wideband noise consists of thermal noise, shot noise, and partition noise. Thermal noise arises in all resistors and is thermodynamic in origin. Chapter 10 discusses this type of noise in more detail. Shot and partition noise components arise due to the discrete character of electrical carriers and become a problem at small current levels. Interference is man-made noise that can be reduced by proper circuit layout, shielding, and grounding techniques, as discussed in the next section.

The $1/f$ noise component, which exists in all natural processes, shows the accumulated effects of small changes. For example, resistors and transistors are constantly changing in their microstructure due to heat, cosmic ray damage, vibration, and a myriad of other effects. These combine to produce *drift*, that is, a slow meandering of signal values. Since the signal normally has a $1/f$ component of its own, which is part of its information content, the $1/f$ noise component cannot

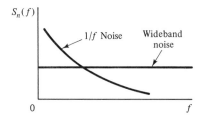

Figure 9.12 Two unavoidable noise components are $1/f$ and wideband noise.

Data-Acquisition Systems   Chap. 9

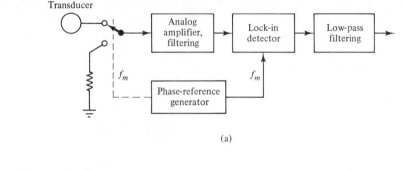

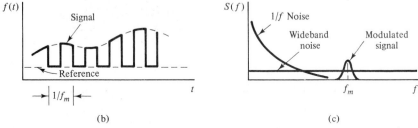

(a)

(b)   (c)

Figure 9.13 (a) The lock-in detection system "chops" the signal at a frequency f_m and detects the information at that frequency; (b) the slow variations in the signal are modulated by the chopping; (c) the effect of the modulation is to move the information to a higher frequency so that $1/f$ noise can be eliminated by the detection and filtering process.

be eliminated by straightforward filtering. An important technique to minimize this noise component is the lock-in modulation/detection system shown in Fig. 9.13a. The information from the transducer is *chopped* or modulated by switching between the transducer output and a reference signal, as shown in Fig. 9.13b. After amplification and analog filtering, the output component at the modulation frequency is detected by a phase-selective detector. The benefit of this scheme is that $1/f$ noise contributed by the electronic system is reduced because only these signal components at the modulation frequency are detected. In effect, the modulation moves the information to a high frequency so that the $1/f$ noise can be filtered, as shown in Fig. 9.13c.

Active filters. An active filter combines amplification with filtering. The *RC* filters investigated in Chapter 8 are passive filters because they provide only filtering. An active filter uses an op amp to furnish gain but has capacitors added to the input and feedback circuits to shape the filter characteristics.

We derived earlier the gain characteristic of an inverting amplifier in the time domain. In Fig. 9.14 we show the frequency-domain version. We may easily translate the earlier derivation into the frequency domain:

$$v_{\text{in}} \Rightarrow \underline{V}_{\text{in}}(\omega), \qquad v_{\text{out}} \Rightarrow \underline{V}_{\text{out}}(\omega)$$

$$A_V = -\frac{R_F}{R_1} \Rightarrow \underline{F}_V(\omega) = -\frac{\underline{Z}_F(\omega)}{\underline{Z}_1(\omega)}$$

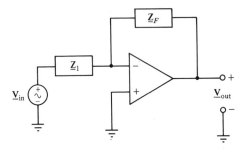

Figure 9.14 Active filter circuit.

The filter function, $\mathbf{F}_v(\omega)$, is thus the ratio of the two impedances, and in general will give gain as well as filtering. We could have written the minus sign as $1/180°$, for in the frequency domain the inversion is equivalent to a phase shift of 180°.

Low-pass filter. Placing a capacitor in parallel with R_F (Fig. 9.15) will at high frequencies tend to lower $\mathbf{Z}_F$ and hence reduce the gain of the amplifier; consequently, this capacitor converts an inverting amplifier into a low-pass filter with gain. We may write

$$\mathbf{Z}_F(\omega) = R_F \parallel \frac{1}{j\omega C_F} = \frac{1}{(1/R_F) + j\omega C_F} = \frac{R_F}{1 + j\omega R_F C_F} \qquad (9.21)$$

Thus the gain would be

$$\mathbf{F}_V = -\frac{R_F}{R_1} \frac{1}{1 + j\omega R_F C_F} = A_V \frac{1}{1 + j(\omega/\omega_c)} \qquad (9.22)$$

where $A_V = -R_F/R_1$ is the gain without the capacitor, and $\omega_c = 1/R_F C_F$ is the cutoff frequency. The gain of the amplifier is approximately constant until the frequency exceeds ω_c, after which the gain decreases with increasing ω. The Bode plot of this filter function is shown in Fig. 9.16 for the case where $R_F = 10$ kΩ, $R_1 = 1$ kΩ, and $C_F = 1$ μF. The shape is identical to that of the low-pass filter in Fig. 8.26, except for the increase in gain due to the op amp.

Figure 9.15 Low-pass filter circuit.

Figure 9.16 Bode plot for active low-pass filter.

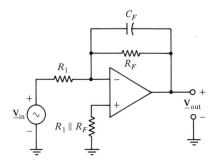

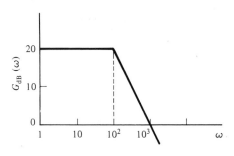

Data-Acquisition Systems   Chap. 9

High-pass filter. The high-pass filter shown in Fig. 9.17 uses a capacitor in series with R_1 to reduce the gain at low frequencies. The details of the analysis will be left to a problem. The gain of this filter is

$$\underline{F}_V(\omega) = -\frac{R_F}{R_1}\frac{j(\omega/\omega_c)}{1 + j(\omega/\omega_c)} = A_V\frac{j(\omega/\omega_c)}{1 + j(\omega/\omega_c)}$$

where $A_V = -R_F/R_1$ is the gain without the capacitor and $\omega_c \rightleftharpoons 1/R_1C_1$ is the cutoff frequency, below which the amplifier gain is reduced. The Bode plot of this filter characteristic is shown in Fig. 9.18. This Bode plot is identical to that of the high-pass filter given in Fig. 8.28 except for the increase in gain due to the op amp.

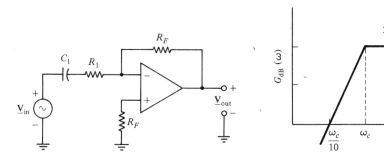

Figure 9.17 High-pass filter circuit.

Figure 9.18 Bode plot for active high-pass filter.

Low-pass, two-pole Butterworth filter. The low-pass Butterworth filter characteristic is

$$|\underline{F}(f)| = \frac{1}{\sqrt{1 + (f/f_c)^{2n}}} \tag{9.23}$$

where n is the order of the filter and f_c is the cutoff frequency. For $n = 1$, we have the characteristic of the low-pass filter discussed in Chapter 8. The filter characteristics for $n = 1$ and $n = 2$ are shown in the asymptotic Bode plots of Fig. 9.19b. Note that the second-order filter characteristic drops at -40 dB/decade after the critical frequency is passed. The circuit shown in Fig. 9.19a gives a second-order Butterworth filter characteristic. This circuit combines positive and negative feedback, and its principles of operation are not obvious. We may confirm the characteristic by using the op-amp analysis techniques developed in Chapter 8.

Analysis of Butterworth filter circuit. Application of op-amp principles requires that the input voltage to both op-amp inputs be essentially $\underline{V}_{\text{out}}$. Thus the current $\underline{I}$ shown in Fig. 9.19a is

$$\underline{I} = j\frac{\omega C}{\sqrt{2}}\underline{V}_{\text{out}} \tag{9.24}$$

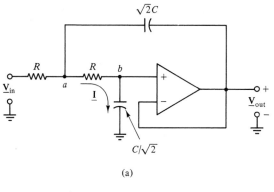

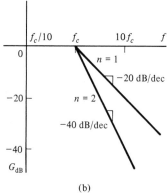

(a)

(b)

Figure 9.19 (a) Second-order Butterworth low-pass filter circuit; (b) Bode plots for first- and second-order Butterworth low-pass filters.

Hence the voltage at point a is

$$\underline{V}_a = \underline{V}_{out} + R\underline{I} = \underline{V}_{out}\left(1 + j\frac{\omega RC}{\sqrt{2}}\right) \tag{9.25}$$

We can now write KCL for node a, which is a standard nodal equation:

$$\frac{\left(1 + j\frac{\omega RC}{\sqrt{2}}\right)\underline{V}_{out} - \underline{V}_{in}}{R} + \frac{\left(1 + j\frac{\omega RC}{\sqrt{2}}\right)\underline{V}_{out} - \underline{V}_{out}}{1/j\sqrt{2}\omega C} + j\frac{\omega C}{\sqrt{2}}\underline{V}_{out} = 0 \tag{9.26}$$

Equation (9.26) has the solution

$$\underline{V}_{out} = \frac{\underline{V}_{in}}{1 - (\omega RC)^2 + j\sqrt{2}\omega RC} \tag{9.27}$$

For the Bode plot, we require the square of the magnitude of the filter gain:

$$|\underline{F}(\omega)|^2 = \frac{|\underline{V}_{out}|^2}{|\underline{V}_{in}|^2} = \frac{1}{[1 - (\omega RC)^2]^2 + (\sqrt{2}\omega RC)^2} = \frac{1}{1 + (\omega RC)^4} \tag{9.28}$$

which is the required Butterworth characteristic with

$$f_C = \frac{1}{2\pi RC} \tag{9.29}$$

Example. We will design a second-order low-pass Butterworth filter with a 3-dB frequency of 8 Hz. The output impedance of the source is less than 100 Ω.

The design requires two choices: the critical frequency in Eq. (9.29) and the value of either R or C. The 3-dB frequency occurs when the output power is one-half that of the passband, which requires in Eq. (9.28) that the term raised to the fourth power be unity. Thus the critical frequency given by Eq. (9.29) is also the 3-dB frequency. We may ensure that the filter has insignificant loading of the signal source by making R much greater than 100 Ω. Accordingly, we choose $R = 10$ kΩ and Eq. (9.29) determines the value of C to be

$$C = \frac{1}{2\pi R f_C} = \frac{1}{2\pi(10^4)8} = 1.99 \ \mu F \qquad (9.30)$$

This is not the value of either capacitor in the circuit in Fig. 9.19a but is rather the geometric mean between the two capacitors. The capacitors are thus approximately 2.8 μF and 1.4 μF, and the required circuit is shown in Fig. 9.20.

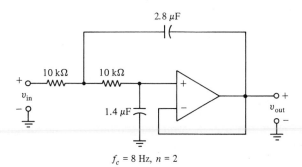

$f_c = 8$ Hz, $n = 2$

Figure 9.20 Second-order low-pass Butterworth filter circuit.

Other filters. Standard designs exist for high-pass and band-pass Butterworth filters, as well as several other filters having superior characteristics to the Butterworth in some aspects. Handbooks and application literature give practical circuits for many active filter circuits. Filter manufacturers produce filters of prescribed characteristics for high-frequency, extremely narrowband band-pass, and narrowband band-reject filters.

9.2.3 Cabling, Grounding, and Shielding Techniques

Sources of interference. Unwanted, man-made electrical signals constitute a class of noise called *interference*. Interference may couple into a transducer circuit via an electric field or a magnetic field. Electric-field interference arises from high-voltage signals and couples into high-impedance portions of electronic circuits. Magnetic fields induce signals by coupling to loops of wire in an electronic circuit. Both types of interference may be reduced by proper circuit layout, shielding, and grounding techniques.

Interference-prone locations for low-level electronic equipment are in the vicinity of high-power electrical equipment such as motors, welders, and transformers, near radio and TV stations, or near power-electronic equipment such as motor controllers and large power supplies. To reduce interference, avoid such locations if possible and shield electronic circuits with grounded metallic cabinets and power lines with metal conduit. Portions susceptible to magnetic coupling may be shielded by magnetic foil. Low-level signals should be protected by twisted-pair conductors that are shielded by a braided outer conductor.

Grounding. The *signal ground* is a network of wires, all tied together, to provide a reference potential of zero volts to all parts of the circuits. Grounding problems arise because currents flow in ground wires, and since the ground wires have resistance and inductance, voltage differences are created between different portions of the grounding system. Such differences can create false signals at low-level inputs. One precaution is to connect all grounds within a system to a common point through separate wires so that ground currents do not share wires, but this is usually unnecessary. A more practical technique is to provide a grounding system for high-level signals and for signals that have rapid transitions that is separate from the grounding system for low-level instrumentation signals.

Long cable runs between instruments or between a transducer and its analog electronics can cause special problems in grounding. When connected through the cable in an attempt to provide a common ground between two separate systems, as shown in Fig. 9.21, as much as 100 mV of voltage can exist between ground *a* and ground *b* due to interference signals. This connection is unacceptable unless signals are immune to that level of interference.

The cabling connection shown in Fig. 9.22a allows separate grounds. The input to the receiving chassis is floating, with the cable shield as one input. The connection in Fig. 9.22b uses a common ground, but the signal is carried by a twisted-pair shielded cable and thus floats. The signal-carrying wires are twisted to reduce magnetic coupling.

9.2.4 Check Your Understanding

1. Grounding one part of a floating input produces a single-ended input. (T/F?)
2. The voltages of a floating input are $+1.5$ V and -0.8 V. Find the common-mode and difference-mode components.
3. A difference amplifier has a voltage gain of 15,000 and a common-mode rejection ratio of 65 dB. What is the gain for the common-mode component of the input?
4. What would be the gain of the programmable attenuator in Fig. 9.11 if all switches were closed?

Figure 9.21 Separated electronic chassis with grounds connected through the connecting cable. This approach can cause interference problems.

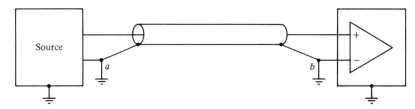

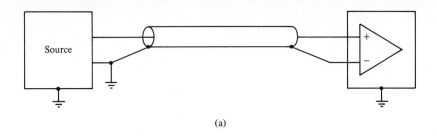

(a)

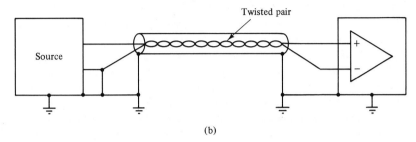

(b)

Figure 9.22 (a) No common ground is provided in this connection; (b) here a common ground is provided, but no signal current flows in the ground connection.

5. Show that the product of the voltage gain and the bandwidth of the low-pass filter in Fig. 9.15 is independent of the resistor in the feedback path.

6. The grounding system used in Fig. 9.21 would be acceptable for logic signals with a 5-V amplitude. (T/F?)

7. Twisted-pair cables are used to reduce electric-field or magnetic-field interference. (Which?)

   Answers. 1. True; 2. $V_{cm} = 0.35\,\text{V}, V_{dm} = 2.3\,\text{V}$; 3. 8.435; 4. $0.00629 = -44\,\text{dB}$; 5. gain $\times$ BW $= 2\pi/R_1 C_F$; 6. true; 7. magnetic-field interference.

9.3 DIGITAL SIGNAL PROCESSING

9.3.1 Analog-to-Digital Conversion

Definitions. An analog-to-digital converter (ADC) converts the analog signal to a digital signal for processing in digital form. Figure 9.23a shows an ADC with analog input, a GO input to initiate conversion, outputs to indicate the status of the conversion, and three output bits. If the ADC is linear, the digital output will be related to the analog input as indicated in Fig. 9.23b. The *precision* of the ADC is the number of output bits, in this case three. The *range* of the ADC is given by the maximum and minimum input voltages that can be converted with at most a one-half least significant bit (LSB) error. If, for example, the full-scale (FS) input voltage of the ADC were 10 V, the range of the ADC would be from 625 mV (1/16 FS) to 9.375 V (15/16 FS). Voltages below this range would give an output of 000, indicating underrange, and voltages above this range would give an output of 111 plus an overrange indication. The under- and overrange indi-

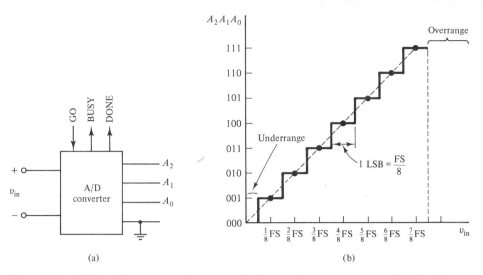

Figure 9.23 (a) The A/D converter takes an analog input and converts to a parallel digital output; (b) the output binary word depends on the range of the analog input.

cations can be used to control the gain of the analog channel to keep the analog voltage within the range of the ADC. The *resolution* of the ADC is the FS voltage divided by the number of output states, in this case $10 \text{ V}/2^3 = 1.25 \text{ V}$, and corresponds to 1 LSB. Typical commercial ADCs have 8- and 12-bit precision, indicating 256 and 4096 states.

Two-bit flash A/D converter.   Figure 9.24a shows a 2-bit flash ADC. The resistors in series set up four reference voltages, and the analog input voltage is compared simultaneously with all references, with the results shown in the truth table in Fig. 9.24c. The lowest comparator indicates underrange (or underflow) and the highest indicates overrange (or overflow). The middle three comparators indicate which of the four possible ranges contains the input, and the logic circuit in the box converts this information to a binary output code. The A/D conversion takes place almost instantaneously. This type of ADC, also called a parallel encoder, is available commercially in 4- and 8-bit versions. Note from Fig. 9.24b that the ranges for this converter are set up with an offset from a linear function passing through the origin, unlike the ideal in Fig. 9.23b.

Single-slope ramp analog-to-digital conversion.   Many types of single-slope converters are possible. The basic idea is shown in Fig. 9.25a, in which an op amp is used to integrate an input reference voltage. This gives an output ramp function. While the ramp is increasing, the AND gate allows the counter to count clock pulses, but when the comparator indicates that the ramp voltage has exceeded the analog input, the count will stop and the A/D conversion will be complete, as shown in Fig. 9.25b. The capacitor would then be shorted electronically, and the ADC would be ready to begin its conversion cycle again.

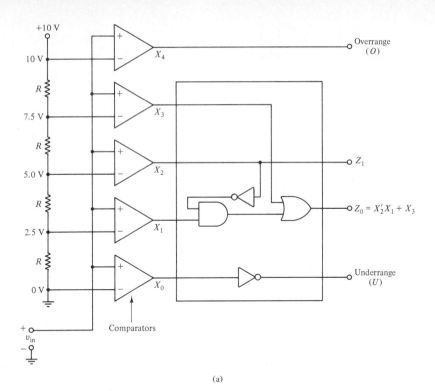

(a)

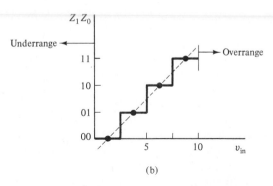

(b)

	X_4	X_3	X_2	X_1	X_0	Z_1	Z_0	O	U
$v_{in} < 0$	0	0	0	0	0	X	X	0	1
$0 < v_{in} < 2.5$	0	0	0	0	1	0	0	0	0
$2.5 < v_{in} < 5.0$	0	0	0	1	1	0	1	0	0
$5.0 < v_{in} < 7.5$	0	0	1	1	1	1	0	0	0
$7.5 < v_{in} < 10.0$	0	1	1	1	1	1	1	0	0
$10.0 < v_{in}$	1	1	1	1	1	X	X	1	0

(c)

Figure 9.24 (a) The two-bit flash A/D converter compares the input with ranges of voltage; (b) the output–input characteristic of the ADC; (c) the truth table of the ADC, including over- and underrange indications.

Sec. 9.3   Digital Signal Processing

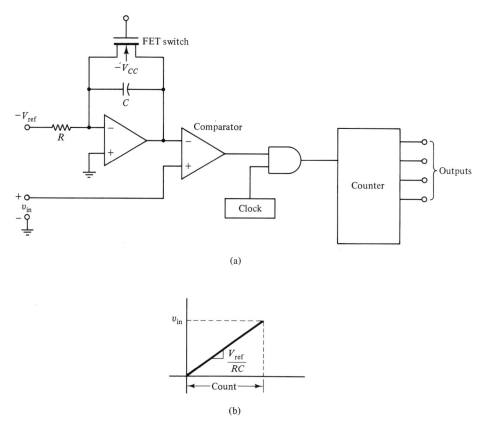

Figure 9.25 (a) The single-slope converter counts clock pulses until an increasing ramp voltage exceeds the input voltage; (b) the count will be proportional to the input voltage.

Dual-slope ramp conversion. A dual-slope conversion technique has several advantages over the single-slope converter. As shown in Fig. 9.26a, the input to the integrator is connected first to the unknown analog input voltage, and the output voltage builds proportionally for a fixed amount of time. Then the input is switched to a known reference voltage of opposite polarity, and clock pulses are counted until the integrator voltage passes through zero, as shown in Fig. 9.26b. The advantage of this scheme is that it measures the unknown relative to the voltage reference, independent of variations in the clock frequency. This is the method employed in most digital voltmeters.

9.3.2 Sampling

Typically, transducers monitoring physical processes produce data rates well below the capability of a computer. Thus many inputs can be monitored simultaneously. Numerous analog inputs may be sampled at discrete times sequentially through a multiplexer. A sample-and-hold circuit may be required to arrest change in an input while A/D conversion is being performed. Finally, the discrete samples of the time-varying inputs may be processed as a time sequence by the computer

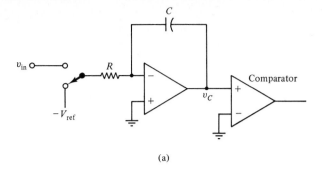

(a)

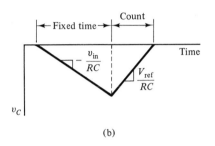

(b)

Figure 9.26 (a) Dual-slope ramp conversion input stage; (b) the number of counts while the voltage is coming to zero indicates the analog input voltage.

by techniques known as digital filtering. Processes associated with the sampling of the analog inputs are discussed in this section.

Analog multiplexers. The analog multiplexer (AMUX) shown in Fig. 9.27a accepts eight analog inputs. A 3-bit address $B_2B_1B_0$ selects the input to connect to the output. By cycling through all inputs, the ADC can serially convert

Figure 9.27 (a) This multiplexer has eight analog inputs. The input selected by the incoming address is connected to the sample-and-hold circuit for analog-to-digital conversion. (b) The FET switch is an open circuit except when its gate is at $+15$ volts, connecting input to output.

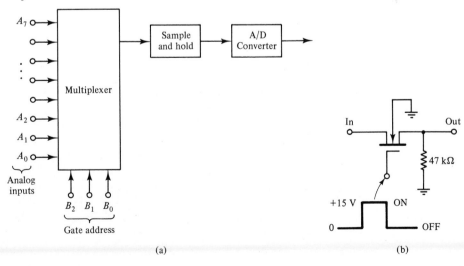

(a)

(b)

all inputs to digital form for processing by the computer. Figure 9.27b shows how the AMUX uses a FET switch to connect and disconnect an input from the output. output. The FET is used as a voltage-controlled switch.† With zero voltage on the gate, the drain–source resistance is thousands of megohms and the input voltage is blocked. With sufficient voltage applied to the gate, the FET is placed in the ohmic region, and the drain–source resistance is 25 to 100 Ω, allowing the analog input to appear across the 47-kΩ resistor and thus the output. The internal logic in the AMUX decodes the channel address and connects one channel at a time with a "break-before-make" connection to ensure that no two input channels are shorted together through ON FET's.

Sample-and-hold circuit. Figure 9.28 shows a sample-and-hold circuit, which is used to arrest time variations in the input signal during ADC processing. The circuit uses two voltage-follower amplifiers to buffer input and output. An FET switch is activated long enough for the capacitor to charge to the input voltage, and then the capacitor holds the voltage while A/D conversion is taking place. The capacitor is sized as a compromise between the requirements for rapid charging (small C) and long voltage-retention time (large C). If, for example, the FET ON resistance is 25 Ω and the requirement is for charging to 99% of the input value in 1 μs, about five time constants are required. Thus the time constant must be less than 0.2 μs, and the required value of C cannot be larger than 8 nF. This in turn places a limit on how long the capacitor can hold the voltage to a prescribed tolerance, considering the leakage current through the FET and into the op amp. We leave the remaining details for a homework problem.

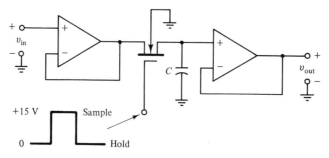

Figure 9.28 The sample-and-hold circuit uses a capacitor to remember the voltage after the FET switch is gated ON briefly.

Sampling and digital filtering. Let us consider that we have a data-acquisition system with a number of inputs. We use an AMUX to cyclically sample each input, with the sample-and-hold circuit holding the sampled voltages long enough for A/D conversion. The end result of this operation is a sequence of samples representing the time structure of each input, as shown by Fig. 9.29a. The discrete sampling of the time structure of the input raises two questions:

† For relatively slow multiplexing rates, mechanical switches of the reed-relay type may be used.

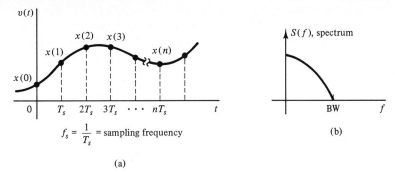

$$f_s = \frac{1}{T_s} = \text{sampling frequency}$$

(a)

Figure 9.29 (a) The continuous waveform is sampled at a frequency f_s, producing a sequence of discrete values, $x(0)$, $x(1)$, $x(2)$, . . . ; (b) the spectrum of the signal waveform before sampling. The *Nyquist criterion* states that the sampling frequency must be at least twice the signal bandwidth to avoid loss of information.

"What is an appropriate sampling rate?" and "How should we process the samples to improve the information content of the data?"

The question of how often to sample is answered by the *Nyquist criterion*, which states that no information is lost through the sampling process if the sampling frequency f_s satisfies the following criterion:

$$f_s > 2 \, \text{BW} \tag{9.31}$$

where BW is the bandwidth of the waveform being sampled, as shown in Fig. 9.29b. If the Nyquist criterion is violated, information is lost and the information content of the sampled values becomes confused. For a wideband signal, the bandwidth prior to sampling should be limited to alleviate this confusion.

The second question of how to process the sampled information raises questions of *digital filtering*. Of course, the samples may be used without digital filtering, but often benefits are gained by processing of the samples. To illustrate this process, we will consider three sample-processing schemes for the filter indicated by Fig. 9.30.

The first is called a finite impulse response (FIR, or nonrecursive) filter, because it calculates the output of the "digital filter," $y(n)$, from previous values of the input, $x(n)$, $x(n - 1)$, ... For example, consider the FIR filter represented by the calculation

$$y(n) = \frac{x(n) + x(n - 1)}{2} \tag{9.32}$$

Figure 9.30 The digital filter accepts discrete input samples and uses a numerical procedure to produce output discrete data. The "filter" is a software algorithm in the computer.

$x(n) = x(1), x(2), \ldots$ → Digital filter → $y(n) = y(1), y(2), \ldots$

We identify the output as the running unweighted average of the previous two samples of the input. By contrast, an infinite impulse (IIF, or recursive) filter calculates $y(n)$ from previous inputs $(x(n), x(n-1), \ldots)$ and previous outputs $(y(n-1), y(n-2), \ldots)$. An example would be

$$y(n) = \frac{y(n-1) + x(n)}{2} \tag{9.33}$$

We may compare the response of the FIR and the IIR filters described by Eqs. (9.32) and (9.33) by examining their responses to a sudden increase of the input. Figure 9.31b shows the responses calculated in the table in Fig. 9.31a for the filter algorithms given in Eqs. (9.32) and (9.33). We have filled in between the samples with dashed and dotted lines to aid in identification of the responses. The IIF algorithm has a greater smoothing effect because it has a longer memory of the past history of the input.

	n	0	1	2	3	4	5
Input	$x(n)$	0	0	1	1	1	1
FIR	$y_1(n)$	0	0	0.5	1	1	1
IIR	$y_2(n)$	0	0	0.5	0.75	0.88	0.94

(a)

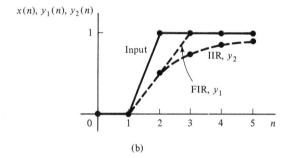

(b)

Figure 9.31 (a) Table showing input and output of a digital filter with the filter algorithms given in Eqs. (9.32) and (9.33). The input is a sudden increase in the input value. (b) Plots of input and output samples.

Digital filters and their effects are an established technique of digital-signal processing. Digital filters exist to simulate analog filters such as the Butterworth filters and can be designed to eliminate discrete frequencies (60-Hz hum, for example).

Summary. In this section we have discussed the effects of sampling information at discrete intervals. We have introduced the Nyquist criterion to define the minimum sampling rate. We have given two examples of digital filtering.

9.3.3 Check Your Understanding

1. The increment of voltage corresponding to the least significant bit for an 8-bit ADC would be one-half the increment of voltage for a 4-bit ADC. (T/F?)
2. The underrange signal on the ADC in Fig. 9.24a would be a digital one for a negative input signal. (T/F?)

3. The dual-slope ramp conversion ADC is immune to minor variations in the clock frequency. (T/F?)
4. A sample-and-hold circuit may not be required if the ADC is fast enough in its conversion speed. (T/F?)
5. If the output signal from a temperature-indicating device contains no significant information beyond 1 Hz in its output spectrum, what is the maximum sampling period that can be used for this device without losing information?
6. A digital filter filters digital signals. (T/F?)

Answers. 1. False; 2. true; 3. true; 4. true; 5. 0.5 s; 6. false.

9.4 DIGITAL CONTROL SYSTEMS

9.4.1 Digital-to-Analog Converters

The digital data-acquisition system shown in Fig. 9.1 monitors and processes data from several transducers. We have included the possibility of an analog output from the computer, however, because often the purpose of the data-gathering function is to produce output signals to control the processes being monitored. For example, the temperature of a chamber might be monitored by a system to maintain a prescribed temperature in the chamber through heaters. In this section we discuss digital-to-analog converters (D/A converters or DACs) and simple control functions.

The digital-to-analog converter (DAC) accepts a digital input and produces an analog output. The input–output characteristics are the same as those shown in Fig. 9.23b for the ADC, except that the vertical axis represents the input and the horizontal axis represents the output. Depending on details of the construction (and specifications), the output might be offset from zero in the manner of Fig. 9.24b. We will show circuits for two representative 3-bit DACs.

The voltage-summing circuit shown in Fig. 9.32 uses a binary-weighted sum of inputs controlled by three single-pole, double-throw switches. The principle of operation of a summing amplifier circuit is discussed on page 365. Here each input is either $-V_{ref}$ or 0 V, depending on the switch setting. The switch positions shown correspond to the binary word $B_2B_1B_0 = 101$, and the output will be $1.25V_{ref}$.

Figure 9.32 This 3-bit digital-to-analog converter circuit operates by summing voltages.

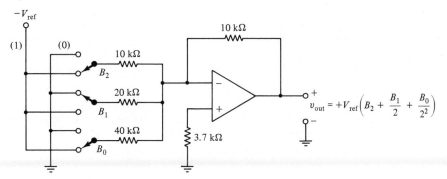

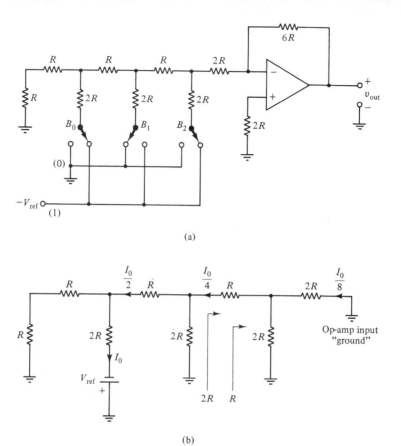

Figure 9.33 (a) This digital-to-analog converter is based on the current-division properties of the R-$2R$ ladder circuit; (b) the R-$2R$ ladder network divides the current by a factor of 2 at each node.

Clearly, 000 produces zero output and 111 produces $1.75V_{ref}$, the least-significant bit being $0.25V_{ref}$ for this DAC.

The circuit shown in Fig. 9.33a is based on the properties of the R-$2R$ ladder network, which are indicated in Fig. 9.33b. We have used the principle of superposition to examine the effect of the LSB input (B_0). Note that the equivalent resistance of the ladder is R at each parallel node and $2R$ into the series resistors. Thus the equivalent resistance values are the same, regardless of the length of the ladder. At each input, the input resistance is $3R$, so the input current from the source at B_0 is $-V_{ref}/3R$. The R-$2R$ ladder network halves the current at each node. By the time the input current at B_0 reaches the op-amp input, it is halved three times, a factor of 8, such that the current into the op-amp is $-V_{ref}/24R$. As explained on page 362, this current goes through the feedback resistor of $6R$ and produces an output voltage of $+V_{ref}/4$. The input of each higher-order bit produces a larger voltage by a factor of 2 because the current is halved one less time. Thus the input at B_2 produces an output voltage of $+V_{ref}$, and generally the ladder can be extended to a larger number of bits for an 8- or 12-bit DAC.

9.4.2 Digital Control Systems

Block diagram. Figure 9.34 represents a digital control system to control a physical process represented by the variable $x(t)$. The portion represented by the right branch we recognize as a data-acquisition system: sensor(s), analog amplification and filtering, multiplexing, sample-and-hold circuit, analog-to-digital conversion, and digital filtering within the computer. The control-system portion of the system consists in comparing the sampled variable $x'(t)$ with a prescribed value x^* and an error criterion ϵ, and generating a control signal $u(t)$ through a control algorithm. We will consider three control algorithms: bang-bang, incremental, and proportional control.

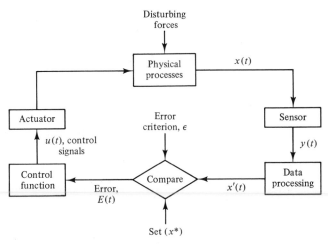

Figure 9.34 A digital control system.

Bang-bang control. The operation of a bang-bang system is represented by Fig. 9.35. Within a dead zone about the desired value, no drive is applied by the control system because the output is acceptable. If the control variable wanders below the dead zone, full positive drive is applied to correct the error, and likewise full negative drive is applied if $x'(t)$ becomes too high. The ON–OFF nature of the drive characterizes the bang-bang algorithm.

Figure 9.36 shows a bang-bang system to control the temperature of a chamber. The temperature of the chamber is monitored by a thermocouple. Its output

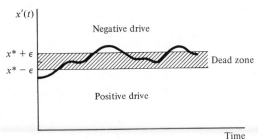

Figure 9.35 A bang-bang control algorithm drives the process to remain within a dead zone about the set value of the signal.

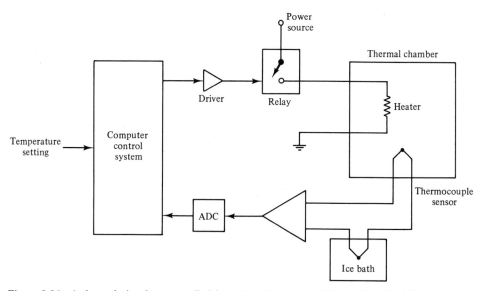

Figure 9.36 A thermal chamber controlled by a bang-bang control loop. The "bang" consists of turning on the heater in the chamber.

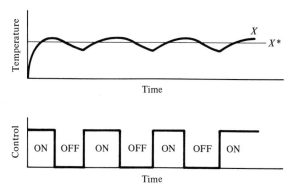

Figure 9.37 The bang-bang control algorithm turns the heaters on when the indicated temperature falls below the desired value.

voltage is amplified and converted to digital form. The computer compares the measured temperature with the desired temperature and activates the heater in the chamber through a driver and relay. The chamber temperature variations and system operation are indicated in Fig. 9.37.

In this system, the positive drive is provided by the heater/relay, the negative drive is provided by the ambience, assumed cooler than the chamber, and the dead zone is provided by the time delays inherent in the heater and thermocouple. The extent of the temperature variations in the chamber depend on the thermal insulation of the chamber and the thermal inertia of the loop.

Incremental control. The algorithm for incremental control is

$$U_{n+1} = U_n + \Delta U, \quad \text{if} \quad x' < x^* - \epsilon$$
$$= U_n - \Delta U, \quad \text{if} \quad x' > x^* + \epsilon \tag{9.34}$$

where U_{n+1} is the new drive level, U_n is the previous drive level, and ΔU is a fixed increment of drive. Thus, if the temperature is too cool, the drive is increased by a fixed increment, and if the temperature is too hot, the drive is decreased by the fixed increment. If the temperature is within the acceptable region, the drive is not changed.

Figure 9.38 shows one means for implementing incremental control for the temperature chamber. By electronic control, the heater is connected to the power source during entire half-cycles of the ac frequency. Figure 9.38a shows the heater connected for two of the five half-cycles shown. Figure 9.38b shows the logic of a control algorithm, based on an overall period of sixteen half-cycles. If the tem-

Figure 9.38 (a) Incremental control is implemented by turning ON the heater during an integral number of ac half-cycles; (b) the logic of the incremental control algorithm. Depending on the temperature comparison, the count of ON cycles is increased, decreased, or left unchanged.

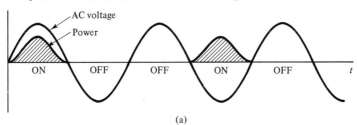

(a)

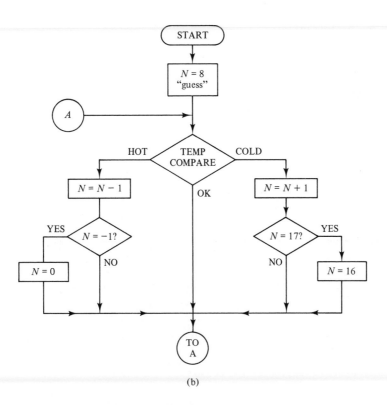

(b)

perature is out of range on the cold side, the number of heating cycles is increased by one, and if the temperature is out of range on the hot side, the count is decreased by one. If the temperature is within the acceptable range, the count of ON half-cycles is not changed. Additional tests must be applied to ensure that the ON count remains between zero (minimum heating) and sixteen (maximum heating).

Control by the incremental algorithm is clearly more steady than bang-bang control because it allows for steady heating and also adjusts the heating in small increments instead of full ON or OFF. A proportional control algorithm allows even smoother control.

Proportional control.   The control algorithm for proportional control is

$$U_{n+1} = U_n + k(x^* - x') \tag{9.35}$$

where k is a gain constant. The heater drive is thus changed proportional to the difference between the desired and the measured temperature. Such control can be accomplished by the method shown in Fig. 9.39. Through the techniques of power electronics, the ac waveform can be applied to the heater for a prescribed fraction of the ac cycle. The control algorithm could be accomplished by logic similar to that shown in Fig. 9.38b, except that the number of increments would be large and would correspond to the fraction of the ac cycle during which the heaters are ON. For example, we could divide the $\frac{1}{120}$-s (8333-μs) half-cycle into 8333 one-microsecond increments. The count would correspond to the number of microseconds that the heaters are ON. Thus, this algorithm amounts to incremental control with 8333 increments (not all equal), but is essentially continuous in the heating effect.

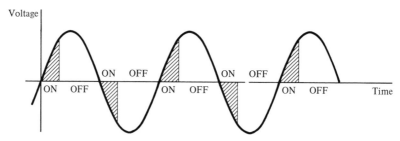

Figure 9.39   For proportional control, the heaters are turned ON for a portion of each cycle. Such control can be achieved by techniques of power electronics.

Summary.   We have introduced digital-control systems by describing three means for controlling the temperature of a thermal chamber. The bang-bang system is full ON or OFF, depending on the temperature. The incremental system controls the level of thermal drive, and the proportional system uses yet finer control.

9.4.3 Check Your Understanding

1. What is the increment in output voltage corresponding to a 1-bit change in the least significant bit for a 8-V full-scale ADC with a precision of 8 bits?

2. Show that the op-amp bias current sees the same impedance at the + and − input terminals in Fig. 9.33a.

3. The type of control system that is inactive some fraction of the time is the bang-bang, incremental, or proportional control. (Which?)

4. Can the bang-bang control be unstable?

   Answers.   1. 31.25 mV;   2. $6R \parallel (2R + R) = 2R$;   3. bang-bang;   4. yes, if the zone of acceptability is too small.

PROBLEMS

Section 9.1: Introduction to Data-acquisition Systems

P9.1. A thermistor is used in the bridge circuit shown in Fig. P9.1. The characteristic of the thermistor is

$$\ln\left(\frac{R}{R_0}\right) = 4000(T^{-1} - T_0^{-1})$$

where R is the resistance at temperature T in kelvin, and R_0 is the resistance at temperature T_0, in this case 4 kΩ at 20°C. Find and plot the output voltage of the bridge as a function of temperature in the range $0 < T < 100°C$.

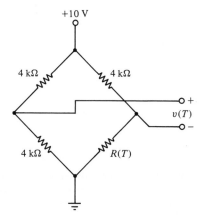

Figure P9.1

P9.2. The circuit in Fig. P9.2 represents an unbonded strain gage. All four resistors are affected by changes in the length of the device, Δd, with opposite resistors changed in opposite directions. Determine the output voltage as a function of Δd.

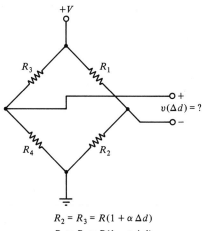

$$R_2 = R_3 = R(1 + \alpha \Delta d)$$
$$R_1 = R_4 = R(1 - \alpha \Delta d)$$

Figure P9.2

P9.3. The potentiometer transducer in Fig. P9.3 gives an output voltage proportional to the displacement x only if unloaded; otherwise, loading effects make the output nonlinear. Determine the ratio of the load resistance, R_L, to the potentiometer resistance, R_p, such that the maximum error is 1% of full scale. *Hint:* Find the maximum value of the output resistance and make this 1% of the load resistance.

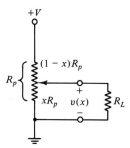

Figure P9.3

P9.4. A voltage in a first-order transient circuit decays from 10 V to 0 V with an RC time constant. If the resistance can vary by 5% and the capacitor by 10%:
  (a) Find the percent of variation in the time constant using a linearized analysis.
  (b) Find the percent of variation in the voltage at $t = RC$, where RC is the nominal time constant. Use the exact analysis.

Answers. (a) $\pm 15\%$; (b) $\pm 15.1\%$

P9.5. The resistive voltage divider in Fig. P9.5 is built with 10% tolerance resistors.
  (a) Determine the exact limits of the output voltage if $R_2 = 2R_1$ (nominal values).
  (b) Explain why the limits do not give a 20% error due to each 10% error.

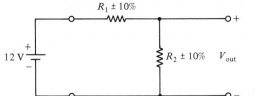

Figure P9.5

P9.6. Assume that you have a number of measurements of the same physical quantity, all of comparable accuracy. Show that the standard deviation of the average is reduced by $\sqrt{n}$ from the standard deviation of a single measurement, where n is the number of samples that are averaged. To improve the accuracy by a factor of 10, how many independent measurements must be made?

Section 9.2: Analog Signal Processing

P9.7. For the instrumentation amplifier in Fig. 9.9, assume that the two R_3's are not identical but differ by a normalized error of ϵ_R. That is, assume $R_{3(\text{bottom})} = R_{3(\text{top})}(1 \pm \epsilon_R)$. Using the high-gain technique from Chapter 8, determine the error in the output due to this asymmetry. Does this affect the common-mode rejection ratio or just the difference-mode gain of the amplifier?

P9.8. Determine the output of the noninverting amplifier in Fig. 8.48 if the amplifier has finite gain (A_d) and a finite common-mode rejection ratio ($A_c = A_d/\text{CMMR}$). Express your answer is terms of A_d, CMMR, R_1, and R_F.

P9.9. An amplifier has a gain of 50 dB for the difference input and a common-mode rejection ratio of 68 dB. The input signals are 1.012 V and 1.006 V. What is the percent of error in the output signals due to the common-mode component?

  Answer. 6.7%

P9.10. The circuit in Fig. 9.10 has a gain of unity for the signal only if the loading due to the potentiometer is ignored. Find the ratio R_p/R such that the gain of the signal channel is always between 0.98 and 1.00.

P9.11. Design an op-amp-based amplifier with the following characteristics:
  • Amplifies the difference between two signals
  • Filters the signal with the filter function shown in Fig. P9.11
  • Has an input impedance of 1 kΩ at both inputs
  • Uses one 10-μF capacitor
  • Uses no more than two op amps

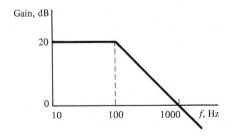

Figure P9.11

P9.12. The sound energy in the voice has a power spectrum with most of the energy between 500 and 3000 Hz. Design an active filter to pass only this band of frequencies and also give a gain of 12 dB. Use no capacitor larger than 10 μF. Combine the first-order high-pass and low-pass characteristics.

P9.13. For the active filter circuit shown in Fig. P9.13:
    (a) Find the filter function $\mathbf{F}(j\omega)$.
    (b) If the input voltage is 2 V dc, what is the output voltage?
    (c) Is there a critical frequency? If so, what is it in hertz, and does the gain begin to decrease or increase at the critical frequency?

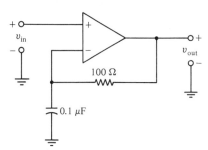

Figure P9.13

P9.14. The Bode plot of the gain of a single-ended active filter is shown in Fig. P9.14.
    (a) Draw an active filter circuit to give this characteristic.
    (b) If the circuit capacitor is 10 μF, find the resistors.
    (c) At what frequency is the amplitude of the output voltage equal to the input voltage if the signal is a sinusoid?
    (d) If $v_{in}(t) = 5\cos(2000\pi t)$ is the input signal, which is the output signal in the time domain?

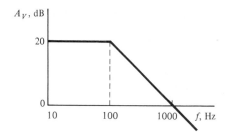

Figure P9.14

P9.15. For the op-amp circuit shown in Fig. P9.15:
    (a) If a 1-V battery is connected from ground to the input, with the + of the battery connected to ground, what is the voltage of points a and b relative to the ground?
    (b) Now instead of a battery the input is a 1-V pulse. What is the duration of the shortest pulse that the amplifier can pass without serious distortion?

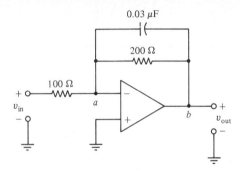

Figure P9.15

P9.16. Design an active filter having a gain of $+15$ dB, a phase shift of 180°, an input resistance of 500 Ω in the passband, and a high-pass characteristic with cutoff frequency of 1250 rad/s.

P9.17. Derive the filter function, $\mathbf{F}(\omega) = \mathbf{V}_{out}/\mathbf{V}_{in}$, for the high-pass filter shown in Fig. 9.17. Confirm the formula for the cutoff frequency, ω_c. What would be the input impedance in the band of frequencies above the cutoff frequency, ω_c?

P9.18. Design an active high-pass filter with a cutoff frequency of 500 Hz, an input impedance of 1000 Ω in the high-frequency region, and a gain of $+10$ in the passband. *Hint:* Use a passive high-pass filter at the input of a noninverting amplifier.

P9.19. An op-amp amplifier is required to pass pulses of 1 ms width but to eliminate unwanted frequencies higher than those required for the pulse. The gain must be -2. Assume that you have available a 0.003-μF capacitor. Give the circuit and component values.

P.20. In an audio system an amplifier is required that will give a gain of 10 ($+$ or $-$, it does not matter) and have upper and lower cutoff frequencies to exclude signals outside the normal audio region, 30 to 20,000 Hz. Combine high-pass and low-pass active filters and design a circuit that will accomplish this. Assume that the largest capacitor you have available is 100 μF.

P9.21. Design a second-order Butterworth filter with a cutoff frequency of 500 Hz. The largest capacitor may be no more than 10 μF. Determine the gain (amplitude and phase shift) at 100, 500, and 2000 Hz.

Section 9.3: Digital Signal Processing

P9.22. An 8-bit ADC has a range from 0 to 10 V.
(a) What increment in input voltage causes a 1-bit change at the output?
(b) What range of input voltages corresponds to the output of 1001? Use the no-offset characteristic similar to that shown in Fig. 9.23b.

P9.23. Verify the logic for Z_0 and Z_1 in the 2-bit flash ADC in Fig. 9.24a by making Karnaugh maps for Z_0 and Z_1 with X_1, X_2, and X_3 as inputs.

P9.24. Redesign the flash ADC in Fig. 9.24a to have a characteristic passing through the origin. In other words, what would be the voltage-reference levels (and the resistors in the voltage-divider circuit) such that the transition to 01 is at $v_{in} = 0.125V_{ref}$, to 10 at $v_{in} = 0.375V_{ref}$, and so on.

P9.25. An 8-bit ADC of the type shown in Fig. 9.25a takes 100 ms to make its conversion.
   (a) What would be the clock frequency?
   (b) If the input range is 0 to 10 V and the reference voltage is 5 V, what should be the RC product for the integrator?

P9.26. The sample-and-hold circuit in Fig. 9.28 was designed in the text to have an 8-nF capacitor. If the combined leakage current due to the FET, op-amp input bias current, and capacitor leakage were 1 nA, how long does the capacitor hold the voltage to 99% of a 10-V capacitor voltage when the FET is turned OFF?

P9.27. The FET switches used in the multiplexer in Fig. 9.27a, one of which is shown in Fig. 9.27b, have a resistance of 25 Ω when ON and a resistance of 10 MΩ when OFF. Assume the inputs A_0 through A_7 come from sources with a 100-Ω output impedance. Determine the coupling in decibels from the channel that is ON to the other seven channels that are OFF. *Note:* The eight channels are switched by eight FETs, all of which connect to the same 47-kΩ resistor for an output.

P9.28. The sample-and-hold circuit in Fig. 9.28 uses an input voltage-follower amplifier with an output impedance of 40 Ω and an output voltage-follower amplifier with an input impedance of 20 MΩ. The FET resistance is 25 Ω when ON and 10 MΩ when OFF. The FET is ON for 1 μs and OFF for 1 μs for the ADC to make its conversion. Under these conditions, what is the optimum capacitor to maximize the voltage at the end of the 2-μs cycle, and what is the final voltage as a percent of the input voltage to the sample-and-hold circuit?

Answers. $C = 1.33$ nF and output is 99.988% of input

P9.29. Determine the response of the two digital filters shown in Fig. 9.31 to the sudden increase at the input if the filters use three terms: for the FIR filter, $y(n) = [x(n) + x(n - 1) + x(n - 2)]/3$, and for the IIR filter, $y(n) = [x(n) + y(n - 1) + y(n - 2)]/3$.

P9.30. Show that the FIR digital filter in Eq. (9.32) will remove a 60-Hz component in the data if the sampling period is $\frac{1}{120}$ s.

P9.31. A transducer signal is sampled every 5 s. The output spectrum is sufficiently broad to cause confusion if the transducer bandwidth is not limited prior to sampling. Determine the critical frequency of a simple low-pass filter to ensure that 99% of the power to the sampler lies within the Nyquist limit. Assume a worst-case transducer spectrum that is constant.

Answer. $f_c = 1.57$ MHz

10 Communication Systems

Introduction to communication systems   As a pipeline moves liquid from one location to another, so a communication system moves information from one location to another. A communication system consists of a source of information with some sort of encoder, a recipient of information with the appropriate decoder, and a medium connecting the two. Some examples are obvious: the telephone system, commercial TV broadcasting, computer networks for airline reservations. Some are not so obvious: a police radar for measuring speed, a system for verifying credit cards, an internal bus in a computer.

There are several reasons for using electrical signals in such communication systems. Electrical signals travel with speeds approaching the speed of light. Electrical signals in the form of radio waves go almost anywhere. At certain broadcast frequencies, the radio waves follow the earth's curvature and also reach distant points by bouncing off the ionosphere. Electrical signals lend themselves to a variety of ingenious coding schemes such as amplitude modulation, frequency modulation, and various digital codes.

Contents of this chapter.   In this section, we focus on the system used in commercial AM radio. First we look at how information is encoded onto and decoded from radio signals. The explanation requires expansion of our understanding of the frequency domain to include nonlinear effects. Then we consider the various forms of electromagnetic waves such as are used in radio systems.

425

We examine several common types of antennas, both as transmitting and as receiving devices. Finally we consider the details of two communication systems.

10.1.1 General Principles of Nonlinear Devices in the Frequency Domain

Linear circuits. One of the important principles we used in our analysis of ac circuits was that linear circuits create no new frequencies. In Chapter 4 we showed that if you excite a circuit at a certain frequency it will respond only at that frequency. This principle allowed us to simplify greatly our analysis of ac circuits; indeed, the approach using phasors and impedance is founded entirely on this principle.

Nonlinear circuits with one input frequency. The reverse principle is that nonlinear circuits create new frequencies. We have already discovered in Chapter 6 that a diode, a nonlinear device, can rectify a signal. Earlier when we discussed the spectrum of a half-wave rectified sinusoid (Fig. 8-6) we noted that it contains, in addition to the frequency of the input sinusoid, frequency components at dc and all the even harmonics of the input frequency. These new frequencies are created by the nonlinear action of the diode.

The spectrum of a rectified sinusoid is one instance of the general principle suggested in Fig. 10.1. For a nonlinear circuit with a sinusoidal input, the output of the circuit will in general contain all the harmonics of the input. The amplitudes of the various output harmonics depend on the details of the circuit, but in principle all harmonics can be present in the output of a nonlinear circuit.

This property of nonlinear circuits finds direct application in many practical devices. One familiar example is a power supply, where the output of interest is the zeroth harmonic of the input. Another application occurs in the generation of stable high-frequency sinusoids for precision communication systems or spectroscopic measurement systems. Often high-frequency signals of the required stability are generated by taking the output of a highly stable oscillator with a low frequency, say, 5 MHz, and putting this signal through a chain of frequency multipliers until it reaches the required high frequency, say, 3240 MHz. In such a scheme, the output of the multiplier chain will be harmonically related to the low-frequency source and hence will exhibit the same relative stability.

On the other hand, the creation of new frequencies may be undesirable. A good audio amplifier should have low *harmonic distortion*, meaning that it should not generate harmonics of the audio signal. The amplifier should also have low *intermodulation distortion*. This term refers to the creation of new frequencies through the interaction of separate frequencies in the auto spectrum. Nonlinearities in the amplifier cause both of these undesirable effects.

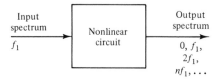

Figure 10.1 Nonlinear circuits create new frequencies.

Nonlinear circuits with two input freqiencies.   The creation of harmonics in a nonlinear circuit, suggested in Fig. 10.1, is a special case of a more general property of nonlinear circuits, as suggested by Fig. 10.2. Here we show a nonlinear circuit with two input frequencies. The figure indicates that the output will in general contain all the harmonics of both input frequencies *plus* all the sum and difference frequencies of those harmonics. For example, if we had inputs of 30 and 100 Hz, the output would contain not only the harmonics of 30 (60, 90, 120, . . .) and the harmonics of 100 (200, 300, . . .), but the output would also contain the various sum and difference frequencies, such as $100 - 30$, $100 + 30$, $200 - 30$, $200 + 30$, $100 - 2 \times 30$, $100 + 2 \times 30$, $200 - 2 \times 30$, ... Radio systems use such new frequencies to encode communication signals.

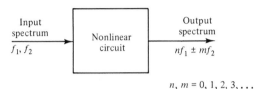

Figure 10.2 Nonlinear circuits create all harmonics plus all sum and difference frequencies.

Importance of filtering.   Although a multitude of frequencies can appear in the output of a nonlinear circuit, not all these frequencies will be useful. The circuit designer must ensure that only the desired frequencies are strong in the output and that undesired components are minimized. Filters are used to remove unwanted frequency components in the output spectra of nonlinear devices. Only through careful control of the filtering properties of such circuits can the benefits of nonlinear action be achieved.

The design and operation of nonlinear circuits is complicated by the creation of new frequencies and the filtering of these frequencies to enhance desired and diminish unwanted components. In the following we will examine the applications of a few nonlinear circuits, emphasizing especially their role in communication systems. Our goal is to develop a sufficient basis for discussing the operation of the common AM radio.

10.1.2 Modulation

Amplitude modulation.   In communication engineering, *modulation* is the process by which information at a low frequency is coded as part of a high-frequency signal. The higher frequency is called the *carrier* because it carries the information signal. In this section we will explain the amplitude modulation (AM) scheme that is widely used in commercial broadcasting of radio and TV signals. Initially, we will explain the nature of amplitude modulation, emphasizing the frequency-domain viewpoint throughout. Then we will discuss why modulation is required in communication systems. Finally, we will present a circuit that accomplishes amplitude modulation.

The audio range covers frequencies from about 30 Hz to about 15 kHz. In a communication system, we do not broadcast audio frequencies directly; rather we use high-frequency carriers. The AM radio band uses carriers in the range from 540 to 1600 kHz.

Let us first consider the nature of amplitude modulation and how modulation affects the spectrum of the carrier. Equation (10.1) gives the equation of a modulated carrier in the time domain.

$$v_{AM}(t) = \underbrace{V_c(1 + m_a \cos \omega_m t)}_{V_e(t),\text{ the envelope}} \cos \omega_c t \qquad (10.1)$$

Here ω_m is the angular frequency of the modulating sinusoid, ω_c is the frequency of the modulated carrier, and V_c is its amplitude. The parameter m_a, the modulation index, indicates the degree of modulation. The modulation index is proportional to the amplitude of the modulating signal, which is indicated by V_m in Fig. 10.3, subject to the restriction that m_a never exceeds unity; that is, $m_a \leq 1$.

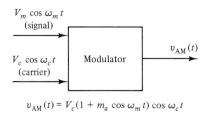

$$v_{AM}(t) = V_c(1 + m_a \cos \omega_m t) \cos \omega_c t$$

Figure 10.3 A modulator circuit has two inputs and one output.

The character of the modulated carrier and the role of the modulation index are pictured in Fig. 10.4, where we have shown modulated carriers for $m_a = 0.5$ and $m_a = 1.0$. Let us examine the role of the various factors characterizing the AM signal. The carrier is the sinusoid whose amplitude is being modulated. The carrier frequency controls the spacing of the zero crossings and individual peaks of the composite waveform. If we were to keep everything else the same yet double the carrier frequency, everything would look the same except there would be twice as many peaks and zero crossings. In other words, the hills and valleys of the amplitude would look the same, but the underlying sinusoid would have a higher frequency.

The amplitude of the carrier, V_c, characterizes the overall strength of the signal. With a broadcast signal, the amplitude of the carrier depends on the power of the station transmitter, the distance to the receiver, and the transmitting and receiving antennas.

The frequency of the modulating signals, ω_m, determines the time between the hills and valleys of the envelope, $V_e(t)$, as defined in Eq. (10.1). If for example the modulating frequency were doubled, the envelope of the carrier would vary up and down twice as fast.

Finally, the modulation index controls the degree of modulation, the ratio of hills to valleys in the envelope. This parameter is limited to values less than unity, so the envelope factor in Eq. (10.1), $V_e(t)$, never goes negative. The modulation index is proportional to the amplitude of the modulating signal. The design of the modulator must ensure that the maximum amplitude of the modulating signal can be accommodated without exceeding the allowed value of unity.

$$m_a = KV_m, \quad \text{but} \quad m_a \leq 1 \qquad (10.2)$$

where K is a constant.

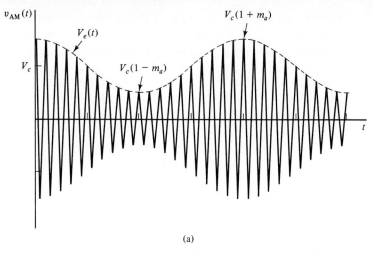

(a)

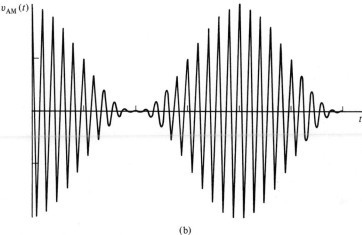

(b)

Figure 10.4 Two AM waveforms: (a) $m_a = 0.5$; (b) $m_a = 1.0$.

AM modulator circuit. The circuit in Fig. 10.5 will act as a simple modulator circuit. Ignore for the moment the RC filter between a-a' and b-b', and consider the part of the circuit that consists of two generators, a diode, and a resistor. The generator marked $v_c(t)$ is the carrier and has a high frequency. The generator marked $v_m(t)$ is the modulating signal, and it has a lower frequency and a smaller amplitude than the carrier.

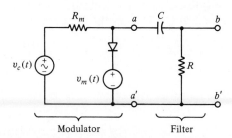

Figure 10.5 AM modulator circuit.

429

Unfiltered output.   Consider now the action of the diode on the signal at $a\text{-}a'$. When the instantaneous carrier voltage, $v_c(t)$, exceeds the instantaneous modulating voltage, $v_m(t)$, the diode will turn ON and current will flow clockwise around the circuit. The output voltage in this case will be equal to the instantaneous modulating voltage plus a small voltage required to turn ON the diode. On the other hand, when the instantaneous carrier voltage is smaller than the instantaneous modulating voltage, the diode will be OFF and no current will flow. During this time, there will be no voltage drop across the resistor and hence the voltage at $a\text{-}a'$ will be equal to the carrier voltage. The result is shown in Fig. 10.6.

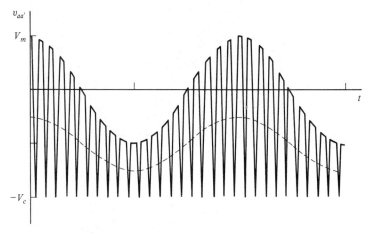

Figure 10.6   Unfiltered output of AM modulator.

Filtered output.   The signal at $a\text{-}a'$, shown in Fig. 10.6, differs from a pure AM signal in two respects: (1) only the top is modulated, the bottom being flat, and (2) the carrier waveform is not round on top and is thus not a pure sinusoid with varying amplitude. We can solve the first problem by filtering out the dc and the component at ω_m (the dashed line in Fig. 10.6). These components we can eliminate with a high-pass filter that passes the carrier while rejecting the modulating frequency. This is the function of the high-pass filter between $a\text{-}a'$ and $b\text{-}b'$.

The other problem, that the tops of the modulated carrier are distorted, implies that some higher harmonics of the carrier are also created in the modulator circuit. These also can be eliminated with a filter. With proper filtering, the output of the modulator in Fig. 10.6 would look like the ideal AM waveform in Fig. 10.4.

Spectrum of an AM signal.   The spectrum of an AM signal can be determined through expansion of Eq. (10.1) with the trigonometric identity

$$\cos a \cos b = \tfrac{1}{2}[\cos(b + a) + \cos(b - a)] \tag{10.3}$$

When we expand Eq. (10.1) and apply Eq. (10.3), letting $a = \omega_m t$ and $b = \omega_c t$, we obtain

$$v_{AM}(t) = V_c \cos \omega_c t + m_a V_c \cos \omega_m t \cos \omega_c t \tag{10.4}$$

$$= V_c \cos \omega_c t + \frac{m_a V_c}{2} \left[\underbrace{\cos(\omega_c + \omega_m)t}_{\text{sum freq.}} + \underbrace{\cos(\omega_c - \omega_m)t}_{\text{diff. freq.}} \right]$$

This expression suggests a spectrum with three frequency components: the carrier at ω_c, an *upper sideband* at $\omega_c + \omega_m$, and a *lower sideband* at $\omega_c - \omega_m$. The spectra before and after modulation are shown in Fig. 10.7. We have broken the frequency scale in Fig. 10.7 because the carrier frequency is normally much higher than the modulating frequency. For example, the emergency broadcast warning tone of 853 Hz, broadcast on a carrier of 1200 kHz, would produce upper and lower sidebands at 1,200,853 Hz and 1,199,147 Hz, respectively.

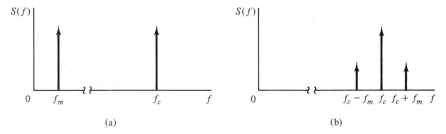

Figure 10.7 Modulation in the frequency domain: (a) before modulation; (b) after modulation.

Modulation with a random signal.   If your AM radio received the AM signals in Fig. 10.4, and if the modulating frequency were in the audio band, you would hear a steady tone from the radio speaker, and you might suppose that a civil defense test was underway. In normal broadcasting, the program material would be music and voice signals, not steady tones. The modulated signal would thus be of the form

$$v_{AM}(t) = V_c[1 + m_a s(t)] \cos \omega_c t \tag{10.5}$$

where $s(t)$ represents the program material and hence would be a random function. In Eq. (10.5), $s(t)$ must remain smaller than unity so that the product $m_a s(t)$ never exceeds unity at any time.

When the program information consists of a random signal such as music or voice, the individual frequencies comprising the spectrum of the audio signal are shifted to the sidebands. Let us investigate the case of normal AM broadcasting. By law, the spectrum of the audio signal is limited to frequencies between 100 and 5000 Hz. The effect of modulating with such a signal is indicated by Fig. 10.8. The sidebands now become continuous spectra. The upper sideband is identical in shape with the audio signal, whereas the lower sideband is reversed. The bandwidth of the radio-frequency (RF) spectrum is twice the audio bandwidth, and hence 10 kHz of RF bandwidth is required for each AM station.

If you could examine the spectrum of radio waves received by an AM radio antenna, you would see carriers for each station in your area, each bracketed by its upper and lower sidebands. Since the AM broadcast band covers frequencies

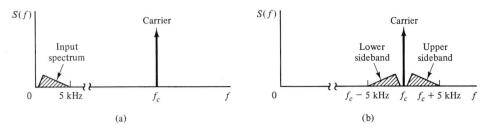

Figure 10.8 Modulation with a spectrum: (a) before modulation; (b) after modulation.

from 535 to 1605 kHz, and since carriers can be spaced at 10-kHz intervals, there are channels for 106 AM stations, although obviously most channels are unoccupied in geographic area.

Standard AM and single-sideband modulation. Because AM detectors are simple, standard AM modulation is used for commercial broadcasting. This AM system is, however, inefficient in its use of the available bandwidth and the transmitter power. Bandwidth is wasted because, although both sidebands and carrier are broadcast, all the information is contained in one sideband. Consequently, the power used in transmitting the carrier and one of the sidebands is wasted because these carry no information not already contained in the other sideband. For these reasons, sophisticated radio communication systems, such as a telephone company uses for long-distance messages, employ single-sideband modulation for economy in both power and bandwidth. Single-sideband modulation, as the name suggests, transmits only one of the sidebands over the communication path, and the receiver must generate the carrier and missing sideband to detect the information.

Need for modulation. Modulation shifts the information from the audio band to the RF radio band. We modulate to solve antenna problems, filtering problems, and confusion problems. The last problem is the easiest to understand: clearly, we cannot allow all radio stations to broadcast their signals in the same frequency band because then our radios would receive all stations simultaneously. Hence, modulation allows us to assign different portions of the RF spectrum to different radio stations, and they can place their signals neatly into their allotted frequency bands. We choose stations by tuning our radio receivers to these bands.

Discussion of the antenna problems solved by modulation will be deferred to the section on antennas. We will consider how modulation solves filtering problems later in this section, after we have discussed AM detectors and mixers. Having shown how the information in the audio band is shifted to higher frequencies for broadcast, we turn now to the inverse operation, how your radio shifts the signal back to the audio band so that you can hear it.

10.1.3 Radio Receivers

AM detector. Figure 10.9 shows a simple circuit that will extract the audio information from the AM signal envelope. We have studied this circuit in Chapter

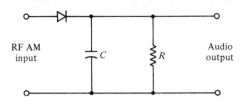

RF AM
input

C

R

Audio
output

Figure 10.9 AM detector circuit.

6 (page 219) as a power-supply circuit. In that application the input is a simple sinusoid. The diode allows the capacitor to charge to the peak voltage of the input, and the resistor represents the load. To work well as a power supply, the circuit employs an *RC* time constant much larger than the period of the input sinusoid. This long time constant ensures that the output voltage of the power supply will remain essentially at the peak voltage of the input.

To function as an AM detector, this circuit requires an *RC* time constant intermediate between the period of the carrier and that of the audio information, that is, $2\pi/\omega_c < RC < 2\pi/\omega_m$. With this provision, the detection of the envelope of the carrier takes place as suggested by Fig. 10.10. The output of the detector follows the peaks of the input and hence approximates the envelope of the AM signal. We have drawn the period of the carrier relatively large, resulting in a jagged appearance for the output; but in practice the period of the carrier is so short that the output is quite smooth. The small amount of jaggedness, which is inevitable, constitutes a small amount of high-frequency noise in the output, easily removed with a low-pass filter. Disregarding this noise component, we see that the effect of the detector in the frequency domain is to shift the information in the sidebands back to the audio band, as shown in Fig. 10.11.

Examining the relationship between the spectra in Figs. 10.8 and 10.11, we see that the detector inverts the modulation process. For this reason the word *demodulation* is often used to describe this process. You might suppose that we now have all the makings of a radio: we can shift the audio information to the sidebands of a carrier for broadcast and we can shift the information back to the audio region for conversion to acoustic waves with loudspeakers. In a sense you

Figure 10.10 Output of AM detector.

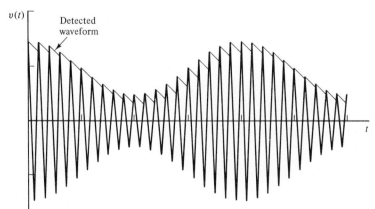

$v(t)$

Detected
waveform

t

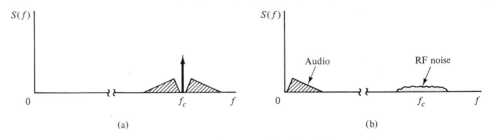

$S(f)$

Audio   RF noise

f_c f

(a)   (b)

Figure 10.11 Effect of AM detector in the frequency domain: (a) detector input spectrum; (b) detector output spectrum.

suppose correctly, for a simple communication system can be made with direct demodulation, but the usual radio utilizes *mixing*, another nonlinear process, to shift the spectrum of the received signal to an intermediate frequency.

Mixer. A *mixer*, like a modulator, is a nonlinear circuit with two inputs and one output, as suggested by Fig. 10.12a. Indeed, modulators and mixers operate on similar principles, yet differ in their function in communication systems. Figure 10.12b shows a circuit that will act as a mixer. The input called V_2 at ω_2 is the carrier and sidebands, and its frequency is called the radio frequency (RF). The input $\underline{V}_1$ at ω_1 is the local oscillator (LO) signal, to be explained below. The mixer output has a frequency intermediate between the RF and the audio, and hence is called the intermediate frequency (IF). We choose the inductor and capacitor to be resonant at the intermediate frequency; hence the capacitor's impedance will be so small at RF that the RF current will be controlled by the diode characteristic [Eq. (6.6), assume $\eta = 1$].

$$i_D = I_o(e^{v_D/V_T} - 1) = I_o \left[1 + \frac{v_D}{V_T} + \frac{1}{2}\left(\frac{v_D}{V_T}\right)^2 + \cdots - 1 \right]$$

$$= a_1 v_D + a_2 v_D^2 + \cdots$$

(10.6)

Figure 10.12 (a) Mixer input and output frequencies; (b) mixer circuit. The *LC* filter presents a small impedance at the radio frequency but a large impedance at the intermediate frequency.

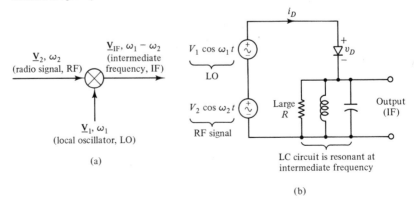

(a)

(b)

TABLE 10.1 SOME OF THE FREQUENCIES CREATED IN THE MIXER

	Frequency Component
$i_D = a_1(V_1 \cos \omega_1 t + V_2 \cos \omega_2 t) + a_2(V_1 \cos \omega_1 t + V_2 \cos \omega_2 t)^2 + \cdots$	
$\quad = a_1 V_1 \cos \omega_1 t$	ω_1
$\qquad + a_1 V_2 \cos \omega_2 t$	ω_2
$\qquad + a_2 V_1^2 \cos^2 \omega_1 t\, [= a_2(V_1^2/2)(1 + \cos 2\omega_1 t)]$	dc, $2\omega_1$
$\qquad + 2a_2 V_1 V_2 \cos \omega_1 t \cos \omega_2 t\, [= a_2 V_1 V_2 \cos(\omega_1 t + \omega_2 t)$	$\omega_1 + \omega_2$
$\qquad\qquad\qquad\qquad\qquad\qquad + a_2 V_1 V_2 \cos(\omega_1 t - \omega_2 t)]$	$\omega_1 - \omega_2$
$\qquad + a_2 V_2^2 \cos^2 \omega_2 t\, [= a_2(V_2^2/2)(1 + \cos 2\omega_2 t)]$	dc, $2\omega_2$

where $a_1, a_2, \ldots$ are constants. Because the impedance of the LC filter at RF is small, the voltage across the diode will be the sum of the two voltage sources,

$$v_D = V_1 \cos \omega_1 t + V_2 \cos \omega_2 t \qquad (10.7)$$

and therefore the power-series form of Eq. (10.6) will have many terms involving the two input frequencies, of which we will investigate but a few. In Table 10.1, we have shown some of the terms that will appear in the mixer output, and we have entered in the last column the frequency components that will be produced. The output provides an example of the principle suggested in Fig. 10.2, that is, that a nonlinear circuit with two input frequencies will in general create all the harmonics of those input frequencies, plus the sums and differences of those harmonics. Had we extended our power series in Eq. (10.6) to higher-order terms and used trigonometric identities to expand these terms, we would have verified the presence of all higher harmonics, with their sums and differences.

Of the various frequencies in the mixer output, only one is desired; the rest are noise to be eliminated by filters. We require the difference frequency, $\omega_1 - \omega_2$, listed in the second column in Table 10.1. This component represents a shift of the information in the carrier and sidebands to the intermediate frequency. Figure 10.13 shows the effects of mixing and demodulation. The higher frequency (f_1) is called the local oscillator (LO) frequency, because this frequency is generated within the radio by an oscillator circuit. The mixer combines this frequency with the carrier and sidebands, acting on each frequency component separately, and shifts the entire RF spectrum to the intermediate frequency (IF). The IF signal is amplified and filtered and then detected by a normal AM detector; the audio signal is thus recovered. This audio signal is further amplified, low-pass filtered, and furnished to a loudspeaker.

Figure 10.13 Mixing and demodulation in the frequency domain.

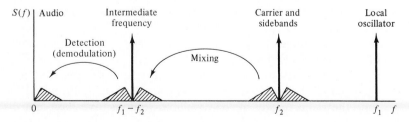

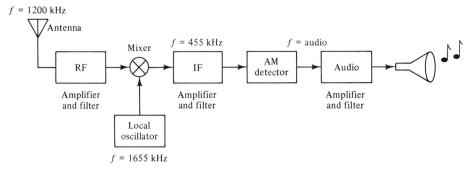

Figure 10.14 Basic AM radio system.

Example: AM system. Consider an AM radio station broadcasting at 1200 kHz. The standard IF for an AM radio is 455 kHz; hence the local oscillator frequency must be 1200 + 455 = 1655 kHz. The resulting AM radio system is pictured in Fig. 10.14.

Reasons for using an intermediate frequency. We will now give the reasons for shifting the information to an IF rather than using direct demodulation of the RF signal. As a preliminary we review the filtering that we have already encountered in the circuit. We require low-pass filtering at the output of the AM detector to separate the audio signal from the detector noise. This filtering is usually incorporated into the audio amplifier circuit and is relatively easy to accomplish because of the wide difference in frequency between the audio spectrum and the noise spectrum.

Similarly, we need filtering at the mixer output to eliminate all but one of the frequencies created by the nonlinear action of the mixer diode. This filtering is likewise easily accomplished because again there is a large difference between the desired and the undesired frequency components in the mixer output. As we will see, this filtering is accomplished by the IF amplifier.

But the principal filtering task of the radio, and the most demanding, we have so far neglected to mention. We must filter the RF spectrum to eliminate all the other radio stations that are broadcasting simultaneously with the station of interest. If we were to portray the true effect of mixing in Fig. 10.13, we would have put the signals from many AM radio stations side by side in our RF spectrum, and our radio has to select one and eliminate the rest. How is this filtering to be accomplished?

The simple answer does not work well for practical reasons. The obvious solution would be to put a filter in the RF amplifier to select the desired station and reject the rest. The two reasons why this is impractical are as follows: the requirements on such a filter would strain the limits for practical inductors and capacitors; and the requirement for a *tunable* filter further complicates the design of an effective RF filter.

Let us consider the second difficulty first. To construct a radio, we need to filter out all the radio stations except one; but, in addition, we need to be able to tune the radio to any of the available stations. We have not discussed tunable filters, hitherto, nor will we here except to assert that they are difficult to design. The other problem is that, even without the requirement that the filter be tunable, it would be challenging to design an RF filter selective enough to reject all but one of the available radio stations. The combination of these two difficulties has driven radio designers to use intermediate frequencies, because mixing to an IF lessens both difficulties.

In Fig. 10.14, we have followed the mixer with an IF amplifier/filter. In the normal AM radio, this amplifier would have high gain only for frequencies within ± 5 kHz of the IF frequency of 455 kHz and hence would act as a filter eliminating all frequencies out of this band of frequencies. The IF amplifier is "fixed-tuned," that is, its bandpass remains the same as you tune the radio; and since its operating frequency is relatively low, it can provide effective filtering and high gain. Consequently, not only does the IF amplifier/filter eliminate the undesired frequencies created by the mixer, it eliminates the radio stations adjacent to the desired station. Thus it performs the same function as an RF filter, but it filters at a lower frequency and does not need to be tunable.

Selecting stations. The radio is tuned by varying the local oscillator (LO) frequency, f_1. When you turn the tuning knob on your radio, the principal effect inside the radio is to vary the frequency of the LO. We saw before that to receive a radio station broadcasting at 1200 kHz we had to set the LO frequency to 1655 kHz. Clearly, if we increase the LO to 1665 kHz, the station broadcasting at 1210 kHz would now be mixed into the passband of the IF amplifier/filter, and the station at 1200 kHz would be eliminated. By this technique, one radio station is selected and all others are rejected.

Image rejection. To the previous filtering actions of the radio circuit, we must add yet one more. In Fig. 10.14, you will note that the box representing the RF amplifier is also called a filter. This does not contradict our earlier assertion that practical problems prevent us from designing an effective RF filter. This filter is not required to select the radio station, but RF filtering is necessary to eliminate the *image band* of the mixer. To explain about the image of the mixer and why we need one more filter to get rid of it, we must refer back to our explanation of mixer operation. You will note in Table 10.1 that the term relating to the mixer action was of the form $\cos(\omega_1 - \omega_2)t$. This is the logical way to write this term when ω_1 (the LO frequency) is higher than the RF frequency and when only one RF frequency comes into the mixer. But if the RF frequency were higher than the LO frequency, we would write the mixer term as $\cos(\omega_2 - \omega_1)t$. So if, for example, our LO were tuned to 1655 kHz and there were an RF input at $1655 + 455 = 2110$ kHz, then this frequency would also be mixed into the IF passband. We conclude that the mixer would not distinguish between RF frequencies above and below the LO frequency and hence would mix both into the IF amplifier/filter passband. Without additional filtering, the radio would receive two stations simultaneously, one above and one below the LO frequency. This second, undesirable frequency is called the *image band* of the mixer, and hence the RF amplifier/filter is required to eliminate any station broadcasting in the image band before it reaches the mixer. Thus the passband of the RF amplifier/filter can be rather broad and often does not have to be tunable, although the RF amplifier/filter is tunable in most radios.

Summary. The superheterodyne radio combines amplification, filtering, and spectrum shifting. Obviously, much amplification is required because of the small signals picked up by the antenna, and we have indicated amplification in Fig. 10.14 at RF, IF, and audio frequencies. Spectrum shifting is performed by the mixer, to lower the spectrum of the information from the RF frequency to the IF frequency; and spectrum shifting is also accomplished by the AM detector (or demodulator) to move the information back into the audio frequency band. Filtering is done at RF to get rid of the station broadcasting at the mixer image frequency; filtering is done by the IF amplifier to eliminate unwanted mixer frequencies and other radio stations; and filtering is done at the audio frequencies to eliminate noise created by the detector.

The radio provides an excellent example of the importance of the frequency-domain viewpoint in electrical engineering. Of the three essential processes involved in radio circuits, two (filtering and spectrum shifting) are accomplished in the frequency domain.

We mention in closing this section that there are obviously other schemes for effecting modulation, with their corresponding detection techniques. Frequency modulation (FM) provides a well-known example. Numerous additional modulation techniques are also used, particularly techniques well adapted to the transmission of digital information. But all communication systems, from the common telephone system to the most advanced space communication network, utilize amplification, spectrum shifting, and filtering.

10.1.4 Noise in Receivers

Importance of noise. When we spoke of op amps in Chapter 8, we asserted that gain is cheap. With radio receivers, gain is not as cheap as with op amps, but we can still achieve as high a gain as we wish by adding more stages of amplification. We might suppose, therefore, that we can successfully detect any input RF signal, no matter how weak, simply by having adequate gain in the receiver. This is not true, however, because the sensitivity of a radio receiver is limited by noise. Noise, by definition, is any undesired signal in the system, whether natural or man-made. Man-made noise is called interference; Chapter 9 showed techniques to reduce its effects.

Sources of natural noise. Natural noise in communication systems comes from two sources. Some noise comes into the receiver from the antenna; for example, in an AM receiver we can hear during storms a sporadic crackling noise from lightning. Other sources of noise that enter the system through the antenna are atmospheric gases (thermal noise) and extraterrestrial sources such as the sun and our galaxy (cosmic noise).

The second source of noise is the receiver itself. Just as any body radiates *blackbody radiation* due to the thermal motion of its atomic constituents, so also resistors radiate their own type of thermal noise into the circuits of which they are a part. We will investigate resistor noise below. Transistors add, in addition to thermal noise, *shot noise* due to the discrete charges that carry the current across the *pn* junctions, and *flicker* noise due to slow changes in device properties from aging, temperature, and other physical changes.

Thermal noise. The thermal radiation of resistors plays an important role in the description of noise because resistors provide a simple calibration for noise signals. Figure 10.15a shows a resistor, R, and indicates the noise voltage generated by the thermal motion of the charges in the resistor. The Thévenin equivalent circuit in Fig. 10.15b separates the noise voltage from the resistance, which here is noise-free. The power spectrum of the noise voltage in volts²/hertz is shown in Fig. 10.15c, with k = Boltzmann's constant of 1.38×10^{-23} J/K. We note the following:

- The noise spectrum is *white* noise, meaning that all frequencies are present equally. This results from the extremely short collision times for thermal motion, as discussed in Chapter 6, page 225. Thus a resistor will give to a circuit a total power that is proportional to the bandwidth of the circuit.
- The height of the power spectrum is proportional to the temperature of the resistor. For sensitive amplifiers, we may measure the noise of the system by comparing system noise to the noise from a resistor of known temperature.
- The height of the power spectrum is proportional to the resistance. For this reason, large values of resistance are avoided in sensitive electronic amplifiers.

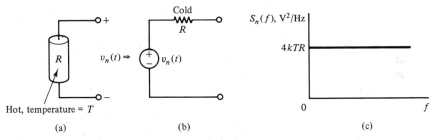

Figure 10.15 (a) A noisy resistor; (b) Thévenin equivalent circuit of a noisy resistor; (c) power spectrum of thermal noise.

The *available power* out of a circuit is the maximum power that the circuit is capable of delivering into a matched load (see Chapter 5, page 172). The available power spectrum from the circuit in Fig. 10.15b would be

$$S_{av}(f) = \frac{S_n(f)}{4R} = kT \quad \text{W/Hz} \tag{10.8}$$

where S_{av} is the available power spectrum in watts/hertz. Thus the total power a noisy resistor will contribute to a matched circuit is

$$P = kTB \quad \text{W} \tag{10.9}$$

where B is the bandwidth in hertz. Equation (10.9) is useful in determining the signal-to-noise ratio in a communication circuit, once we have defined the noise figure and system temperature of an amplifier.

System temperature. The noise originating in the resistors and semiconductors in a radio amplifier is generally broadband or white noise, at least over the bandwidth of the radio. Thus we may represent the noise *as if* it originated in a resistor at the input of the radio. Figure 10.16a shows the true situation: an amplifier with a power gain of G and a bandwidth of B has internal noise sources that produce N_s watts of noise at its output. In Fig. 10.16b, we show an equivalent amplifier, assumed to be noise-free, with the same noise at its output *attributed* to a noisy resistor at the input. The *system temperature* is defined to be the temperature the resistor would have to have in Fig. 10.16b to produce the same amount of output noise as in Fig. 10.16a. Using Eq. (10.9), we can express the output noise as

$$N_s = kT_sGB \Rightarrow T_s = \frac{N_s}{kGB} \ \text{K} \tag{10.10}$$

where T_s is the system temperature. All the internally generated noise is *referred to the input* of the receiver and then expressed as a temperature. The system temperature thus describes the sensitivity of the amplifier; the lower the system temperature, the more sensitive the amplifier.

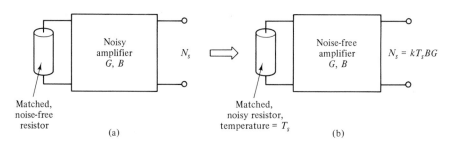

Figure 10.16 (a) Amplifier with output noise; (b) the system temperature is the temperature of the input resistor to give the same output noise.

Noise figure. The operating noise figure is defined as the input signal-to-noise ratio divided by the output signal-to-noise ratio. Figure 10.17 shows an amplifier with input signal and noise, S_{in} and N_{in}, and output signal and noise, S_{out} and N_{out}. The *operating noise figure* is defined to be

$$F = \frac{S_{in}/N_{in}}{S_{out}/N_{out}} = \frac{N_{out}}{N_{in}G} = \frac{N_{in}G + N_s}{N_{in}G} = 1 + \frac{N_s/G}{N_{in}} \tag{10.11}$$

where F is the operating noise figure of the amplifier and is often expressed in decibels: $F_{dB} = 10 \log(F)$. The first form on the right side of Eq. (10.11) is the definition of operating noise figure. The second form introduces the gain, $G = S_{out}/S_{in}$, and the third form breaks the output noise into two components, the amplified input noise and the internally generated noise. The last form shows that

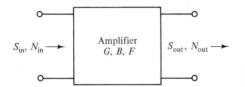

Figure 10.17 Amplifier with signal and noise at input and output.

the noise figure depends on the internally generated noise, referred to the input of the amplifier, compared with the input noise. Thus the operating noise figure depends in part on the input noise and not on the amplifier only.

The operating noise figure is useful in performing signal-to-noise calculations in a communication system. For example, if the signal-to-noise ratio for a TV signal from a satellite antenna were $+50$ dB, and the operating noise figure of the amplifier in the radio receiver were 20 dB, then the output signal-to-noise ratio would be $+50$ dB $-$ 20 dB $=$ $+30$ dB. The operating noise figure is useless for comparing amplifiers, however, until a standard noise environment is established. For purposes of comparing amplifiers, we use the *standard noise figure*, in which the input noise is assumed to be the thermal noise of a matched resistor at a temperature of 290 K. For this case, $N_{in} = k290GB$; and from Eqs. (10.10) and (10.11), we can express the standard noise figure as

$$F = 1 + \frac{T_s}{290} \tag{10.12}$$

Example. One amplifier has a standard noise figure of 1.25, and another has a system temperature of 75 K. Which is the more sensitive amplifier? To compare the amplifiers, we must express their noise contributions in the same terms. The first amplifier has a system temperature of

$$T_s = 290(F - 1) = 72.5 \text{ K} \tag{10.13}$$

and thus produces slightly less noise than the second amplifier.

We assume the first amplifier is used in a receiver with a gain of 120 dB, a bandwith of 4 MHz, and an output impedance of 50 Ω. What would be the output noise power due the amplifier and the peak-to-peak noise voltage into a 50-Ω load? The output power can be determined from Eq. (10.10) as

$$N_s = kT_sGB = 1.38 \times 10^{-23} \times 72.5 \times 10^{(120 \text{ dB}/10)}$$

$$\times 4 \times 10^6 = 4.00 \times 10^{-3} \text{ W} \tag{10.14}$$

This would be the power into a matched load of 50 Ω, and thus the rms voltage would be

$$\frac{V_{rms}^2}{50} = 4.00 \text{ mW} \Rightarrow V_{rms} = 0.447 \text{ V} \tag{10.15}$$

The output voltage would be a random signal with a correlation time of $\tau = 1/B$ $= 0.25$ μs. A good rule of thumb for such random voltages is that the peak-to-peak voltage is five to six times the rms value; thus we would expect a peak-to-peak voltage of about 3 V.

10.1.5 Check Your Understanding

1. If the input spectrum to a circuit contains a fundamental and third harmonic and the output contains in addition a second harmonic, is the circuit linear or nonlinear?

2. The input of a nonlinear system contains frequencies of 30 and 100 Hz. Which of the following can be in the output: 0, 70, 95, 100, 115, 210, 270 Hz?

3. The most critical filtering in a superheterodyne radio is done by the RF amplifier, mixer, IF amplifier, or audio amplifier. (Which?)

4. In a communication system, a carrier of 500 kHz is amplitude-modulated with a 5-kHz tone. What is the required radio-frequency bandwidth to pass the signal?

5. In a radio, the RF amplifier is required to remove the LO, IF, or image band of the mixer. (Which?)

6. If in a radio the RF frequency is 22 MHz, the IF frequency is 5 MHz, and the LO frequency is 27 MHz, what is the image frequency?

7. An amplifier has a standard noise figure of 6 dB. What is its system temperature?

8. List four sources of noise in communication systems.

9. Calculate the available noise power from a 300-K, 5-kΩ resistor in the bandwidth from 0 to 10 MHz.

Answers. 1. Nonlinear; 2. 0, 70, 100, 210, 270 Hz; 3. IF amplifier; 4. 10 kHz; 5. image band of the mixer; 6. 32 MHz; 7. 865 K; 8. man-made interference, atmospheric noise such as lightning, thermal noise, cosmic noise, shot and flicker noise; 9. 4.14×10^{-14} W.

10.2 ELECTROMAGNETIC WAVES

10.2.1 Introduction

History.   In early experiments in electricity, a connection between light, electric phenomena, and magnetic phenomena was unanticipated. It was known that moving charges produce a magnetic field (Ampere's circuital law) and that a changing magnetic flux produces an electric field (or a voltage, which is Faraday's law).† But no one related these phenomena to light. One of the major triumphs of mathematical physics occurred when James Clerk Maxwell (1831–1879) realized that mathematical consistency required that a changing electric field produce a magnetic field. When Maxwell added the needed term to the then-known equations, thus formulating Maxwell's equations, he showed that coupled electric and magnetic fields exist in the form of waves. Moreover, his predicted velocity of these waves corresponded to the known velocity of light. Thus Maxwell unified electrical science, and showed that light consists of electromagnetic waves. Maxwell's predictions were soon confirmed experimentally by Heinrich Hertz. Maxwell discovered what we call the electromagnetic spectrum, summarized in Fig. 10.18.

Figure 10.18 incorporates the relationship between frequency and wavelength:

---

† These fields and laws will be discussed in Chapter 12.

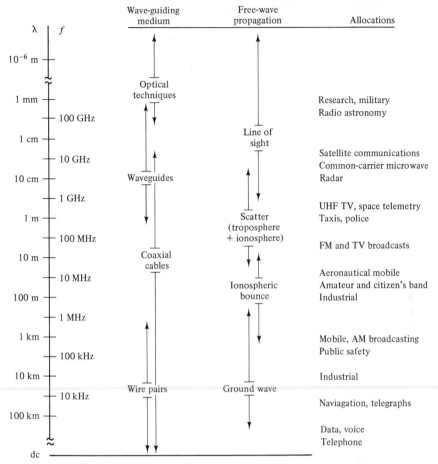

Figure 10.18 The electromagnetic spectrum.

$$\lambda = \frac{c}{f} = \frac{3 \times 10^8}{f} = \frac{300}{f_{\mathrm{MHz}}} \qquad (10.16)$$

where λ is the wavelength in meters, c is the velocity of light in meters/second, f is the frequency in hertz, and f_{MHz} is the frequency in megahertz. Figure 10.18 shows the manner in which such electromagnetic waves are guided from point to point. We will discuss wave-guiding structures below. Also shown are the ways in which free waves are affected by earth and ionosphere for long-distance communication, which we will also discuss. The last columnn shows some of the applications of electromagnetic waves in the various frequency-wavelength ranges.

Free and guided electromagnetic waves. Figure 10.18 distinguishes between free waves and guided waves. Guided waves are one-dimensional waves such as waves that are guided as shown in Fig. 10.18. Free waves are launched by antennas and spread out in space. We must investigate both types of waves to understand communication systems. We begin with guided waves.

10.2.2 Transmission Lines

Introduction.   Along many rural roads you will see wires strung on high poles. The power line is usually on top, with the telephone line below. These are two examples of *transmission lines* and exemplify two major applications, communication and distribution of electric power. The theory of transmission lines is the same for both types of lines, but in this section we will survey the major concepts of transmission lines used in communication systems.

A transmission line guides waves of electric energy from source to load. Such lines may be parallel wires, as described above, or coaxial cables such as distribute TV and other communication services in urban areas, or an internal bus in a computer, to give common examples. Any "circuit" in which the distributed capacitance and inductance of the conductors become a factor in circuit performance is a transmission line. Put another way, any circuit whose physical dimensions are not greatly smaller than a wavelength at the highest significant frequency must be considered a transmission line. The analysis of waves on transmission lines fits nicely on the foundation of circuit that we have laid, but space limitations prohibit development here.

The following are properties of transmission lines:

- *Wave velocity* on an overhead transmission line is the velocity of light, 3×10^8 m/s in air. In coaxial cables the waves are slowed down by the plastic that separates inner and outer conductors. The formula for the wave velocity is

$$v = \frac{1}{\sqrt{LC}} \quad \text{m/s} \tag{10.17}$$

  where L is the distributed inductance of the line in henrys/meter and C is the distributed capacitance of the line in farads/meter. For example, a common coaxial cable (RG-58) has $L = 0.253$ μH/m and $C = 101$ pF/m, so the wave velocity is 1.98×10^8 m/s. The wavelength at any frequency can be calculated from Eq. (10.16) provided we use the velocity on the transmission line for c.

- *Loss* on the line is caused by distributed resistance. For example, the loss of the coaxial cable given above is 5 dB/100 ft at 100 MHz.

- *Dispersion* describes the tendency of waves at different frequencies to travel at different velocities on the line. Dispersion causes distortion in the spectra of communication signals sent over transmission lines, but amplifiers at the receiving end can compensate partially.

- The *characteristic impedance* gives the ratio of the voltage to the current waves that move along the line. The characteristic impedance of the transmission line is given by

$$Z_o = \sqrt{\frac{L}{C}} \tag{10.18}$$

For example, the characteristic impedance corresponding to the values of L and C given above is 50.0 Ω, and hence this coaxial cable is "50-Ω coax."

- *Reflection* occurs when the impedance of the load at the receiving end of the line differs from the characteristic impedance of the line. Reflections are avoided if possible because power is lost to the load, and often the reflected waves interfere with the source of the signal. When the load has the same impedance as the characteristic impedance of the line, the load is said to be "matched to the line," and the no reflection occurs.

10.2.3 Free Electromagnetic Waves

What we mean by "free."   We deal in this section with electromagnetic waves that are free of man-made guiding structures. Such unconfined waves spread out in space. They may spread out in three dimensions, like waves radiated from a satellite antenna, or they may spread out in two dimensions, like waves guided by the surface of the earth. The distribution of energy flow in such waves is affected by the source of the energy, usually an antenna, and by the matter that the waves encounter, for example, by reflection from the ionosphere.

Free electromagnetic waves are used in almost all communication systems except local telephone and cable TV systems. For this reason, we need to understand the character of such waves, how they are launched, how they interact with earth, obstacles, and ionosphere, and how they are received. In this section we will focus on the character of the waves; in the next section, we look at how the waves are launched and received by antennas.

Spherical waves.   The simplest case is that of waves that travel away in all directions from a source. Such waves consist of coupled electric and magnetic fields traveling near the speed of light. We may describe the waves by their polarization and power density, the power per unit area in the wave, normally in watts per square meter. We begin with the isotropic (all-directional) antenna. Such an antenna is defined† to radiate equally in all directions. Since the energy travels outward in straight lines, and since energy is conserved, the power passing a spherical surface of radius R must account for all the power radiated by the antenna.‡ Thus the power density is

$$S_{\text{iso}}(R) = \frac{P_{\text{rad}}}{4\pi R^2} \quad \text{W/m}^2 \tag{10.19}$$

where P_{rad} is the total power radiated, $4\pi R^2$ is the area of a sphere of radius R, and $S_{\text{iso}}(R)$ is the power density at a distance R from the origin of the radiation. The power density is a vector quantity, having a direction in space directed away from the source. Equation (10.19) indicates that spherical waves weaken due to spreading as they travel away from their source.

---

† It has been proved that an isotropic antenna is incompatible with Maxwell's equations and thus can never by constructed. It is a useful concept, however.

‡ There will be a delay, of course, between the time of radiation and the time of passing the sphere. Equation (10.19) assumes steady power levels.

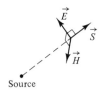

Source

Figure 10.19 Geometry of a wave traveling away from a source. The wave direction and power density are indicated by $\vec{S}$. The electric and magnetic fields are indicated by $\vec{E}$ and $\vec{H}$, respectively.

Polarization. Electromagnetic waves consist of electric and magnetic fields that, at any point, also have directions in space. Figure 10.19 shows the situation: the wave direction and power density are indicated by $\vec{S}$; the electric field ($\vec{E}$) and the magnetic field ($\vec{H}$) are directed at right angles to the direction of $\vec{S}$ and at right angles to each other. Thus, at a location, we can describe a radio wave in terms of its direction of travel and its polarization, for example, E-field vertical relative to the earth's surface.

Polarization is important because most antennas produce and receive only one polarization. Thus, if a satellite transmits vertical polarization and your local antenna receives only horizontal polarization, the radio waves arriving at your antenna will not be collected by the antenna. Commercial AM broadcast signals are vertically polarized, but FM and TV signals are required to use horizontal polarization. Circular polarization uses both vertical and horizontal polarizations, with a 90° phase shift between the two. Due to multiple reflection, however, transmitted signals can become in part depolarized, so the orientation of the receiving antenna is often not critical. At microwave frequencies, matching antenna and wave polarization is very important.

Surface waves. At lower frequencies such are used for AM broadcasting, radio waves can be guided by the surface of the earth. These surface waves follow the curvature of the earth and do not depend upon line-of-sight between transmitting and receiving antennas. In addition to the loss due to the spreading of the wave, a surface wave also diminishes due to resistive losses in soil.

Plane waves. Radio waves originate from finite sources and hence have curvature. However, for practical purposes their radius of curvature is usually large enough to treat them as plane waves. In determining the intensity of such waves, we must consider the distance to the source, but in considering the local nature of the waves, we usually consider them plane waves.

Things that can happen to radio waves. Earlier we mentioned two things that can happen to radio waves: they can be reflected and they can become depolarized. Among the more important effects on radio waves due to their interaction with matter are the following:

Figure 10.20 A microwave relay station using reflectors.

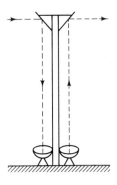

- *Reflection* can occur from the earth's surface, bodies of water, buildings, and, at low frequencies, the ionosphere. Reflection is used in radar to locate the range and direction of a target. Sometimes reflectors are elevated on towers in microwave relay stations, as shown in Fig. 10.20. At low frequencies, the reflection of radio waves from the ionosphere allows long distance communication.
- *Depolarization* occurs when a wave loses its pure polarization in the horizontal or vertical direction and becomes a partial mix of the two.
- *Refraction* refers to the bending of a wave in the direction of denser matter. For example, radio waves in the lower atmosphere are refracted toward the earth's surface because the atmosphere becomes less dense with increasing

height. This has the effect of extending the "radio horizon" slightly and is beneficial in line-of-sight microwave transmission systems.

- *Diffraction* occurs when waves spread into a shadow region behind an obstacle. Diffraction allows radio waves to be received beyond the radio horizon.
- *Scattering* occurs when radio waves bounce off a multitude of small objects and "scatter" in all directions. For example, a wave that was reflected from the surface of the ocean would be in part scattered by the irregular surface.
- *Doppler shift* occurs when a wave reflecting from a moving target is shifted in frequency. This effect allows a radar to measure the speed of a target.
- *Attenuation* refers to the absorption of electromagnetic energy by matter. Atmospheric attenuation is a problem at high microwave frequencies.

Summary. We have investigated the behavior of free electromagnetic waves. We have shown how waves diminish through spreading as they travel away from sources, and we have described the various effects that can influence their travel through the atmosphere. We now describe the means for launching and receiving the waves.

10.2.4 Antennas

What is an antenna?   An antenna is a structure that couples between a guided and a free electromagnetic wave. Most antennas can be used for both transmitting and receiving waves. The transmitting and receiving properties of an antenna are closely related but are described in different terms. Figure 10.21 represents an antenna as a transmitting device. It receives power (P_{in}) from a circuit and radiates power in the form of radio waves, which we describe by their polarization and their power density as a function of distance from the antenna, $S(R)$. As a circuit element, the antenna is described by its input impedance, $\mathbf{Z}_{in}$. The radiation efficiency, η, is the ratio of the radiated power to the input power:

$$\eta = \frac{P_{rad}}{P_{in}} \tag{10.20}$$

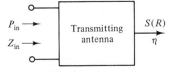

Figure 10.21   A transmitting antenna receives power from a circuit and radiates an electromagnetic wave.

Antenna gain.   Parabolic reflectors, such as shown in Fig. 10.22a, are used in radar and satellite communication systems because they focus the power into a narrow region of space. Simple wire antennas, such as shown in Fig. 10.22b, focus the power only slightly and are used for broadcast applications. The focusing ability of an antenna is described by the *antenna gain*, which is defined as

$$G = \eta \frac{S(R)}{S_{iso}(R)} \tag{10.21}$$

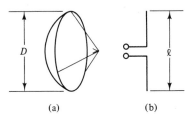

Figure 10.22 (a) A parabolic reflector-type antenna; (b) a wire-type antenna.

where G is the antenna gain. The antenna gain involves an efficiency factor and a focusing factor. The efficiency factor describes how much of the power delivered to the antenna is radiated. The focusing factor is the ratio of the power density radiated by the antenna divided by the power density of an isotropic antenna. The gain thus is the power density of the antenna compared to that of a lossless antenna that does not focus at all. A simple wire antenna has a maximum gain in the range 1.5 to 2.0. A large parabolic reflector may have a gain exceeding 1000. The antenna is normally positioned to focus the energy in a preferred direction, and thus we deal usually with the maximum gain of the antenna.

We may combine Eqs. (10.19), (10.20), and (10.21) to obtain an important equation relating power density to distance from the antenna:

$$S(R) = \frac{P_{\text{in}}G}{4\pi R^2} \tag{10.22}$$

Equation (10.22) assumes spherical-wave spreading of the waves and also the absence of reflection, attenuation, and the like.

Receiving antennas. Figure 10.23 represents a receiving antenna. The input to the antenna is a plane wave of a given polarization and power density. As a circuit element, the antenna is characterized by its Thévenin equivalent circuit. Although in practice the equivalent circuit is not easy to determine, we may work directly with the available power from the antenna.

As mentioned above, the available power of a circuit is the maximum power that the circuit is capable of delivering into a matched load. The load on the antenna is the input impedance of the radio receiver attached to the antenna, which must be $\underline{Z}_{\text{out}}^* = R - jX$ to receive the available power from the antenna. The ability of the antenna to collect power from a radio wave and deliver it to a circuit is characterized by the *effective area* of the antenna

$$A_{\text{eff}} = \frac{P_{\text{av}}}{S} \tag{10.23}$$

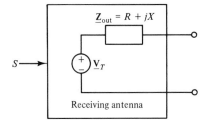

Figure 10.23 A receiving antenna. The input is a plane wave and the output is power furnished to a circuit.

where P_{av} is the available power in watts, S is the incident power density in watts/(meter)^2, A_{eff} is the effective area of the antenna in square meters, and we have assumed that the antenna is correctly oriented to receive the polarization of the incident radio wave. It can be shown† that the effective area of any antenna is related to its gain by

$$A_{eff} = \frac{\lambda^2}{4\pi} G \qquad (10.24)$$

where λ is the wavelength in meters. Equations (10.22), (10.23), and (10.24) permit us to calculate the coupling of transmitter and receiver through a radio link.

Communication equation. We now consider the communication link pictured in Fig. 10.24. A transmitter delivers P_{in} of power to a transmitting antenna. A receiving antenna at a distance R accepts a portion of the radio wave and delivers a power P_{rec} to a receiver with a matched input impedance. We consider only the line-of-sight wave and thus neglect reflection, scattering, and the like. Combining Eqs. (10.22) and (10.23), we determine the received power to be

$$P_{rec} = \frac{P_{in}G_t}{4\pi R^2} \times A_{eff} \qquad (10.25)$$

where G_t is the gain of the transmitting antenna and A_{eff} is the effective area of the receiving antenna. We may change the form of Eq. (10.25) by use of Eq. (10.24), with the result

$$P_{rec} = \frac{P_{in}G_tG_r}{(4\pi R/\lambda)^2} \qquad (10.26)$$

where G_r is the gain of the receiving antenna. The coupling between transmitter and receiver depends therefore on the gains of the transmitting and receiving antennas and the distance between antennas measured in wavelengths. The denominator in Eq. (10.26) is often called the *space loss* in the link and is often given in decibels. Equation (10.26) was presented and illustrated in Chapter 8, page 340. We further illustrate its use below after we have discussed two common types of antennas.

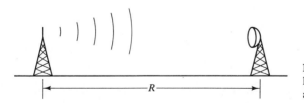

Figure 10.24 A radio transmission link with transmitting and receiving antennas.

---

† The proof of Eq. (10.24) from Maxwell's equations is difficult. A relatively simple proof can be obtained from thermodynamic equilibrium by considering resistors attached to each antenna. For the resistors to reach thermodynamic equilibrium through radiative heat transfer, Eq. (10.24) must be true. In other words, if Eq. (10.24) were not true, a perpetual motion machine theoretically could be constructed out of resistors and antennas.

Reflector-type antennas. Figure 10.22a shows the reflector-type antenna commonly used in microwave communication systems. The metallic reflector is parabolic in shape, and a "feed" at the focus of the parabola radiates or receives the radio wave. Power is coupled between the transmitter (or receiver) and the feed through a transmission line. In the transmit mode, a spherical wave is radiated by the feed in the direction of the reflector, and the reflector redirects the spherical wave into one direction. In the receive mode, the incoming plane wave is reflected from the parabolic reflector and converted into a spherical wave that converges on the feed. The feed collects power from the spherical wave.

A reflector antenna is most naturally described in terms of its effective area. A good rule of thumb is that about 50% of the incident energy that strikes the reflector is collected by the feed. Since the losses of a reflector antenna are quite low, the effective area is about

$$A_{\text{eff}} = \frac{1}{2} A_{\text{geo}} = \frac{1}{2} \frac{\pi}{4} D^2 = \frac{\pi D^2}{8} \tag{10.27}$$

where A_{geo} is the geometric area of the reflector and D is its diameter. Equation (10.24) gives the gain of the reflector-type antenna as

$$G = \frac{4\pi}{\lambda^2} A_{\text{eff}} = \frac{\pi^2}{2} \left(\frac{D}{\lambda}\right)^2 \tag{10.28}$$

Thus the gain of the antenna depends on the diameter measured in wavelengths. Equation (10.28) allows estimation of the gain of a microwave antenna based on its size and operating frequency. For example, a 6-ft-diameter antenna operating at 500 MHz ($\lambda = 0.6$ m) would have a gain of 45.8 (16.6 dB).

Wire-type antennas. Figure 10.22b shows a common wire-type antenna called a *dipole*. This antenna is used on TV receivers and is closely related to the telescoping antenna used on automobiles. For good antenna properties, the length of the antenna ℓ should be either one-half or a full wavelength of the electromagnetic wave.† This explains why audio signals must be modulated to higher frequencies for effective radiation, since the half-wave dipole is the smallest efficient antenna. The half-wave dipole has a gain of about 1.5, depending on the losses, and an input resistance of about 50 Ω. Thus the effective area of this antenna is

$$A_{\text{eff}} = 1.5 \frac{\lambda^2}{4\pi} \tag{10.29}$$

which is considerably larger than the geometric cross section of the antenna.

Other antenna types. Multiple dipole antennas are often used in tandem arrangements called *arrays*. The common house-top TV antenna is an array antenna of several dipole antennas with slightly different lengths to increase the gain

---

† For practical reasons, the length of an auto antenna is much less than the wavelength, which is about 300 m for AM signals. For this reason, auto antennas are inefficient.

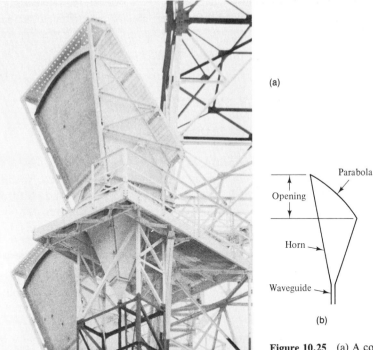

(a)

Parabola

Opening

Horn

Waveguide

(b)

Figure 10.25 (a) A cornucopia horn; (b) antenna cross section.

and bandwidth of the antenna. Often an AM station will use an array of two or more dipoles on separate towers to direct their signals to populated areas.

An external antenna is avoided in an AM radio by using a small, multiturn coil wound on a ferrite rod. A horn may be used for a microwave antenna, much as a megaphone is used by a cheerleader to focus his or her voice. The *cornucopia* antennas used in microwave relay systems combine a horn with a parabolic reflector, as indicated in Fig. 10.25b.

Summary. We have now discussed the components of a communication system. Transmitting antennas radiate waves that are received by an antenna, which furnishes power to a radio receiver. The receiver amplifies and filters the signal, plus the inevitable noise, and delivers the signal to the ultimate user of the information. We next consider two representative communication systems.

10.2.5 Check Your Understanding

1. A 0.5-μs pulse is transmitted down a RG-58 coaxial cable. How much space on the line does the pulse occupy at any given time?
2. One way that radio waves get bent is through refraction. (T/F?)
3. What is the wavelength of a radio wave with a frequency of 15 MHz?
4. An isotropic source radiates 10 W. What is the power density at a distance of 1 mile?
5. A wave with a power density of 2×10^{-9} W/m^2 falls on an antenna with an effective area of 10 m^2. How much power is available from the antenna to its receiver?

6. What is the space loss in decibels between two antennas that are 10^4 wavelengths apart?

7. What is the gain at 3000 MHz of an reflector-type antenna that has a diameter of 3 ft?

Answers. 1. 99.0 m; 2. true; 3. 20.0 m; 4. 0.307 μW/m²; 5. 2 × 10⁻⁸ W; 6. 63.3 × 10⁻¹² or −102.0 dB; 7. 412.6 or 26.2 dB

10.3 EXAMPLES OF COMMUNICATION SYSTEMS

10.3.1 FM Broadcast Station

General description. Frequency modulation uses a wide bandwidth to achieve its remarkable noise-repression characteristics; specifically, an FM station uses 200 kHz of RF bandwidth for an audio bandwidth of 50 to 15,000 Hz. The University of Texas at Austin operates an FM radio station, KUT, on a carrier frequency of 90.5 MHz. The station radiates approximately 15 kW of RF power.

Antenna system. The station employs an array of 12 helical antennas, radiating circular polarization. The Federal Communication Commission requires horizontal polarization for FM, as stated above, but permits circular polarization. Circular polarization is desirable because orientation of the receiver antenna becomes unimportant.

Frequencies exceeding about 20 MHz do not normally bounce off the ionosphere; hence reception is limited to line-of-sight, extended somewhat by refraction and by diffraction over the radio horizon. The array of transmitting antennas directs power out toward the horizon. The gain of the array, considered as a single antenna, is about 6.4 (8.1 dB). The transmitter output is about 17 kW, of which 80% is radiated by the antenna; thus the radiated power is 13.6 kW. The station announcers claim "100,000 watts," which presumably takes account of the antenna gain and normal media exaggeration.

Coverage. The antenna is located west of Austin on a tower that is 528 ft high. Assuming a perfect sphere, we calculate through simple geometry the distance, D, to the line-of-sight horizon to be

$$D = \sqrt{2R_e h} \qquad (10.30)$$

where h is the antenna height and R_e is the equivalent† radius of the earth, 1.3 × 3956 miles. Substitution into Eq. (10.30) yields a distance of about 32 miles to the radio horizon. The antenna tower is about 8 miles west of Austin on a high hill, which extends the radio horizon somewhat.

We may with Eq. (10.26) calculate the received power at the radio horizon. The wavelength is 300/90.5 m, and we assume a gain of 1.5 for the receiving antenna. Under these assumptions, the received power is 3.4 × 10⁻⁶ watts. Some noise would be broadcast with the signal; the FCC requires a 60-dB or better signal-to-noise ratio; some noise would also be added to the signal due to thermal

---

† The earth's radius is increased by about 30% in such calculations to account for refraction.

radiation of the earth and atmosphere, but the main source of noise would be the receiver. We assume a standard noise figure of $F = 5$; hence the noise power in the 200-kHz bandwidth of the receiver would be

$$N = k(F - 1)T_0B = 1.38 \times 10^{-23}(5 - 1)290$$
$$\times 200 \times 10^3 = 3.2 \times 10^{-15} \quad \text{W} \quad (10.31)$$

where T_0 is the standard temperature (290 K) and the noise is referred to the input of the receiver. Thus the signal-to-noise ratio would be about 90.3 dB for a receiver at the radio horizon. This signal is much stronger than required for line-of-sight operation. The station is therefore extending coverage beyond the horizon.

10.3.2 Speed-Measuring Radar

Radar concepts. Radar is an acronym for RAdio Direction And Range. A conventional radar sends out a pulse of power that reflects from a target such as an airplane and returns to the radar site. The direction to the target is determined from the directionality of the antenna, and the range to the target is determined from the time delay between transmission and reception of the pulse. Such radars are used for airplane traffic control, weather detection, and a variety of military and space applications.

A police radar measures neither direction nor range but speed. A continuous microwave signal is radiated from a directional antenna, bounces off a vehicle, and returns to the point of origin. The return signal is received with the same antenna and diverted to a receiver. The motion of the vehicle Doppler-shifts the frequency of the return signal, allowing accurate determination of speed. In this section we will perform some representative calculations for such a radar.

Radar equation. Figure 10.26 pictures the situation. The power density at the vehicle would be given by Eq. (10.22), repeated below:

$$S = \frac{PG}{4\pi R^2} \quad \text{W/m}^2 \qquad (10.32)$$

where G is the gain of the antenna, P is the power radiated, and R is the distance. The reflection from the target back toward the transmitter is described by the *radar cross section*, σ, which has the units of square meters. The radar cross section includes two factors: the total power reflected, and the directionality of that reflection toward the transmitter. Hence the equivalent power reflected back toward the transmitter, P_{ref}, is

$$P_{\text{ref}} = \frac{PG\sigma}{4\pi R^2} \quad \text{W} \qquad (10.33)$$

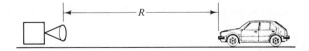

Figure 10.26 A speed-measuring radar.

This power weakens due to spreading and produces a power density at the transmitter/receiver of

$$S_{\text{rec}} = \frac{PG\sigma}{(4\pi R^2)^2} \quad \text{W/m}^2 \tag{10.34}$$

where S_{rec} is the power density at the transmitter/receiver. The power available from the antenna is

$$P_{\text{av}} = \frac{PG\sigma A_{\text{eff}}}{(4\pi R^2)^2} = \frac{\lambda^2 PG^2\sigma}{(4\pi)^3 R^4} \tag{10.35}$$

where A_{eff} is the effective area of the transmitter/receiver antenna. The second form of Eq. (10.35) uses the relationship between effective area and gain from Eq. (10.24). Equation (10.35) is called the *radar equation* and relates the available power to the transmitted power, the radar cross section of the target, and the gain of the radar antenna.

Received power calculation. Modern police radars use a frequency of 24,150 MHz, which corresponds to a wavelength of 1.242 cm. The antenna is a horn with a 4-in. diameter. The gain of the antenna from Eq. (10.28) is about 330. The power source in the radar is a solid-state device, and the radiated power would be a minimum of 10 mW. The radar cross section of an automobile, being an irregularly shaped object, would be a statistical quantity because the incident energy would reflect from many points on the surface and combine in a random fashion in various directions. We will use a nominal value of 0.01 m². We assume a range of 100 yards (91.4 m). Substitution of these numbers into the radar equation, Eq. (10.35), yields a received power of

$$P_{\text{rec}} = \frac{\lambda^2 PG^2\sigma}{(4\pi)^3 R^4} = \frac{(0.0124)^2 (10 \times 10^{-3}) (330.1)^2 (0.01)}{(4\pi)^3 (91.4)^4}$$

$$= 1.212 \times 10^{-14} \quad \text{W} \tag{10.36}$$

Not much power, but we must estimate the noise in the system to see if this is detectable.

Signal-to-noise estimation. To determine the noise level in the receiver, we must know the noise figure and the bandwidth. Because the noise power out of the receiver is proportional to bandwidth, the minimum bandwidth would be used. The minimum bandwidth depends in turn on the maximum speed that the system would expect to encounter in practice. We will assume 150 mph (67.0 m/s) and use the Doppler equation

$$\Delta f = f \times \frac{2v}{c} = 24.15 \times 10^9 \times \frac{2(67.0)}{3 \times 10^8} = 10.8 \text{ kHz} \tag{10.37}$$

This would be the bandwidth of our receiver. The noise in the receiver depends on the input noise from the antenna and the receiver noise figure, with the major contribution coming from the receiver. A worst-case standard noise figure for a receiver of the type used is $F = 10$. Using Eqs. (10.13) and (10.9), we estimate the noise, referred to the receiver input, to be

$$N = kT_0(F - 1)B = 1.38 \times 10^{-23} \times 290 \times (10 - 1)$$
$$\times 10.8 \times 10^3 = 3.89 \times 10^{-16} \quad \text{W} \tag{10.38}$$

and hence the signal-to-noise ratio is about 31.2. This is not very impressive, and speeding tickets would be easy to discredit if this were the end of the story.

Equation (10.38) gives the signal-to-noise if only one sample of the speed were obtained, but the radar would be able to collect many samples in a short time. Let us assume that the radar collects samples over 0.5 s before displaying the speed. This would mean that a counter would count the frequency of the return for 0.5 s, and then the results would be displayed, scaled to indicate miles per hour. The correlation time of the receiver noise would be approximately $1/B = 0.1$ ms, and hence the number of independent samples would be approximately 5000. A well-known rule of statistics is that accuracy improves as the square root of the number of independent samples, which would be about 70 in this case. Thus the signal-to-noise of the speed measurement is about 2200 or 33.4 dB. This gives an accurate measurement of target speed.

Summary. In this chapter we have considered the components of communication systems. We have shown how nonlinear devices are used to shift communication signals in the frequency domain and how a variety of such spectrum shifts are employed in radio circuits. We have described the properties of free and guided radio waves. We have described how such waves are radiated and received by antennas. Finally, we have examined the details of two communication systems.

PROBLEMS

Section 10.1: Radio Principles

P10.1. A nonlinear circuit has input frequencies at 300 and 200 Hz. How many output frequencies are there below 1000 Hz, and what are these frequencies?

Answer. 10 frequencies, counting dc but not counting 1 kHz

P10.2. The input of a *linear* system contains frequencies of 40 and 100 Hz. Which of the following can be in the output: 0, 70, 100, 120, 210, 270 Hz? Which of these can be in the output for a *nonlinear* system of the same input frequencies?

P10.3. In a superheterodyne radio receiver, the receiver bandwidth is 100 to 101 MHz and the image bandwidth is 120 to 121 MHz. What is the LO frequency?

Answer. 110.5 MHz

P10.4. The AM radio band of carrier frequencies is 540 to 1600 kHz and the standard IF is 455 kHz. Calculate the image frequency when the radio is tuned to the bottom of the band to receive a station broadcasting with a carrier of 540 kHz. Is this image in the AM band?

P10.5. For the simplest design of the local oscillator (LO) in a radio, the ratio of the maximum to minimum LO frequencies should be as low as possible (that is, the percent tuning range of the LO should be minimum). To show why the LO is placed above the RF band, calculate the ratio f_{max}/f_{min} for the LO both above and below the RF in the standard AM radio.

P10.6. For a standard AM superheterodyne radio receiver tuned to the station with a carrier of 1600 kHz:
  (a) What is the LO frequency?
  (b) What is the IF frequency?
  (c) What is the IF bandwidth?
  (d) What is the maximum RF bandwidth?
  (e) Where does most of the gain occur?
  (f) Where does most of the filtering occur?
  (g) How is the receiver tuned to another station?

P10.7. A superheterodyne radio receiver has the following characteristics: the antenna receives an RF spectrum of 10.016 MHz $\pm$ 3.2 kHz, with a voltage level of 8 μV; the RF amplifier has a gain of $+15$ dB; the mixer has a gain of -12 dB; the IF amplifier has a center frequency of 1.5 MHz and a gain of $+120$ dB; the second detector has a gain of -16 dB (based on audio voltage versus IF voltage); and the audio amplifier has a gain of $+20$ dB. Find:
  (a) LO frequency, assuming that the LO is above the RF frequency
  (b) IF bandwidth required
  (c) Output voltage at audio
  (d) Maximum bandwidth of the RF amplifier

Answers. (a) 11.516 MHz; (b) 6.4 kHz; (c) 17.91 V; (d) about 6 MHz

P10.8. The FCC has allocated the band from 88 to 108 MHz for FM broadcasting, with 200 kHz for each station. The intermediate frequency amplifier in an FM receiver has a center frequency of 10.7 MHz and a bandwidth of 200 kHz.
  (a) What is the number of FM stations that can be assigned different carrier frequencies in the total FM band?
  (b) What range is required for the frequency of the local oscillator in an FM receiver, assuming that the LO is located above the RF frequency?
  (c) What is the maximum frequency that might get mixed into the IF passband from the image band of the mixer?

P10.9. An FM radio receiver receives carriers at frequencies of 88.1, 88.3, . . . , 107.7, 107.9 MHz. The RF bandwidth required by each station for the FM information is 200 kHz, the IF is 10.7 MHz, and the audio bandwidth is 50 to 15,000 Hz.
  (a) Find the lowest and highest LO frequency. State assumptions.
  (b) Find the image frequency of the first detector when the LO is tuned to its lowest frequency.
  (c) What would be an appropriate bandwidth for the IF amplifier?
  (d) What would be highest frequency amplified by the audio amplifier?

P10.10. An amplifier with a 290-K matched resistor at its input produces a certain amount of noise at its output. When the resistor is cooled to 77 K, the output noise is observed to diminish by 15%. What is the standard system temperature of the amplifier?

Answer. 1130 K

P10.11. A communication system has a bandwidth of 5 kHz and an output power spectrum of 2×10^{-4} (V)^2/Hz into a matched 100-Ω-resistor. What is the rms current in the resistor?

Section 10.2: Electromagnetic Waves

P10.12. The distributed inductance and capacitance of a lossless vacuum-filled coaxial transmission line are

$$L = \frac{\mu_o}{2\pi} \ln \frac{b}{a} \quad \text{H/m} \quad \text{and} \quad C = \frac{2\pi\epsilon_o}{\ln(b/a)} \quad \text{F/m}$$

where b = inner radius of outer conductor and a = outer radius of the inner conductor; $\mu_o = 4\pi \times 10^{-7}$ H/m, and $\epsilon_o = 8.854 \times 10^{-12}$ F/m.
(a) Find the wave velocity in the cable.
(b) Determine the characteristic impedance if $b = 4a$.

P10.13. A certain 60-Hz power transmission line has an inductive reactance of 1.4 Ω/mile and a capacitive susceptance of 2.9 $\mu\mho$/mile.
(a) Convert these to H/m and F/m, as required for calculation of the wave velocity and characteristic impedance.
(b) Determine the wave velocity and characteristic impedance of this transmission line.

P10.14. Consider the sun as an isotropic radiator of energy. The power density at the surface of the earth is about 1000 W/m². Determine the total power radiated by the sun, assuming 92.8 million miles distance between sun and earth.

P10.15. An antenna with a gain of 12 dB radiates 10 W of power. What is the power density at 1-km distance?

Answer. 12.6 μW/m²

P10.16. The effective area of an antenna at 3000 MHz is 5 m². What is its gain at that frequency?

P10.17. Two half-wave dipoles, each having a gain of 1.3, are separated by 100 m. A 10-W signal at 100 MHz is transmitted by one antenna.
(a) What is the available power at the receiving antenna?
(b) What is the open-circuit voltage at the receiving antenna, assuming a 50-Ω output impedance?

P10.18. A 10-ft-diameter satellite TV receiving antenna operates over a frequency range of 3700 to 4200 MHz. What is the maximum and minimum gain in decibels over this range of frequency?

P10.19. An antenna produces at a certain point a wave intensity of 10^{-7} W/m², whereas an isotropic antenna would give a wave intensity of 10^{-8} W/m². What is the gain of the antenna?

P10.20. The beamwidth of a reflector-type antenna describes the angular width of the region in space into which it focuses its radiated energy. Using Eq. (10.28) and conservation of energy, show that the conical beamwidth (θ) of a reflector-type antenna is approximately

$$\theta = \frac{4}{\pi} \sqrt{2} \left(\frac{\lambda}{D}\right) \quad \text{rad}$$

where D is the antenna diameter and λ is the wavelength.

P10.21. Using Eq. (10.28), show that the gain of a 6-ft-diameter reflector antenna at 500 MHz is 16.6 dB.

Section 10.3: Examples of Communication Systems

P10.22. For the FM station example, calculate the received power at the radio horizon (32 miles) and confirm the S/N of 90.3 dB, as stated in the text, page 453.

P10.23. If a ship-borne radar must detect any target out to 10 miles, how high does the radar antenna have to be relative to the level of the sea?

P10.24. It is desired to increase the range of an air-control radar by a factor of 50%.
    (a) If only the transmitted power is increased, what is the required percent of increase in power?
    (b) If, rather than increasing the power, the reflector-type antenna is replaced by a larger antenna, what is the percent of increase in antenna diameter?

P10.25. What would be the signal-to-noise ratio for the police radar if the device measured the speed for 0.3 s instead of 0.5 s?

P10.26. Confirm that the gain of the police radar antenna described on page 454 is about 330.

11 Linear Systems

11.1 COMPLEX FREQUENCY

11.1.1 Introduction to This Chapter

What is a system? A system consists of several components that interact together to accomplish some purpose. An automobile, for example, has a motor, steering mechanism, lights, padded seats, radio, and more, operating together to give safe and pleasant transportation. Likewise, a stand-alone ac generator requires a control system to regulate the frequency and voltage of its output.

Often systems are modeled by linear equations. The analysis of such linear systems has furnished a powerful language for system description. This method of system description builds on our earlier study of the frequency domain. This chapter introduces system models and explores basic techniques of linear system analysis.

Contents of this chapter. We begin by generalizing the concept of frequency. We then introduce the language of system notation by deriving the generalized impedance of electrical circuits. From this impedance we determine the natural frequencies and natural response of electrical circuits, and we then investigate the transient response of first- and second-order circuits. Next, this viewpoint is applied to a system to control the temperature of an oven. Finally, we control the oven with a feedback system and investigate the transient response and dynamic stability of the feedback system.

11.1.2 Definition and Meaning of Complex Frequency

Definition.   We continue our exploration of the frequency domain through the following definition of complex frequency, $\underline{s}$: a time-domain variable (say, a voltage) is said to have a complex frequency $\underline{s}$ when it has the form given in Eq. (11.1)

$$v(t) = \text{Re}\{\underline{V}e^{\underline{s}t}\} \tag{11.1}$$

where $\underline{V}$ is a complex number (a phasor), and

$$\underline{s} = \sigma + j\omega \tag{11.2}$$

where $\underline{s}$, σ, and ω all have units of $(\text{second})^{-1}$. Equation (11.1) is a slightly modified version of, say, Eq. (4.15), except that frequency is now a complex number.

Relation to earlier material.   Before we explore the implications of complex frequency, we wish to relate Eq. (11.1) to prior methods and results. When the complex frequency in Eq. (11.1) is *zero*, $\underline{s} = 0$, the voltage in Eq. (11.1) is a constant, or a dc voltage. When the complex frequency $\underline{s}$ is *real*, $\underline{s} = \sigma$, the time-domain voltage is

$$v(t) = Ve^{\sigma t} \tag{11.3}$$

For negative σ, the voltage is a decreasing exponential function such as we encountered in Chapter 3 in first-order transient problems. Specifically, the time constant would be the negative of the reciprocal of σ. If σ is positive in Eq. (11.3), the voltage is an increasing exponential, which we have not encountered hitherto. Figure 11.1 shows the time functions that result from *zero* and *real* complex frequencies. Thus, a complex frequency that is real includes the response that we studied as a transient solution in Chapter 3, except that we now can consider growing as well as decaying exponentials. We will discover that a growing exponential represents a possible response of an unstable system.

Complex frequency that is pure imaginary, $\underline{s} = j\omega$.   When the complex frequency is pure *imaginary*, Eq. (11.1) takes the form of an sinusoidal function, such as we studied in Chapter 4. In this case,

$$v(t) = \text{Re}\{\underline{V}e^{j\omega t}\} = V_p \cos(\omega t + \theta) \tag{11.4}$$

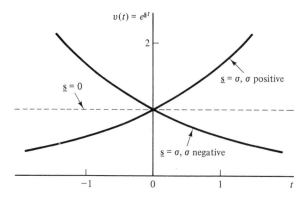

Figure 11.1   Time-domain responses for complex frequencies that are real.

where $\underline{V} = V_p e^{j\theta}$, with V_p the peak value and θ the phase of the sinusoidal function. Imaginary complex frequencies are used in the analysis of ac circuits.

From another point of view, we may consider that a sinusoidal function is represented by two complex frequencies. This is implied by the *real part of* operation in Eq. (11.4), because an alternate way to express the real part is through the identity in Eq. (11.5)

$$\text{Re}\{\underline{z}\} = \tfrac{1}{2}(\underline{z} + \underline{z}^*) \qquad (11.5)$$

where $\underline{z}^*$ is the complex conjugate of $\underline{z}$. Thus Eq. (11.4) can be expressed in the form

$$V_p \cos(\omega t + \theta) = \text{Re}\{\underline{V}e^{j\omega t}\} = \frac{\underline{V}}{2} e^{j\omega t} + \frac{\underline{V}^*}{2} e^{-j\omega t} \qquad (11.6)$$

When we compare Eq. (11.6) with Eq. (11.1), we see that two complex frequencies, $\underline{s} = j\omega$ and $\underline{s} = -j\omega$, express a sinusoidal function. In our consideration of complex frequency, we will find this second point of view to be useful.

We have a slight semantic problem with this case, however. Although the complex frequency $\underline{s}$ is *imaginary,* the frequency (ω) is said to be *real.* Thus a sinusoidal function has a real frequency, but is described by a complex frequency that is imaginary. As shown above, a complex frequency that is real corresponds to an exponential function.

General interpretation of complex frequency.   We now consider the meaning of $e^{\underline{s}t}$, with $\underline{s} = \sigma + j\omega$. In general

$$v(t) = \text{Re}\{\underline{V}e^{\underline{s}t}\} = \text{Re}\{\underline{V}e^{(\sigma+j\omega)t}\}$$
$$= e^{\sigma t} \text{Re}\{\underline{V}e^{j\omega t}\} = V_p e^{\sigma t} \cos(\omega t + \theta) \qquad (11.7)$$

Equation (11.7) expresses a time function that combines sinusoidal behavior with the exponential behavior that we hitherto have associated with transients. Equation (11.7) can also be considered a sinusoidal function in which the peak value of the sinusoid changes exponentially with time. Here we define it simply to be the time function associated with a complex frequency $\underline{s} = \sigma + j\omega$. Figure 11.2 shows the character of the time function in Eq. (11.7).

Figure 11.2   The function $\text{Re}\{e^{(\sigma+j\omega)}\}$ for (a) σ positive and (b) σ negative.

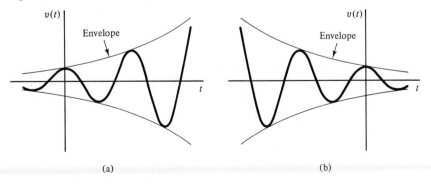

(a)

(b)

The s-plane.   A complex number, s, can be represented by a point in the complex plane. In Fig. 11.3 we show such an s-plane and identify the regions of complex frequency corresponding to possible time responses. The origin corresponds to a constant. Complex frequencies on the real axis correspond to growing and decaying exponential functions. Complex frequencies on the imaginary axis correspond to sinusoidal functions. The region to the right of the vertical axis, the right-half plane, corresponds to sinusoids that are growing exponentially. The left-half plane corresponds to sinusoids that are decreasing exponentially.

Note the patterns of the time functions. To the right, we have increasing functions, and the closer we come to the vertical axis, the slower the rate of increase. To the left, we have decreasing functions, and the closer we come to the vertical axis, the slower the rate of decrease. On the vertical axis, we have steady sinusoidal functions. On the horizontal axis, we have functions that do not oscillate. As we move away from the horizontal axis, we have oscillating functions whose frequency increases as we move further away from the horizontal axis.

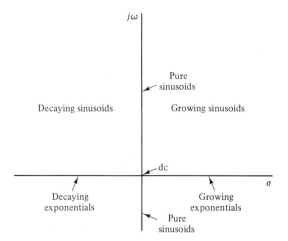

Figure 11.3   The s-plane and associated time functions.

Time domain and frequency domain.   The introduction of complex frequency expands our concept of the frequency domain to cover a wider class of time-domain behavior. We now have a frequency-domain representation of exponentially growing or decreasing time functions. As hinted above, this expansion will allow frequency-domain techniques to be applied to transient problems as well as sinusoidal steady-state problems such as we studied in Chapter 4. Furthermore, these new functions allow the study of a wider class of transient problems.

Importance of e^{st}.   This entire chapter explores the importance of e^{st}. At this point, we wish to look ahead and anticipate later results. We will find that all linear systems are characterized by certain complex frequencies. Once we determine these frequencies for a given system, say, a feedback amplifier, we can determine from them the system response in the time domain or the frequency domain. In the following section we show how to determine these characteristic frequencies and how to derive from them the time-domain response of the system.

Summary. In this section we have generalized frequency to include growing or decreasing exponentials and exponentially growing or decreasing sinusoids. Such complex frequencies allow frequency-domain techniques to be applied to a wide class of problems. The characteristics of linear systems are often described in terms of complex frequency.

11.1.3 Generalized Impedance

Basic concept. In this section, we use complex frequency to generalize the concept of impedance. We will use the technique presented in Chapter 4, where frequency was real. As before, we argue that $e^{\underline{s}t}$ is a function that is indestructible to linear operations such as addition, differentiation, and integration. To determine the impedance, therefore, we excite a circuit with a voltage $\text{Re}\{\underline{V}(\underline{s})e^{\underline{s}t}\}$ and calculate a response, $\text{Re}\{\underline{I}(\underline{s})e^{\underline{s}t}\}$. We illustrate this technique with an inductor, as shown in Fig. 11.4. The equations are

$$v(t) = L\frac{d}{dt}i(t) \Rightarrow \underline{V}(\underline{s})e^{\underline{s}t} = L\frac{d}{dt}\underline{I}(\underline{s})e^{\underline{s}t} = \underline{s}L\underline{I}(\underline{s})e^{\underline{s}t} \qquad (11.8)$$

where we have omitted the "real part of" for simplicity. The differentiation is performed only on the $e^{\underline{s}t}$ function because this is the only function of time. Thus the generalized impedance is

$$\underline{Z}_L(\underline{s}) = \frac{\underline{V}(\underline{s})e^{\underline{s}t}}{\underline{I}(\underline{s})e^{\underline{s}t}} = \underline{s}L \quad \Omega \qquad (11.9)$$

In like manner we can establish the generalized impedances of resistors and capacitors:

$$\underline{Z}_R(\underline{s}) = R, \qquad \underline{Z}_C(\underline{s}) = \frac{1}{\underline{s}C} \qquad (11.10)$$

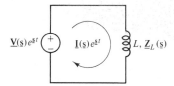

Figure 11.4 An inductor excited by $\underline{V}e^{\underline{s}t}$

Note that the impedance is established by replacing time derivatives with $\underline{s}$. All this would seem to add little to what we did in Chapter 4 for real frequencies except for a substitution of $\underline{s}$ for $j\omega$. This is true mathematically, but not true for the interpretation of the results. With complex frequency, we are implicitly considering the response of the circuit to exciting voltages such as $7e^{-3t}$, or to sinusoids with increasing amplitudes, or to sinusoids with constant amplitude as in Chapter 4. This leads to new insights in circuit behavior, one of which we illustrate in the next two sections.

Application of generalized impedance. We begin with the RL circuit shown in Fig. 11.5. This is a series circuit; hence the impedances of the resistor and inductor add, as in Eq. (11.11).

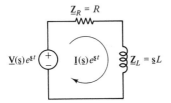

Figure 11.5 An R-L circuit excited by $\underline{V}(s)\,e^{\underline{s}t}$.

$$\underline{Z}_{RL}(\underline{s}) = \frac{\underline{V}(s)e^{\underline{s}t}}{\underline{I}(\underline{s})e^{\underline{s}t}} = R + \underline{s}L \qquad (11.11)$$

The impedance function, $\underline{Z}_{RL}(\underline{s})$ offers several potential uses. With $\underline{s} = j\omega$, we have the impedance for an ac signal at a real frequency of ω. Or we could excite the circuit with an exponential function like $5e^{-5t}$ and calculate the response. But in the next section we will use the impedance function to determine the *natural frequencies* of the circuit.

11.1.4 Check Your Understanding

1. The half-life of the exponential decay of carbon 14 is 3730 years. What complex frequency would describe this process?

2. What complex frequency describes an RC transient with a dc source if $R = 100\ \Omega$ and $C = 10\ \mu F$.

3. What would be the time between zero crossings for a function described by the complex frequency $\underline{s} = -2 + j10$?

4. A pure sinusoid may be described by a complex frequency that is pure imaginary, two complex frequencies that are pure imaginary, either, or neither. (Which?)

5. Complex frequencies near the origin in the $\underline{s}$-plane describe functions that vary slowly with time. (T/F?)

6. What is the impedance of a 100-μF capacitor at a complex frequency of $\underline{s} = -100\ s^{-1}$?

Answers. 1. -1.86×10^{-4} year^{-1}; 2. $-1000\ s^{-1}$; 3. 0.314 s; 4. either, depending on context; 5. true; 6. $-100\ \Omega$.

11.2 NATURAL FREQUENCIES AND TRANSIENT BEHAVIOR OF LINEAR SYSTEMS

11.2.1 Natural Frequencies

Forced and natural responses. When we excite a circuit with a voltage or current source, we force a certain response from the circuit. For example, if we apply a dc source, we expect a dc response. But, as we saw in Chapter 3, part of the transient response of the circuit takes a form that is natural to, and determined by, the circuit itself. For example the transient response of the circuit in Fig. 11.5 would be controlled by its time constant, $\tau = L/R$, and by initial and final conditions. (You may wish to review Section 3.2.2 to refresh your understanding of such transient problems.) The *natural response* of the circuit is always of the form $e^{\underline{s}_n t}$, where $\underline{s}_n$ is a natural frequency of the circuit. The natural frequency is related simply to the time constant for first-order circuits.

Forced response. The forced response of the circuit may be determined directly from the impedance function when the forcing function (input current or voltage) is of the form $e^{\underline{s}t}$. For example, if the voltage source in Fig. 11.5 were a dc source, we would set $\underline{s} = 0$ in the impedance function to determine the forced (steady-state) current, which would also be dc. Or if the voltage were

$$v(t) = 10e^{-4t} \tag{11.12}$$

then $\underline{s} = -4$ and the current would be

$$i(t) = \mathrm{Re}\left\{\frac{10e^{-4t}}{R + (-4)L}\right\} = \frac{10e^{-4t}}{R - 4L} \tag{11.13}$$

where the numerical values of R and L would be substituted to yield the amplitude of the current. If $R - 4L = 0$ in the denominator, that would indicate that we have excited the circuit at its natural frequency. Generally, exciting a circuit at one of its natural frequencies leads to an undefined response. When the exciting function is not of the form $e^{\underline{s}t}$, the forced response must be determined by other methods, such as Laplace transform theory.

Natural response. The circuit in Fig. 11.5 has only one natural frequency, which may be determined by setting the impedance to zero:

$$\underline{Z}_{RL}(\underline{s}) = 0 \Rightarrow \underline{s}_n = -\frac{R}{L} \tag{11.14}$$

where $\underline{s}_n$ is the natural frequency for the circuit. The natural frequency in this case is the negative of the reciprocal of the time constant, and you may confirm that this is also true for a simple RC circuit.

Why does setting the impedance function to zero give the natural frequency of the circuit? This may be understood by examining Eq. (11.11). The natural response of the circuit is the response of which it is capable *when the forcing function is turned* OFF. This means in Eq. (11.11) that $\underline{V}(\underline{s})$ is zero, but $\underline{I}(\underline{s})$ is nonzero. Equation (11.11) can be satisfied under these conditions only if $\underline{Z}_{RL}(\underline{s})$ is zero. In Fig. 11.5, we excited the circuit with a voltage source; hence, turning OFF $\underline{V}(\underline{s})$ is equivalent to replacing the voltage source with a short circuit. Thus, we have determined the *short-circuit* natural frequency of the circuit.

Open-circuit and short-circuit natural frequencies. In Fig. 11.6 we show a circuit containing one capacitor and two resistors. The impedance function is

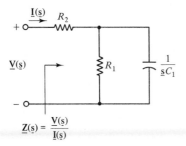

$$\underline{Z}(\underline{s}) = \frac{\underline{V}(\underline{s})}{\underline{I}(\underline{s})}$$

Figure 11.6 An R-C circuit.

$$\underline{Z}(\underline{s}) = R_2 + \frac{1}{(1/R_1) + \underline{s}C_1} = R_2 \frac{\underline{s} + 1/(R_1 \parallel R_2)C_1}{\underline{s} + 1/R_1 C_1} \tag{11.15}$$

where the natural frequencies may be easily recognized in the second form. Note that the value of $\underline{s}$ that makes the impedance function go to zero corresponds to the natural frequency with the input short-circuited. That is, if the input were short-circuited, the capacitor would see an equivalent resistance of $R_1 \parallel R_2$; thus, the time constant of the circuit would be $(R_1 \parallel R_2)C$, which is the negative of the reciprocal of the value of $\underline{s}$ that makes the impedance function go to zero.

But additional information can be derived from the value of $\underline{s}$ that makes the impedance function go to infinity. Infinite impedance would correspond to the input current $\underline{I}(\underline{s})$ going to zero with the input voltage $\underline{V}(\underline{s})$ nonzero. Note that setting the denominator to zero gives $\underline{s}_n = -1/R_1 C_1$, which is the negative of the reciprocal of the time constant of the circuit with the input open-circuited. This corresponds to exciting the circuit with a current source, which is turned OFF for the natural frequency.

Summary. We note that the value of $\underline{s}_n$ that sets the impedance function to zero corresponds to the short-circuit natural frequency, and the value of $\underline{s}_n$ that makes the impedance function go to infinity corresponds to the open-circuit natural frequency of the circuit. These natural frequencies are normally called the *zeros* and *poles* of the impedance function.

Poles and zeros. The impedance function of a complicated circuit generally takes the form of Eq. (11.15), except that the numerator and denominator are polynomials in $\underline{s}$. Thus the function can always be factored into the form

$$\underline{Z}(\underline{s}) = K \frac{(\underline{s} - \underline{z}_1)(\underline{s} - \underline{z}_2)}{(\underline{s} - \underline{p}_1)(\underline{s} - \underline{p}_2)(\underline{s} - \underline{p}_3)} \tag{11.16}$$

where K is a constant (in ohm $-$ s) and we have assumed a second-order polynomial in the numerator and a third-order polynomial in the denominator. In Eq. (11.16), $\underline{z}_1$ and $\underline{z}_2$ are called the *zeros* of the function because these are the values of $\underline{s}$ at which the function goes to zero. Likewise, $\underline{p}_1$, $\underline{p}_2$, and $\underline{p}_3$ are called the *poles* of the function because these are the values of $\underline{s}$ at which the function goes to infinity. As we have shown above, the zeros correspond to the short-circuit natural frequencies and the poles correspond to the open-circuit natural frequencies of the circuit. Except for a multiplicative constant, the poles and zeros of the impedance function fully establish the behavior of the circuit. They give the natural frequencies, as we have illustrated above, and they also give the steady-state (forced) response to excitations of the form $e^{\underline{s}t}$, such as dc ($\underline{s} = 0$) or ac ($\underline{s} = j\omega$).

Impedance or admittance. A circuit may be described by an impedance function or an admittance function or a transfer function (introduced in the next section.) The impedance and admittance functions are reciprocals; hence, the zeros of the impedance function are the poles of the admittance function, and *vice versa*. The zeros of the admittance function correspond, therefore, to the open-circuit natural frequencies of the circuit, and the poles correspond to the short-circuit natural frequencies. This can be confusing unless one returns to the basic definitions. For example, the definition of the admittance function is

$$\underline{Y}(\underline{s}) = \frac{\underline{I}(\underline{s})e^{\underline{s}t}}{\underline{V}(\underline{s})e^{\underline{s}t}} = K' \frac{(\underline{s} - \underline{z_1})(\underline{s} - \underline{z_2})}{(\underline{s} - \underline{p_1})(\underline{s} - \underline{p_2})} \qquad (11.17)$$

where K' is a constant (in mhos) and we have assumed second-order polynomials in numerator and denominator. Here the zeros correspond to $\underline{I}(\underline{s}) = 0$ and hence are the open-circuit natural frequencies because $\underline{I}(\underline{s}) = 0$ means that the input is open-circuited.

Second-order circuit. Figure 11.7 shows a circuit identical to that in Fig. 11.6 except that a second capacitor has been added. The impedance of the circuit is

$$\underline{Z}(\underline{s}) = R_2 + \frac{1}{\underline{s}C_2} + \frac{1}{(1/R_1) + \underline{s}C_1} = \frac{(\underline{s}R_2C_2 + 1)(\underline{s}R_1C_1 + 1) + R_1C_2\underline{s}}{\underline{s}C_2(\underline{s}R_1C_1 + 1)} \qquad (11.18)$$

One of the two poles is identical to that in Eq. (11.15), corresponding to the time constant of the parallel R_1 and C_1; the other pole occurs at $\underline{s} = 0$. The circuit permits $\underline{I}(\underline{s}) = 0$ with $\underline{V}(\underline{s}) \neq 0$ at dc ($\underline{s} = 0$) because C_2 blocks the current.

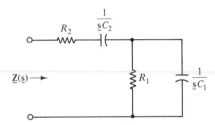

Figure 11.7 A second-order circuit.

The zeros of $\underline{Z}(\underline{s})$, determined from the numerator of Eq. (11.18), lead to the quadratic equation

$$R_1R_2C_1C_2s^2 + (R_1C_1 + R_2C_2 + R_1C_2)s + 1 = 0 \qquad (11.19)$$

For example, with $R_1 = 20 \ \Omega$, $R_2 = 10 \ \Omega$, $C_1 = 0.5 \ \mu F$, and $C_2 = 1 \ \mu F$, the roots of the equation are $\underline{s_1} = -50 \times 10^3 \ s^{-1}$ and $\underline{s_2} = -200 \times 10^3 \ s^{-1}$. These correspond to time constants of 20 μs and 5 μs, respectively. These time constants do not correspond to any RC combination in the circuit but are influenced by the interactions of the entire circuit.

Circuits with either two independent† capacitors or two independent inductors are characterized by two time constants, which are determined through solution of a quadratic equation. The roots are always real and negative; hence the response of this type of circuit is generally

$$i(t) = i_f + Ae^{\underline{s_1}t} + Be^{\underline{s_2}t} \qquad (11.20)$$

where i_f is the forced response and A and B are constants to be determined from the initial conditions of the circuit.

† Two capacitors in series or parallel are not independent but may be replaced by a single equivalent capacitor. The same is true for two inductors in series or parallel.

Nonelectrical example. A thermocouple has a linear output with a time constant of 0.4 s. It is placed in an oven at an ambient temperature of 75°F, and both oven and transducer are in equilibrium. The oven begins to increase in temperature at a rate of 10°F/s. Determine the temperature indicated by the thermocouple in response to this oven temperature.

Let $T(t)$ represent the temperature indicated by the thermocouple and $T_o(t)$ be the temperature of the oven. The specification of a time constant implies that the indicated temperature would respond to a hypothetical sudden change in oven temperature as shown in Fig. 11.8. Such a response would be described by the differential equation

$$\frac{dT(t)}{dt} + \frac{T(t)}{\tau} = \frac{T_o(t)}{\tau} \tag{11.21}$$

where $\tau = 0.4$ s for the thermocouple. Equation (11.21) would have a transform in the frequency domain, but use of the frequency domain is not particularly helpful in this case because the oven temperature has the form $T_o = 75 + 10t$, which cannot be expressed in the form e^{st}. Thus the form of the solution is

$$T(t) = T_f(t) + ae^{-t/\tau} \tag{11.22}$$

where $T_f(t)$ is the forced solution, a is a constant, and the natural solution follows from Eq. (11.21).

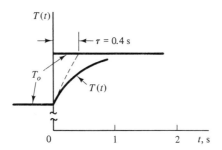

Figure 11.8 Response of the thermocouple to a hypothetical sudden increase in temperature.

The forced function will, like the oven temperature, be a polynominal of the form $T_f(t) = b + ct$. When we substitute this form into Eq. (11.21) with the proper form for the oven temperature, we determine the constants to be $c = 10$ and $b = 71$. When we impose the initial condition that $T(0) = 75$, we determine $a = 4$; hence the total solution is

$$T(t) = 71 + 10t + 4e^{-t/0.4} \tag{11.23}$$

which is shown in Fig. 11.9a. The principal effect is a delay of the thermocouple response by one time constant.

The response of the thermocouple was determined without use of frequency-domain techniques. We note from Eq. (11.21) (if transformed into the frequency domain) that the thermocouple acts in the frequency domain as a low-pass filter, as indicated in Fig. 11.9b.

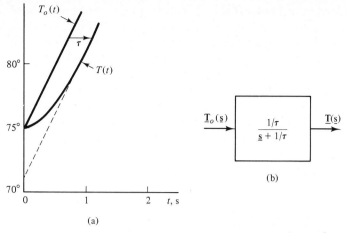

(a)

(b)

Figure 11.9 (a) Response of the thermocouple to a linear increase in temperature. The principal effect is a delay equal to the time constant. (b) The thermocouple as a low-pass filter in the frequency domain.

11.2.2 Natural Response of *RLC* Circuits

Transfer function of an *RLC* circuit. The circuit in Fig. 11.10 introduces two new features into this development. Instead of the impedance, we will derive the *transfer function* of the circuit, which describes the relationship between the input and output voltages. Second, we must consider an oscillatory response because the circuit contains an inductor and capacitor.

This circuit can be analyzed as a voltage divider:

$$\underline{\mathbf{T}}(\underline{s}) = \frac{\underline{\mathbf{V}}_{out}(\underline{s})e^{\underline{s}t}}{\underline{\mathbf{V}}_{in}(\underline{s})e^{\underline{s}t}} = \frac{R \parallel (1/\underline{s}C)}{\underline{s}L + R \parallel (1/\underline{s}C)} = \frac{1}{LC}\frac{1}{\underline{s}^2 + (\underline{s}/RC) + (1/LC)} \quad (11.24)$$

where $\underline{\mathbf{T}}(\underline{s})$ is the transfer function. The poles of the transfer function correspond to an output with no input, and hence are the natural frequencies of the circuit with the input short circuited. These we will investigate in detail because they reveal the various types of behavior that can result with an *RLC* circuit.

We may determine the poles with the quadratic formula

$$\underline{s}^2 + \frac{\underline{s}}{RC} + \frac{1}{LC} = 0 \Rightarrow \underline{s} = -\frac{1}{2RC} \pm \sqrt{\left(\frac{1}{2RC}\right)^2 - \frac{1}{LC}} \quad (11.25)$$

There are four possibilities, depending on the relative values of R, L, and C. The possibilities are undamped, underdamped, critically damped, and overdamped responses.

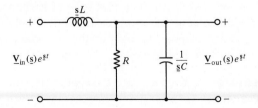

Figure 11.10 An *RLC* circuit with input and output.

Undamped behavior. First we consider the case where the resistance is infinite, which is equivalent to removing the resistor from the circuit. Equation (11.25) gives imaginary roots:

$$\underline{s}_1 = +j\omega_o \quad \text{and} \quad \underline{s}_2 = -j\omega_o \tag{11.26}$$

where $\omega_o = 1/\sqrt{LC}$ rad/s. Thus, with no resistance, the short-circuit natural frequencies are imaginary, meaning that the natural response of the circuit is a sinusoid of constant amplitude. This may be understood as a lossless resonance, with ω_o the resonant frequency. With no resistor, any energy imparted to the circuit will produce a sinusoidal output as it alternates between the inductor and capacitor. This is called *undamped* behavior because the oscillations do not diminish with time.

Underdamped behavior. For large but finite resistance, the discriminant will be negative, and the roots of the quadratic will be complex:

$$\underline{s} = -\alpha \pm \sqrt{\alpha^2 - \omega_o^2} \quad \text{or} \quad \underline{s}_1 = -\alpha + j\omega \quad \text{and} \quad \underline{s}_2 = -\alpha - j\omega \tag{11.27}$$

where $\alpha = 1/2RC$ s^{-1} and $\omega = \sqrt{\omega_o^2 - \alpha^2}$ s^{-1}. The two roots are complex conjugates and correspond to an exponentially decreasing sinusoid. Thus the natural response of the circuit is of the form

$$v_{\text{out}}(t) = Ae^{-\alpha t}\cos(\omega t + \theta) \tag{11.28}$$

where A and θ are constants determined by the initial and final values. (We will work out an example below.) The constant α is called the *damping constant* and is the reciprocal of the time constant of the dying oscillation. The frequency of the oscillation, ω, is less than the resonant frequency, ω_o, because the exchanges of energy between inductor and capacitor are slowed down by the loss in the resistor.

The exponentially decreasing oscillation is called an *underdamped* response. The condition for an underdamped response is

$$\alpha < \omega_o \quad \text{or} \quad R > \frac{1}{2}\sqrt{\frac{L}{C}} \tag{11.29}$$

Thus for large values of resistance, the response of the circuit is dominated by the resonance of the inductor and capacitor, but the oscillations die out due to the loss of the resistor.

Critical damping. When $\alpha = \omega_o$, the roots of Eq. (11.25) become real and equal. This is interesting mathematically but unimportant in practice because it exists only for one exact value of resistance, given by Eq. (11.29) with an equality sign. Even if we wished to produce this type of response in a physical circuit, we would be able to achieve the required resistance only with great care or good fortune. We consider critical damping as the boundary between underdamped oscillations and overdamped behavior.

Overdamped behavior. When $\alpha > \omega_o$, the roots of the quadratic equation are real, negative, and unequal.

$$\underline{s}_1 = -\alpha + \sqrt{\alpha^2 - \omega_o^2} \quad \text{and} \quad \underline{s}_2 = -\alpha - \sqrt{\alpha^2 - \omega_o^2} \qquad (11.30)$$

Thus the *overdamped response* has the form

$$v_{\text{out}}(t) = Ae^{\underline{s}_1 t} + Be^{\underline{s}_2 t} \qquad (11.31)$$

where $\underline{s}_1$ and $\underline{s}_2$ are real, negative numbers. This response consists of two time constants, which are the negatives of the reciprocals of $\underline{s}_1$ and $\underline{s}_2$. Thus, for small resistance, the loss eliminates the resonance between the inductor and capacitor.

Summary. The natural frequencies of the *RLC* circuit are derived from the roots of a quadratic equation. The roots may be imaginary, complex, real and equal, or real and unequal. Imaginary roots indicate a undamped oscillation. Complex roots indicate a damped oscillation. Real, unequal roots indicate two time constants but no oscillation. Real, equal roots represent the mathematical boundary between oscillatory and nonoscillatory behavior.

11.2.3 *RLC* Circuit Transient Behavior

Forced response. We now investigate the response of the circuit shown in Fig. 11.11, in which a dc voltage is applied to an *RLC* circuit. The output voltage consists of two components, a natural response and a forced response due to the dc input. We may determine the forced response by the methods presented in Chapter 3, that is, by treating the capacitor as an open circuit and the inductor as a short circuit. Alternately, we may consider the input as $10e^{\underline{s}t}$, with $\underline{s} = 0$, and derive the forced response from the transfer function given in Eq. (11.24) with $\underline{s} = 0$. Thus the forced component of the output voltage is

$$V_{\text{out}}(0) = \underline{T}(0) \times 10e^{0t} = 10 \text{ V} \qquad (11.32)$$

This indicates a dc output of 10 V due to the battery. The total response will therefore be

$$v_{\text{out}}(t) = 10 + v_n(t) \quad \text{V} \qquad (11.33)$$

where $v_n(t)$ is the natural response.

Natural response: undamped case. Several forms are possible for the natural response, depending on the resistance. We consider first the undamped response because the undamped circuit is used for "resonant charging" of capacitors. The response is

$$v_{\text{out}}(t) = 10 + A\cos(\omega_o t + \theta) \qquad (11.34)$$

where A and θ must be determined from the initial conditions.

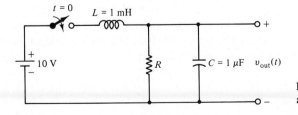

Figure 11.11 An *RLC* circuit with a source.

Initial conditions. We may establish the initial conditions of the output voltage through the techniques presented in Chapter 3. The capacitor acts initially like a short circuit; hence the output voltage must be zero at $t = 0^+$.

$$0 = 10 + A \cos(\theta) \tag{11.35}$$

The inductor acts initially like an open circuit. Consequently, the initial current through the capacitor is also zero, and the derivative of the output voltage must also be zero at $t = 0^+$.

$$\left. \frac{dv_{\text{out}}(t)}{dt} \right|_{t=0^+} = 0 \Rightarrow 0 = -\omega_o A \sin\theta \tag{11.36}$$

so $\theta = 0$ and Eq. (11.35) yields $A = -10$ V. Hence the response is

$$v_{\text{out}}(t) = 10 - 10 \cos(\omega_o t) \tag{11.37}$$

This response is shown in Fig. 11.12.

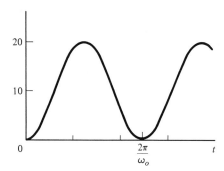

Figure 11.12 Response of the circuit in Fig. 11.11 with $R = 0$. The peak voltage is twice the input voltage, a property of the circuit that finds many applications.

The voltage on the capacitor reaches twice that of the input because the inductor gives a momentum to the current to continue charging past the equilibrium point. If a diode is placed in series with the inductor to prevent the discharge of the capacitor, the addition of the diode and inductor results in a fourfold increase in stored energy in the capacitor. This technique is known as *resonant charging* of the capacitor, and applications are found in radar and power electronics.

Damped response. For the inductance and capacitance values given in Fig. 11.11, the value of the resistance for critical damping is given by Eq. (11.29) with an equality sign:

$$R_c = \frac{1}{2}\sqrt{\frac{L}{C}} = \frac{1}{2}\sqrt{\frac{10^{-3}}{10^{-6}}} = 15.8 \ \Omega \tag{11.38}$$

where R_c is the resistance for critical damping. We consider first the underdamped response, which implies a resistance larger than the critical value. We assume $R = 5R_c = 79.1 \ \Omega$. The damping coefficient and resonant frequency are

$$\alpha = \frac{1}{2RC} = 6324 \ s^{-1} \quad \text{and} \quad \omega_o = \frac{1}{\sqrt{LC}} = 31{,}623 \ \text{rad/s} \quad (5033 \ \text{Hz}) \tag{11.39}$$

and thus the natural frequencies, determined from Eq. (11.27), are

$$\underline{s}_1 = -6324 + j30{,}984 \quad \text{and} \quad \underline{s}_2 = -6324 - j30{,}984 \qquad (11.40)$$

Note that ω is slightly less than the resonant frequency calculated in Eq. (11.39). The total response is

$$v_{out}(t) = 10 + Ae^{-6324t}\cos(30{,}984t + \theta) \quad \text{V} \qquad (11.41)$$

The initial conditions are the same as for the undamped response. Hence at $t = 0^+$

$$0 = 10 + A\cos\theta \qquad (11.42)$$

and

$$\left.\frac{dv_{out}(t)}{dt}\right|_{t=0} = 0 \Rightarrow 0 = A[-6324\cos\theta - 30{,}984\sin\theta] \qquad (11.43)$$

Equations (11.42) and (11.43) yield $\theta = -11.5°$ and $A = -10.21$ V. The solution is therefore

$$v_{out}(t) = 10 - 10.21e^{-6324t}\cos(30{,}984t - 11.5°) \quad \text{V} \qquad (11.44)$$

which is shown in Fig. 11.13.

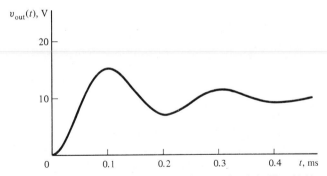

Figure 11.13 Underdamped response for the circuit in Fig. 11.11.

Overdamped response. To obtain an overdamped response, we will use a resistance one-fifth the critical value given by Eq. (11.38): $R = R_c/5 = 3.16\ \Omega$. With this resistance, $\alpha = 158{,}114$, and Eq. (11.30) gives the natural frequencies as

$$\underline{s}_1 = -3195\ s^{-1} \quad \text{and} \quad \underline{s}_2 = -313{,}033\ s^{-1} \qquad (11.45)$$

Thus the output voltage is

$$v_{out}(t) = 10 + Ae^{-3195t} + Be^{-313{,}033t} \quad \text{V} \qquad (11.46)$$

where A and B are constants. The first term is the forced response, which is the same as before. The second term has a time constant of 0.313 ms and the third term a time constant of 3.19 μs. The initial conditions are the same as before; hence we may solve for A and B from the equations

$$0 = 10 + A + B \quad \text{and} \quad 0 = -3195A - 313{,}033B \qquad (11.47)$$

which yield $A = -10.10$ V and $B = 0.10$ V. Thus the overdamped response is

$$v_{\text{out}}(t) = 10 - 10.10e^{-3195t} + 0.10e^{-313{,}033t} \quad \text{V} \qquad (11.48)$$

This response is shown in Fig. 11.14. The third term in Eq. (11.48) has a brief influence to give the zero derivative at the origin, but otherwise the response is that of a first-order transient of the RL part of the circuit.

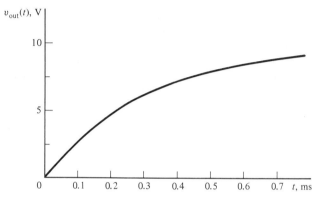

Figure 11.14 Overdamped response of the circuit in Fig. 11.11.

Series or parallel _RLC_ circuit? The circuit in Fig. 11.10, when excited by a voltage source, is called a parallel RLC circuit. With the source turned OFF, the resistor, inductor, and capacitor are connected in parallel. In the circuit shown in Fig. 11.15, the resistor, inductor, and capacitor are connected in series. The analysis of the series RLC circuit for the generalized impedance, natural frequencies, and transient response follows the same lines as for the parallel circuit, but the results differ in detail. We leave the analysis of the series RLC circuit for a homework problem.

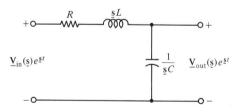

Figure 11.15 This is a series RLC circuit when driven by a voltage source.

Relationship with Laplace transform theory. The Laplace transform gives a strong mathematical foundation to the techniques used in this chapter and is closely associated with linear system theory. Although the Laplace transform yields the total response of a circuit or system, including initial conditions, its primary importance lies in furnishing a means for describing linear systems.

The functions such as $\underline{V}(\underline{s})$ and $\underline{I}(\underline{s})$, which we have been treating as constants, can be considered _transforms_ of the time-domain voltage and current. These transforms can be interpreted as generalized spectra of the time-domain functions. Similarly, generalized impedances and transfer functions are frequency-domain transforms of the linear differential equations that describe the circuit or system component in the time domain.

In this book we introduce the notation and application of the system notation, with a minimum of mathematical sophistication. Books on linear system theory normally include a mathematical section on Laplace transform theory.

Summary. In this section we have used complex frequency to investigate transient behavior of first- and second-order circuits. Natural responses are decreasing exponentials or exponentially decreasing sinusoids. The natural frequencies and the character of the natural response may be determined from the poles and zeros of the impedance or transfer functions. In the next section, we will illustrate how the transfer function, as a function of complex frequency, describes the properties of a linear-system component.

11.2.4 Check Your Understanding

1. An impedance consists of $R = 100 \ \Omega$ and $C = 10 \ \mu F$ in series. What impedance does this present to an input voltage $v(t) = 10e^{-500t}$?
2. A 10-Ω resistor is connected in series with a 100-μF capacitor. At what complex frequency does this combination look like a short circuit?
3. The impedance function can be used to determine the forced response of a circuit if the input excitation is of the form e^{st}. (T/F?)
4. An impedance with a zero at $\underline{s} = 0$ will pass a dc current. (T/F?)
5. The open-circuit natural frequencies of a circuit correspond to the zeros of the impedance function. (T/F?)
6. A circuit with two inductors must have natural frequencies that are real. (T/F?)
7. Describe the largest number and the character of the natural frequencies of the following circuits:
   (a) Two resistors and a capacitor
   (b) One resistor and two capacitors
   (c) Three resistors and two inductors
   (d) Two resistors, a capacitor, and an inductor
8. Determine the natural frequencies of the circuit in Fig. 11.11 if the resistor has the value for critical damping.

Answers. 1. $-200 \ \Omega$; 2. $-1000 \ s^{-1}$; 3. true; 4. true; 5. false; 6. true; 7. (a) one natural frequency, real and negative; (b) one natural frequency, real and negative; (c) two natural frequencies, real and negative; (d) two natural frequencies, either complex with negative real part or both real and negative; 8. $\underline{s} = -31,623 \ s^{-1}$ (repeated)

1.3 SYSTEM ANALYSIS

Introduction. In this section we will introduce the notation and techniques commonly used to analyze linear systems. Our example will be a thermostatically controlled oven. We will determine its response when controlled by a feedback control system. Although heating of the oven is inherently nonlinear, we will linearize the problem to use the techniques of the frequency domain. We will show that a second-order system (two independent modes of energy storage) is unconditionally stable, although the system response may be unacceptable. We then show that a third-order system can become unstable for large loop gain.

Problem description. In Chapter 8, page 353, we discussed a system for controlling an oven with a feedback control system. We described the system in general terms and mentioned briefly the dynamic stability of the system. We now have the tools to investigate the system stability.

Figure 11.16 shows the oven-control system. The oven is heated by a resistive element driven by an amplifier. The oven temperature is monitored by a sensor, which we assume produces a voltage proportional to its temperature. The op amp is driven by the difference between an input set voltage and the feedback signal from the sensor.

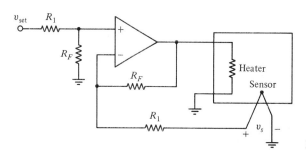

Figure 11.16 Oven-control system.

11.3.1 System Functions

We will now express the properties of the sensor in the frequency domain. Equation (11.21) presented the differential equation describing the oven temperature and the indicated sensor temperature. Here we add one factor: the output signal of the sensor is a voltage and the input signal is a temperature. Thus we introduce a constant K_s V/°F describing its conversion of temperature to voltage. Thus the output voltage of the sensor in the time domain is

$$\frac{dv_s}{dt} + \frac{v_s}{\tau_s} = K_s \frac{T_o(t)}{\tau_s} \tag{11.49}$$

Figure 11.17 System function representing the sensor. The input and output are frequency-domain transforms of the oven temperature and sensor voltage, respectively.

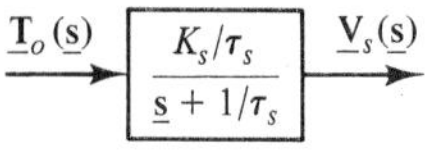

$$\underline{T}_o(\underline{s}) \quad \boxed{\dfrac{K_s/\tau_s}{\underline{s} + 1/\tau_s}} \quad \underline{V}_s(\underline{s})$$

where $v_s(t)$ is the sensor voltage, τ_s is the time constant of the sensor, and $T_o(t)$ is the oven temperature. We may transform Eq. (11.49) into the frequency domain by assuming input temperature and output voltage to be of the form $e^{\underline{s}t}$. The result is

$$\left(\underline{s} + \frac{1}{\tau_s}\right) \underline{V}_s(\underline{s}) = \frac{K_s}{\tau_s} \underline{T}_o(\underline{s}) \Rightarrow \underline{V}_s(\underline{s}) = \frac{K_s/\tau_s}{\underline{s} + (1/\tau_s)} \times \underline{T}_o(\underline{s}) \tag{11.50}$$

where $\underline{V}_s(\underline{s})$ is the transform of the sensor voltage and $\underline{T}_o(\underline{s})$ is the transform of the oven temperature. Thus we may represent the sensor in the frequency domain by the system function shown in Fig. 11.17.

The oven temperature will be controlled by the temperature of the heater element. We will assume a time constant of τ_o for this process. Thus the heater temperature and oven temperature are related by a system function

$$\underline{T}_o(\underline{s}) = \frac{1/\tau_o}{\underline{s} + (1/\tau_o)} \times \underline{T}_H(\underline{s}) \tag{11.51}$$

where τ_o represents the oven time constant and $\underline{T}_H(\underline{s})$ is the transform of the heater temperature.

Although the heater is nonlinear, we will deal only with small changes in the signal levels and hence may assume that a linear increment in voltage will produce a linear increment in power in the heater, and hence a linear increment in the heater temperature. There will be a delay associated with the thermal mass of the heater element, but we will ignore this effect in the present analysis. Thus we assume that changes in heater voltage produce proportional and instantaneous changes in the temperature of the heater element. These changes are represented in the time domain and frequency domain by

$$T_H(t) = K_H v_H(t) \Rightarrow \underline{T}_H(\underline{s}) = K_H \underline{V}_H(\underline{s}) \tag{11.52}$$

where $\underline{V}_H(\underline{s})$ is the transform of the heater voltage and K_H °F/V is a constant relating heater temperature to heater voltage.

System diagram. We may represent the interaction of the various system variables by the system diagram shown in Fig. 11.18. The amplifier is represented by a summer (differencer) and a gain A.

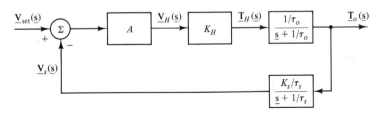

Figure 11.18 System diagram for the oven-control system.

Summary. In this section we have introduced the system functions of the various components of a feedback control system. Each variable is represented by its frequency-domain transform. Each component is represented by its system function. We may now use feedback theory to determine the overall system function of the controller. From the natural frequencies of this system function, we may determine the natural frequencies of the system and from these the transient response of the system in the time domain. The next section continues this investigation.

11.3.2 Dynamic Stability of Feedback Systems

Feedback analysis. The gain with feedback of an "amplifier" with feedback is given in Eq. (8.45) as

$$A_f = \frac{A}{1 - L} \qquad (11.53)$$

where A_f is the gain with feedback, A is the gain without feedback, and L is the loop gain. Here we are expressing the system properties in the frequency domain, so our signals and "gains" are complex functions of $\underline{s}$. The gain without feedback in Fig. 11.18 is the product of the system functions between input and output:

$$\underline{A}(\underline{s}) = A \times K_H \times \frac{1/\tau_o}{\underline{s} + (1/\tau_o)} \qquad (11.54)$$

and the loop gain is the product of the system functions around the loop:

$$\underline{L}(\underline{s}) = -AK_H \times \frac{1/\tau_o}{\underline{s} + (1/\tau_o)} \times \frac{K_s/\tau_s}{\underline{s} + (1/\tau_s)} = + \frac{L_0/\tau_o\tau_s}{[\underline{s} + (1/\tau_o)][\underline{s} + (1/\tau_s)]} \qquad (11.55)$$

where L_0 is the dc loop gain (at $\underline{s} = 0$) and is a dimensionless quantity that indicates the importance of feedback in establishing the system properties.

System natural frequencies. Our investigation of the dynamic behavior of the system does not require that we substitute Eqs. (11.54) and (11.55) into Eq. (11.53). We are interested in the natural response of the system, when the input is zero but the output is nonzero. This can occur only at the poles of the system function, that is, when $1 - \underline{L}(\underline{s}) = 0$. Thus we may determine the natural frequencies from the quadratic equation

$$1 - \frac{L_0/\tau_o\tau_s}{[\underline{s} + (1/\tau_o)][\underline{s} + (1/\tau_s)]} = 0 \Rightarrow \underline{s}^2 + \left(\frac{1}{\tau_o} + \frac{1}{\tau_s}\right)\underline{s} + \frac{1 - L_0}{\tau_o\tau_s} = 0 \qquad (11.56)$$

We will continue our analysis with a time constant of 5 min for the oven and a time constant of 30 s for the sensor. The roots of Eq. (11.56) for dc loop gains of -1, -10, and -50 are given in Fig. 11.19a. The associated dynamic responses shown in Fig. 11.19b are derived in the next section.

Figure 11.19 (a) Natural frequencies for dc loop gains of -1, -10, and -100; (b) system response to a sudden increase in input for the various dc loop gains.

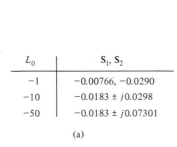

L_0	S_1, S_2
-1	$-0.00766, -0.0290$
-10	$-0.0183 \pm j0.0298$
-50	$-0.0183 \pm j0.07301$

(a)

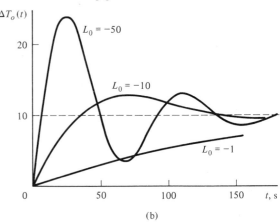

(b)

Dynamic response: overdamped case. We may determine the dynamic response of the circuit from the natural frequencies, which are the poles of the overall system function with feedback, and from the initial and final conditions of the output. We assume in this section that the input voltage is increased suddenly to cause eventually a 10°F increase in the oven temperature.

For $L_0 = -1$, the roots of Eq. 11.56 are $\underline{s}_1 = -0.00766$ and $\underline{s}_2 = -0.02901$; hence these are the natural frequencies of the system. The response is overdamped, and the total response will be of the form

$$\Delta T_o(t) = A + Be^{-0.00766t} + Ce^{-0.02901t} \tag{11.57}$$

where $\Delta T_o(t)$ is the change in the oven temperature and A, B, and C are constants to be determined from the initial and final conditions. The final condition is that $\Delta T_o(t)$ should approach $+10°F$ as time gets large; hence $A = +10$.

The initial conditions can be reasoned from the time delays in the system. Because it takes time for the oven to heat up, the initial value of $\Delta T_o(t)$ is zero. The initial value of the derivative will not be affected by the feedback effects because of the time delay in the sensor. Therefore, the initial value of the derivative of $\Delta T_o(t)$ will be $\Delta T_o'$ °F/300 s, where $\Delta T_o'$ is the value that the temperature would reach if there were no feedback. In other words, $\Delta T_o'$ is proportional to the increment in input voltage required to produce a 10°F temperature increment in the oven. The proof that this is $10(1 - L_0)$ is required in a homework problem.

With these two initial conditions, we can readily determine the constants B and C in Eq. (11.57). The initial condition on the temperature increments

$$0 = 10 + B + C \tag{11.58}$$

and the initial condition on the derivative of $\Delta T_o(t)$ for $L_0 = -1$ follows from the derivative of Eq. (11.57) as

$$\frac{10}{300} \times [1 - (-1)] = 0 - 0.00766B - 0.02901C \tag{11.59}$$

Simultaneous solution of Eqs. (11.58) and (11.59) yields $B = -10.466$ and $C = +0.466$. Hence the response of the system with a loop gain of -1 to a sudden increase in input voltage is

$$\Delta T_o(t) = 10 - 10.466e^{-0.00766t} + 0.466e^{-0.02901t} \tag{11.60}$$

which is shown in Fig. 11.19b. We note that the transient is dominated by the longer time constant of $1/0.00766 = 130.5$ s. This is the time constant of the oven, sped up by a factor of approximately 2 by the feedback.

Dynamic response: underdamped case. For dc loop gains greater than 2.025, the roots of Eq. (11.56) are complex, indicating underdamped response. For $L_0 = -10$, the roots are given in Fig. 11.19a as $\underline{s}_{1,2} = -0.0183 \pm j0.0298$. These natural frequencies indicate underdamped behavior. The form of the response can be written in several ways, but the form given in Eq. (11.61) is most convenient.

$$\Delta T_o(t) = A + e^{-0.0183t}[B \cos(0.0298t) + C \sin(0.0298t)] \tag{11.61}$$

where again A, B, and C are constants to be determined from the initial and final conditions. The final value (A) is 10, as before; the initial condition on $\Delta T_o(t)$ is the same as before, which allows B to be determined by inspection as -10 since the sine term vanishes at $t = 0$. The initial value of the derivative is discussed above; differentiation of Eq. (11.61)

$$\frac{d\Delta T_o(t)}{dt} = \frac{10}{300} \times [1 - (-10)] = 0 - 0.0183B + 0.0298C \quad (11.62)$$

which yields $C = -6.159$. Hence the response for a dc loop gain of -10 is

$$\Delta T_o(t) = 10 + e^{-0.0183t}[-10\cos(0.0298t) - 6.159\sin(0.0298t)] \quad (11.63)$$

This response is shown in Fig. 11.19b, where we have also shown the response for a dc loop gain of -50.

Summary of feedback effects for second-order systems. Figure 11.19b shows the effects of feedback on the second-order system, having delays in both oven and sensor. As the magnitude of the dc loop gain is increased, the response goes from overdamped to underdamped. The character of the various responses is shown in Fig. 11.19b, which shows the system speeding up with increasing feedback. However, the use of excessive feedback causes a severe overshoot problem and is unacceptable. The second-order system is unconditionally stable, meaning that growing oscillations cannot occur, regardless of the amount of feedback. As we will see in the next section, a third-order system can become dynamically unstable if the loop gain is too high.

Control theory. Figure 11.19b shows that large loop gain degrades the dynamic response of the system by causing a "hunting" type of response. But with low loop gain, we lose the benefits of feedback. One goal of control theory is to achieve an acceptable compromise between these two effects. Improvement may result from modifying the system properties with electrical filters.

Dynamic analysis of third-order system. We now repeat the analysis of the oven-control system considering the delay in the heater element. When the power is applied to the heater element, its temperature heats up with a time constant τ_H. This changes the system function of the heater element in Fig. 11.18 to that shown in Fig. 11.20. The analysis of the feedback system proceeds as

Figure 11.20 System diagram for oven-control system with an additional time constant due to the heater element.

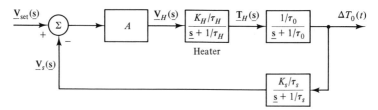

before, and the natural frequencies are the roots of $1 - \underline{L}(\underline{s}) = 0$. This results in the cubic equation

$$\left(\underline{s} + \frac{1}{\tau_H}\right)\left(\underline{s} + \frac{1}{\tau_o}\right)\left(\underline{s} + \frac{1}{\tau_s}\right) + \frac{L_0}{\tau_H\tau_o\tau_s} = 0 \qquad (11.64)$$

We have determined the roots of Eq. (11.64) with time constants of 30 s, 300 s, and 30 s for the time constants of heater, oven, and sensor, respectively. Figure 11.21a tabulates the roots for various values of dc loop gain.

Root-locus plot. Figure 11.21b shows the motion of the natural frequencies in the complex plane as the magnitude of the dc loop gain is increased. We note that for a loop gain of -1 all three roots are real, indicating exponential behavior in the time domain. But for loop gain magnitudes greater than about 1.3,

L_0	Roots
−1	−0.0430, −0.0167, −0.0103
−2	−0.0464, −0.0118 ± j0.0100
−5	−0.0527, −0.00865 ± j0.01862
−10	−0.0591, −0.00545 ± j0.0257
−20	−0.0674, −0.0013 ± j0.0340
−30	−0.0732, +0.0016 ± j0.0396

(a)

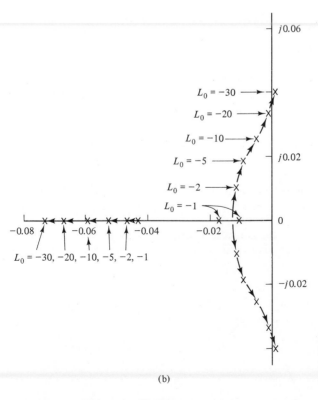

(b)

Figure 11.21 (a) Natural frequencies for the system in Fig. 11.20; root plot for third-order system with increasing loop gain.

two roots become complex and begin to move toward the real axis as the oscillation frequency increases. For loop-gain magnitudes greater than about 25, the roots move into the right half of the complex plane, indicating oscillations that increase with time, similar to the response shown in Fig. 11.2a. This type of response indicates dynamically unstable behavior.

Control theory. The dynamic performance of this system is unsatisfactory for many applications, even if the poles are kept in the region of stability. Control theory addresses questions of stability and dynamic response in such systems. We have pursued this example to illustrate applications of complex frequency and to introduce the notation and concerns of system theory.

11.3.3 Check Your Understanding

1. What would be the system function for an integrator?
2. Give the system function of a low-pass filter with a critical frequency of 100 Hz.
3. A second-order feedback system can never be dynamically unstable. (T/F?)
4. Natural frequencies in the left half of the complex plane represent systems that are stable or unstable? (Which?)
5. System stability becomes an increasing problem as (1) the magnitude of the loop gain is increased, (2) the order of the system is increased, or (3) both. (Which?)

Answers. 1. $1/\underline{s}$; 2. $200\pi/(\underline{s} + 200\pi)$; 3. true; 4. stable; 5. both.

PROBLEMS

Section 11.1: Complex Frequency

P11.1. A decaying sinusoid has a complex frequency $\underline{s} = -1 + j2$ and has its maximum value at $t = 0$. What is the angle of the phasor representing this signal in the frequency domain?

P11.2. A decaying sinusoid crosses zero every 10 ms and each positive peak is 90% of the previous positive peak. What complex frequency describes this function?

Answer. $-5.27 + j314.2$

P11.3. A time-domain function is shown in Fig. P11.3. The function is of the form

$$v(t) = A + Be^{\sigma t}\cos(\omega t)$$

(a) From the graph, determine A, B, σ, and ω.
(b) What complex frequency (or frequencies) is (are) involved in this function?

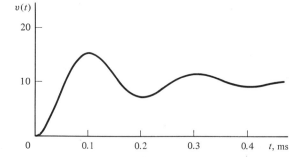

482

Figure P11.3

Section 11.2: Generalized Impedance

P11.4. Consider a resistor, R, in series with a capacitor, C. Determine the generalized input impedance as a function of complex frequency and, from that, the pole and zero of the circuit. What is the interpretation of the pole? What is the interpretation of the zero?

P11.5. The circuit shown in Fig. P11.5 is excited by a voltage source that is zero for negative time and exponential for positive time, as shown.
(a) Determine the impedance of the circuit, $\underline{Z}(\underline{s})$.
(b) What is the complex frequency of the source?
(c) What is the natural frequency of the circuit?
(d) Determine the forced response of the current.
(e) Determine the total response of the circuit, including the initial condition.

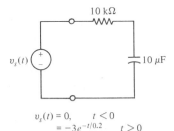

$$v_s(t) = 0, \quad t < 0$$
$$= -3e^{-t/0.2} \quad t > 0 \qquad \textbf{Figure P11.5}$$

P11.6. For the RLC circuit in Fig. 11.11, what would be the natural frequencies of the network if R and L were exchanged? What would be the initial value and initial derivative of the output response to the sudden application of an input voltage?

P11.7. For the series RLC circuit in Fig. 11.15, with $L = 1$ mH and $C = 1$ μF:
(a) Determine the transfer function, $\underline{T}(\underline{s})$.
(b) From the transfer function, determine the natural frequencies of the circuit if excited by a voltage source.
(c) What is the value of resistance for critical damping of the circuit? Is this different from the critical value given in Eq. (11.38) for the parallel RLC circuit?
(d) For resistances one-half and twice the critical value established in part (c), determine the natural frequencies of the circuit and write the corresponding time responses with unknown constants.
(e) Consider that a 10-V source is suddenly applied to the input in the manner shown in Fig. 11.11. What are the initial value and initial derivative of the output voltage? Work out the complete response for the overdamped case calculated in part (d).

P11.8. A circuit has the impedance function

$$\underline{Z}(\underline{s}) = 10^4 \frac{\underline{s}^2 + 3\underline{s} + 1}{\underline{s}(\underline{s} + 1)} \ \Omega$$

(a) What is the impedance at dc?
(b) What are the natural frequencies of the circuit if the input is shorted?
(c) What are the natural frequencies if the input is open-circuited?
(d) In addition to resistors, this circuit has one inductor, or one capacitor, or two inductors, or two capacitors, or one capacitor and one inductor? Which combinations are possible (may be more than one)? Explain your answer.

(e) If the circuit is excited with a current source of value $-2e^{-2t}$ A that is suddenly turned on at $t = 0$, what is forced response of the voltage at the input to the circuit?

(f) What is the total response for the input voltage if the initial value of the voltage and its rate of change are zero?

P11.9. Determine the open-circuit and short-circuit natural frequencies for the circuits shown in Fig. P11.9 and interpret these in terms of the time constants of the circuits.

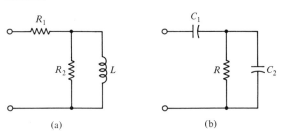

(a)   (b)   **Figure P11.9**

P11.10. For the circuit shown in Fig. P11.10:
   (a) Find the input impedance, $\mathbf{Z}(\mathbf{s})$.
   (b) Determine the open-circuit natural frequencies.
   (c) What value (range) of resistance corresponds to behavior that is (1) undamped, (2) underdamped, (3) critically damped, and (4) overdamped?

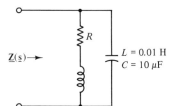

Figure P11.10

P11.11. Figure P11.11a shows the input function to the low-pass filter shown in Fig. P11.11b. Determine the equation of the output voltage in the time domain.

Answer. $v_{\text{out}}(t) = 2.2e^{319t} \cos(393t - 7.6°)$

Figure P11.11

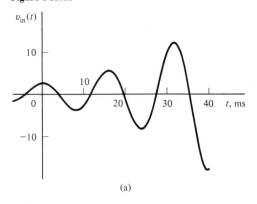

(a)

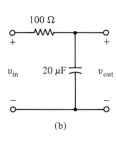

(b)

P11.12. A circuit has the pole-zero pattern shown in Fig. P11.12 and has an impedance magnitude of 30 Ω at $\underline{s} = +1 \text{ s}^{-1}$. The circuit is excited by a voltage source that is zero for negative time and has a value of

$$v(t) = 5 \cos(2t)$$

for positive time. Determine the current into the circuit for positive time, assuming zero for the initial value and initial derivative of the current.

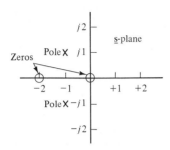

Figure P11.12

P11.13. The circuit shown in Fig. P11.13 is excited by a voltage source that is zero for negative time but

$$v(t) = 100e^{-10000t} \text{ V}$$

for positive time. Determine the current for positive time, assuming zero current at $t = 0$.

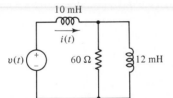

Figure P11.13

P11.14. Figure P11.14 shows a circuit with a disconnected source.
  (a) Find the generalized impedance of the circuit, $\underline{Z}(\underline{s})$.
  (b) What is the impedance at $\underline{s} = 0$?
  (c) What is the impedance at $\underline{s} = \infty$?
  (d) If the switch is closed, what are the natural frequencies of the circuit?
  (e) What is the character of the response (undamped, underdamped, critically damped, overdamped)?
  (f) What is the initial value of the current in the resistor?
  (g) What is the final value of the current in the resistor?

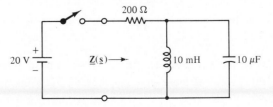

Figure P11.14

P11.15 The impedance of the circuit in Fig. P11.15 has one zero at $\underline{s} = 0$ and one pole at $\underline{s} = -1 \text{ s}^{-1}$. At $\underline{s} = +1 \text{ s}^{-1}$, the impedance into the circuit has a value of 15 Ω. The circuit is excited by a current source that is zero for negative time and has a value of

$$i(t) = 0.5e^{-2t} \text{ A}$$

for positive time. Determine the input voltage, as shown, assuming $v(0) = +2$ V.

Answer. $v(t) = 30e^{-2t} - 28$

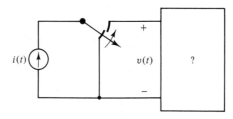

Figure P11.15

P11.16. An *RC* network has the impedance function

$$\underline{Z}(\underline{s}) = 100 \frac{\underline{s}^2 + 30\underline{s} + 100}{\underline{s}(\underline{s} + 10)} \; \Omega$$

(a) Does the circuit allow a dc current?
(b) Find the impedance of the circuit at a frequency of 5 Hz.
(c) If the circuit is excited with a current source, what is (are) the natural frequency (frequencies) of the circuit?
(d) If the circuit source is of the form $I_o e^{-10t}$, what input voltage results?
(e) What are the short-circuit natural frequencies of the network?

P11.17. Figure P11.17 shows an undamped *LC* circuit that will "resonant charge" the capacitor to twice the power-supply voltage, and the diode will hold the voltage until the capacitor is discharged. Determine the inductance and capacitance to store 2 J of energy in the capacitor in 1 ms from the time of switch closure. Assume an ideal diode.

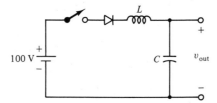

Figure P11.17

Section 11.3: Linear System Analysis

P11.18. A simplified analysis of an electric-oven control system includes the following effects:

- When the heater element in turned on, it heats with a time constant of 15 s.
- If the heater element were fully hot, the air in the oven would heat with a time constant of 3 min.

- The time constant of the thermostat is negligible.
- If the heater element were left on a long time, the oven would heat to 600°F relative to ambient temperature.

Figure P11.18 shows a block diagram of the system. The output of the thermostat is compared to the oven temperature setting, and the output of the comparator is represented as a dashed line to the switch.

(a) Determine the form of the transfer functions for the heater and oven-temperature blocks.
(b) Derive the system function from the input of the heater to the output of the thermostat.
(c) What is the form of the output signal in the time domain, given that the oven is excited with a constant voltage at the heater element?
(d) Determine the time required to heat the oven to 350°F and compare that time with the time required if there were no heat-up delay in the heater element. Assume an ambient temperature of 75°F.

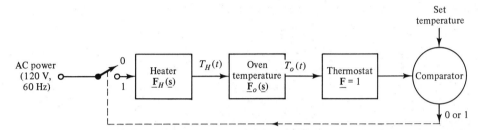

Figure P11.18

P11.19. A feedback system is shown in Fig. P11.19.
(a) What is the system function with feedback?
(b) Find A for critical damping, that is, to put the system on the boundary between under- and overdamped behavior.
(c) For a value of A ten times that computed in part (b), what is the form of the natural response of the system?

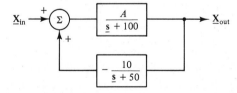

Figure P11.19

P11.20. Show that, in the absence of feedback, the final oven temperature used as an example in Section 11.3.1 will reach $10(1 - L_0)$, where L_0 is the dc loop gain and the final change in the oven temperature is 10°F. *Hint:* Determine the increment in input temperature to produce a 10°F change in the oven temperature with feedback operating.

P11.21. Determine from Eq. (11.56) the value of dc loop gain to give critical damping. Show that this is -2.025 for $\tau_o = 300$ s and $\tau_s = 30$ s, as stated in the text.

P11.22. Show from Eq. (11.56) that the second-order system cannot become dynamically unstable, regardless of the dc loop gain.

P11.23. A linear system is described by the differential equation

$$\frac{d^2 x_{out}}{dt^2} + 5\frac{dx_{out}}{dt} + 16 x_{out} = 5\frac{dx_{in}}{dt}, \qquad x_{out} = \text{output}, \; x_{in} = \text{input}$$

A fraction of the output, call it β, is fed back to the input and subtracted from the input in a negative feedback system.

(a) Draw a block diagram of the system in the frequency domain.

(b) What range of β's gives transient behavior that is overdamped?

P11.24. The block diagram for a linear process is shown in Fig. P11.24.

(a) Determine the maximum value of B for an overdamped response.

(b) Find the transfer function for $B = 0.1$.

(c) What value(s) of $\underline{s}$ allow output with no input?

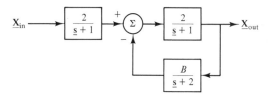

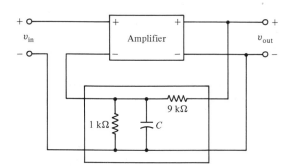

Figure P11.24

P11.25. The circuit shown in Fig. P11.25 has a main amplifier and a feedback network. The main amplifier has a gain of 500, infinite input impedance, zero output impedance, and a -3-dB point at 5 kHz (one pole only). Find C such that the poles are real and negative and differ by a factor of 2.

Answer. 8.0 μF

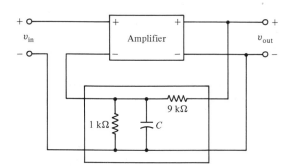

Figure P11.25

P11.26. When we borrow money and make payments at a rate $R(t)$, the principal is figured from the equation

$$P(t + \Delta t) = P(t)(1 + i\Delta t) - R(t)\Delta t$$

where $P(t + \Delta t)$ is the principal after the payment, $P(t)$ is the principal before the payment, Δt is the period of time covered by the payment (usually one month), $R(t)$ is the annual rate of payment (usually constant at 12 times the monthly payment), i is the annual rate of interest, and t is time in years.

(a) Approximate this by a differential equation and draw a block diagram describing this process in the frequency domain.

(b) If we borrow P_0 at $t = 0$, determine the debt as a function of time using the techniques in Chapter 11.

12 The Physical Basis of Electromechanics

2.1 INTRODUCTION TO ELECTROMECHANICS

Importance of electrical energy conversion. Electric motors are such an important part of modern civilization that we hardly notice them. But if you look, you see them everywhere: in your car (windshield wipers, starter motor, window lifts); in the office (electric typewriters, floor buffers); in the home (hair dryers, clocks, large appliances); and of course in the factory.

Electromechanical devices are also required in the generation of electrical power from basic energy sources and in the distribution of such power to individual users. Also important are the electrical transformer, which operates on principles similar to those of electromechanical devices, and other magnetic devices such as electrical relays, bells (doorbells, telephone bells), solenoids (electric locks, starter solenoids on cars), magnets, inductors used in electronics and fluorescent lights.

Purpose and scope of Part IV on electromechanics. Our primary purpose in Part IV is to explain the principles of the common types of electric motors and generators, and to describe their characteristics and typical applications. We also will discuss the related topics of electrical transformers, power-distribution systems, and many of the miscellaneous electromechanical devices named above. We end with a study of the electronic control of electrical power.

The electrical power industry employs a mature technology. Few engineers outside the major manufacturers will design a new electrical motor. Thus, we do not emphasize the design of electromechanical equipment; rather, we aim at imparting a physical understanding of electromechanics and the practical knowledge that would guide one to choose the right device for a specific need. But we emphasize our intellectual goal of fostering understanding. Electromechanics is an interesting area of study, in part because it centers on the interaction between electrical and mechanical systems. Electromechanics is also mathematically challenging because it employs vector, phasor, and field concepts.

Contents of this chapter. In Chapter 12 we review the electrical physics underlying electromechanics. Our emphasis falls on defining electric and magnetic fields and stating the basic laws that operate in electromechanical devices. Magnetic field concepts are emphasized because virtually all electromechanical devices utilize magnetic phenomena.

12.2 ELECTRIC FORCES AND ELECTRIC FIELDS

12.2.1 Forces between Charges

Two experimental results. Figure 12.1a shows an experiment that one can perform with a battery and two sheets of metal. The battery puts opposite charges on the sheets, and the sheets exhibit an attractive force, such that they would pull together unless restrained. The force can be explained as an attraction between the positive and negative charges on the sheets. We call the force an *electrostatic* force because it is associated with the relative location of the charges; and the energy associated with this type of force is called electrostatic energy, or more simply electric† energy. Electrostatic forces are responsible for lightning, because charges become separated through the breakup of water droplets in clouds, and electrostatic forces are used in photocopy machines to form images on a selenium surface with charged bits of dry ink. As we will see, electrostatic forces indicate the presence of an electric field.

Figure 12.1b shows another simple experiment, where currents exist in long, parallel wires. The wires exhibit a force of attraction and will move together if not restrained. This type of force requires *moving* charges; thus, such a force

Figure 12.1 Two simple experiments.

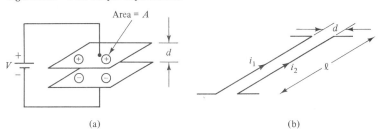

(a)   (b)

† "Electric" is an adjective appropriate to all electrical phenomena but in the present context is contrasted with "magnetic."

exists between currents. This we call magnetic force, and energy associated with this type of force we call magnetic energy. Such magnetic forces operate in electric motors and magnets and also can be used to deflect beams of electrons in TV tubes. As we will see presently, magnetic forces indicate the presence of a magnetic field.

Forces and fields.   It is convenient to think of the forces described above in terms of electric and magnetic fields. That is, rather than thinking of the forces produced directly between stationary or moving charges, we assert that one set of charges, due to its presence and motion, sets up electric and magnetic fields, and that other charges experience force through interacting with, or responding to, these fields.

12.2.2 Electric Fields

Definition of an electric field.   An electric field is defined as the vector force on a stationary charge divided by that charge; thus

$$\vec{E} = \frac{\vec{f}}{q} \quad \text{N/C} \tag{12.1}$$

where $\vec{E}$ is the vector electric field in newtons/coulomb and $\vec{f}$ is the vector force in newtons on the charge q in coulombs. A single charge produces an electric field directed outward from the charge.† We will deal mathematically with cases where the direction of the electric field is evident; hence we will minimize vector mathematics. Figure 12.2a shows the electric field pattern from an isolated positive charge; Fig. 12.2b shows the electric field pattern of a dipole, that is, an associated pair of equal but opposite charges; and Fig. 12.2c shows the electric field for a capacitor.

Figure 12.2   Electric field distributions for (a) single charge, (b) an electric dipole, and (c) a capacitor.

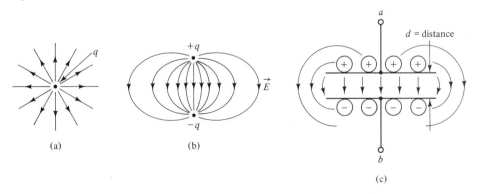

(a)   (b)   (c)

---

† The electric field is directed outward only if the charge is positive. The direction of the electric field is the direction a second positive charge would tend to move, which would be away from q if it were positive.

Dimensions of an electric field. The units of an electric field are newton/coulombs (N/C), but this is equivalent to volt/meter, as shown by the following:

$$\frac{N}{C} = \frac{N \times meter}{C \times meter} = \frac{joule}{C} \times \frac{1}{meter} = \frac{volt}{meter} \qquad (12.2)$$

because a joule/coulomb is a volt. Thus, we can interpret an electric field as a voltage per unit distance; or alternately, we can interpret a voltage as the integral of an electric field. We now investigate this interpretation.

Relationship with voltage. Recall that the voltage from point a to point b is defined as the work done by the electrical system in moving a charge from a to b, divided by the magnitude of the charge: $V_{ab} = W_{ab}/q$. Let us consider the charged capacitor of Fig. 12.1a, shown in cross section in Fig. 12.2c. Opposite charges spread over the top and bottom plates of the capacitor. The electric field is directed from the positive to the negative charges because that is the direction a positive charge would tend to move. Between the plates of the capacitor, the electric field is essentially constant in magnitude and direction, but the electric field lines outside the plates spread out, which is called *fringing*. The work done by the electrical system in moving a test charge q from a to b would be qV_{ab}. We can express this work mechanically in terms of force and displacement, with the integral of the vector dot product of the force and displacement

$$W_{ab} = \int_a^b \vec{f} \cdot \vec{d\ell} = q \int_a^b \vec{E} \cdot \vec{d\ell} \quad J \qquad (12.3)$$

where $\vec{d\ell}$ is the vector increment of distance along the path of the line integral. Two comments are required before we continue to investigate the relationship between voltage and electric field.

1. The relationship between electric field and voltage that follows from Eq. (12.3) results from defining voltage as the voltage (or potential) drop from point a to point b.
2. We use the notation of vector calculus in writing the equations describing electric (and magnetic) fields because these are required to express the relationships mathematically. For the benefit of students who have not studied this branch of mathematics, we will describe in words the mathematical relationships. We apply the equations only in situations where, due to symmetry, we do not require application of vector calculus. In the present instance, for example, we will perform the line integral only in the region where the electric field is straight, and we will follow a path aligned with the electric field. Thus the line integral becomes an ordinary integral.

The integral in Eq. (12.3) describes the summation of the electric field component times the increment in distance in the direction of the path followed. For the capacitor, we will perform the integral near the middle of the capacitor, where the electric field goes directly from the top to the bottom plate, and we will move the charge in a straight line from top to bottom. Thus the line integral reduces to the ordinary integral

$$W_{ab} = \int_a^b f\, dx = \int_a^b q\,|\vec{E}|\, dx \Rightarrow V_{ab} = \frac{W_{ab}}{q} = \int_a^b |\vec{E}|\, dx \quad \text{V} \quad (12.4)$$

where dx is the incremental displacement of q in moving from a to b. The electric field, like a gravitational field, is a conservative field, which means that the work done, and hence the voltage, is independent of the path taken from a to b. Equation (12.4) shows that the voltage between two points can be expressed as the integral of the electric field or, alternatively, that a voltage indicates the presence of an electric field.

Electric field in the capacitor. We may apply Eq. (12.4) to the capacitor shown in Fig. 12.2c to determine the electric field between the plates of the capacitor, given the voltage. In this region, the electric field is uniform, so we take a path directly from the top to the bottom plate, following the lines of the electric field, to obtain for the electric field

$$V_{ab} = \int_a^b |\vec{E}|\, dx = |\vec{E}| \int_a^b dx = |\vec{E}|\, d \Rightarrow |\vec{E}| = \frac{V_{ab}}{d} \quad \text{V/m} \quad (12.5)$$

where d is the distance between the plates. Equation (12.5) suggests the interpretation we mentioned earlier, that the electric field is a voltage per unit distance in space.

Summary. Electric fields are produced by charges and exert force on other charges. An electric field can be considered a voltage stretched over space. The electric fields of importance in electromechanics, and the associated voltages, are produced by changing magnetic fluxes.

12.2.3 Check Your Understanding

1. Give the units for an electric field.
2. An electron accelerates downward under the influence of an electric field. What is the direction of the field?

   Answers. 1. Newtons/coulomb or volts/meter; 2. upward.

12.3 MAGNETIC FORCES AND MAGNETIC FIELDS

12.3.1 Currents and Magnetic Forces

Force equation. Magnetic force is produced when one set of moving charges interacts with another set of moving charges; or more simply, magnetic forces exist between currents. In Fig. 12.1b we show two parallel wires of length ℓ separated by a distance d. The total force of attraction (F) acting on the wires is described by Ampere's force law:

$$F = \mu_o \frac{i_1 i_2 \ell}{2\pi d} \quad \text{N} \quad (12.6)$$

where μ_o is the *permeability* of free space and has a numerical value of $4\pi \times 10^{-7}$ henries/meter in the mks system. In Eq. (12.6), i_1 and i_2 are referenced in the same direction.

Our approach here, as for electrostatics, is to assert that one current produces a *magnetic field* in its vicinity, and the other current interacts with the field and experiences a force. Here we will consider the force on the wire carrying i_2 due to the magnetic field produced by i_1. Thus we will divide the problem into two parts: first the relationship between i_1 and its magnetic field, H_1; and second the interaction of i_2 with H_1 to produce a force on the wire carrying i_2.[†]

Currents and their magnetic fields. Currents produce magnetic fields just as stationary charges produce electric fields. Magnetic fields are vector fields but, unlike electrostatics, their source currents are also vectors. In many important instances the mathematical difficulties associated with vectors can be avoided because of symmetry or because the physical structures control the direction and distribution of the fields. We can gain a sufficient understanding of the nature of magnetic fields and of many important devices using magnetic fields, such as electric motors and transformers, with a minimum of vector mathematics.

Because the current produces a magnetic field, we anticipate a theory for determining the field, given the current. Such a theory exists, but is mathematically difficult to apply. There is, however, a relatively simple relationship known as *Ampere's circuital law* that allows us to compute the current, given the magnetic field; this is what we will be using, through symmetry and other means, to determine magnetic fields.

Ampere's circuital law. Unlike electric fields, which start on positive charges and end on negative charges, magnetic field lines encircle their source currents. Figures 12.3a and 12.3b show magnetic field configurations from a long straight wire and a coil of wires, respectively. Ampere's circuital law is a conservation principle that relates the integral of the magnetic field around a closed contour to the current passing through the area enclosed by the contour.

Figure 12.3 The magnetic field encircles the current.

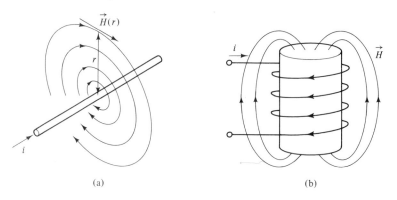

(a) (b)

[†] There will be an equal and opposite force on the wire carrying i_1.

The Physical Basis of Electromechanics   Chap. 12

$$\oint_C \vec{H} \cdot \vec{d\ell} = i \qquad (12.7)$$

where $\vec{H}$ is the magnetic field in amperes/meter, $\vec{d\ell}$ is a vector element of distance, C is the contour of integration that follows the magnetic field, and i is the current passing through the area bounded by C. This gives the picture of magnetic field lines circling currents, but always in such a way that, if you follow a magnetic field line, the spatial integral of the magnetic field strength (in A/m) around the path equals the current encircled. If the path is further away from the current, the field will be weaker because the integration path length is longer.

Magnetic field of a long straight wire. We will use Ampere's circuital law to calculate the magnetic field due to an infinitely long wire, shown in Fig. 12.3a. Here the magnetic field lines must by symmetry encompass the current in circular paths, and hence the magnitude of the magnetic field $|\vec{H}(r)|$ at a radius r from the wire, times the circumference of the path, $2\pi r$, must by Ampere's circuital law equal the current in the wire. Thus the magnetic field is given by

$$\oint \vec{H} \cdot \vec{d\ell} = |\vec{H}(r)| \oint |\vec{d\ell}| = |\vec{H}(r)| \times 2\pi r = i \qquad (12.8)$$

or

$$H(r) = |\vec{H}(r)| = \frac{i}{2\pi r} \qquad (12.9)$$

Thus the magnetic field is proportional to the current and, in this case, inversely proportional to the distance from the wire.

Magnetic forces. We return to Fig. 12.1b and the force equation in Eq. (12.6). We may now rearrange the equation into the form

$$f = \frac{F}{\ell} = \mu_o \left(\frac{i_1}{2\pi d} \right) i_2 = \mu_o H_1(d) i_2 \quad \text{N/m} \qquad (12.10)$$

where f is the force in newtons per meter. Note that the force per meter on the wire carrying i_2 depends on the magnetic field due to i_1 at the wire and the current in the wire. Later we will reformulate Eq. (12.10) in vector form and elaborate on the importance of the quantity $\mu_o H_1(d)$. For now, we have accomplished our goal of separating the force into two terms: the magnetic field produced by i_1, and the force between that field and the wire carrying i_2.

Magnetomotive force (MMF). Ampere's law can be applied to a coil, as shown in Fig. 12.3b. Here the left side of Eq. (12.7) is impossible to evaluate because we do not know the exact paths of the magnetic field lines. However, the right side of Eq. (12.7) must be ni, where n is the number of turns in the coil, and i is the current in the wire, because the magnetic field lines encircle n times the current in a single wire. The product of current and number of turns in a coil is the magnetomotive force (MMF $= ni$) and usually is given the descriptive unit ampere-turns (A-t). As we will see in Chapter 13, magnetomotive force acts as a source for magnetic systems involving coils of wire and iron structures.

12.3.2 Direction of the Magnetic Field

Right-hand rule. Ampere's circuital law, as we have stated it, gives only the magnitude of the magnetic field. To determine the direction of the field, we use the right-hand rule. Actually, there are several forms of the right-hand rule that relates the directions of current and magnetic field. In Fig. 12.3a, the convenient form is: take the wire in your right hand with your thumb in the reference direction of the conventional† current; your fingers will then encircle the wire in the direction of the magnetic field.

For a coil, as in Fig. 12.3b, the following is convenient: take the coil in your right hand with the fingers pointing in the direction of the conventional current. Then your thumb will point in the direction of the magnetic field *inside* the coil.

Magnetic poles. The direction of the magnetic field as given by the right-hand rule is consistent with the definition based on the magnetic pole concept, which came first historically. The early theory of magnetism was developed along the same lines as electrostatics through the concept of magnetic poles. A compass was said to consist of a north and a south magnetic pole, with the north pole of the compass indicating the direction of north. The direction of the magnetic field is defined as the direction a compass would indicate as northerly. Since opposites attract, it follows that the north geographic pole contains a south magnetic pole, as indicated in Fig. 12.4a. Put another way, a surface with emerging magnetic field lines contains a north magnetic pole, and a surface with entering magnetic field lines contains a south magnetic pole. Thus Fig. 12.4a shows the magnetic field lines emerging from the north magnetic pole at the south

Figure 12.4 (a) Terrestrial magnetic field; (b) a current loop has a magnetic moment; (c) magnetic dipole.

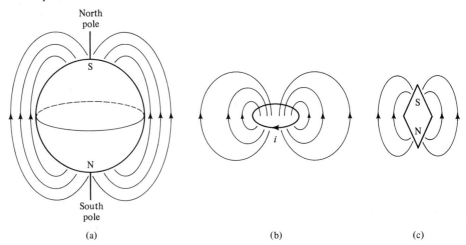

(a) (b) (c)

† The conventional current is the direction in which positive charges move. The right-hand rule gives the *reference direction* for the magnetic field. If *i* were numerically negative, the magnetic field would be numerically negative in the direction indicated by the right-hand rule.

geographic pole and disappearing into the south magnetic pole located at the north geographic pole.

A compass is analogous to an electrostatic dipole and is called a magnetic dipole. A small loop of current has a magnetic moment as shown in Fig. 12.4b, and Fig. 12.4c shows a dipole created by a pair of magnetic poles. Although isolated magnetic poles (magnetic charges) have not been detected by physicists, some basic atomic particles possess a magnetic moment.

12.3.3 Magnetic Effects in Matter

Magnetic fields interact with matter through several effects. The important effect from a practical point of view is ferromagnetism. This effect arises out of the inherent magnetic moment associated with the spin of the orbiting electrons (Fig. 12.5a). In certain materials, notably iron, the magnetic moments of the orbital electrons interact with each other to produce a region of magnetic coherence over an extended region of the material, a *magnetic domain*. A magnetic domain can be thought of as a microscopic magnet and is capable of interacting significantly with an external field. Under normal conditions, the magnetic domains are randomly oriented, as suggested by Fig. 12.5b and produce no net effect. However, under the influence of an external magnetic field, the boundaries of the domains move such that the magnetic domains in the direction of the applied field grow in size at the expense of those in other directions, and the net effect is a considerable enhancement of the magnetic flux. This effect, called *ferromagnetism*, establishes large magnetic fluxes as required for electromechanical devices.

(a) (b)

Figure 12.5 (a) Electron spin; (b) magnetic domains in iron.

12.3.4 Magnetic Flux Density

Definition. Consider the state of matter that is magnetized by an external field, such that the magnetic domains produce a net magnetic moment. We define the magnetic dipole density, $\vec{M}$, as the vector sum of all the magnetic moments over a volume of space, divided by that volume. The magnetic dipole density characterizes the effect of the magnetic field on the magnetic state of the matter. The magnetic flux density, $\vec{B}$, combines the applied and the induced magnetism, as defined in Eq. (12.11).

$$\vec{B} = \mu_o(\vec{H} + \vec{M}) \tag{12.11}$$

where $\mu_o = 4\pi \times 10^{-7}$ henries/meter in the mks system. The units of magnetic flux density are webers/m², which has been given the honorary unit tesla (T) after Nikola Tesla (1856–1943). Later we show a close relationship between these unfamiliar units and the more familiar units of volts, joules, coulombs, and henries of inductance.

Magnetic properties of iron. Because iron is important to electromechanics, we concentrate on the magnetic properties of iron. For moderate applied fields, the magnetic dipole density is proportional to the applied field, and the magnetic flux density is proportional to the magnetic field. Thus Eq. (12.12) applies.

$$\vec{M} \propto \vec{H} \Rightarrow \vec{B} = \mu\vec{H} = \mu_o\mu_r\vec{H} \tag{12.12}$$

where μ is called the *permeability* of the iron. If we think of $\vec{H}$ as the cause of the magnetization, and $\vec{B}$ as the effect, then the constant μ in Eq. (12.12) indicates the magnitude of the induced effect. The constant μ_r is called the relative permeability of the material and is essentially unity for all materials except ferromagnetic materials, notably iron, for which the value of the relative permeability lies typically in the range 1,000 to 10,000. Thus for iron, a small magnetic cause ($\vec{H}$) creates a large magnetic effect ($\vec{B}$).

Magnetic saturation and hysteresis. A more complete picture of magnetic effects in iron must consider the effects of magnetic saturation and hysteresis, as shown in Fig. 12.6. Here we have shown the effect of applying an external magnetic field to unmagnetized iron. The curve starts at the origin and increases linearly as the field magnetizes the iron. The slope of the curve in this region is $\mu = \mu_r\mu_o$, as indicated by Eq. (12.12). Eventually, however, all the magnetic domains align with the applied field, and the curve flattens out as the iron becomes magnetically saturated. If the applied field is then reduced to zero, the flux density follows a different curve because the iron tends to retain its magnetized state. The thermodynamic forces that disorient the magnetic domains in the iron do not

Figure 12.6 A magnetic hysteresis curve for iron.

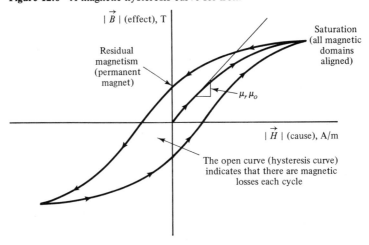

fully overcome the order imposed beforehand by the external field, and the iron retains a residual magnetism. We thus have produced a permanent magnet. As we reverse the applied field, the iron will eventually become magnetized in the reverse direction until it again saturates. If we continue to cycle the magnetic state of the iron (by applying ac current to the coil creating the applied field), the curve continues to follow an S-shaped curve, which is called a *hysteresis* curve. Later we will interpret the area enclosed by the hysteresis curve as the energy loss per unit volume per cycle. This loss will heat the iron and is one reason why electric motors and transformers get hot.

Air and iron paths for magnetic flux.   An iron path has a strong effect on the magnetic flux density, as we will illustrate for the hypothetical experiment shown in Fig. 12.7. Figure 12.7a shows a coil energized by a current. The magnetic field, $\vec{H}$, has the shape shown and conforms to Ampere's law. Because $\vec{M}$ is zero for air, the magnetic flux density is given by $\vec{B} = \mu_o \vec{H}$ and is relatively small. An identical coil wound around an iron core, however, would produce a rather different flux pattern. The magnetic field, $\vec{H}$, is distorted by the presence of the iron. The magnetic fields in the region around the coil have roughly the same magnitude as before, because Ampere's law must still be satisfied. Within the iron, however, the magnetic flux density, B, is much larger than before (by the magnitude of μ_r, say, 5000). The result is that a large magnetic flux follows the iron path.

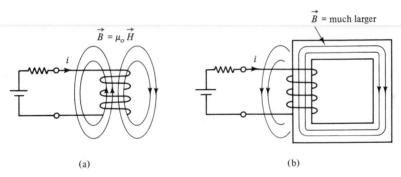

Figure 12.7   (a) A coil in air; (b) same coil, iron path for flux.

Interaction with current.   We return now to the original experiment in Fig. 12.1b that we used to introduce magnetic forces. The force given in Eq. (12.6) has a direction in space. A general form of the force equation that gives the direction is the vector cross product of current and flux density,

$$\vec{f}_2 = \vec{i}_2 \times \vec{B}_1 \tag{12.13}$$

where $\vec{f}_2$ is the vector force in newtons per meter on the wire carrying $\vec{i}_2$ and $\vec{B}_1$ is the vector flux density due to $\vec{i}_1$. Equation (12.13) is a form of Ampere's force law.

We also have a right-hand rule relating current, magnetic flux density, and force from Eq. (12.13). In this case, point the fingers of your right hand in the direction of the current; then bend them in the direction of the magnetic flux.

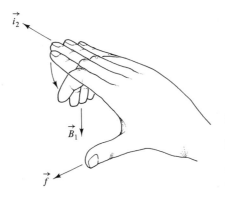

Figure 12.8 (a) For the direction of $\vec{i}_2 \times \vec{B}_1$, put the fingers of the right hand in the direction of $\vec{i}_2$; then swing to the direction of $\vec{B}_1$. The thumb points in the direction of force.

Your thumb will indicate the direction of the force as shown in Fig. 12.8. This is, of course, consistent with the definition of a vector product in a right-hand system, as Eq. (12.13) requires.

Let us work out the directions of the flux and force for the case shown in Fig. 12.1b using the right-hand rule of Sec. 12.3.2. Putting the thumb of the right hand in the direction of i_1, we determine from the right-hand rule that B_1 is downward at the position of the wire carrying i_2 and has a magnitude from Eqs. (12.9) and (12.12) ($\mu = \mu_o$) of $\mu_o i_1 / 2\pi d$. Crossing $\vec{i}_2$ into $\vec{B}_1$, we see that the force on the wire carrying i_2 is directed toward the wire carrying i_1. Similarly, we could confirm a force of repulsion between parallel wires carrying currents in opposite directions.

Summary. Iron enhances greatly the magnetic flux and controls its distribution in space. Large magnetic flux densities produce correspondingly large forces on conductors carrying currents. Thus electromechanical devices normally use magnetic structures made out of iron.

12.3.5 Magnetic Flux (Φ) and Magnetic Flux Linkages (λ)

Magnetic flux density is an important quantity because magnetic flux, of which $\vec{B}$ gives the spatial distribution, plays an important role in electromechanical energy conversion. In this section we will discuss the properties of magnetic flux and then begin our examination of the role played by magnetic flux in electromechanical energy conversion. Subsequent chapters apply these principles in a variety of practical devices.

Properties of magnetic flux. Magnetic flux lines encircle the currents that generate them. Thus magnetic flux lines exist only in closed loops. This means that magnetic flux is conserved in any closed spatial region: what goes in must come out. This is expressed by

$$\oint_S \vec{B} \cdot \hat{n} \, da = 0 \tag{12.14}$$

where S is a closed surface, $\hat{n}$ is a unit vector at right angles to the surface, and da is a differential element of area. Magnetic flux lines behave therefore like the flow lines of an incompressible fluid. It follows that the amount of magnetic flux passing through a loop of wire having an area (A) enclosed by the contour of the wire (C) is a uniquely defined quantity, as pictured in Figure 12.9.

$$\Phi = \int_A \vec{B} \cdot \vec{n} \, da \qquad (12.15)$$

Equation (12.15) gives a mathematical expression for calculating the magnetic flux, Φ, passing through such an area. The magnetic flux, a scalar quantity, has the units of webers, after Wilhelm Weber (1804–1891).

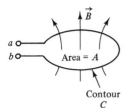

Figure 12.9 A one-turn coil.

Flux linkage. For a coil of wire, the number of turns in the coil becomes important. The flux passing through the coil is the product of the number of turns (n) and the flux passing through a single turn (Φ). This product is called the magnetic flux linkage of the coil, λ:

$$\lambda = n\Phi \qquad (12.16)$$

The definition of flux linkage in Eq. (12.16) assumes that every turn of the coil has the same flux passing through it. Faraday's law, to be considered in the next section, shows that the concept of flux linkage, broadly construed, is not founded on this assumption.

12.3.6 Check Your Understanding

1. Give the units for a magnetic field.
2. A wire is stretched around the equator, closed into a loop, and excited by a current traveling westward. Does the magnetic field from the current aid or oppose the earth's magnetic field above the surface of the earth?
3. At the south geographic pole, which direction (N, E, S, W, up, or down) would a magnetic compass indicate as "north"?
4. What would be the value of the integral of the magnetic field around the electric cord of a lighted 120-V, 60-W electric light?
5. Find the magnetic flux density at the surface of a No. 12 wire (0.00808-in. diameter) carrying 20-A dc current.
6. A cosmic ray with a positive charge would tend to be deflected in what direction by the earth's magnetic field?

Answers. 1. Amperes/meter; 2. aid; 3. up; 4. zero because the currents in the two wires add to zero; 5. 0.0395 Wb/m²; 6. eastward.

12.4 DYNAMIC MAGNETIC SYSTEMS

12.4.1 Faraday's Law

In 1831, Michael Faraday published his law of electromagnetic induction relating voltage to time-varying magnetic flux:

$$v_t = \pm \frac{d\Phi}{dt} \tag{12.17}$$

where v_t is the voltage per turn induced by a time-varying magnetic flux, Φ, through a coil. We will show below how the polarity of the voltage is determined through Lenz's law. A coil having n turns constitutes a series connection of the individual turns; hence, the total induced voltage would be

$$v = nv_t = \pm n \frac{d\Phi}{dt} = \pm \frac{d\lambda}{dt} \tag{12.18}$$

Equation (12.18) shows that the induced voltage is the time derivative of the flux linkages of a coil. The last form of the equation is valid, and indeed serves as a definition of the flux linkage, when all the magnetic flux does not pass through every turn of a coil. For example, if we crumpled a length of wire into a random tangle, we would not have a coil with well-defined turns, but the flux linkages would still be a meaningful quantity and would still be related through Eq. (12.18) to the induced voltage. In other words, the tangle of wire would still have an inductance. We investigate the relationship between flux linkage and inductance after we consider the sign of the induced voltage.

12.4.2 Lenz's Law

Consider the situation shown in Fig. 12.9 and assume that the magnetic flux is changing with time. According to Faraday's law, a voltage v_{ab} will be induced. We can determine the sign of the induced voltage through Lenz's law, which states that, if the loop were closed (a connected to b), the current would flow in the direction to produce a flux opposing the original flux *change*. For example, let us say that the flux is increasing in the coil in Fig. 12.9. If we connected a resistor between a and b, then current would flow to generate a reaction magnetic flux opposing the increase; in this case, the reaction flux would be downward. This means that current will flow *through the external connection* from b to a. Thus the voltage v_{ba} will be positive; that is, the external circuit will see b as $+$ and a as $-$ because current flows out of the $+$ and into the $-$ *from the viewpoint of the external connection*. Thus, if $\vec{B}$ were increasing, v_{ab} would be numerically negative.

This reasoning is based on the fact that *for a source,* the current flows out of the $+$ terminal, as shown in Fig. 12.10. The voltage described by Faraday's law acts as a source for the external circuit; thus the $+$ polarity marking should be put at the terminal where the current flows out. Of course, if the coil is open-circuited, no current will flow, but the $+$ voltage is produced nevertheless.

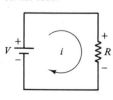

Figure 12.10 The current flows *out of* the $+$ on the source (battery) and *into* the $+$ on the load.

If the magnetic flux in Fig. 12.9 were *decreasing* and *a* and *b* were connected, the current would flow through the circuit so as to oppose the decrease of the flux; that is, it would generate a flux in the upward direction. For this case, *a* would be + and *b* would be −, since the current would flow from *a* to *b through the external connection*, and v_{ab} would be numerically positive.

12.4.3 Definition of Inductance

Figure 12.11
Reference directions for an inductor.

The circuit-theory concept of inductance is related to the flux linkages through Faraday's law. We have but to compare Faraday's law in Eq. (12.18) with the circuit-theory definition of inductance in Eq. 3.1, (see Fig. 12.11):

$$v = \frac{d}{dt}(Li) \qquad (12.19)$$

to identify

$$\lambda = Li \qquad (12.20)$$

Four comments follow:

1. Equation (12.20) is the definition of inductance. To calculate the inductance of a coil, assume a current, determine the flux linkages, and use Eq. (12.20).
2. Although we normally write the definition of inductance with the inductance treated as a constant, and hence bring *L* outside the derivative in Eq. (12.19), the form shown above is required when the inductance can change with time, as would be the case, for example, if the coil were part of a moving system. Thus Eq. (12.19) is the more general form relating voltage and current in an inductor.
3. The ambiguity of sign in Eq. (12.18), which may be resolved through application of Lenz's law, does not exist in the circuit-theory case because we have a standard sign convention. We use a load set convention for inductors, which fixes the relationship between the voltage polarity and the current reference direction (and hence the direction of the magnetic flux). We leave for a homework problem the demonstration that Lenz's law is compatible with the standard polarity convention for an inductance.
4. In a linear magnetic system, the flux linkages are proportional to the current, and hence the inductance defined in Eq. (12.20) is not a function of the current or the flux linkage but depends only on the geometry of the coil and the magnetic properties of the surroundings. However, if iron is involved in the inductor, then the inductance can be a function of the current because the *B–H* (Fig. 12.6) curve is not linear.

12.4.4 Stored Energy in a Magnetic System

Consider a lossless magnetic system. We may compute the energy stored in the system by integrating the input power, $p = vi$. The incremental energy input is then

$$dW_m = p \, dt = vi \, dt = \frac{d\lambda}{dt} i \, dt = i \, d\lambda \qquad (12.21)$$

where dW_m is the incremental increase in the stored magnetic energy in time dt. The total stored energy is thus

$$W_m = \int_0^{\lambda_f} i \, d\lambda \qquad (12.22)$$

where λ_f is the final value of the flux linkage. The integral in Eq. (12.22) corresponds to the shaded area in Fig. 12.12.

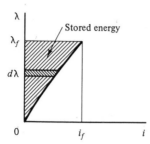

Figure 12.12 The stored energy corresponds to the shaded area.

Linear magnetic systems. When the magnetic system is linear, the inductance, $L = \lambda/i$, and Eq. (12.22) can be evaluated as shown in Eq. (12.23).

$$W_m = \int_0^{\lambda_f} \frac{\lambda}{L} \, d\lambda = \frac{\lambda_f^2}{2L} = \frac{1}{2} Li_f^2 \qquad (12.23)$$

where i_f is the final value of the current. Thus for a linear inductor the stored energy can be related to the final state of the current or flux linkage and to the inductance, which characterizes the geometry of the system.

Since this chapter explores the physical basis of electromechanics, we are entitled to ask where in physical space the magnetic energy is stored. This question will be addressed in the next chapter, where we will analyze some specific magnetic structures.

Relating λ and i to $\vec{B}$ and $\vec{H}$. Because the flux linkage, λ, is proportional to the magnetic flux density, $\vec{B}$, and the magnetic field, $\vec{H}$, is proportional to the current, it follows that the λ–i graph, as shown in Fig. 12.12, and the B–H plot, as exemplified by Fig. 12.6, would be similar; that is, the two would be scaled versions of each other. This proportionality has several implications:

- Since the shaded area in Fig. 12.12 represents the stored energy, the corresponding area in the B–H curve must relate to energy. Indeed, we can show that the dimension of area in the B–H diagram is energy per unit volume, J/m³ in the mks system. Thus we would logically expect that the area in the B–H diagram corresponding to the shaded area in Fig. 12.12 would represent the stored energy per unit volume in the magnetic system.

$$w_m = \int_0^{B_f} H \, dB \quad \text{J/m}^3 \tag{12.24}$$

where w_m is the energy density, and H and B are the magnitude of the magnetic field and flux density, respectively. The total energy in the system may be computed from the integral of the energy density over the entire volume of the magnetic system.

$$W_m = \int_{\text{volume}} w_m \, dv \quad \text{J} \tag{12.25}$$

when dv is a differential volume element. Such integrals as Eq. (12.25) are difficult to perform except in certain magnetic structures.

- If the magnetic system contains iron, the λ–i curve will possibly be nonlinear and exhibit hysteresis, as shown in Fig. 12.6. When there is hysteresis, the magnetic system returns less energy to the external system than it received when energized. Thus there is magnetic loss resulting physically from the reorientation of the magnetic domains by the external source. The loss per unit volume is the area inside the hysteresis curve in the B–H diagram. The energy lost to the electrical system through hysteresis loss appears as thermal energy in the iron.

- Equations (12.24) and (12.25) give us our first clue to the question raised earlier: where in space is the energy stored? Although we will not make direct use of them in our discussion of this question in the next chapter, these equations underlie our final conclusion.

12.4.5 Moving Conductors

Faraday's law for moving conductors. Faraday's law in Eq. (12.18) is valid for coils in which the wire is moving through the magnetic field, provided the rate of flux change includes the effect of the motion. An alternate approach is to separate the voltage induced by a changing flux from the voltage induced in a moving conductor.

The calculation of the voltage induced in a moving conductor may be complicated, but the case of a straight wire in orthogonal motion through a uniform magnetic flux field is easily stated:

$$e = \ell \, | \, \vec{u} \times \vec{B} \, | = \ell \, | \, \vec{u} \, | \, | \, \vec{B} \, | \sin \theta \quad \text{V} \tag{12.26}$$

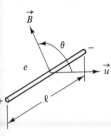

Figure 12.13 A conductor moving through magnetic flux generates a voltage called the electromotive force (emf).

where ℓ is the length of the wire, $\vec{u}$ is its velocity, $\vec{B}$ is the uniform flux density, and θ is the angle between the motion and the flux direction. The voltage e is called the electromotive force (emf). If the two ends of the bar were connected through a resistor, current would flow in the direction of $\vec{u} \times \vec{B}$, out of the end marked $+$ and into the end marked $-$ in Fig. 12.13. Thus the emf is positive in the direction indicated by the vector product $\vec{u} \times \vec{B}$.

The generation of a voltage by a magnetic flux, whether through time rate of change or through motion of a conductor, indicates the presence of an electric field in the wire. This electric field integrated through the conductor path produces the voltage (emf).

Sec. 12.4 Dynamic Magnetic Systems

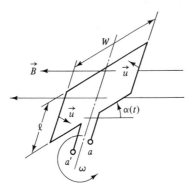

Figure 12.14 A rectangular loop of wire turning in a magnetic flux density will generate an ac voltage.

The rectangular loop of wire shown in Fig. 12.14 illustrates Faraday's law. The loop has a length of ℓ, a width W, and is turning with an angular velocity ω. The plane of the coil makes an angle $\alpha(t)$ relative to the direction of the magnetic flux density. The emf is produced in the parts of the coil that are parallel to the axis of rotation; hence the linear velocity of the wire is $|\vec{u}| = \omega W/2$. The angle between $\vec{u}$ and $\vec{B}$ is $90° - \alpha(t)$; hence the emf from one wire is

$$e = \ell |\vec{u}||\vec{B}| \sin[90° - \alpha(t)] = \ell\omega \frac{W}{2} |\vec{B}| \sin(90° - \alpha(t))$$

$$= \frac{\omega\Phi_m}{2} \cos \alpha(t) \tag{12.27}$$

where $\Phi_m = |\vec{B}| W\ell$ is the maximum flux in the coil [when $\alpha(t) = 90°$]. An equal voltage is induced in the other wire. The voltages are in series, so the total motion-induced voltage is

$$e_{aa'} = \omega\Phi_m \cos \alpha(t) = \omega\Phi_m \cos \omega t \tag{12.28}$$

since $\alpha(t) = \omega t$.

This result can be derived directly from Faraday's law. The flux through the coil is

$$\Phi(t) = \Phi_m \sin \alpha(t) = \Phi_m \sin \omega t \tag{12.29}$$

The emf by Faraday's law is

$$e_{aa'} = \pm \frac{d\Phi(t)}{dt} = \pm\omega\Phi_m \cos \omega t \tag{12.30}$$

The sign determined from Lenz's law is the same for both methods.

12.4.6 Electromechanical Energy Conversion

Introduction. We have now stated the physical laws underlying electromechanical energy conversion. We have asserted that electrical currents in the presence of magnetic flux produce mechanical force on wires. We have also asserted that mechanical motion of a conductor through a magnetic flux produces an electrical voltage. In this section we will show how these effects combine to

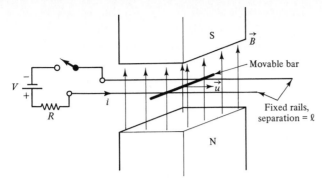

Figure 12.15 A simple transducer.

transform energy from electrical form (voltage and current) to mechanical form (force and velocity), and vice versa.

Electromechanical transducer. We will investigate the energy conversion process for the transducer shown in Fig. 12.15. We show a magnetic flux furnished by an external magnet. The electrical circuit consists of a battery (V), a resistor (R), two stationary rails, and a movable bar that can roll or slide along the rails with electrical contact. We will consider that the bar is stationary and has an active length ℓ between the rails. We close the switch and consider the sequence of effects that follow:

1. Current does not start immediately due to the inductance of the circuit (it is a one-turn coil). The time constant of L/R is very small, however.
2. The current quickly reaches the value V/R.
3. A force will be exerted on the bar due to the interaction between the current and magnetic flux. The magnitude of this force, by Eq. (12.13), is $F = iB\ell$ to the right, and the bar will begin to move with a velocity u. The instantaneous mechanical power output is Fu. This energy comes from the electrical circuit, ultimately from the battery, and would accelerate the bar or do work against an external force, or both.
4. The motion of the bar will produce an electromotive force. This is the condition pictured in Fig. 12.13 with $\theta = 90°$; thus, by Eq. (12.26), $e = uB\ell$. The polarity of the emf is positive where the current enters the moving bar. The equivalent circuit, shown in Fig. 12.16, models the electrical aspects of the system. The moving bar generates a "back" emf that opposes the imposed voltage and reduces the current.

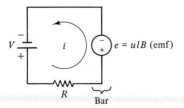

Figure 12.16 The moving bar presents a back emf to the circuit.

5. The instantaneous electrical power into the bar will be ei, or $uB\ell i$, and the instantaneous mechanical power output of the bar would be Fu or $iB\ell u$. The electrical input and the mechanical output powers are equal.

6. The bar has an equilibrium speed of $u = V/\ell B$ at which the current, and hence generated force, is zero. If moved faster by an external mechanical force, the bar becomes a generator, reverses the current direction, and supplies electrical power to the resistor and battery. In this circumstance, we could easily show that the input mechanical power is equal to the electrical output power supplied by the bar to the resistor and battery.

We draw four conclusions:

1. The electrical input power to the bar (ignoring resistive losses) is transformed to mechanical output power with 100% efficiency.

2. The energy lost to the electrical circuit via the emf appears as mechanical energy. The emf voltage source represents electrically, therefore, the coupling between electrical and mechanical systems.

3. This structure is a two-way transducer between the electrical and mechanical systems. The bar may represent an electrical load or source, depending on its velocity.

4. Magnetic flux, though an agent in the electromechanical coupling, does not participate directly in the energy exchange.

Whether motor or generator, the bar exchanges energy between electrical and mechanical systems with 100% efficiency. We have ignored mechanical losses such as friction and electrical losses associated with the resistance of the bar, but these are easily modeled in their respective systems.

Summary.   Through an analysis of a linear transducer we have established several important concepts. We have shown that currents in wires moving through magnetic flux effect electromechanical energy conversion. The moving conductor generates an electromotive force. If the current flows against the electromotive force, electrical energy is converted to mechanical energy; hence, the system acts as a motor. If the current is driven by the electromotive force, mechanical energy is converted to electrical energy, and the system acts as a generator. In both cases, an emf voltage source in the equivalent circuit represents the exchange of energy between the electrical and mechanical systems.

12.4.7 Chapter Summary

This chapter reviews the basic laws underlying electromechanical devices. In adopting a modest mathematical level, we have presented the physics in a piecemeal manner that obscures the unity and elegance of the subject. In some cases we have given only a special case of a physical law or treated a vector equation as if it were a scalar equation. Many details and subtleties were omitted. Our goal was to introduce the important concepts and equations that are used in the following chapters, not to give a full exposition of electrical physics.

The definition of electric field is important because electric fields and voltages go together. The section on dynamic magnetic systems is important because Faraday's law, stored magnetic energy, and electromotive forces generated by moving conductors are the heart of electromechanics. The analysis of a transducer, which can act as either motor or generator, introduces the main ideas that will be explored in the next four chapters.

12.4.8 Check Your Understanding

1. Figure 12.17a shows a one-turn coil. The magnetic flux is decreasing. Mark the physical polarity (+ and −) on the coil terminals according to Lenz's law.
2. For the rotating coil shown in Fig. 12.17b, determine the numerical sign of v_{ab} at the moment and angle shown.
3. What are the units of $\vec{u} \times \vec{B}$ in terms of volts, amperes, meters, and/or seconds? (All are not required.)
4. The deflection of the electron beam in a TV tube is accomplished with magnetic flux. If the TV set is facing south and the beam is deflected west, what is the direction of the magnetic flux established by the deflection coils?

Answers. 1. b is $+$; 2. v_{ab} is negative; 3. volts/meter; 4. down.

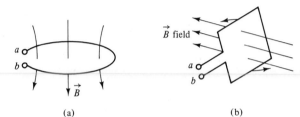

(a) (b)

Figure 12.17 (a) A one-turn coil; (b) rotating coil.

PROBLEMS

Section 12.3: Magnetic Forces and Magnetic Fields

P12.1. An infinite straight wire carries 10 A dc, as shown in Fig. P12.1. A nearby rectangular loop carries 5 A dc.
   (a) Find the flux passing through the rectangular loop due to the 10-A current. Count downward as positive.
   (b) Find the total force on the rectangular loop. Give the direction of the force relative to the infinite wire. *Hint:* The forces on the radial sides of the loop cancel by symmetry.

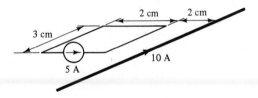

Figure P12.1

P12.2. An auto battery is being charged with a current of 35 A. *Estimate* the magnetic flux density at the top of the battery, neglecting any effect due to the iron in the vicinity. Draw a picture showing the battery and the direction of the flux.

    Answer. $(2.5 - 6) \times 10^{-5}$ T

P12.3. Two wires, infinite in the y-direction, are located at $x = 0$ and $x = 1$ cm, as shown in Fig. P12.3. The wires carry dc currents of 4 A and 2 A, as shown.
    (a) Find the magnetic field at $x = 0.4$ cm. Give direction and magnitude.
    (b) What is the force per meter on the wire labeled b? The field required in the $\vec{i} \times \mu_o \vec{H}$ is that due to the current in wire a.
    (c) If the magnetic field were integrated in a circular path at radius $r = 10$ meters from the y-axis, what would be the magnitude of the result?

P12.4. For the two wires shown in Fig. P12.3:
    (a) At what point on the x-axis between the wires is the flux density a minimum?
    (b) At what noninfinite point on the x-axis is the flux density zero?

    Answers. (a) 5.86 mm; (b) 2 cm

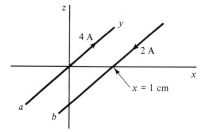

                                            Figure P12.3

P12.5. Figure P12.5 shows two infinite wires aligned with the x- and y-axes. The wires carry 1 A dc and 2 A dc, as shown.
    (a) Where in the x–y plane is the magnetic field zero?
    (b) In what regions of the x–y plane is the magnetic flux density out of the paper (in the positive z-direction)?
    (c) Calculate the torque on the conductor carrying 1 A due to the field from the conductor carrying 2 A. Calculate either total torque or torque per meter, whichever is appropriate.

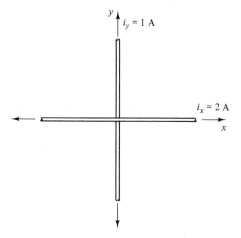

                                            Figure P12.5

Section 12.4: Dynamic Magnetic Systems

P12.6. For the circuit and coil in Fig. P12.6:
   (a) Draw the magnetic flux pattern.
   (b) Would a compass needle inside the coil indicate north to be up or down?
   (c) If the inductance of the coil were $L = 0.02$ H, find its magnetic flux linkages. Note that the wire in the coil and the rest of the circuit has a resistance of 1 Ω, in addition to the 4 Ω in the lumped resistance.
   (d) As the battery weakens and the current decreases, an induced voltage would appear across the coil. Would the + of the induced voltage be at the top or bottom of the coil?

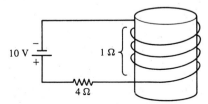

Figure P12.6

P12.7. The earth's magnetic field at the equator has a strength of about 30 μT. Consider an aircraft flying east over the equator at 600 mph. The wing span is 100 ft and the fuselage 15 ft in diameter.
   (a) To what part of the aircraft would the conduction electrons tend to move?
   (b) What would be the maximum emf created by the aircraft's motion?

P12.8. Figure P12.8 shows a lossless, nonlinear inductor that is energized with a battery. After the switch is closed, the current is observed to increase as $i(t) = 0.05t^{1.2}$ A. Determine the flux linkages as a function of the current $\lambda(i)$.

Answer. $\lambda(i) = 121i^{+0.833}$

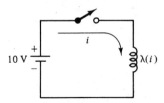

Figure P12.8

P12.9. Figure P12.9 shows a nonlinear, lossless inductor with a flux linkage relationship $\lambda(i) = 5\sqrt{i}$. The inductor is energized by a battery through a switch that is closed at $t = 0$.
   (a) Determine the time after switch closure when the power into the inductor is 12 W.
   (b) At what time is the stored energy 100 J?

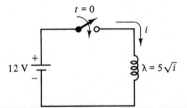

Figure P12.9

P12.10. Two parallel infinite wires, each carrying a current i in the same direction, set up a magnetic field. The wires are a distance d apart, as shown in cross section in Fig. P12.10. A conducting bar of length ℓ moves upward midway between the wires at a velocity $\vec{u}$. Derive a formula for the emf generated in the bar as a function of time, counting $t = 0$ when the bar passes the plane of the wires.

Answer. $\mu_o \ell i u^2 t / \pi[(ut)^2 + (d/2)^2]$

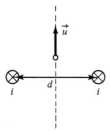

Figure P12.10

P12.11. Figure P12.11 shows a permanent magnet that has a loop of wire around one of its poles.
(a) Draw the magnetic flux pattern, including the direction of the flux.
(b) As the loop of wire is lifted vertically, is the voltage v_{ab} positive or negative?

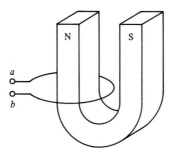

Figure P12.11

P12.12. An infinite wire carrying a current i puts flux through a rectangular loop, as shown in Fig. P12.12. Note that the area of integration is $w\,dr$, where w is the width of the loop and r is the distance from the wire.
(a) Determine the flux in the loop for $w = 10$ cm and $i = 1$ A.
(b) Determine the voltage v_{ab} if $di/dt = 100$ A/s.

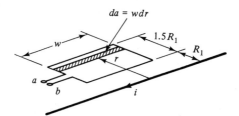

Figure P12.12

P12.13. Figure P12.13 shows a *homopolar* generator, which consists of a conductor rotated in the presence of a magnetic flux. In this case, sliding contacts are made at the outer radius, r_o, and the radius of the axis, r_i. Determine the voltage v_{ab} as a function of the magnetic flux density, B, the angular velocity, ω, r_i, and r_o.

The Physical Basis of Electromechanics   Chap. 12

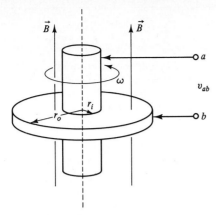

Figure P12.13

P12.14. The relationship between the flux linkages and current for a magnetic system is plotted in Fig. P12.14 and is approximated by the equation $\lambda = 2 \ln(i + 1)$.
(a) What is the inductance at $i = 0.2$ A?
(b) Calculate the stored energy for $i = 0.4$ A.

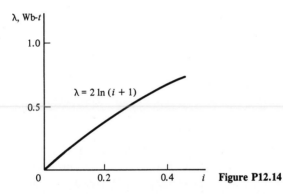

Figure P12.14

P12.15. Figure P12.15 shows the magnetic flux linkages of a coil as a function of the current. The curve is nonlinear due to saturation of the iron in the system.
(a) What is the inductance of the coil for currents below 2 A?
(b) If the current starts at zero and increases at a rate of 100 A/s, determine the resulting voltage as a function of time during the first 40 ms. Plot the voltage during this time period.
(c) Determine the stored energy of the system for a current of 4 A. Assume that the stored energy is a unique function of the current.

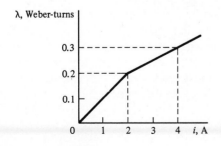

Figure P12.15

13 Magnetic Structures and Electrical Transformers

13.1 INTRODUCTION TO MAGNETIC STRUCTURES

Magnetic structures. A magnetic structure† is an iron structure that increases the amount of magnetic flux and controls its distribution in space. Figure 13.1 shows several types of magnetic structures. The transformer in Fig. 13.1a changes electrical voltage, current, and impedance levels. Transformers find application in every area of electrical technology, but most notably in electrical power-distribution systems. Relays, Fig. 13.1b, are electrical actuators used in switches, locks, and the like. In electric motors, Fig. 13.1c, magnetic structures are vital to controlling the magnetic fluxes that produce torque and motion. Transducers, such as the loudspeaker in Fig. 13.1d, form yet another class of magnetic structures. Other transducers are phono pickups, tape heads, and tachometers. Finally, we should mention the ordinary inductors used as circuit elements, many of which require magnetic structures.

Although we stated above that magnetic structures are made out of iron, some use other magnetic materials. Materials research has produced ferrite ceramic materials with excellent magnetic properties, and a special class of electric motors, not to mention other devices, employ such materials in their magnetic structures. However, most magnetic structures are made of iron because of its low cost and excellent magnetic properties.

† Also called magnetic circuit.

514

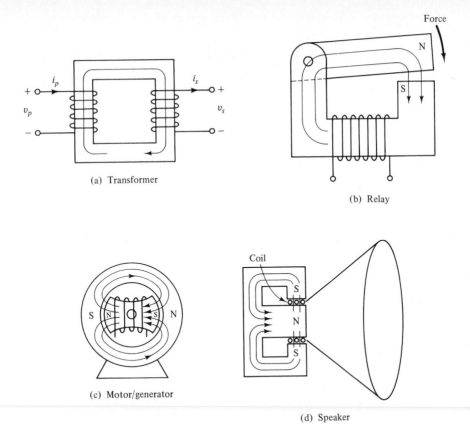

(a) Transformer

(b) Relay

(c) Motor/generator

(d) Speaker

Figure 13.1 Typical magnetic structures.

Contents of Chapter 13. First we will demonstrate methods of analysis for magnetic structures, continuing the themes of Chapter 12. We then apply this analysis to transformers, beginning from the definition of an ideal transformer and then developing models to describe real transformers. The transformer section concludes with a discussion of transformer applications in three-phase systems. Finally, we study forces generated by magnetic structures, using conservation of energy as a basis for calculating force and torque in electromechanical systems.

3.2 ANALYSIS OF MAGNETIC STRUCTURES

13.2.1 Toroidal Ring

Figure 13.2 shows the magnetic structure we will analyze, a toroidal iron ring with a gap. The magnetomotive force (mmf) is created by a coil of wire carrying dc current. Our goal is to determine the magnetic field, flux density, and flux in the toroid and air gap.

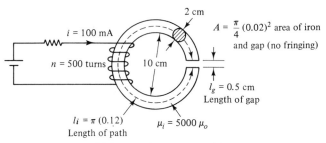

Figure 13.2 A toroidal magnetic structure with a gap.

Gaps in magnetic structures. You may wonder why we have included a gap in the iron ring, since its effect will reduce the flux. We include the gap because many electromechanical devices require such a gap. Look back, for example, at Fig. 13.1. Note that the relay, motor, and loudspeaker must have gaps. The magnetic flux in such devices couples the electrical and mechanical systems, and the gap is required to permit motion.

Assumptions. Our analysis of the iron ring is based on several assumptions. We assume that the relative permeability of the iron is large and constant. In other words, we neglect magnetic saturation and hysteresis. Second, we assume that the significant magnetic flux remains in the iron except at the gap, where the flux must pass through air. Third, we assume that the flux density is constant in magnitude, meaning that the flux density does not vary across the cross section of the toroid and does not vary around the ring. Finally, we assume that fringing in the gap is negligible. This means that the flux density of the gap does not vary over its cross section and is circumferential, as in the iron. These assumptions are reasonable in view of the discussion of the effect of the iron structure, as discussed on page 499.

Analysis. We begin with Ampere's circuital law:

$$\oint \vec{H} \cdot \vec{d\ell} = \int_{\text{iron}} \vec{H}_i \cdot \vec{d\ell} + \int_{\text{gap}} \vec{H}_g \cdot \vec{d\ell}$$

$$= |\vec{H}_i| \, \ell_i + |\vec{H}_g| \, \ell_g = ni \quad \text{A-t} \quad (13.1)$$

where $\vec{H}_i$ and $\vec{H}_g$ are the magnetic fields within the iron and gap, respectively, and ℓ_i and ℓ_g are the circumferential distances within the iron and gap. The path of integration is the center of the toroid. The integral in Ampere's circuital law has contributions from both iron and gap. In each medium the magnetic field is constant by assumption, leading to the last form of Eq. (13.1). Because the field is constant along the path, the field can be brought outside the integrals. The magnetic fluxes in the iron and gap are also easily derived:

$$\Phi_i = \int \vec{B}_i \cdot \hat{n} \, da = |\vec{B}_i| \, A_i \quad \text{and} \quad \Phi_g = |\vec{B}_g| \, A_g \quad (13.2)$$

where $\hat{n}$ is a unit circumferential vector, $\vec{B}_i$ and $\vec{B}_g$ are the magnetic flux densities, A_i and A_g the cross-section areas, and Φ_i and Φ_g the fluxes in the iron and gap, respectively. Conservation of flux requires equal fluxes:

$$\Phi_i = \Phi_g = \Phi \qquad (13.3)$$

where Φ is the magnetic flux in the magnetic structure. The magnetic fields and flux densities are related by the permeabilities of the iron (μ_i) and gap (μ_o):

$$H_i = \frac{B_i}{\mu_i} \quad \text{and} \quad H_g = \frac{B_g}{\mu_o} \qquad (13.4)$$

where the magnetic fields (H_i and H_g) and the magnetic flux densities (B_i and B_g) are here and hereafter the magnitudes of the vector fields. Because the direction of these vector fields is constrained by the magnetic structure, we can treat them as scalar quantities.

Analytical result. Combining Eqs. (13.1) through (13.4), we can eliminate all unknown quantities except the flux, with the result

$$\Phi = \frac{ni}{(\ell_i/\mu_i A_i) + (\ell_g/\mu_o A_g)} \quad \text{Wb} \qquad (13.5)$$

Magnetomotive force, reluctance, and flux. Equation (13.5) can be cast into the form

$$\Phi = \frac{ni}{\mathcal{R}} = \frac{\text{mmf}}{\text{reluctance}} \qquad (13.6)$$

The mmf is the cause and the flux is the effect in the magnetic structure; the reluctance describes the geometry and magnetic properties of the magnetic structure.

Numerical results. When we substitute the dimensions of the structure into Fig. 13.2 and assume that the relative permeability of the iron is 5000,† we obtain the numerical result

$$\Phi = \frac{500(0.1)}{1.91 \times 10^5 + 1.27 \times 10^7} = 3.89 \times 10^{-6} \quad \text{Wb} \qquad (13.7)$$

We can now determine the values of the various field quantities

$$B_i = B_g = \frac{\Phi}{A} = \frac{3.89 \times 10^{-6}}{3.14 \times 10^{-4}} = 1.24 \times 10^{-2} \quad \text{T}$$

$$H_i = \frac{B_i}{\mu_i} = \frac{1.24 \times 10^{-2}}{5000(4\pi \times 10^{-7})} = 1.97 \quad \text{A/m} \qquad (13.8)$$

$$H_g = \frac{B_g}{\mu_o} = 9851 \quad \text{A/m}$$

As final matters, we can determine the flux linkages, $\lambda = n\Phi = 1.95 \times 10^{-3}$ weber-turns, and the inductance of the structure, $L = \lambda/i = 19.4$ mH. In the next

† That is, $\mu_i = 5000\mu_o$.

section we discuss and generalize this analysis, using the numerical calculations to justify certain approximations.

13.2.2 Principles of Analysis of Magnetic Structures

Magnetomotive force. Examination of the results of the previous section indicates that the magnetic flux depends on several factors. Notice first the role of the magnetomotive force (mmf $= ni$). The formal units of mmf are amperes, but ampere-turns (A-t) is commonly used to distinguish mmf from current. Magnetic systems are energized by current-carrying coils, and the mmf acts as the magnetic source for the system. The "polarity" of the mmf is given by the right-hand rule. If there were more coils contributing to the flux, the mmfs would add or subtract according to this polarity. Equation (13.1) can be written in the form

$$ni - H_i\ell_i - H_g\ell_g = 0 \qquad (13.9)$$

which suggests that mmf is gained in the coil and lost to the path around the toroid. For example, the mmf lost to (or required to magnetize) the gap is $H_g\ell_g$. In fact, the magnetic field can be considered as the loss in mmf per unit length along the field lines.

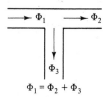

$$\Phi_1 = \Phi_2 + \Phi_3$$

Figure 13.3
Conservation of magnetic flux at a junction.

Conservation of magnetic flux. Another important principle, although it applied trivially in our example, is conservation of magnetic flux. Had our magnetic structure contained branches and parallel paths, such as suggested in Fig. 13.3, we would have required conservation of magnetic flux at each such junction.

Reluctance. Other factors that appear in Eq. (13.5) are the *reluctances* of the iron and gap:

$$\mathcal{R}_i = \frac{\ell_i}{\mu_i A_i} \quad \text{and} \quad \mathcal{R}_g = \frac{\ell_g}{\mu_o A_g} \quad \text{A-t/Wb} \qquad (13.10)$$

Reluctance measures the mmf required to produce a given amount of flux in a magnetic structure. The longer the path, the more mmf required; the larger the cross section, the less mmf required; and the larger the permeability, the less mmf required. These factors are reflected in Eq. (13.10). Because the iron and gap are in series (have the same flux), their reluctances add to give the total reluctance of the path. The coil mmf is divided by this total reluctance to give the flux in the system. In Eq. (13.7), note that the reluctance of the iron path is small relative to the reluctance of the gap because of the large permeability of the iron. The total reluctance is effectively that of the air gap, and hence the magnetic flux of the system is in effect determined by the dimensions of the air gap. This is typical: the iron does such an effective job in guiding the magnetic flux that the magnetic system is limited by the air gap. In other words, the mmf of the coil is used largely to magnetize the gap, and very little mmf is required to magnetize the iron.

Combining reluctances. Reluctances in series or parallel may be combined like resistances in series or parallel. Equation (13.5) illustrates a series addition. Parallel magnetic paths are sometimes used in transformers (see Fig. 13.14).

Nonlinear effects. When saturation of the iron becomes a factor, several changes occur in our analysis. The equations become nonlinear and hence require a numerical or graphical solution. Also, when the iron becomes saturated, the reluctance of the iron part of the path increases and can become a factor in limiting the magnetic flux. This is often the case in practice, for practical devices often are operated partially saturated. However, our study of magnetic structures will usually exclude the effects of saturation.

Inductance. The inductance (L) is closely related to the reluctance of the system.

$$L = \frac{\lambda}{i} = \frac{n\Phi}{i} = \frac{n}{i} \times \frac{ni}{\mathcal{R}} = n^2/\mathcal{R} \quad \text{H} \qquad (13.11)$$

Note that small reluctance corresponds to large inductance. This definition of inductance applies when flux linkages are proportional to current, that is, when the magnetic system is linear.

Summary. In this section we have introduced the concept and function of a magnetic structure. We have shown the role of magnetomotive force and reluctance in the analysis of magnetic structures. We now apply these concepts in the study of electrical transformers. In the next chapter we introduce cylindrical magnetic structures, which are used in motors.

13.2.3 Check Your Understanding

1. A magnetic system is shown in Fig. 13.4.
   (a) For $i > 0$, mark the direction of the magnetic flux in the iron.
   (b) If i were decreasing at a rate of 20 A/s, find v_{ab}.
2. The magnetic field direction is the direction a compass needle would indicate as north. (T/F?)
3. Magnetic flux comes out of a N magnetic pole. (T/F?)

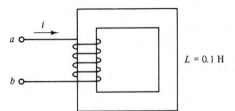

$L = 0.1$ H

Figure 13.4

4. In a one-loop magnetic structure the flux is 0.1 Wb. The air gap has an area of 0.1 m²
and a length of 1 mm.
(a) What is the flux density in the air gap?
(b) What is the reluctance of the air gap?
(c) What mmf is required to force this flux across the air gap?

Answers. 1. (a) Down in coil side, up in other side, (b) -2 V; 2. true;
3. true; 4. (a) 1 T; (b) 7.96×10^3 A-t/Wb; (c) 796 A-t.

13.3 ELECTRICAL TRANSFORMERS

13.3.1 Introduction

An electrical transformer consists of two or more coils (or windings), tightly coupled by magnetic flux that is guided by a magnetic structure. In Chapter 5 we
showed how transformers are used for voltage transformation, current transformation, and impedance transformation. Transformers find an extremely wide
range of applications in all sorts of electrical systems. Our treatment of transformers in this chapter is oriented toward power transformers.

Contents of this section. We begin with a brief review of ideal transformer relationships, and then analyze the transformer as a magnetic structure.
From this analysis, supplemented by physical considerations, we develop equivalent circuit models for circuit calculations. We conclude with a discussion of
transformers in three-phase systems.

13.3.2 Ideal Transformer

In Chapter 5, we introduced the ideal transformer with the symbol shown in Fig.
13.5a plus Eqs. (13.12) and (13.13):

$$\frac{\mathbf{V}_p}{n_p} = \frac{\mathbf{V}_s}{n_s} \tag{13.12}$$

$$n_p \mathbf{I}_p = n_s \mathbf{I}_s \tag{13.13}$$

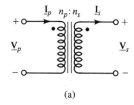

(a)

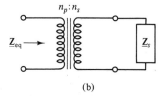

(b)

Figure 13.5 (a) Circuit symbol for
an ideal transformer; (b) impedance
transformation.

Magnetic Structures and Electrical Transformers   Chap. 13

Division of Eq. (13.13) into Eq. (13.12) shows that the transformer changes impedances by the square of the turns ratio, Fig 13.5b,

$$\mathbf{Z}_{eq} = \frac{\mathbf{V}_p}{\mathbf{I}_p} = \left(\frac{n_p}{n_s}\right)^2 \frac{\mathbf{V}_s}{\mathbf{I}_s} = \left(\frac{n_p}{n_s}\right)^2 \mathbf{Z}_s \qquad (13.14)$$

Our purpose in the following section is to show the physical basis for these relationships and to investigate the degree with which they describe real transformers. The analysis is based on the physical laws from Chapter 12 plus the concepts of magnetic structures from the previous section.

13.3.3 Analysis of a Transformer as a Magnetic Structure

The structure to be analyzed is shown in Fig. 13.6. We will assume sinusoidal steady state, and hence use phasors to represent voltage, current, and flux quantities. The voltage relationship in Eq. (13.12) follows from Faraday's law (p. 502) if we assume the same flux in primary and secondary windings. In our analysis of a real transformer, we will not make this assumption, and hence the ideal voltage relationship of Eq. (13.12) does not apply strictly to a real transformer. However, in a well-designed magnetic structure, only a small amount of flux escapes from the iron to become stray (or leakage) flux; hence the ideal voltage relationship is very nearly obeyed in a real transformer. In the following, we assume that all flux couples primary and secondary coils, and at the end we insert small series inductors into our equivalent circuit to account for leakage flux.

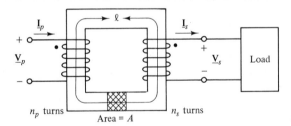

n_p turns   Area = A n_s turns

Figure 13.6 Transformer conventions.

Faraday's law. From Faraday's law, we gain an important relationship between voltages and flux values. Applied to the primary, Faraday's law is

$$v_p(t) = n_p \frac{d\Phi(t)}{dt} \Rightarrow \mathbf{V}_p = n_p(j\omega)\mathbf{\Phi} \qquad (13.15)$$

where the first form is time domain and the second a phasor relationship in the frequency domain.

Important change in phasor notation. In Eq. (13.15) and henceforth, the magnitude of a phasor quantity will be the rms value of the corresponding sinusoid. This is the custom in the study of electromechanics and power systems. Thus formulas for power and energy quantities will not contain a factor of $\frac{1}{2}$ as in Eq. (5.22). However, we will retain the peak values for phasor diagrams because these relate intimately to the sinusoids of the time domain. The quantity $\mathbf{\Phi}$ in Eq.

(13.15) is the phasor representation of the flux, rms amplitude and phase. The j factor on the right side indicates that the flux lags the voltage by 90° in phase. Since the maximum value of the flux is frequently used, Eq. (13.15) is often written

$$V_p = \frac{\omega n_p}{\sqrt{2}} \Phi_{max} \tag{13.16}$$

Φ_{max} is the maximum flux and V_p is the rms magnitude of the primary voltage. Clearly, Eqs. (13.15) and (13.16) could also have been written for secondary voltages. Equation (13.12) results when the common factors are eliminated between Eq. (13.15) and the corresponding equation involving the secondary voltage.

Polarities and the dot convention. The relative polarities between transformer windings are indicated by the dots in Figs. 13.5 and 13.6. The dot on the primary is assigned arbitrarily, and a load set of voltage and current reference directions is assigned with the + on the voltage corresponding to the dot. The + on the secondary voltage is then determined by the right-hand rule and Lenz's law in the following manner. Assume current flows into the dot on the primary and use the right-hand rule to establish the flux direction. Then take the secondary coil in your right hand with the thumb *opposite* to the flux from the primary. Your fingers then indicate the direction current would flow if allowed. The dot is placed on the end of the secondary coil where current would exit.† The dot corresponds to the + on the secondary voltage. The secondary acts as a source to its load; hence a source set of voltage and current reference directions is used. To summarize, the dots indicate the polarity of the voltages of the windings. The primary current reference direction is into the dot, but the secondary reference direction is out of the dot. Thus the primary is represented as a load, and the secondary is represented as a source.

Ampere's circuital law. The currents in primary and secondary are related through Ampere's circuital law

$$\oint \underline{\vec{H}} \cdot \vec{d\ell} = n_p \underline{I}_p - n_s \underline{I}_s \tag{13.17}$$

where $\underline{\vec{H}}$ is the phasor magnetic field in the magnetic structure. The minus sign of the $n_s\underline{I}_s$ term in Eq. (13.17) follows from Lenz's law. That is, we placed the dots in Fig. 13.6 such that the flux from $\underline{I}_s$ would *oppose* the flux created by $\underline{I}_p$; hence the mmfs of the two series coils subtract. We can change the left-hand side of Eq. (13.17) by introducing the concepts of magnetic structures:

$$\oint \underline{\vec{H}} \cdot \vec{d\ell} = |\underline{\vec{H}}| \ell_i = \frac{|\underline{\vec{B}}| \ell_i}{\mu_i} = \frac{\underline{\Phi}\ell_i}{\mu_i A} = \underline{\Phi}\mathcal{R}_i \tag{13.18}$$

where $\mathcal{R}_i$ is the reluctance of the iron (no gap here), and $\underline{\Phi}$ is the coupling flux. Thus, Eq. (13.17) can be rearranged to the form

---

† A consequence of these conventions is that currents into the dots give flux in the same direction.

$$\underline{I}_p = \frac{n_s}{n_p} \underline{I}_s + \frac{\Phi \mathcal{R}_i}{n_p} \tag{13.19}$$

We can eliminate the flux from Eq. (13.19) by introducing Eq. (13.15), with the result

$$\underline{I}_p = \frac{n_s}{n_p} \underline{I}_s + \frac{\mathbf{V}_p}{j\omega L_i} \tag{13.20}$$

where $L_i = n_p^2/\mathcal{R}_i$, from Eq. (13.11). Equation (13.20), interpreted as Kirchhoff's current law, suggests the equivalent circuit shown in Fig. 13.7. Subject to the assumptions we have made, the real transformer is represented by an ideal transformer plus an inductor. The inductor accounts for the magnetic energy of the flux; thus in this circuit model of the transformer, real power is conserved but reactive power is not because we have more input than output reactive power.

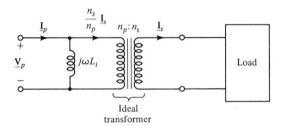

Figure 13.7 The inductor represents the energy stored in the magnetic flux.

13.3.4 Transformer Equivalent Circuits

In developing an equivalent circuit to predict the operation of a real transformer, we need to account for the energy storage and loss of the transformer, as well as the transformation properties represented by an ideal transformer. We have shown above that, when applied to the transformer geometry, Ampere's circuital law introduces an inductor in the equivalent circuit. Similarly, the other significant effects in the transformer are represented by circuit elements in the equivalent circuit.

Iron losses.   The losses in the magnetic structure are represented by a resistor (R_i) in parallel with the inductance in Fig. 13.7. These losses are due to hysteresis, as discussed in Chapter 12, page 498, and to *eddy currents*. Eddy currents, caused by the changing flux in the iron, heat the resistance of the metallic conductor. These losses may be reduced but not eliminated by fabricating the magnetic structure out of thin, insulated laminations.

Copper losses.   The losses in the resistance of primary and secondary coils are called copper losses.† These losses are represented in the equivalent

---

† Although aluminum and occasionally other types of metals are used for wires in transformers and other electrical apparatus, originally only copper was used. For this reason, resistive losses are traditionally called "copper losses," regardless of the metal used.

circuit by resistances (R_p and R_s) in series with the primary and secondary of the ideal transformer.

Leakage flux.   Although the iron provides a low-reluctance path for the magnetic flux, some flux escapes the iron path and thus fails to couple primary and secondary, as shown in Fig. 13.8. The energy associated with this flux is modeled by inductors (L_p and L_s) in series with the primary and secondary resistances. In the ac equivalent circuit, these inductances become primary and secondary leakage reactances (X_p and X_s).

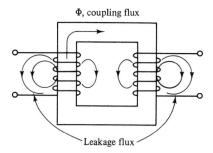

Φ, coupling flux

Leakage flux

Figure 13.8   Some leakage flux escapes the iron path.

Equivalent circuit.   We have now considered the significant losses and energy-storage effects in a real transformer. An equivalent circuit that accounts for these various effects is given in Fig. 13.9.

In Fig. 13.9 we distinguish between $\underline{V}_p$, the primary voltage, and $\underline{E}_p$, the voltage induced in the primary coil by the changing flux. The $\underline{E}$s in primary and secondary obey the ideal transformer relationship. The *exciting current*, $\underline{I}_e$, accounts for the energy required to magnetize the iron and supply its magnetic and eddy-current losses. Note that if the secondary were an open circuit the only current flowing in the primary would be the exciting current; thus, the exciting current would be the no-load current into the primary.

Simplified equivalent circuit.   The equivalent circuit in Fig. 13.9 models a real transformer. The circuit in Fig. 13.10 is fully equivalent; the secondary coil resistance and leakage reactance have merely been transformed into the primary. Normally, the exciting current is small compared with the total input current.

Figure 13.9   The resistors account for losses. The inductors account for stored magnetic energy.

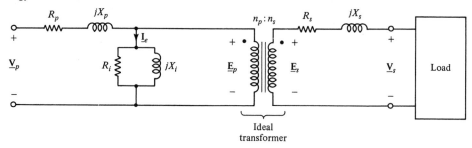

Ideal
transformer

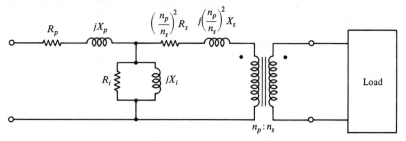

Figure 13.10 Impedance transformation is used to bring all impedances into the primary.

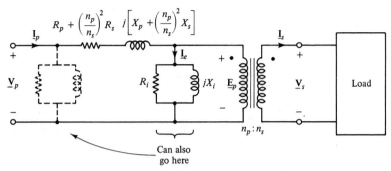

Figure 13.11 Combining the series elements causes insignificant error.

Hence, we may move R_i and X_i to either position shown in Fig. 13.11 without significant error. This change is attractive because it combines primary and secondary resistances and leakage reactances and hence simplifies circuit calculations, as in the following example.

Example Transformer. A 10-kVA, single-phase transformer is rated 2400/240 V. The equivalent circuit parameters are $R_p = 3\ \Omega$, $X_p = 15\ \Omega$, $R_s = 0.03\ \Omega$, $X_s = 0.15\ \Omega$, $R_i = 30\ \text{k}\Omega$, and $X_i = 20\ \text{k}\Omega$. If the load is 10 kVA at 0.9 power factor (lagging), find the primary voltage ($\underline{V}_p$), primary current ($\underline{I}_p$), exciting current ($\underline{I}_e$), percent regulation (% reg), and efficiency (η). We will use the equivalent circuit in Fig. 13.11, but also give in parentheses the results from the exact equivalent circuit in Fig. 13.10 to show that the two give almost identical results.

First we must interpret the transformer specifications. The secondary voltage will be the rated voltage, $240\underline{/0°}$ V, and the ratio 2400/240 is interpreted to give the turns ratio as 10:1. The primary voltage will be greater than 2400 voltage due to real and reactive power requirements of the transformer. With this interpretation, the secondary current will be

$$\underline{I}_s = \frac{\text{kVA}}{V_s}\underline{/-\cos^{-1}(\text{PF})} = \frac{10 \times 10^3}{240}\underline{/-\cos^{-1}(0.9)} = 41.67\underline{/-25.8°}\ \text{A}$$

$$(13.21)$$

For a 10:1 ratio of turns, the primary voltage of the ideal transformer is $\underline{E}_p = 2400\underline{/0°}$, and its current is $4.167\underline{/-25.8°}$.

The exciting current is most easily calculated by considering separately the magnetizing current in the inductor X_i and the loss current in the resistor R_i. The magnetizing current required to establish the magnetic flux is

$$\underline{I}_\Phi = \frac{\underline{E}_p}{jX_i} = \frac{2400\underline{/0°}}{j20 \times 10^3} = 0.120\underline{/-90°} \quad \text{A} \tag{13.22}$$

and the loss current required to supply the iron losses is

$$\underline{I}_i = \frac{\underline{E}_p}{R_i} = \frac{2400\underline{/0°}}{30 \times 10^3} = 0.080\underline{/0°} \quad \text{A} \tag{13.23}$$

As shown in Fig. 13.12, the exciting current is

$$\underline{I}_e = \underline{I}_i + \underline{I}_\Phi = 0.080 - j0.120 = 0.144\underline{/-56.3°} \quad \text{A} \tag{13.24}$$
$$(0.147\underline{/-55.1°})$$

The total current in the primary will therefore be

$$\underline{I}_p = \frac{n_s}{n_p} \underline{I}_s + \underline{I}_e = \frac{41.67}{10} \underline{/-25.8°} + 0.144\underline{/-56.3°} = 4.292\underline{/-26.8°} \quad \text{A} \tag{13.25}$$
$$(4.295\underline{/-25.8°})$$

Finally, we can determine the primary voltage:

$$\underline{V}_p = \underline{E}_p + \underline{I}_p \left[R_p + \left(\frac{n_p}{n_s}\right)^2 R_s + jX_p + j\left(\frac{n_p}{n_s}\right)^2 X_s \right] \tag{13.26}$$
$$= 2400\underline{/0°} + 4.292\underline{/-26.8°} (6 + j30) = 2483\underline{/2.38°} (2481\underline{/2.36°}) \quad \text{V}$$

The percent regulation is defined as

$$\% \text{ reg} = \frac{V_s \text{ (no load)} - V_s \text{ (loaded)}}{V_s \text{ (loaded)}} \times 100\% \tag{13.27}$$

where V_s (no load) is the magnitude of the secondary voltage with no load and V_s (loaded) is the magnitude of the secondary voltage under the given load conditions. For no load, we can ignore the voltage loss in the series impedance and

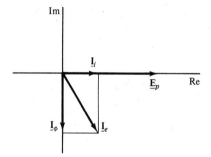

Figure 13.12 The exciting current consists of an in-phase component to supply iron losses and an out-of-phase component to supply energy stored in the magnetic structure.

use the ideal transformer relationship: V_s (no load) $= 2483/10 = 248.3$ V. Thus the percent regulation is

$$\% \text{ reg} = \frac{248.3 - 240.0}{240.0} \times 100\% = 3.46\% \tag{13.28}$$

Efficiency calculation. For an efficiency calculation, we consider only real power. The output power is 10 kVA $\times$ (0.9) = 9000 W, and the input power is

$$P_{in} = V_p I_p \text{ PF} = 2483 \times 4.292 \cos(26.8° + 2.38°) = 9304.5 \ (9305) \quad \text{W} \tag{13.29}$$

The efficiency is therefore $\eta = 9000/9304.5 \times 100\% = 96.7 \ (96.7)\%$. Comparison of the results for the exact (in parentheses) and the approximate equivalent circuits shows that the latter gives good accuracy.

Had we wanted to know only the efficiency, we could have performed a simpler calculation by estimating the losses. The iron losses are approximately

$$P_i = \frac{(2400)^2}{30 \text{ k}\Omega} = 192 \text{ W} \tag{13.30}$$

and, if we ignore the exciting current, the copper losses are

$$P_r = (4.167)^2(6 \ \Omega) = 104.2 \quad \text{W} \tag{13.31}$$

Thus the efficiency is approximately

$$\eta = \frac{P_{out}}{P_{out} + P_i + P_r} = \frac{9000}{9000 + 192 + 104.2} = 96.8\% \tag{13.32}$$

Clearly, this method gives accurate results.

Transformer measurements. The equivalent circuit parameters in Fig. 13.11 may be determined by an open-circuit/short-circuit test. In an open-circuit test, the high-voltage side is open circuited, and the transformer is excited at rated voltage on the low-voltage side, while voltage, current, and input power are measured. The measured current is the exciting current, and the measured power is the iron losses. Copper losses are not a factor because current levels are small. From these measurements, the parameters R_i and X_i may be determined. All measured quantities are referred to the low-voltage side.

In a short-circuit test, the low-voltage side is short circuited and a reduced voltage is applied to the high-voltage side. The voltage is reduced to give rated current in the windings. Input voltage, current, and power are measured. The input power is the copper losses only because the iron losses for the reduced voltage are negligible. From the copper losses and the current, the combined resistance of the two coils may be determined. The input voltage and current yield the magnitude of the input impedance, and from this and the resistance the combined leakage reactance may be determined. Thus the iron and copper losses are measured directly,† and the full equivalent circuit in the form of Fig. 13.11 may be established from the open-circuit/short-circuit test.

† Although not at the same time.

13.3.5 Why Apparent Power (kVA) Rating Is Important

There are important reasons why transformers are rated for a specific apparent power (kVA). A transformer is limited by the amount of heat it can dissipate into the environment and hence, for a given environment, it is limited by its losses. The magnetic structure is designed for a specific maximum flux, which implies a specific voltage rating since the frequency and number of turns are constant. To operate the transformer at a voltage higher than the design value would give excessive losses and invite eventual failure. To operate at a lower voltage would underutilize the transformer. The other loss component comes from the copper losses in the primary and secondary windings; hence the maximum current is limited likewise by the amount of heat the transformer can dissipate. For these reasons, the transformer is rated according to the product of rated voltage and rated current, or apparent power.

We may distinguish two apparent power limits:

1. The rated apparent power is the operating level that may be sustained under a worst-case thermal environment. For example, a small pole-hung transformer would be rated for a hot summer day, in full sun and with no wind. This would be the most severe conditions expected, and the transformer would be expected to perform satisfactorily under such conditions.

2. The actual limit for the transformer would depend on the operating conditions. For example, a transformer rated for conditions in Texas could be operated at higher levels of apparent power in Alaska.

In a given application, the operating level of the transformer depends on the load and may be safely below the rated or actual limits of the transformer.

Example. When transformers have multiple primary and/or secondary windings, the kVA rating is independent of the way in which the transformer is connected. Consider, for example, a 480:240/240:120 transformer with a 8.8-kVA rating. The voltage rating implies that both primary and secondary have two identical windings, which may be connected in series or parallel. Figure 13.13a shows the windings in series. The primary voltage is 480 V, the series connection of two 240-V windings, and the secondary voltage is 240 V. From the kVA rating, we calculate the primary current as 18.3 A and the secondary current as 36.7 A.

Figure 13.13b shows the parallel connection of both windings. From the kVA calculation, we now have twice the current in primary and secondary, but with parallel connections this current is divided between two windings to give the rated current in each winding. The magnetic structure will operate at the same flux level in both connections, and the currents in the individual windings are unchanged; hence the losses for the two cases are identical. As we will see in the next section, we have a similar result for three-phase connections. Consequently, the transformer is rated according to its apparent power; the power factor and hence the real power flowing through the transformer are not directly a limiting factor.

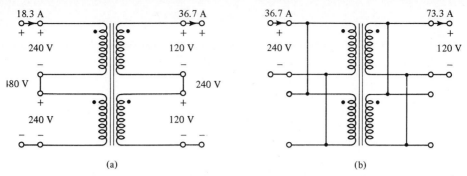

Figure 13.13 (a) Windings in series; (b) windings in parallel.

13.3.6 Transformers in Three-phase Circuits

Three-phase systems constitute the overwhelming majority of all power generation and distribution systems. Transformation of three-phase power from one voltage level to another may be accomplished by three single-phase transformers or by one three-phase transformer. Figure 13.14 shows one possible configuration for a three-phase transformer.

A three-phase transformer is cheaper, smaller, and more efficient than three single-phase transformers, but the latter configuration is more versatile. For example, should a single transformer fail, only one transformer would have to be replaced, and in certain cases the system could continue operation at reduced load until the replacement arrived.

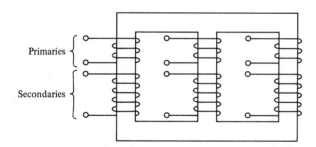

Figure 13.14 A three-phase transformer wound on a common core.

Three-phase transformer connections. In three-phase transformation, primaries and/or secondaries can be connected in delta (Δ) or wye (Y). This results in four possible combinations: Y–Δ, Δ–Y, Δ–Δ, and Y–Y.

Example. Figure 13.15 shows a Y–Δ connection in circuit and in schematic representation. We assume three 2400/240, 24-kVA (each) single-phase transformers, so the allowed currents in the primary and secondary windings are 10 A and 100 A, respectively. Operated at rated voltage and current, the primary line voltage and current are $\sqrt{3} \times 2400$ V and 10 A, for an apparent power of $\sqrt{3}\ VI = 72$ kVA. The secondary line voltage is 240 V and the rated current is $\sqrt{3} \times 100$ A. Hence the apparent power in the secondary is also 72 kVA. In both

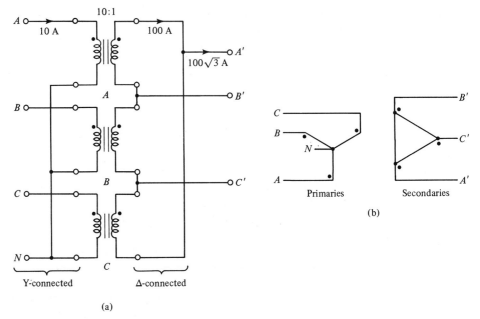

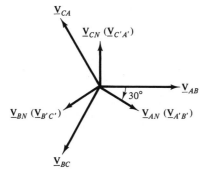

Figure 13.15 Three single-phase transformers in a Y–Δ connection: (a) circuit diagram; (b) schematic diagram.

Figure 13.16 Phasor diagram for a Δ–Y transformer connection.

cases the rating of the three-phase connection is three times the rating of the single-phase transformers. This would be true for all four possible connections.

The phasor diagram for the Y–Δ connection is shown in Fig. 13.16. Note that the secondary voltages are one-tenth the primary line-to-neutral voltage, as required, but 30° out of phase. This suggests that care should be used in designing and installing such systems because if multiple paths exist, and they do in most power systems, phases must be correct.

13.3.7 Per-phase Representation

In a three-phase power system, four combinations exist for source (transformer secondaries) and load connections. Each has its practical advantages, depending on the application. But the case where both source and load are in the Y connection has conceptual advantages because, with the neutral in place, the Y–Y connection is equivalent to three single-phase circuits sharing a common neutral. Thus a single-phase circuit can represent a balanced three-phase circuit, provided that line-to-neutral voltage is used and also that power-type calculations are multiplied by three. This is called a *per-phase equivalent circuit* and its use simplifies three-phase circuit diagrams and calculations. In the per-phase circuit, line current appears at its true value, line voltage is reduced by $\sqrt{3}$, and powers are reduced by 3 relative to the three-phase circuit. Wye-connected impedances appear at their true values, but delta-connected impedances are divided by 3.†

Example.   As an example of a per-phase representation, consider a source with 600 V (line-to-line) and a load of three 30-Ω resistors connected in delta as shown in Fig. 13.17a. We will give the per-phase circuit and calculate the power in the load. The per-phase voltage is $600/\sqrt{3}$ because we must use line-to-neutral voltage. The load resistance is one-third of 30 Ω because this is the value that, Y-connected, would draw the same current as a 30-Ω delta load. Thus the per-phase equivalent circuit is that shown in Fig. 13.17b. The power in the load would be $(600/\sqrt{3})^2/10 = 12.0$ kW per phase, or 36.0 kW for the three-phase load.

When a per-phase circuit is given for a transformer connection, it should not be inferred that Y-connections are being used. The per-phase equivalent circuit is merely a single-phase model of the three-phase circuit.

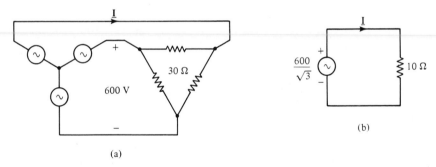

(a)

(b)

Figure 13.17   (a) A three-phase circuit; (b) its per-phase equivalent.

13.3.8 Check Your Understanding

1. The basic laws describing the transformer are Coulomb's law, Ampere's circuital law, Faraday's law, and/or Ohm's law. (Which two?)
2. For an ideal transformer, many turns on a winding go with high or low voltage?
3. A transformer has 500 turns on the side connected to the power source and 25 turns on the side connected to the load. If the load requires 120 V, what should be the voltage of the power source?
4. A single-phase transformer has 300 turns on the side connected to the source of power and 27 turns on the side connected to the load. If the load requires 100 A, what should be the current capacity on the other side?

† See Delta-Wye Conversions, page 191.

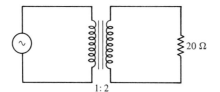

1:2   **Figure 13.18**

5. In the transformer circuit shown in Fig. 13.18, an ammeter measures 2 A in the primary. What would the following measure?
   (a) An ammeter in the secondary?
   (b) A voltmeter in the primary?
   (c) A voltmeter in the secondary?
   (d) A wattmeter in the primary?
   (e) A wattmeter in the secondary?

6. A 240-V, delta-connected resistive three-phase load requires a power of 3000 W. What is the per-phase resistance?

7. A three-phase transformer bank (three transformers) has 12,000 V on the primary (line-to-line) and a line current of 10 A. If the secondary voltage is 320 V line-to-neutral, what is the secondary line current?

8. A 560-V rms, 20-kVA three-phase transformer operates fully loaded with a power factor of 0.85, lagging.
   (a) What is the per-phase voltage?
   (b) What is the per-phase current?

   Answers. 1. Ampere's circuital law and Faraday's law; 2. high voltage; 3. 2400 V; 4. 9.0 A; 5. (a) 1 A, (b) 10 V, (c) 20 V, (d) and (e) 20 W; 6. 19.2 Ω; 7. 217 A; 8. (a) 323 V, (b) 20.6 A.

13.4 FORCES IN MAGNETIC SYSTEMS

13.4.1 Magnetic Pole Approach

Consider the device shown in Fig. 13.19. This is a magnetically driven mechanical actuator, which might operate a lock or ring a bell. The magnetic structure is iron, with a hinged member. The electrical input is produced by a coil with n turns. From our experience with such devices, we expect that a magnetic force will be produced. Our goal in this section is to develop means for determining the magnitude and direction of this force.

Everyone has handled permanent magnets and knows about the magnetic compass. The properties of these devices can be described in terms of magnetic poles. We can use the concept of magnetic poles to build an intuitive understanding of the force in our actuator and to determine the direction of the force.

Relationship between magnetic poles and flux. We have already discussed terrestrial magnetism on page 496 in defining the direction of a magnetic field. We stated that a north magnetic pole is attracted in a northerly direction, and vice versa, and that the direction of the magnetic field is the direction that a north magnetic pole tends to move. It follows that a north magnetic pole acts as

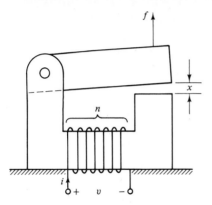

Figure 13.19 A magnetic actuator.

a *source* of magnetic flux, and a south magnetic pole acts as a *sink* of magnetic flux.

The pole is inside the surface. In this viewpoint, we assume that we have a physical body, a piece of iron or perhaps the earth as shown in Fig. 13.20a, and that magnetic flux is leaving or entering its surface. If magnetic flux is leaving a surface, as at the south geographic pole, we attribute this flux to the presence of a north magnetic pole within the surface, which is acting as a source for the flux. Alternatively, if flux lines are entering a surface, as at the north geographic pole, we attribute this flux to the presence of a south magnetic pole within the surface.

These ideas are useful in determining the direction of forces and torques. Consider again the device in Fig. 13.19. If i is positive, then by the right-hand rule the direction of the magnetic flux is as shown in Fig. 13.20b. The top member thus contains a north magnetic pole because *from the viewpoint of the gap* it produces the magnetic flux. Likewise, the bottom of the gap contains a south magnetic pole because the flux enters the bottom surface. Since opposites attract, we conclude that a force of attraction exists between the north and south magnetic poles, tending to close the gap. We would also expect that the force would increase

Figure 13.20 (a) The earth's magnetic poles and flux; (b) the actuator with magnetic poles shown.

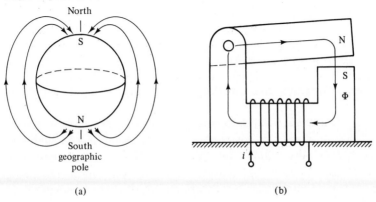

(a)

(b)

as the gap becomes smaller because the magnetic poles are closer and increase in magnitude. It is easily shown that the force is independent of the direction of the current.

Conclusion. This viewpoint gives us a qualitative understanding of the magnetic forces in the system but no quantitative information. In Section 13.4.3, we develop a method, based on conservation of energy, that gives both qualitative and quantitative information, albeit at the cost of some abstraction.

13.4.2 Analysis from Current–Flux Interaction

In Eq. (12.13), we presented the expression for the force per meter on a current-carrying conductor in a magnetic flux:

$$\vec{f} = \vec{i} \times \vec{B} \quad \text{N/m} \tag{13.33}$$

This expression may be integrated to determine the force on a system of current-carrying conductors, as in a motor, if the magnetic flux density and currents are known. However, magnetic structures such as shown in Fig. 13.20b do not permit force calculation because the force is produced by magnetic dipoles in the magnetic material.

13.4.3 Analysis from Energy Considerations

A general approach to electromechanic interactions may be founded on the principle of conservation of energy. The full analysis is presented in advanced texts; only the results are given here.

A magnetic system with a movable member, such as that shown in Fig. 13.20b, will have a stored energy function that depends on the electrical excitation and the mechanical displacement. This energy function is given by

$$W_m(\lambda_f, x) = \int_0^{\lambda_f} i(\lambda, x)\, d\lambda \tag{13.34}$$

where $W_m(\lambda_f, x)$ is the stored magnetic energy, λ_f is the flux linkages under which the force is to be calculated, and x is the mechanical displacement. Conservation of energy and other assumptions to be itemized below then require that the force $f(\lambda_f, x)$ exerted by the magnetic system on the mechanical system be given by

$$f(\lambda_f, x) = -\left.\frac{\partial W_m(\lambda_f, x)}{\partial x}\right|_{\lambda_f\,=\,\text{constant}} \tag{13.35}$$

The assumptions that must be satisfied are:

- The magnetic system must be lossless. Since most magnetic systems do have some loss, this assumption weakens the analysis slightly, meaning that the accuracy of the results is compromised by the magnetic losses.
- The stored energy must be zero if the current or flux linkages are zero. This assumption rules out forces due to permanent magnets.

- The stored-energy function must be a single-valued state function. This requires that the stored energy depend only on the final value of λ and x, and not on how the system is energized. Since most magnetic systems exhibit some hysteresis, this assumption also weakens the analysis.

Example. Let us apply Eqs. (13.34) and (13.35) to the system shown in Fig. 13.20b. The total flux will be

$$\Phi = \frac{ni}{\mathcal{R}_i + \mathcal{R}_g(x)} \quad \text{Wb} \tag{13.36}$$

where the reluctances of iron and gap are

$$\mathcal{R}_i = \frac{\ell_i}{\mu_i A_i} \quad \text{and} \quad \mathcal{R}_g(x) = \frac{x}{\mu_o A_g} \tag{13.37}$$

Notice that the mechanical displacement enters the analysis through its effect on the reluctance of the gap. The flux linkages will therefore be

$$\lambda = n\Phi = \frac{n^2 i}{\mathcal{R}_i + \mathcal{R}_g(x)} \quad \text{Wb-t} \tag{13.38}$$

Since Eq. (13.34) requires the current as a function of flux linkages, we rearrange Eq. (13.38) to the form

$$i(\lambda, x) = \frac{1}{n^2} [\mathcal{R}_i + \mathcal{R}_g(x)]\lambda \tag{13.39}$$

and integrate:

$$W_m(\lambda_f, x) = \int_0^{\lambda_f} i(\lambda, x) \, d\lambda = \frac{1}{n^2} [\mathcal{R}_i + \mathcal{R}_g(x)] \int_0^{\lambda_f} \lambda \, d\lambda$$
$$= \frac{1}{2n^2} [\mathcal{R}_i + \mathcal{R}_g(x)]\lambda_f^2 \tag{13.40}$$

This is the required energy state function. At this point we may drop the f subscript from λ, since the energizing process is conceptually accomplished.[†] To compute the developed force, Eq. (13.35) requires that we take the partial derivative of $W_m(\lambda, x)$ with respect to x, with λ held constant.

$$f_{\text{dev}} = -\left. \frac{\partial W_m(\lambda, x)}{\partial x} \right|_{\lambda = \text{constant}} = -\frac{\partial}{\partial x} \frac{1}{2n^2} \left[\mathcal{R}_i + \frac{x}{\mu_o A_g} \right] \lambda^2 = -\frac{\lambda^2}{2n^2 \mu_o A_g} \tag{13.41}$$

Equation (13.41) gives the force on the mechanical system as a function of the flux linkage. If we wish to have the force as a function of the current, we may reintroduce Eq. (13.39) to yield

$$f_{\text{dev}} = -\frac{(ni)^2}{[\mathcal{R}_i + \mathcal{R}_g(x)]^2} \frac{1}{2\mu_o A_g} \tag{13.42}$$

† In Eq. (13.40), λ is a dummy variable of integration.

The minus sign means that the force is directed opposite to the $+x$ direction and tends therefore to decrease x, that is, to close the gap. The force depends on the square of the mmf. To see how the force depends on the width of the gap, we will assume that the reluctance of the iron is negligible compared with the reluctance of the gap. Under this assumption, Eq. (13.42) reduces to

$$f_{dev} = -\frac{1}{2}\frac{(ni)^2}{(x/\mu_o A_g)^2}\frac{1}{\mu_o A_g} = -\frac{(ni)^2 \mu_o A_g}{2x^2} \qquad (13.43)$$

and hence the force varies as the inverse square of the gap width.

Zero gap? Although Eq. (13.43) leads to infinite forces as x approaches zero, this is unrealistic for two reasons. For one, the gap cannot have zero width because of surface roughness; hence a minimum effective value of x would exist. But, more important, the finite permeability of the iron would limit the force. If we were to seek to determine the maximum force, we would have to consider carefully the properties of the iron, both the finite value of permeability and also the saturation effects.

Torques. A rotational system, such as shown in Fig. 13.21, would require slight modification of the above development. Specifically, $W_m(\lambda, x)$ becomes $W_m(\lambda, \theta)$, and the torque is

$$T_{dev}(\lambda, \theta) = -\frac{\partial W_m(\lambda, \theta)}{\partial \theta}\Bigg|_{\lambda = \text{constant}} \qquad (13.44)$$

where $T_{dev}(\lambda, \theta)$ represents the developed torque. In this case, the dependence of the reluctance on θ becomes the important factor.

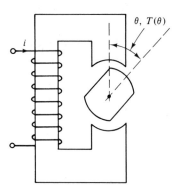

Figure 13.21 A magnetic structure involving rotation.

Analysis based on co-energy. The co-energy of a magnetic system is defined as

$$W'_m(i_f, x) = \int_0^{i_f} \lambda(i, x)\, di \qquad (13.45)$$

where $W'_m(i_f, x)$ is the co-energy function and i_f is the current under which the magnetic force is to be calculated. The force is determined from the co-energy function as

$$f(i_f, x) = + \left. \frac{\partial W'_m(i_f, x)}{\partial x} \right|_{i_f = \text{constant}} \qquad (13.46)$$

Again the f can be dropped after the integration is performed.

Same example, using co-energy. Let us illustrate the use of co-energy in reworking the example in Fig. 13.20b. Our object is to calculate the developed force as a function of current and mechanical displacement. The flux linkages are still given by Eq. (13.38). Thus, the co-energy function is

$$W'_m(i_f, x) = \int_0^{i_f} \lambda(i, x) \, di = \frac{n^2 i_f^2}{2[\mathcal{R}_i + \mathcal{R}_g(x)]} \qquad (13.47)$$

The force is therefore

$$f_{\text{dev}} = + \frac{\partial W'_m(i, x)}{\partial x} = - \frac{(ni)^2}{2[\mathcal{R}_i + \mathcal{R}_g(x)]^2} \frac{\partial \mathcal{R}_g(x)}{\partial x} = - \frac{(ni)^2}{2\mu_o A_g [\mathcal{R}_i + \mathcal{R}_g(x)]^2} \qquad (13.48)$$

We note that this is the same answer as Eq. (13.42).

What is co-energy? Mathematically, co-energy is the function calculated by Eq. (13.45). Graphically, the co-energy is the area under the λ–i curve, as shown in Fig. 13.22a. But these answers do not explain co-energy.

A good starting place is to address the question, "What is energy?" The answer is not obvious; certainly many brilliant minds pondered the physical creation before coming on the concept of energy and postulating its conservation. Our frequent use of the concept of energy has perhaps dulled us to its abstract quality. Whatever we say about energy from a physical point of view, philosophically energy is a concept. The energy function is a descriptor of a system that has useful mathematical properties. Conservation of energy leads to important relationships between variables, sometimes allows shortcuts in an analysis, and usually gives insight into the operation of a physical system. In the present context, for example, conservation of energy allows us to calculate the magnetically generated force on a mechanical system.

Figure 13.22 (a) Co-energy is the area under the curve; (b) for a linear magnetic system, energy and co-energy are equal.

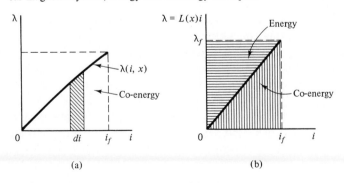

Similarly, co-energy is a mathematical function with useful properties. Here co-energy also allows us to compute the magnetically generated force. But as a basic principle of science, co-energy has minor importance compared with energy.

Energy is the area above the λ–i curve, and co-energy is the area below the curve in Fig. 13.22a. Energy and co-energy are equal when we have a linear relationship between λ and i, as in Fig. 13.22b. This is the case following, where we apply these ideas to circuit theory.

13.4.4 Circuit Approach

Let us continue to speak of the mechanical actuator pictured in Fig. 13.20b. From a circuit point of view, this device would be an inductor. The inductance depends on the reluctance of the magnetic structure [see Eq. (13.11)], which depends on x, the width of the gap. Thus we may write the flux linkages, $\lambda = L(x)i$, where $L(x)$ is the inductance as a function of the mechanical variable, and the co-energy is thus

$$W'_m(i_f, x) = \int_0^{i_f} L(x)i \, di = \frac{1}{2} L(x)i_f^2 \qquad (13.49)$$

The force would be

$$f = \frac{\partial W'_m(i, x)}{\partial x} = \frac{1}{2} i^2 \frac{dL(x)}{dx} \qquad (13.50)$$

Note that for an ac current the time-average force is determined from the time average of the square of the ac current, which is by definition the square of the rms current:

$$\langle f \rangle = \frac{1}{2} \langle i(t)^2 \rangle \frac{dL(x)}{dx} = \frac{1}{2} I_e^2 \frac{dL(x)}{dx} \qquad (13.51)$$

where I_e is the effective value of the current.

Where is the energy? Because the relative permeability for iron is so large, most of the energy is stored in the gap, even though the volume of the gap is relatively small. As Eq. (13.40) implies, the total energy divides between iron and gap in proportion to the reluctances.

You may be surprised to learn that most of the energy is stored in the gap. After all, that is mere empty space. Before you decide this is nonsense, consider a mechanical spring, in which we may store energy by stressing the material. Where in this case is the energy stored? Of course it is in the spring, but we press the point: where in the spring is it stored? After all, the spring is mostly empty space, and presumably the electrons and nuclei do not change physically because the spring is stressed. The stresses are placed on the bonds between the atoms, but these are also mere empty space. In the final analysis, what energy is and where in space it is stored are something of a mystery. The electrical example merely presents this mystery in an unfamiliar form.

Summary. The energy in a magnetic system is stored in the space comprising that system, including the empty space of any gaps. We visualize the magnetic fields as exerting a stress on the material and the space, and hence being the vehicle for force and energy storage. This analysis will be used in the next chapter to determine the torque developed in a cylindrical magnetic structure.

13.4.5 Check Your Understanding

1. In a magnetic system in which the reluctance of the magnetic structure decreases with increasing x, does the magnetic force tend to increase or decrease x?
2. The sum of the energy and co-energy of a magnetic system is constant if the current into the system is constant, even if a mechanical part of the magnetic structure is moved. (T/F?)

   Answers.   1. Decrease;   2. false.

Chapter summary. In this chapter we have applied physical laws from Chapter 12 to magnetic structures. We have analyzed the electrical transformer and presented equivalent circuit models for the transformer based on these laws, supplemented with reasoning based on energy considerations. We have derived expressions for magnetically generated forces and torques in magnetic structures that have movable members. The themes of this chapter continue in the next chapter where we will introduce cylindrical magnetic structures used in electrical motors to produce torque and rotation. Chapters 15 and 16 deal with specific types of electrical motors.

PROBLEMS

Section 13.2: Analysis of Magnetic Structures

P13.1. A toroidal inductor has circular cross section and the dimensions shown in Fig. P13.1. Assume $\mu_r = 1000$ and $H = $ constant in the iron. Use the average radius for computing the path length. Find the inductance.

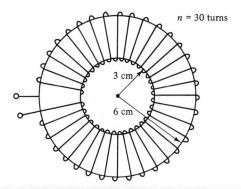

$n = 30$ turns

3 cm

6 cm

Figure P13.1

P13.2. A lossless toroidal inductor with a gap is represented in Fig. P13.2. The reluctances of the iron and air gap are 10,000 and 100,000 A-t/Wb, respectively.

(a) If $i = 1$ A dc, what is the flux in the system?

(b) Under these conditions, an electron passes through the gap traveling outward from the center. What would be the effect on its motion?

(c) If $i = 1$ A (rms) ac at 60 Hz, what would be the rms voltage per turn in the coil?

(d) What is the impedance of the inductor at 60 Hz?

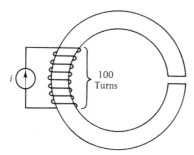

100 Turns

i

Figure P13.2

P13.3. A 12-V (rms), 60-Hz electrical relay has 1000 turns and draws 12 mA (rms) of current. Assume that the relay magnetic system acts as a linear inductor, and neglect resistance.

(a) What is the mmf operating in the magnetic structure?

(b) What is the peak flux in the relay?

(c) What is the reluctance in the magnetic structure?

Answers. (a) 12 A-t (rms); (b) 4.50×10^{-5} Wb; (c) 377×10^3 A-t/Wb

P13.4. An iron ring weighing 3.65 kg and having a relative permeability of 5000 has some wire wrapped around it and is excited by an ac voltage. The current is measured. A small gap is then cut in the ring, and the current is observed to increase by a factor of 10. How much do the filings weigh? The density of iron is 7.65 g/cm³.

Answer. 6.57 g

P13.5. An inductor is shown in Fig. P13.5.

(a) What is the reluctance of the iron path?

(b) A 60-Hz ac voltage of 120 V rms is applied to the coil input. What is the maximum of the magnetic flux in the iron?

(c) What is the rms current required to supply this flux?

(d) The iron hysteresis and eddy-current losses are found to be 30 W. Give a parallel equivalent circuit for the inductor that accounts for the energy storage and loss.

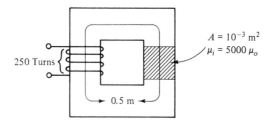

$A = 10^{-3}$ m²
$\mu_i = 5000\,\mu_o$

250 Turns

0.5 m

Figure P13.5

P13.6. An iron ring with a relative permeability of 3000 has a coil with n turns wound on it. The inductance of the structure is measured at 50 mH. Give the new inductance as the following changes are made one at a time (and then restored before the next change).

(a) The number of turns is changed to $n/2$.

(b) The relative permeability increases to 4000.

(c) The entire structure is scaled up in every dimension by 25%, but the number of turns is constant.

(d) The windings are replaced by wires with half the original diameter, but the number of turns remains at n.

(e) We cut a gap in the iron such that 0.5% of the path length is now air gap.

Section 13.3: Electrical Transformers

P13.7. Figure P13.7 shows an ideal magnetic structure ($\mu_i = \infty$), which thus becomes an ideal transformer, except that it has two secondaries, each with a resistor attached. Mark polarity dots on the two secondaries. Determine the equivalent resistance into the primary. *Hint:* Use fundamental laws.

Answer. $R_{eq} = \left(\dfrac{n_p}{n_1}\right)^2 R_1 + \left(\dfrac{n_p}{n_2}\right)^2 R_2$

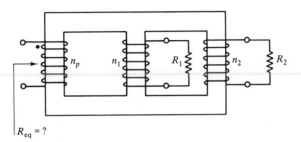

$R_{eq} = ?$

Figure P13.7

P13.8. Figure P13.8 shows an ideal magnetic structure ($\mu_i = \infty$), which thus becomes an ideal transformer, except that it has two secondaries, each with a resistor attached. Mark polarity dots on the two secondaries. Determine the equivalent resistance into the primary. *Hint:* Use fundamental laws.

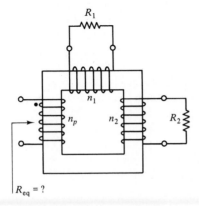

$R_{eq} = ?$

Figure P13.8

P13.9. A transformer has 1000 primary turns and 100 secondary turns. The reluctance of the magnetic structure is $\mathcal{R} = 10^5$ A-t/Wb. Assume no leakage flux, no iron losses, and ignore wire resistance.
- **(a)** If the primary voltage is 120 V, 60 Hz and the secondary is open circuited, how much current will flow in the primary?
- **(b)** If the secondary has a resistance of 30 Ω connected to it, what will be the current in the resistor?
- **(c)** For part (b), what would be the magnitude of the current in the primary?

P13.10. Figure P13.10 shows the approximate model for a nominal 880/220-V transformer with all internal impedances referred to the primary.
- **(a)** What term(s) in the model account for:
  - **i.** Secondary copper losses?
  - **ii.** Magnetic energy stored in the iron?
  - **iii.** Faraday's law of induction?
  - **iv.** Magnetic energy stored in the primary leakage fields?
  - **v.** Core losses due to eddy currents and hysteresis losses?
- **(b)** If the transformer heat exchanger is capable of dissipating 325 W without undue temperature rise, estimate the rated kVA of the transformer.
- **(c)** What would be the approximate efficiency at rated kVA if the output power factor were 0.6?

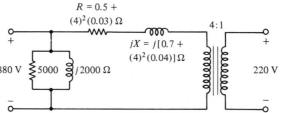

Figure P13.10

P13.11. A 2400/120-V, 30-kVA single-phase transformer draws 2 kVA at 0.2 power factor (lagging) at no load. The winding resistance and reactance due to stray inductance are 3.5 + j5.0 Ω, referred to the primary. Find the input kVA and power factor if this transformer serves a 30-kVA load at 0.9 power factor (lagging).

P13.12. Consider a 60-Hz single-phase transformer with the following specifications:

- Core: area = 0.1 m^2, length 2 m, μ_r = 5000
- Windings: 100 primary and 10 secondary turns
- Coil impedances: primary = 0.5 + j0.8 Ω; secondary = 0.005 + j0.008 Ω
- Allowable losses: 4000 W, divided equally between iron and windings

Also, we know that for 2000-W loss in the iron the maximum magnetic flux density would be B_m = 2.0 T.
- **(a)** Find the allowable kVA for the transformer.
- **(b)** Determine the equivalent circuit for the transformer.

Answer. (a) about 240 kVA

P13.13. A 60-Hz single-phase, 240/120-V transformer with $\mu_i = 4500\ \mu_o$ is shown in Fig. P13.13. The primary resistance is 0.3 Ω and primary leakage reactance is 0.8 Ω. The secondary resistance is 0.075 Ω and secondary leakage reactance is 0.15 Ω.

(a) Find the reluctance of the magnetic structure.

(b) Determine the magnetizing current, $\mathbf{I}_\phi$, required on the high-voltage side to magnetize the iron for the voltages shown.

(c) If the hysteresis and eddy-current losses are 60 W, what is the in-phase current in the primary required to supply these losses? There is no load on the secondary.

(d) What is the no-load current?

(e) Give the equivalent circuit for the transformer with all circuit elements referred to the primary.

(f) If the total losses of the transformer must be kept below 135 W, what should be the kVA rating of the device?

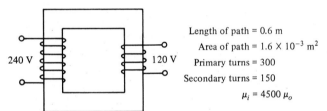

Length of path = 0.6 m

Area of path = 1.6 × 10^{-3} m^2

Primary turns = 300

Secondary turns = 150

$\mu_i = 4500\ \mu_o$

Figure P13.13

P13.14. Figure P13.14 shows the equivalent circuit for a 2400/240-V, 75-kVA transformer. Assume that the transformer is operating at rated secondary voltage and apparent power, with a 0.9 power factor (lagging).

(a) What is the turns ratio?

(b) Find the iron losses.

(c) Find the copper losses.

(d) What is the efficiency?

(e) What would be the exciting current, I_e?

(f) Determine the input power factor. *Hint:* Determine the reactive powers since you know the real powers.

Answers. (a) 10/1; (b) 576   W; (c) 977   W; (d) 97.8%; (e) 0.537   A; (f) 0.888.

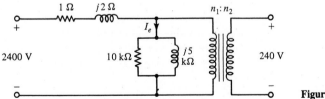

Figure P13.14

P13.15. Figure P13.15 shows the equivalent circuit for a 1-MVA single-phase transformer. The voltages given are rated values.

(a) What current would flow on the high-voltage side if the low-voltage side were open-circuited?

(b) Determine the copper losses at full load.

(c) Determine the iron losses at full load.

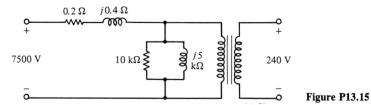

Figure P13.15

P13.16. The equivalent circuit in Fig. P13.16 represents a 240/120-V, 10-kVA single-phase transformer.

(a) What is the exciting current for the transformer referred to the high-voltage side?

(b) Estimate the iron and copper losses if the transformer is operating at 80% of capacity.

(c) If operating at 80% of rated capacity and a power factor of 0.9 lagging, find the input voltage and power to the transformer.

(d) At what percent of rated capacity is the transformer efficiency maximum, and what is the corresponding efficiency?

Answers. (a) $0.72 \underline{/-56.3°}$; (b) 207 W; (c) 245.9 V, 7311 W; (d) $\eta_{max} = 97.2\%$ at 74.4% of rated

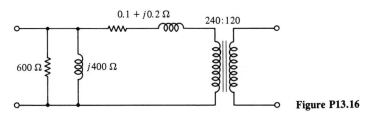

Figure P13.16

P13.17. The equivalent circuit for a 460/230-V, 9.2-kVA transformer is shown in Fig. P13.17a.

(a) Figure P13.17b shows a table to record the results of an open-circuit/short-circuit test. Fill in every slot, even if the quantity would not normally be measured. Assume that the wattmeters are connected to measure the input power on the side supplying the power.

(b) If the secondary voltage is 230 V and current 10 A with unity power factor, find the primary voltage and the efficiency.

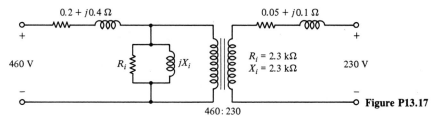

Figure P13.17

(a)

Condition	V_H	I_H	P_H	V_L	I_L	P_L
Open circuit						
Short circuit						

Figure P13.17 (cont.)

(b)

P13.18. A transformer is tested with an open-circuit/short-circuit test at rated voltage and current values, with the results shown in Table P13.18.
  (a) Find the exciting current, as seen from the primary.
  (b) Find the apparent power rating of the transformer.
  (c) Find the efficiency if operated at 80% full load capacity and 0.95 PF.

TABLE P13.18

	Primary			Secondary		
Test	V	I	P	V	I	P
SC	28	20	350	0	N/A	N/A
OC	480	0	N/A	240	1.93	290

P13.19. A transformer operated at unity power factor has a maximum efficiency of 94%, which occurs at a load of 30 kVA. What is its efficiency at a load of 20 kVA, unity power factor?

P13.20. A 240/120-V, 720-VA single-phase transformer has an OC/SC test. The power into the transformer is 25 W in the open-circuit test and 30 W on the short-circuit test.
  (a) Find the current in the low-voltage winding during the short-circuit test, assuming that the voltage is applied to the high-voltage side.
  (b) Find the voltage across the high-voltage winding during the open-circuit test, assuming that the voltage is applied to the low-voltage side.
  (c) Determine the efficiency if operated with a 0.95 power factor and full load.

P13.21. A 50-kVA, 2400/240-V, single-phase, 60-Hz transformer is tested in an open-circuit/short-circuit test, with the results shown in Table P13.21.
  (a) Derive an equivalent circuit for the transformer with all components referred to the primary.
  (b) The load on the transformer is 50 kVA with a power factor of 0.9, lagging. The output voltage is 240 V. Find the efficiency.
  (c) Determine the input voltage required for the condition described in part (b).

TABLE P13.21

	Primary			Secondary		
Test	V	I	P	V	I	P
Open circuit	—	—	—	240	4.22	650
Short circuit	72	20.8	800	—	—	—

P13.22. Figure P13.22 shows (a) the circuit symbol, (b) the wiring connection, and (c) the equivalent circuit for the same transformer connected in the usual mode and as an *autotransformer*. This transformer is a 720-VA, 120/120-V transformer when connected in the normal way. Assume the same flux levels in both connections.

(a) What are the primary and secondary voltages of the autotransformer?

(b) What are the parameters in the equivalent circuit of the autotransformer shown in Fig. P13.22d?

(c) What is the apparent power rating of the autotransformer, assuming the same losses in both cases?

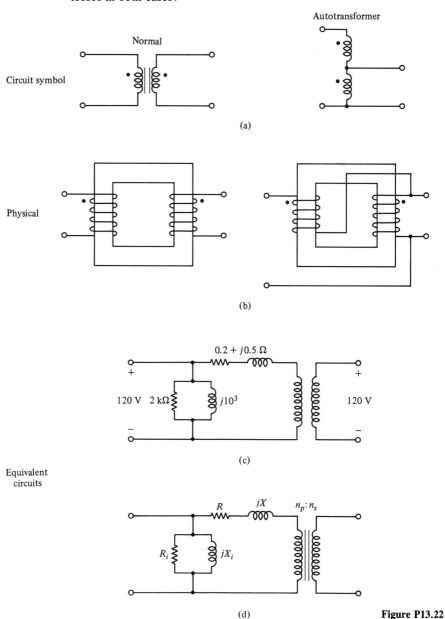

(a)

(b)

(c)

(d)

Figure P13.22

P13.23. Three single-phase 50-kVA transformers, each 600/240-V devices, are used as a three-phase transformer. There are four possible connections, depending on whether primaries or secondaries are placed in delta or wye connections. The system is operated at its rated kVA. Fill in the line voltages and current in Table P13.23. Assume ideal transformers.

TABLE P13.23

	Primary		Secondary	
Connection	Line Voltage	Line Current	Line Voltage	Line Current
Δ–Δ				
Δ–Y				
Y–Y				
Y–Δ				

P13.24. A bank of three single-phase transformers is required to transform 300 kVA of three-phase power from 13,200 V to 440 V. It is decided to use a wye connection on the high-voltage primary and a delta on the low-voltage secondary. Find the required turns ratio for the transformers and calculate the primary and secondary currents in the transformer windings. Assume ideal transformers.

P13.25. A three-phase 60-Hz transformer is used to reduce 4160 V to 240 V. The rating of the transformer is 30 kVA. When operated at rated kVA, the transformer losses are 1150 W, which is 30% iron losses and 35% each primary and secondary copper losses.

(a) At a load of 20 kW, 0.85 power factor, lagging, determine the efficiency of the transformer.

(b) If the reactive power required by the transformer were 2 kVAR for the condition described in part (a), *estimate* the input voltage to the transformer if the output voltage were 240 V exactly.

P13.26. A three-phase transformer is connected Y–Δ and rated 13-kV/480-V, 25 kVA. The efficiency is 95% at full-load, unity power factor, and losses are divided 60% iron, 40% copper. Determine the per-phase equivalent circuit for the transformer, with numerical values for the turns ratio, the resistors in the circuit, and the voltage levels. No values for the inductors are required.

P13.27. A three-phase transformer has the following parameters: 2400/480 V, 60 kVA, primary resistance 1 Ω per phase, primary leakage reactance 1.3 Ω per phase, secondary resistance 0.04 Ω per phase, secondary leakage reactance 0.06 Ω per phase, iron losses 600 W, and no-load current 0.240 A on high-voltage side.

(a) Give the per-phase equivalent circuit for the transformer with all circuit quantities referred to the primary.

(b) For rated output voltage and kVA with 0.9 power factor, lagging, determine the input line voltage.

(c) Determine the efficiency for the conditions in part (b).

Section 13.4: Forces in Magnetic Systems

P13.28. The reluctance of a magnetic system is given by the formula

$$\mathcal{R}(x) = \frac{10^6}{x + 0.001}$$

with x in meters. Find the force generated by the system at 1000 A-t of mmf.

P13.29. The current and flux linkages of a magnetic system are related by the formula

$$i = \lambda(1 + \sin 2|\theta|), \qquad |\theta| < \frac{\pi}{2}$$

with θ in radians. If $i = 1$ A, find the angle at which the torque is -0.3 N-m.

P13.30. An ac relay has a cross-sectional area of 1 cm^2 and a path length of 10 cm in iron with $\mu_i = 5000\mu_o$. With the relay open, the gap is 2 mm, and with the relay energized, the minimum gap is 0.1 mm due to roughness. The relay coil has 5000 turns and operates from 12 V (rms) ac, 60 Hz.
(a) Find the current into the relay with the gap closed.
(b) Find the time – average force on the movable member at maximum gap width.
(c) Find the time – average force on the movable member at minimum gap width.

P13.31. The device shown in Fig. P13.31 is a doorbell ringer, where a magnetic force is operated against a spring (not shown). The ringer is excited by 12 V rms, 60 Hz ac. The inductance of the structure is $L(x) = 2e^{-x/2}$ H, where x is in inches.
(a) At $x = 0$, what is the rms current in the coil?
(b) At $x = 0$, what is the time-average force on the plunger? Consider force positive if in the direction to increase x. *Hint:* Distance must be expressed in meters.

Answers. (a) 15.9 mA (rms); (b) -4.99×10^{-3} N

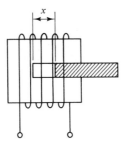

Figure P13.31

P13.32. The magnetic structure shown in Fig. P13.32a has an inductance given by the graph in Fig. P13.32b. Find the force at $d = 0.5$ cm and $i = 1.0$ A.

Figure P13.32

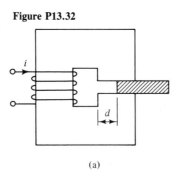

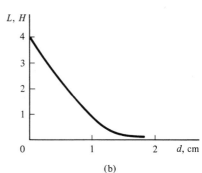

(a)

(b)

P13.33. An electromechanical transducer is shown in Fig. P13.33a. The coil has 1000 turns, and you may neglect stray flux and losses. An experiment is performed in which the coil is excited with 120 V (rms) ac and the current is measured as the gap length is changed. The results are given in Fig. P13.33b. From these data, you are to determine the time-average force developed across the gap at $x = 2$ mm. The frequency is 60 Hz.

Answer.   Force ≈ -105 N

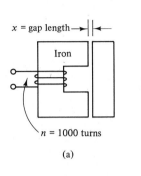

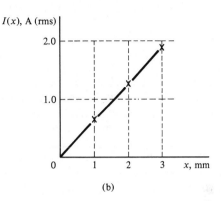

(a)

(b)

Figure P13.33

P13.34. The flux linkages created by a current i are given by the equation $\lambda = (1 + x/10)i$, where x is a mechanical displacement in the system in centimeters.
   (a) Find the inductance at $x = 0$.
   (b) Find the energy required to increase the current from 0 to 2 A for $x = 0$.
   (c) What magnetically generated force operates on the mechanical system in the direction of increasing x for $i = 2$ A?

Answer.   (c) $+20$ N

P13.35. A magnetic structure, with its flux-linkage versus current curve, is shown in Fig. P13.35.
   (a) Find the inductance for $i < 1$ A.
   (b) Determine the energy required to increase the current from 0 to 1 A.
   (c) Determine the energy required to increase the current from 1 to 2 A.
   (d) The current is kept at 2 A and a 1-mm gap is sawed in the iron path. The filings are found to weigh 10 g and the density of the iron is 7.65 g/cm^3.
      i. What is the final flux linkages after the gap is opened?
      ii. As the sawing is going on, does the source maintaining the current receive or give energy to the magnetic system?

Figure P13.35

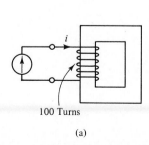

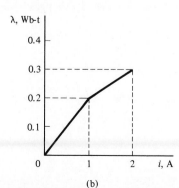

(a)

(b)

549

P13.36. Consider a magnetic structure with a moving mechanical member. Figure P13.36 shows the flux linkage–current characteristic for the structure for two values of mechanical displacement. The areas of several parts of the diagram are given.

(a) *Estimate* the mechanical force generated by the structure at $\lambda = 1.8$ Wb-t, $i = 1.0$ A, and $x = 1.0$ cm.

(b) *Estimate* the stored energy at the same point.

(c) *Estimate* the inductance at $x = 1.0$ cm for small currents.

Answer. (a) 97 N approximately

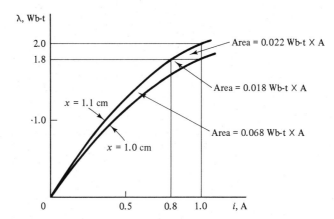

Figure P13.36

P13.37. For the rotational device in Fig. 13.21, the reluctance of the air gap is

$$\mathcal{R}(\theta) = 10^5(1 + |\theta|)$$

where θ is expressed in radians and $\theta < \pi/2$. Neglect mmf losses in the iron.

(a) Find the torque at $\theta = +\pi/4$ for a current of 10 A and 100 turns.

(b) Find the inductance of the system at the same angle.

(c) How much co-energy is stored at this angle and current?

Answers. (a) -1.57 N-m; (c) 2.80 J

P13.38. A 120-V, 60-Hz hair clipper (Fig. P13.38a) has a vibration motor dimensioned in Fig. P13.38b. The spring is adjusted to resonate with the mass of the vibrator for maximum motion. The coil has 5000 turns, the average gap size is about 0.08 in., and the areas are as shown. Ignore mmf losses in the iron.

(a) At what frequency should the spring–mass system resonate?

(b) Estimate the input current required, assuming the real power into the motor is 10 W.

(c) Give a parallel equivalent circuit for the vibrator motor.

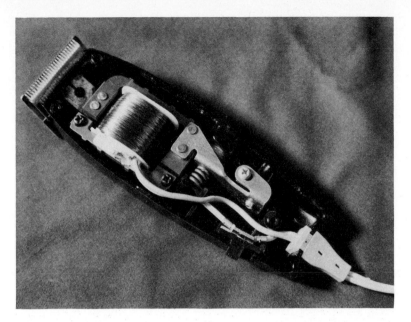

(a)

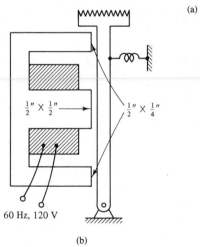

$\frac{1}{2}'' \times \frac{1}{2}''$

$\frac{1}{2}'' \times \frac{1}{4}''$

60 Hz, 120 V

(b)

Figure P13.38

14 Introduction to Motors

14.1 BASIC MOTOR CONCEPTS

Contents of the next three chapters. This chapter begins with a discussion of general considerations that apply to most types of electrical motors. We show how torque is produced by rotor and stator magnetic fluxes. Then we analyze cylindrical magnetic structures and show how magnetic flux is made to rotate in such structures. In Chapter 15 we describe and analyze ac motors. We emphasize primarily the three-phase induction motor but include the single-phase induction motor and the synchronous machine. Chapter 16 deals with the dc motor.

14.1.1 Terminology

Figure 14.1 suggests an electrical motor. It has an electrical input of voltage and current, a mechanical output in the form of torque and rotation, and losses represented as heat. The motor consists *mechanically* of a stator, which does not rotate, a rotor, which can rotate, and an air gap to permit motion. Electrical force (torque) is produced by currents interacting with magnetic flux.

$$\vec{f} = \vec{i} \times \vec{B} \tag{14.1}$$

field — armature

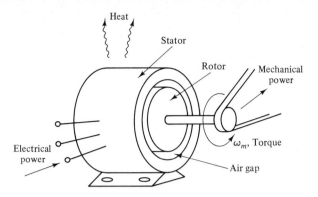

Figure 14.1 A basic motor.

We distinguish *electrically* between the field circuit, which produces a magnetic flux, and the armature circuit, which carries a current. The field usually has many turns of wire carrying relatively small currents, and the armature has few turns of larger wire carrying relatively large currents. In induction motors and dc motors, the field is on the stator and the armature on the rotor. For the synchronous motor, the field is on the rotor and the armature on the stator.

14.1.2 Steady-state Operation of Motors

Electrical input.   We consider first steady-state operation, such as a fan motor turning at constant speed. The electrical quantities of interest in steady state are the input voltage and current and, for ac motors, the power factor. In many cases, analysis of the motor suggests an equivalent circuit that accounts for losses, energy storage, and the conversion of electrical power into mechanical power.

Mechanical output.   The output characteristic is the magnetically generated torque as a function of rotation speed. A mechanical model of the motor should account for mechanical losses, such as friction and losses due to the stirring of the air by the rotor, and the output torque and speed. The operating speed of the motor is jointly determined by the output torque characteristic of the motor and the torque requirement of the load as functions of speed.

Example.   As an example of the factors involved in steady-state operation, consider a 60-Hz, single-phase induction motor with the following nameplate† information: 1 hp, 1725 rpm, 115/230 V, 14.4/7.2 A, service factor = 1.15. The dual voltage/current rating means that the motor has windings that can be placed in series for high-voltage/low-current operation or placed in parallel for low-voltage/high-current operation. The nameplate information reveals that, supplied with

---

† The nameplate of a motor gives the type of motor, name of the manufacturer, the rated voltages, frequency, currents, apparent power, speed, and output power, usually in horsepower. The distinctions made between rated limits, actual limits, and operating levels made on page 528 apply also to motors.

rated voltage (say, 230 V) and at rated output power (1 hp), the current will be 7.2 A(rms) and the speed will be 1725 rpm. At reduced load, the current will be lower and the speed slightly higher. The *service factor* implies that the motor can sustain a 15% overload in output power when operated with rated voltage. However, the nameplate speed and current refer to operation at rated voltage without service factor being considered.

For the motor described above, assume the electrical losses are 100 W and the mechanical losses are 5% of the power converted to mechanical form; that is, the *mechanical* efficiency is 95%. We will find the input electrical power and power factor, the output torque, and the overall efficiency under nameplate conditions.

The voltage, current, and output power are the nameplate values. The output power is 1 hp × 746 W/hp = 746 W, so the developed power is 746/0.95 = 785.3 W. Developed power is the power converted from electrical to mechanical form in the motor. The input power is higher by 100 W: P_{in} = 885.3 W. The input power is VI × PF, where V and I are rms values: 230 V × 7.2 A × PF = 885.3 W. Thus the power factor must be

$$PF = \frac{885.3}{230 \times 7.2} = 0.535 \tag{14.2}$$

The output torque will be

$$T_{out} = \frac{P_{out}}{\omega_m} = \frac{746}{1725(2\pi/60)} = 4.13 \text{ N-m} \tag{14.3}$$

where the mechanical speed has been converted to radians/second. Finally the overall efficiency will be output power divided by input power: η = 746/885.3 = 84.3%.

Motor and load characteristics. Figure 14.2 shows the diverse torque characteristics that can be achieved for several types of electrical motors. Even within a specific motor type, the designer can tailor the motor characteristics

Figure 14.2 Motor torque characteristics.

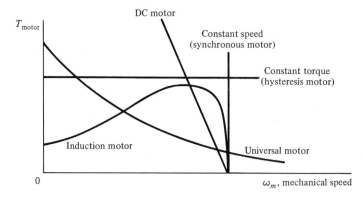

Introduction to Motors   Chap. 14

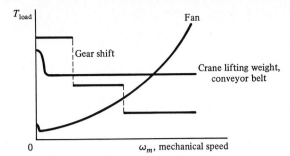

Figure 14.3 Load characteristics.

within broad limits. Figure 14-3 shows torque requirements for various mechanical loads.

System operation is jointly determined by load requirements and motor characteristics. Consider, for example, connecting an induction motor to a fan starting from rest, as shown in Fig. 14.4. For rotation to occur, the starting torque of the motor must exceed the starting-torque requirement of the load. The excess torque, $\Delta T(\omega_m) = T_{motor} - T_{load}$, will accelerate the system to the speed where the two characteristics cross. This crossing determines the steady-state speed of the system.

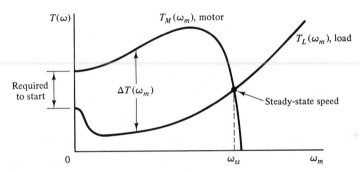

Figure 14.4 Motor and load characteristics. The load is accelerated by the excess torque.

14.1.3 Dynamic Operation

In addition to steady-state operation, we are also interested in characterizing motors for transient, or dynamic, operation. For example, we may need to predict the starting current and the time required to reach the steady-state speed, which is called the runup time. In dynamic operation, the rotor and load moments of inertia become factors, and energy processes are more complicated than for steady-state operation. Although this book deals primarily with steady-state operation, we show below how to calculate the runup time.

Runup time. We can calculate the time required to reach steady-state speed by integrating the equation of motion. In general, for a rotational system,

$$J\frac{d^2\theta_m}{dt^2} + B\frac{d\theta_m}{dt} + K\theta_m = T_{dev} - T_{load} = \Delta T(\omega_m) \tag{14.4}$$

where θ_m represents mechanical angle, J the moment of inertia, B a loss factor that can be combined with the load, T_{dev} the magnetically developed torque, and K a spring constant, which is zero for any system that is free to rotate continuously, as here. Thus we may change variables from θ_m to ω_m, the mechanical angular velocity:

$$J\frac{d\omega_m}{dt} = \Delta T^*(\omega_m) \Rightarrow dt = \frac{J\,d\omega}{\Delta T^*(\omega_m)} \tag{14.5}$$

where $\Delta T^*(\omega_m) = T_{dev} - T_{load} - B\omega_m$. Equation (14.5) integrates to

$$t_{ru} = \int_0^{\omega_{ss}} \frac{J\,d\omega_m}{\Delta T^*(\omega_m)} \approx \Sigma \frac{J\Delta\omega_m}{\Delta T^*(\omega_i)} \tag{14.6}$$

where ω_{ss} is the steady-state speed and t_{ru} is the runup time. We may approximate the runup time as indicated by the second form of Eq. (14.6) and illustrated in Fig. 14.5.

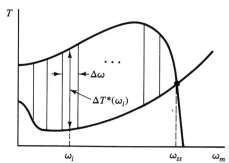

Figure 14.5 Runup time calculation.

Example: runup calculation. Consider a motor with a constant output torque of 5 N-m, driving a load requiring a torque proportional to speed. The steady-state speed is 800 rpm, and the combined moment of inertia of motor and load is 0.02 kg-m². Find the time required to reach equilibrium speed, starting from standstill.

The load requirement, including motor losses, would be $T = C\omega_m$, where C is a constant to be determined. The equilibrium speed would be $\omega_{ss} = 800 \times 2\pi/60 = 83.78$ rad/s; hence the constant is $C = 5/83.78 = 0.05968$ N-m/(rad/s). By Eq. (14.6), the runup time is

$$t_{ru} = 0.02 \int_0^{83.78} \frac{d\omega_m}{5 - 0.05968\omega_m} = \frac{0.02}{-0.05968}\ln(5 - 0.05968\omega)\Big|_0^{83.78} \tag{14.7}$$

Equation (14.7) leads to an infinite result. The difficulty is that the system speed approaches its steady-state asymptotically, so theoretically it never reaches equilibrium. One way out of this difficulty is to assume a "final" speed slightly below the steady-state speed. For Eq. (14.7), changing the upper limit to 98% of the steady-state speed gives a runup time of 1.311 s. Another way to resolve the problem in this case is to solve the differential equation for the speed. We leave for a homework problem the proof that the speed approaches steady-state speed exponentially with a time constant of 0.3351 s.

14.1.4 Conditions for Motor Action

Torque generation. The steady transformation of electrical into rotational mechanical power requires torque and rotation.

$$P_{dev} = T_{dev} \times \omega_m \tag{14.8}$$

where P_{dev} is the power, T_{dev} is the developed torque, and ω_m is the angular speed of the rotor. Let us consider first the means for producing torque. Figure 14.6a shows a compass at an angle θ_m relative to a magnetic flux. At $\theta_m = 0°$, the compass would be in stable equilibrium; no torque would be produced. For θ_m in the first quadrant, as shown in Fig. 14.6a, a torque in the negative θ_m direction would tend to restore the compass to equilibrium. This torque would increase up to $\theta_m = 90°$ and then would decrease to zero at $\theta_m = 180°$. This position has an unstable equilibrium because a small displacement, either way, would produce torque tending to *increase* the displacement. The $T(\theta_m)$ curve shown in Fig. 14.6b is sinusoidal because the lever arm for the torque varies as $\sin \theta_m$.

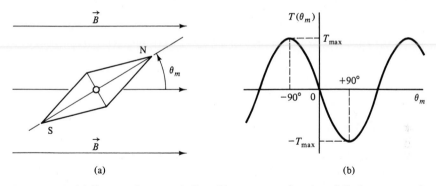

(a)   (b)

Figure 14.6 (a) Compass in magnetic flux; (b) torque as a function of displacement angle.

We consider now the torque generated by the stator and rotor magnetic poles in Fig. 14.7. Here we also will have a stable equilibrium at $\theta_m = 0$, an unstable equilibrium at $\theta_m = 180°$, and a maximum restoring torque around $\theta_m = 90°$. Although it is not obvious, the torque characteristic for this structure is like that of the compass, as will be demonstrated later in this chapter. To summarize, torque can be generated magnetically through a displacement between stator and rotor poles.

Ways to achieve motor action. As indicated by Eq. (14.8), torque *and* rotation are required to effect the steady conversion of electrical energy to mechanical energy. Electrical motors operate from this principle but differ in:

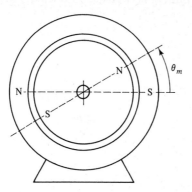

Figure 14.7 Stator and rotor poles.

- How the rotor and stator poles are produced
- Whether the rotor or stator poles cause rotation

We will examine the major types of electrical motors to see how poles are produced and rotated and to determine the motor characteristics that result. To produce torque in all motors, the rotor and stator poles remain in stable but displaced alignment.

Plan for this chapter.   We have now introduced motor terminology, described the electrical and mechanical quantities of concern, and discussed the fundamental principle of electrical motors. In the next section we show by analyzing cylindrical magnetic structures how stator poles can be produced. Then we show how the poles can be made to rotate. The final section derives the torque characteristic for the magnetic structure shown in Fig. 14.7.

14.1.5 Check Your Understanding

1. In a motor the field current is usually larger than the armature current. (T/F?)
2. The output torque of a motor in operation depends in part on the load requirements. (T/F?)
3. The product of the power factor and the efficiency of an ac motor is always equal to the output mechanical power divided by the input electrical apparent power. (T/F?)
4. An induction motor reaches 1780 rpm in 0.6 s without a load. It requires 1.5 s to reach 1750 rpm with a load. Estimate the moment of inertia of the load divided by the moment of inertia of the motor, assuming the slowdown is caused mainly by increased inertia.
5. If the maximum torque in Fig. 14.6b is 5 N-m, determine the work required to move the rotor from its position of stable equilibrium to its position of unstable equilibrium.

Answers.   1. False;   2. true;   3. true;   4. 1.5;   5. 10 J.

14.2 ANALYSIS OF CYLINDRICAL MAGNETIC STRUCTURES

14.2.1 Magnetic Structures for AC Machines

Description of structure.   Alternating-current motors normally have an iron structure with cylindrical symmetry. Figure 14.8a shows an axial view of such a cylindrical magnetic structure with the current-carrying wires on the stator

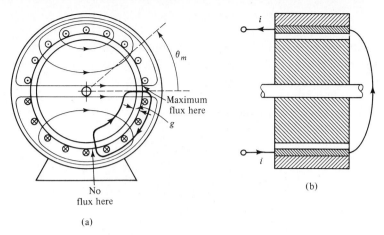

(a)

(b)

Figure 14.8 Cylindrical magnetic structure: (a) axial view; (b) side view. The mechanical angle is θ_m, and the width of the air gap is g.

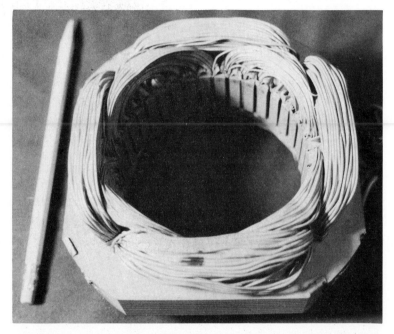

Figure 14.9 Four-pole stator structure for single-phase induction motor. The main windings for the motor are horizontal and vertical. The smaller starter windings are on the diagonals. The windings may be connected in parallel for 120-V operation or in series for 240-V operation.

side of the air gap. The currents in the coils go from front to back on the bottom, cross over to the top at the back, and return on top, as shown in Fig. 14.8b. The field coils lie in slots that have been milled in the inner circumference of the stator. The stator structure for a small motor is shown in Fig. 14.9. The rotor is assumed to be a coaxial iron cylinder in our analysis.

Flux distribution.   Although Fig. 14.8 suggests one conductor in each slot, in practice there would be many, as shown in Fig. 14.9. The mmf of the coil, which is distributed around the circumference, is tapered sinusoidally. This mmf distribution is accomplished by having heavy concentrations in the center of the winding, tapered to few wires† at the edge of the windings. The resulting flux distribution, shown in Fig. 14.8a, has no flux crossing the air gap at the top and bottom, maximum flux density in the air gap in the positive radial direction at θ_m = 0°, and maximum flux density in the air gap in the negative radial direction at θ_m = 180°. This pattern may be interpreted as a distributed south magnetic pole in the stator centered at θ_m = 0° and a distributed north magnetic pole in the stator centered at θ_m = 180°. Thus we are describing a two-pole structure. Because of the sinusoidal taper of the mmf, the flux density will be sinusoidal in form:

$$B(\theta_m) = B_\mathrm{m} \cos \theta_m \tag{14.9}$$

where B_m is the maximum flux density at θ_m = 0°.

Maximum flux density.   We may determine the maximum flux density by applying Ampere's circuital law around the heavily drawn path indicated in Fig. 14.8a. The result is

$$\oint \vec{H} \cdot \vec{dl} = \text{enclosed current} = \frac{n}{2} i \tag{14.10}$$

where n is the number of turns in the coil and i is the current in each turn. The line integral may be expanded into four contributions, two from crossing the air gap twice and two from the paths within the rotor and stator iron. We assume that the iron has a large μ_i, and hence the magnetic field in the iron makes a negligible contribution to the line integral. Because of symmetry, the flux density at the bottom is zero and hence no contribution is made by crossing the air gap at θ_m = −90°. This leaves the contribution from crossing the air gap at θ_m = 0°. Since we are crossing in the same direction as the flux, Eq. (14.10) reduces to

$$H_{gap}g = \frac{B_\mathrm{m}}{\mu_o} g = \frac{n}{2} i$$

and thus the maximum flux in Eq. (14.9) is

$$B_\mathrm{m} = \frac{\mu_o n}{2g} i \tag{14.11}$$

14.2.2 Magnetic Structures for DC Machines

Salient-pole structures.   The magnetic structure for a dc machine is shown in Fig. 14.10a. Cylindrical symmetry is not maintained because the stator magnetic poles do not rotate. The inward protrusions of iron are called *salient*

---

† Don't worry about those partially filled slots; we will later fill them with more coils.

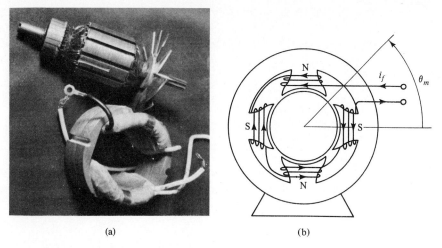

(a)   (b)

Figure 14.10 (a) Small dc motor with two poles; (b) salient-pole magnetic structure with $P = 4$ poles.

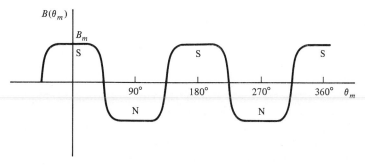

Figure 14.11 Flux density for four-pole dc magnetic structure.

poles,† which are extended in width to leave as little interpole space as practical. The stator coils are wrapped around these poles. The rotor is cylindrical, with slots for wires, as shown in Fig. 14.10a. The flux density distribution approximates a square wave, as shown in Fig. 14.11 for the four-pole structure in Fig. 14.10b.

Analysis. The maximum flux density can be determined from Ampere's circuital law around a path passing through two adjacent poles.

$$\oint \vec{H} \cdot \vec{dl} = ni \Rightarrow B_{\mathrm{m}} = \frac{\mu_o 2ni_f}{2g} \tag{14.12}$$

where ni_f is the mmf per pole and g is the width of the air gap. The factor 2 in the denominator arises because the air gap is traversed twice by the flux, and the factor 2 appears in the numerator because n is the turns on one pole and the path

† In this context, "pole" refers to the mechanical protrusion as well as the magnetic pole associated with it.

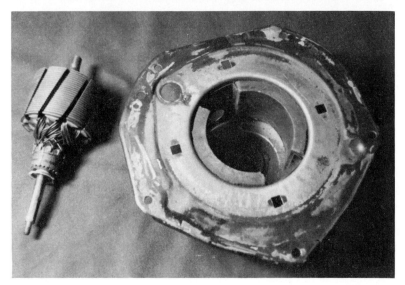

Figure 14.12 Automotive fan motor. Permanent magnets on the stator produce the magnetic field. This stator has four poles.

of integration passes through two poles. The mmf loss in the iron has been neglected.

Permanent magnet fields for DC machines. Figure 14.12 shows an automotive blower motor, which uses permanent magnets on the stator. Each molded ceramic ferrite pole piece has two poles, so the structure has four poles. Larger structures might use Alnico permanent magnets.

14.2.3 Check Your Understanding

1. A cylindrical magnetic structure cannot have an odd number of magnetic poles. (T/F?)
2. In Fig. 14.9, the wires that show are not the wires that produce the motor flux but are the crossover wires between poles. (T/F?)
3. For large flux, the air gap in a machine should be as large or small as possible. (Which?)

   Answers. 1. true; 2. true; 3. small.

14.3 ROTATING MAGNETIC FLUX FOR MOTOR ACTION

14.3.1 Two-phase Rotating Flux

The previous section showed how magnetic flux is produced in a cylindrical magnetic structure. This section shows how the stator flux (or stator magnetic poles, if you like) is (are) made to rotate. Although the three-phase motor is more common than the two-phase, the latter offers a convenient place to start. We therefore consider a stator like that in Fig. 14.8a, but with two coils: an x coil that creates a magnetic flux pattern with its maximum in the horizontal plane (shown), and a y coil that creates a magnetic flux pattern with its maximum in the vertical plane (not shown). The wires/slot distributions for both coils are tapered to produce flux patterns that are sinusoidal in mechanical angle. Hitherto, we have been considering only the x winding, for which the wires are most dense on the bottom and top of the stator; the wires for the y coil would be most dense on the sides of the stator and share intermediate slots with the x coil. The two coils are identical except for their orientation in space. If i_x is the current in the x coil and i_y the current in the y coil, then the magnetic flux densities in the air gap would be, respectively,

$$B_x(\theta_m, t) = Ki_x(t) \cos \theta_m \quad \text{and} \quad B_y(\theta_m, t) = Ki_y(t) \sin \theta_m \quad (14.13)$$

where $K = \mu_o n/2g$, from Eq. (14.11). Both fluxes are radial across the air gap and have their maximum densities at $\theta_m = 0°$ and $\theta_m = +90°$, respectively.

The two-phase currents are

$$i_x(t) = I_p \cos \omega t, \qquad \text{excites } B_m \text{ in } x\text{-direction}$$
$$i_y(t) = I_p \sin \omega t, \qquad \text{excites } B_m \text{ in } y\text{-direction} \qquad (14.14)$$

where I_p is the peak current in each coil, ω is the electrical angular frequency, and a 90° phase shift between the two phases is indicated by the sine and cosine functions.

$$B_x(\theta_m, t) = B_m \cos(\omega t) \cos(\theta_m)$$
$$B_y(\theta_m, t) = B_m \sin(\omega t) \sin(\theta_m) \qquad (14.15)$$

where B_m is the peak flux density created by each coil separately.

Rotating flux. When both coils are excited simultaneously, the flux pattern moves in the counterclockwise direction, as demonstrated mathematically in Eq. (14.16), where we have used the trigonometric identity $\cos(a - b) = \cos(a) \cos(b) + \sin(a) \sin(b)$, identifying a with ωt and b with θ_m:

$$B(\theta_m, t) = B_m[\cos(\omega t) \cos(\theta_m) + \sin(\omega t) \sin(\theta_m)] \qquad (14.16)$$
$$= B_m \cos(\omega t - \theta_m)$$

The flux density has a pattern that is constant in shape, rotating in space with the following properties:

- The flux at all points is radial across the air gap.
- The maximum flux density is constant at B_m.
- The position of maximum flux occurs at the angle where $\omega t - \theta_m = 0$, that is, at $\theta_m = \omega t$. Thus the flux maximum and the entire pattern rotate with a spatial speed of $\omega_s = \omega$ rad/s, where ω_s is called the *synchronous* speed and ω is the electrical radian frequency. For an electrical frequency of 60 Hz, the synchronous speed is $2\pi \times 60$ rad/s, or 60 rev/s, or 3600 rev/min (rpm).
- The flux exhibits angular wave motion. We call this a *rotating wave of flux.*

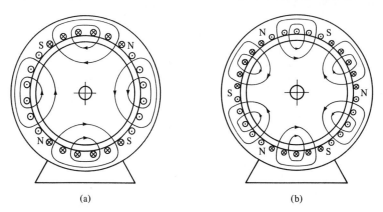

Figure 14.13 (a) Four-pole stator; (b) six-pole stator.

For *P* poles. We consider now a magnetic structure with more than two poles. Figure 14.13a shows the x winding for a stator wound with four poles, and you must imagine the y winding between the x winding. In this case the pattern of the magnetic flux advances one pole pair for each electrical cycle, and hence the synchronous (spatial) speed of the flux pattern is one-half that for two poles. Figure 14.13b shows a stator with $P = 6$ poles. If there are P poles, the flux pattern will advance $2\pi/(P/2)$ in mechanical angle for each electrical period (T), and the synchronous speed, ω_s, will therefore be

$$\omega_s = \frac{2\pi}{T(P/2)} = \frac{\omega}{P/2} \quad \text{spatial rad/s} \tag{14.17}$$

where T is the period of the line frequency, ω is the electrical angular frequency in radians/second, and P the number of stator poles. Table 14.1 tabulates the common synchronous speeds for 60 Hz.

TABLE 14.1 STANDARD VALUES OF SYNCHRONOUS SPEED FOR 60 HZ.

P	ω_s	*rpm*
2	$2\pi \times 60$	3600
4	$2\pi \times 30$	1800
6	$2\pi \times 20$	1200

14.3.2 Three-phase Rotating Flux

Figure 14.14 shows a two-pole magnetic structure wound with three coils separated 120° in space. Note that coil a corresponds to our x coil from before, with the currents entering on the bottom and returning on the top, but coil b currents enter in the first quadrant and those of coil c enter in the second quadrant.† As before, the coils are tapered sinusoidally and share slots between their most dense

† In practice, the three coils would be internally connected in wye or delta, and only three wires would be connected externally to the three-phase source.

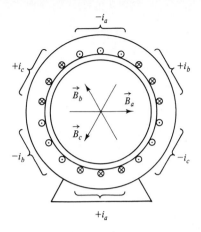

$$B_a = Ki_a(t) \cos \theta_m$$
$$B_b = Ki_b(t) \cos (\theta_m - 120°)$$
$$B_c = Ki_c(t) \cos (\theta_m - 240°)$$

Figure 14.14 A three-phase cylindrical structure.

regions. The equations of the flux patterns from the coils are given in Fig. 14.14. Note that the maximum from coil a lies in the horizontal plane, and the maxima from coils b and c are displaced 120° and 240° in space, respectively.

Physical reasoning. We now excite the three coils with three-phase currents, as shown in Fig. 14.15. We claim that the flux pattern now rotates as a rotating wave of flux, as in the two-phase case investigated above. To see that this is plausible, consider the contribution of each coil at $t = 0$. At this time, i_a is maximum, so its flux (in the horizontal direction) will be maximum. At this time, i_b and i_c are negative and equal; hence, the contribution from coil b will lie in the direction $\theta_m = -60°$, and an equal contribution from coil c will lie in the direction $\theta_m = +60°$. All three make a positive contribution in the horizontal direction, and from symmetry we expect that the combined flux maximum will be at $\theta_m = 0°$. Now consider the situation at $t = T/6$, one-sixth of an electrical cycle later. At this time, i_c has its negative maximum, so its flux has a positive maximum at $\theta_m = +60°$. Simultaneously, i_a and i_b are equal and positive; hence their fluxes also make positive contributions to the flux from i_c, and from sym-

Figure 14.15 The three-phase currents, with three specific times indicated.

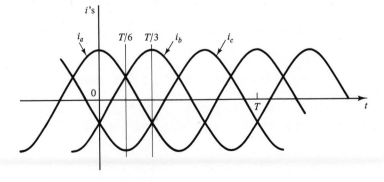

metry we expect that the maximum will have moved up to $\theta_m = +60°$. Similarly, we could show that at $t = T/3$ the maximum will have moved to $\theta_m = +120°$. The flux pattern appears to be rotating.

Mathematical proof. The mathematical proof is straightforward. The three-phase currents are

$$i_a(t) = I_p \cos(\omega t)$$

$$i_b(t) = I_p \cos(\omega t - 120°) \qquad (14.18)$$

$$i_c(t) = I_p \cos(\omega t - 240°)$$

such that, for example, the flux from coil a is

$$B_a(\theta_m, t) = B'_m \cos(\omega t) \cos(\theta_m) \qquad (14.19)$$

where B'_m is the peak flux from one coil alone. The combined flux density will be

$$B(\theta_m, t) = B'_m[\cos(\omega t)\cos(\theta_m) + \cos(\omega t - 120°)\cos(\theta_m - 120°) \qquad (14.20)$$
$$+ \cos(\omega t - 240°)\cos(\theta_m - 240°)]$$

We use the trigonometric identity

$$\cos(a)\cos(b) = \tfrac{1}{2}[\cos(a - b) + \cos(a + b)] \qquad (14.21)$$

For example,

$$\cos(\omega t - 120°)\cos(\theta_m - 120°)$$
$$\qquad (14.22)$$
$$= \tfrac{1}{2}[\cos(\omega t - \theta_m) + \cos(\omega t + \theta_m - 240°)]$$

Thus we obtain the results

$$B(\theta_m, t) = B'_m \{\tfrac{3}{2}\cos(\omega t - \theta_m) + \tfrac{1}{2}[\cos(\omega t + \theta_m)$$
$$\qquad (14.23)$$
$$+ \cos(\omega t + \theta_m - 240°) + \cos(\omega t + \theta_m - 480°)]\}$$

The three bracketed terms add to zero at all times because they are a balanced three-phase set.† Hence we are left with

$$B(\theta_m, t) = \tfrac{3}{2}B'_m \cos(\omega t - \theta_m) = B_m \cos(\omega t - \theta_m) \qquad (14.24)$$

where $B_m = (\tfrac{3}{2})B'_m$ is the maximum flux with all coils operating together. The three coils operating together give a 50% increase in the flux maximum from what each gives individually. Equation (14.24) has the same form as Eq. (14.16) and represents a rotating flux pattern of flux.

Character of the rotating flux. Let us summarize the character of the rotating wave of flux:

- The flux is radially directed across the air gap.

---

† Realize that $-480°$ is the same as $-120°$.

- If time is fixed, the flux is sinusoidal in space with the maximum flux at $\theta_s = \omega t$. Thus the synchronous speed (for two poles) is equal to the electrical angular frequency.
- For P poles, the synchronous speed slows down to $\omega_s = \omega/(P/2)$.
- It is easy to show that changing the three-phase sequence from abc to acb reverses the direction of rotation.
- At a fixed θ_m, the flux magnitude is sinusoidal in time. The time of peak flux is $t_p = P/2\ \theta_m/\omega$.
- Because at every point in the air gap $B(\theta_m, t)$ is sinusoidal in time, it can be represented in the frequency domain by a phasor:

$$B(\theta_m, t) = B_m \cos(\omega t - P/2\ \theta_m) \Rightarrow \underline{B}(\theta_m) = B_m e^{-j(P/2)\theta_m} \qquad (14.25)$$

where $\underline{B}(\theta_m)$ is the phasor representation of the sinusoidal flux density.

Thus, as we move counterclockwise around the stator in real space, the phasor moves clockwise in the complex plane, as shown in Fig. 14.16. This phase change represents delay of the flux maximum for points away from the origin at $\theta_m = 0°$, where the phase is zero.

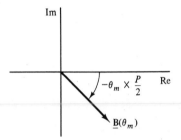

Figure 14.16 The phasor representing the magnetic flux density.

Forward and reverse rotating fluxes. Pure rotation is described in the time domain by the equation

$$B(\theta_m, t) = B_m \cos(\omega t \mp \theta_m) \qquad (14.26)$$

where the minus sign indicates *forward* (counterclockwise) rotation, and the plus sign indicates *reverse* (clockwise) rotation. Equation (14.26) and henceforth assumes two poles. Because the flux density at every point is a sinusoidal function of time, it can be expressed by a phasor

$$\underline{B}(\theta_m) = B_m e^{\mp j\theta_m} \qquad (14.27)$$

The forward and reverse fluxes would be described by

$$\underline{B}(\theta_m) = B_f e^{-j\theta_m} \quad \text{or} \quad B_r e^{+j\theta_m} \qquad (14.28)$$

where B_f and B_r are the peak values of the forward and reverse fluxes, respectively.

In the previous section, we showed how coils are wound in cylindrical structures to produce magnetic flux. In this section we have demonstrated how the flux can be rotated with two- and three-phase currents. These principles can explain the operation of two- and three-phase machines. To explain the single-phase induction motor, we must investigate rotating fluxes in more depth.

14.3.3 Single-phase Expressed as Rotating Flux

Figure 14.17 shows a cylindrical magnetic structure with one winding excited by a single-phase current. The magnetic flux density in the air gap is given by

$$B(\theta_m, t) = B_m \cos \theta_m \cos \omega t \tag{14.29}$$

or as a phasor by

$$\underline{B}(\theta_m) = B_m \cos \theta_m \tag{14.30}$$

Equations (14.29) and (14.30) describe an oscillating flux, not a rotating flux.

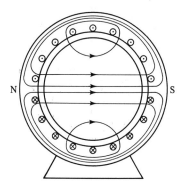

Figure 14.17   Magnetic flux due to a single coil.

Oscillating flux expressed by rotating components.   The single-phase flux density in Eq. (14.30) can be expanded into forward and reverse waves of rotating flux through Euler's theorem, which leads to the identities

$$\cos(x) = \frac{e^{+jx} + e^{-jx}}{2} \quad \text{and} \quad \sin(x) = \frac{e^{+jx} - e^{-jx}}{2j} \tag{14.31}$$

Applying the first of these to Eq. (14.30), we obtain

$$\underline{B}(\theta_m) = B_m \cos \theta_m = \underbrace{\frac{B_m}{2} e^{+j\theta_m}}_{B_r} + \underbrace{\frac{B_m}{2} e^{-j\theta_m}}_{B_f} \tag{14.32}$$

The oscillating flux is the sum of two fluxes of equal magnitude rotating in opposite directions. Figure 14.18 illustrates Eq. (14.32) in terms of magnetic poles. Here we have represented two sets of N–S pole pairs rotating in opposite directions. At $t = 0$, the poles line up and produce a flux to the right, as in Fig. 14.18a. In Fig. 14.18b, the poles cancel after one-fourth of a rotation. Then in Fig. 14.18c, after half a rotation, they again combine to produce a flux to the left. Thus an oscillating ac flux can be resolved into two counterrotating fluxes.

In the next chapter, this representation of a single-phase, oscillating flux into counterrotating fluxes is used to explain the run characteristics of single-phase induction motors.

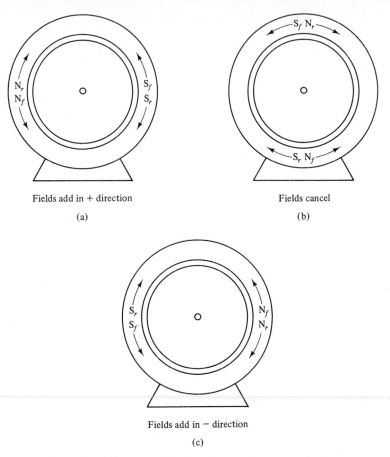

Figure 14.18 The stationary, oscillating poles can be composed of two counterrotating sets of poles: (a) at $t = 0$, the poles add; (b) at $t = T/4$ later, the poles cancel; (c) at $t = T/2$, the poles add in the negative direction.

14.3.4 Generalized Two-phase Rotating Fluxes

To explain how single-phase motors are started, we must expand further our understanding of two-phase rotating fluxes. We have seen two extremes for rotating fluxes produced by two-phase currents. At one extreme, equal and out-of-phase currents produce rotational motion of the flux. This is the most balanced situation. The most unbalanced situation is the single-phase case, which may be considered two-phase currents with one current reduced to zero. This current produces two equal, counterrotating fluxes.

The most general two-phase excitation has unequal amplitudes of the two currents and imperfect phase shift, giving the time-domain fluxes in Eq. (14.33):

$$B(\theta_m, t) = \underbrace{B_h \cos \theta_m \cos \omega t}_{\text{horizontal}} + \underbrace{B_v \sin \theta_m \cos (\omega t - \alpha)}_{\text{vertical}} \qquad (14.33)$$

where B_h is the maximum flux in the horizontal direction, B_v is the maximum flux in the vertical direction, and α is the phase shift between the two fluxes ($\alpha = 90°$ for pure two-phase). It may be shown that these fluxes combine to produce unequal counterrotating fluxes. Equation (14.34) describes the flux in the frequency domain:

$$\underline{\mathbf{B}}(\theta_m) = \underbrace{\tfrac{1}{2}[B_h + jB_v e^{-j\alpha}]e^{-j\theta_m}}_{\underline{\mathbf{B}}_f} + \underbrace{\tfrac{1}{2}[B_h - jB_v e^{-j\alpha}]e^{+j\theta_m}}_{\underline{\mathbf{B}}_r} \tag{14.34}$$

where $\underline{\mathbf{B}}_f$ is the phasor amplitude of the forward-rotating flux and $\underline{\mathbf{B}}_r$ is the phasor amplitude of the reverse-rotating flux.

Generally, the more nearly the currents approach a balanced two-phase set, the more pure the rotation of the flux. Stator circuits for single-phase motors have coils placed 90° apart in space [for $P = 2$ poles, 90°/(P/2) generally], and the currents in the coils are separated in phase as much as is practical. Ideally, the currents should be 90° out of phase and provide equal mmfs for a pure rotational flux to start the motor.

Example. Let $\alpha = 60°$, $B_h = 2$, and $B_v = 1$. Substitution into Eq. (14.34) yields $\underline{\mathbf{B}}_f = 1.455\underline{/9.9°}$ and $\underline{\mathbf{B}}_r = 0.620\underline{/-23.8°}$. Therefore, the amplitude of the forward flux wave is 1.455, and it begins at $\theta_m = +9.9°$ at $t = 0$, and the reverse wave has an amplitude of 0.620 and begins at $\theta_m = +23.8°$. These are shown in Fig. 14.19.

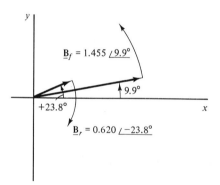

Figure 14.19 Rotating phasors for the example.

Summary of two-phase fluxes. We note the following:

- One oscillating flux can be resolved into two counterrotating flux waves of equal amplitude.
- Two oscillating fluxes, 90° displaced in space, can be resolved into two unequal counterrotating flux waves, provided the fluxes differ in phase.
- A 90° spatial displacement between fluxes is not strictly required. So long as two fluxes are not in the same direction and are not in phase, they can be resolved into counterrotating waves of flux. However, we do not need this general case to understand the single-phase induction motor, and hence we give no proof.

TABLE 14.2 SUMMARY OF TWO-PHASE ROTATING WAVES.

Condition	Result	Application
B_h only	Two equal, counterrotating waves	Single-phase run characteristics
$B_h = B_v$ $\alpha = \pm 90°$	Pure rotation $+90°$ for CCW, $-90°$ for CW	Two-phase induction motor
$B_h \neq B_v$ α arbitrary	Two unequal, counterrotating waves	Starting of single-phase motor

Table 14.2 summarizes the results of this section and anticipates the applications in the next chapter.

14.3.5 Check Your Understanding

1. The magnetic flux in the gap of a machine changes only in magnitude, not in direction. (T/F?)
2. If we change the phase sequence of a three-phase cylindrical structure, the direction of rotation of the magnetic flux will reverse. (T/F?)
3. If we reverse the connections to both windings of a two-phase cylindrical structure, the direction of rotation of the magnetic flux will reverse. (T/F?)
4. For a three-phase rotating magnetic flux, the phase difference between the currents in the coils is 0, 90°, 120°, or 240°. (Which?)
5. Which of the following can be represented by two unequal counterrotating flux waves: (1) a two-phase flux with unequal magnitudes but 90° phase shift; (2) a three-phase flux with equal amplitudes but one phase missing; (3) a single-phase flux; (4) a two-phase flux with equal amplitudes and 90° phase shift?

Answers.   1. True, it is always radial across the air gap; 2. true; 3. false; 4. 120°, but 240° will work also since 480° is the same as 120°;   5. (1) and (2).

14.4 TORQUE DEVELOPMENT BETWEEN ROTOR AND STATOR FLUXES

Introduction.   In this section, we assume that we have both rotor and stator fluxes, that each varies sinusoidally in space, and that they combine to produce the total flux in the air gap. We do not discuss here the means by which the rotor flux is established. In this section we investigate the developed torque and introduce the various power angles that play a role in ac motor characteristics.

During motor operation, the fluxes rotate in synchronism around the cylindrical structure. However, the developed torque depends on the flux magnitudes, the angles between the fluxes, and the geometry of the machine. In the following, we assume for simplicity that the fluxes are stationary.

14.4.1 Combining Fluxes

We assume the stator flux is horizontal:

$$B_s(\theta_m) = B_{\mathrm{m}S} \cos \theta_m \qquad (14.35)$$

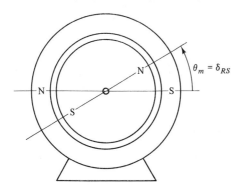

Figure 14.20 Rotor and stator poles are displaced by the rotor–stator power angle, δ_{RS}.

where B_{mS} is the maximum stator-flux density (see Figure 14.20). The rotor flux is displaced from it by a physical angle δ_{RS}, which is called the *rotor–stator power angle*.

$$B_R(\theta_m) = B_{mR}\cos(\theta_m - \delta_{RS}) \tag{14.36}$$

where B_{mR} is the maximum rotor-flux density. The total air-gap flux is the sum of the two, which may be combined *like* phasors to produce

$$B_{RS} = B_{mR}\underline{/-\delta_{RS}} + B_{mS}\underline{/0°} = \underbrace{\sqrt{B_{mR}^2 + B_{mS}^2 + 2B_{mR}B_{mS}\cos\delta_{RS}}}_{B_{mRS}}\underline{/-\delta_S} \tag{14.37}$$

where $\theta_m = \delta_S$ is the angle of the maximum of the combined flux, B_{mRS}. The phasorlike addition of the rotor and stator fluxes is shown in Fig. 14.21. We also define the *rotor power angle, δ_R*, as the angle between the rotor pole and the maximum of the total flux. Since the sum of two spatial sinusoids is still sinusoidal, the total flux is

$$B_{RS}(\theta_m) = B_{mRS}\cos(\theta_m - \delta_S) \tag{14.38}$$

This flux is used to calculate the co-energy of the system, from which the developed torque is derived.

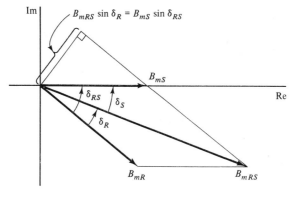

Figure 14.21 The rotor and stator fluxes combine by the law of cosines to produce the total (rotor–stator) flux. The rotor–stator power angle, δ_{RS}, is the angle between the rotor and stator poles. The rotor power angle, δ_R, is the angle between the rotor pole and the total flux.

14.4.2 Energy, Co-energy, and Torque

Energy determination. As shown in Section 13.4.3, we may determine the developed torque by taking the partial derivative with angle of either the energy or the co-energy functions. In the present case we are assuming that the rotor and stator poles (fluxes) are constant in magnitude and vary only in relative position (angle). This requires constant current excitation and hence calls for the co-energy function. However, in this linear magnetic system, the co-energy and the energy are equal and are both proportional to the square of the total flux. When the energy density function is integrated over the volume of the air gap, the result is

$$W'_m(\delta_{RS}) = K_M(B_{mR}^2 + B_{mS}^2 + 2B_{mR}B_{mS}\cos\delta_{RS}) \tag{14.39}$$

where W'_m is the co-energy of the system and K_M is the machine constant that depends on the geometry of the air gap.

Torque. The developed torque is the partial derivative of the co-energy function with respect to rotor–stator power angle, with the rotor and stator flux magnitudes kept constant.

$$T(\delta_{RS}) = +\frac{\partial W'_m}{\partial\delta_{RS}} = -2K_M B_{mR}B_{mS}\sin\delta_{RS} \tag{14.40}$$

Note from Fig. 14.21 the equality of the projections of the total flux and the stator flux:

$$B_{mS}\sin\delta_{RS} = B_{mRS}\sin\delta_R \tag{14.41}$$

This allows an alternative form of Eq. (14.40):

$$T(\delta_R) = -2K_M B_{mR}B_{mRS}\sin\delta_R \tag{14.42}$$

Equation (14.42) will prove useful in the analysis of the synchronous motor.

Note from Eq. (14.40) that the torque is zero for $\delta_{RS} = 0$ and negative if δ_{RS} is positive, meaning the magnetic interaction tends to restore alignment between the rotor and stator poles. Equation (14.40) gives the magnetically generated torque between two sets of poles with a constant angle δ_{RS} between them. The maximum torque depends on the product of the individually contributing fluxes and the geometric factors. The torque characteristic given in Eq. (14.40) is that in Fig. 14.6b, which we asserted from intuition.

Summary. We have now analyzed three requirements for motors: (1) producing magnetic flux in a cylindrical structure, (2) making that flux pattern rotate, and (3) generating torque through having rotating rotor poles that are misaligned with the rotating stator poles. In the following chapter we begin study of induction motors by showing how the rotor poles are produced by transformer action.

14.4.3 Check Your Understanding

1. For maximum torque, the rotor and stator fluxes should be aligned. (T/F?)
2. The rotor–stator power angle is always greater than the rotor power angle. (T/F?)
3. Express the torque in terms of the stator power angle, δ_S, in Fig. 14.21.

Answers. 1. False; 2. true; 3. $T(\delta_S) = -2K_M B_{mS} B_{mRS} \sin \delta_S$.

PROBLEMS

Section 14.1: Basic Motor Concepts

P14.1. A single-phase induction motor has the following nameplate information: 1 hp, 1725 rpm, 115/230 V, 9.2/4.6 A, service factor 1.0, power factor 0.83, starting current 30 A at 115 V, starting torque 9.0 lb-ft.
(a) What is the output torque under nameplate conditions?
(b) Determine the efficiency at nameplate conditions.
(c) Determine the reactive power into the machine under nameplate conditions, assuming lagging current.

P14.2. A 208-V three-phase motor runs 1720 rpm at a load requiring 9.0 N-m torque. The motor draws 5.5 A of current and the power factor is 0.92, lagging.
(a) Find the input power.
(b) Find the losses of the motor.
(c) Find the efficiency of the motor.

Answers. (a) 1823 W; (b) 202 W; (c) 89%

P14.3. A dc motor has an input voltage of 12.6 V and current of 10 A. The overall efficiency is 82%, and the output speed is 640 rpm.
(a) Find the losses of the motor.
(b) Find the output torque of the motor.
(c) Would the torque produced by magnetic interaction in the motor be greater than, less than, or the same as the output torque?

P14.4. A three-phase induction motor has the following characteristics: 25 hp, 1755 rpm, 91.7% efficiency, 230/460 V, 60.0/30.0 A, service factor 1.15.
(a) Find the power factor at nameplate conditions.
(b) Find the torque output at nameplate conditions.
(c) What is the apparent power into the machine at nameplate conditions?
(d) Determine the reactive power into the machine at nameplate conditions, assuming lagging current.

P14.5. A motor has the output torque characteristic shown in Fig. P14.5. The load torque characteristic is

$$T_L(n) = 10 \left(\frac{n}{1200} \right)^2 \text{ N-m}$$

(a) Determine the operating speed.
(b) Determine the output power.
(c) What is the maximum possible output power from the motor?

Answers. (a) 742 rpm; (b) 297 W; (c) 314 W

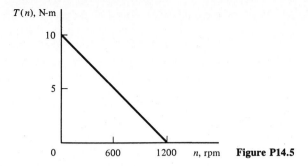

$T(n)$, N-m

Figure P14.5

P14.6. A motor has the parabolic output torque characteristic shown in Fig. P14.6. The load torque characteristic is

$$T_L(n) = 2 + 5 \left(\frac{n}{1800} \right) \text{ N-m}$$

(a) Determine the operating speed.
(b) Determine the output power.
(c) What is the maximum possible output power from the motor?

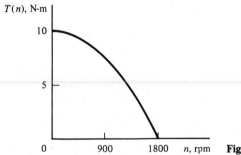

$T(n)$, N-m

Figure P14.6

P14.7. A fan requires a driving torque of the form $T_L(\omega_m) = K\omega_m^2$. The fan requires $\frac{1}{2}$ hp of drive power on the shaft to turn 1600 rpm. The fan is driven by an electrical motor with the output characteristic given in Fig. P14.7.
(a) Find the constant K, with torque and speed expressed in mks units, N-m and radians/second.
(b) Find the speed in rpm at which the fan will operate.
(c) Find the approximate time it takes the fan to reach its final speed. The moment of inertia of the motor-load is $J = 0.06$ kg-m^2.

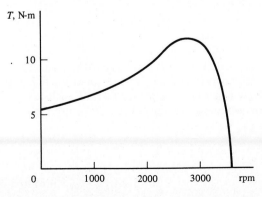

T, N-m

Figure P14.7. Motor torque characteristic.

P14.8. A motor with a moment of inertia $J = 0.5$ kg-m^2 is turning a load at a constant speed of 1160 rpm. When the load is suddenly disconnected, the instantaneous angular acceleration is $+10$ rad/s^2. Find the load torque at 1160 rpm.

P14.9. Write the differential equation for the motor load used in the runup time calculation on page 556. Use the notation J = moment of inertia, K = torque constant, and ω_{ss} = steady-state speed. This system fits the conditions of Chapter 3 transients; hence the initial value, final value, and time constant establish the response. Confirm the time constant given. Calculate the runup time for 98% of equilibrium speed based on the differential equation solution.

P14.10. Calculate the steady-state speed and runup time for the motor and load used in the example on page 556 if the load torque requirement is changed to $T = 0.01\omega^2$ N-m.

P14.11. A motor-load system starts with the characteristic

$$n(t) = 1200(1 - e^{-t/0.5}) \text{ rpm}$$

The motor-load moment is $J = 0.2$ kg-m^2, and the load torque requirement is constant at 10 N-m. Find:
(a) Motor starting torque
(b) Motor torque at 1200 rpm
(c) Motor output power at 600 rpm

Section 14.2: Analysis of Cylindrical Magnetic Structures

P14.12. The cylindrical magnetic structure shown in Fig. P14.12a has 24 slots with tapered windings as shown in the table in Fig. 14.12b. The gap width is 2 mm and the mmf loss in the iron may be neglected. The current is 50 A into the paper on the bottom half and out on the top half. Determine the maximum magnetic flux density in the air gap, and determine the location of the maximum.

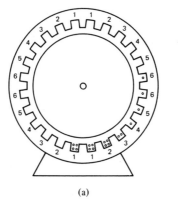

Slot number	Wires
1	15
2	14
3	12
4	9
5	6
6	2

(b)

(a)

Figure P14.12

P14.13. Two magnetic structures with cylindrical geometry are shown in Fig. P14.13. The current i flows in at the cross and out at the dot, with one wire in each slot.
(a) How many stator poles are there in each case?
(b) In each case, determine the direction of the torque on the rotor.

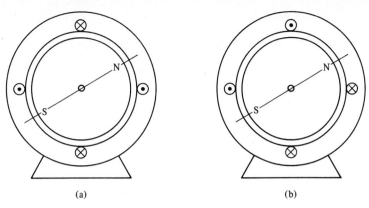

(a)   (b)

Figure P14.13. Two magnetic structures.

Section 14.3: Rotating Magnetic Flux for Motor Action

P14.14. A cylindrical rotor and a three-phase two-pole stator are excited by dc and 60-Hz ac currents, respectively. The rotor is locked and stationary in the position shown in Fig. P14.14. The rotor-flux density is constant with 2.0 T maximum, sinusoidally distributed in space. The stator produces a flux density with 1.0 T maximum, sinusoidally distributed in space and rotating ccw, as shown in Fig. P14.14. The maximum of the flux due to the stator currents is at $\theta_m = 0°$ in space at $t = 0$.

    (a) At $t = \frac{1}{360}$ s, what is the magnitude and direction (in space) of the air-gap flux density at $\theta_m = 0°$?

    (b) At $t = \frac{1}{360}$ s, at what angle in space (θ_m) is the magnetic-flux density a maximum?

    (c) At $t = \frac{1}{360}$ s, what direction would the rotor tend to move if freed?

Answers.   (a) 2.500 T   (b) 19.1°;   (c) ccw

Figure P14.14 Cylindrical structure with rotor locked.

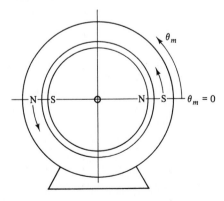

P14.15. A cylindrical magnetic structure is wound with two coils, as shown in Fig. P14.15. One, excited by i_h, produces a horizontal flux, and the other, excited by i_v, produces a vertical flux. The horizontal flux would be

$$B_h(\theta_m, t) = \frac{\mu_o n i_h(t)}{2g} \cos \theta_m$$

and the vertical flux would be

$$B_v(\theta_m, t) = \frac{\mu_o n i_v(t)}{2g} \sin \theta_m$$

where n is the number of turns in each coil and g is the gap width. Find the magnetic flux density at a mechanical angle of 30° in phasor form, that is, find $\underline{B}(\theta_m)$ at $\theta_m = 30°$ for the following conditions. Explain the nature of the flux.
(a) $i_h = 2I_p$ dc and $i_v = -I_p$ dc (At dc the phasor is real.)
(b) $i_v = 0$ and $i_h = I_p \cos(\omega t)$
(c) $i_v = I_p \sin(\omega t)$ and $i_h = 0$
(d) The currents in (b) and (c) at the same time

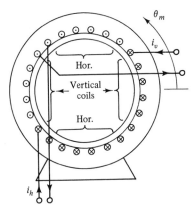

Figure P14.15 A two-phase magnetic structure.

P14.16. Table P14.16 has columns for time-domain and frequency-domain expressions for magnetic flux densities in an air gap, as a function of angle in space, and a description of the field or the source of the field. Fill in the missing information.

TABLE P14.16

Time Domain	Frequency Domain	Description
$1.5 \cos(\omega t + \theta_m)$		
	$e^{j\theta_m} - e^{-j\theta_m}$	
		Pure rotating flux in positive direction

P14.17. Describe in words, being as complete as possible, the nature of the air-gap flux density described by the following. Assume 60 Hz.
(a) $\underline{B}(\theta_m) = 1.2 \, e^{+j2\theta_m}$ T
(b) $B(t, \theta_m) = 0.8 \cos(\omega t) \sin \theta_m$ T

P14.18. Write mathematical expressions for the following air-gap flux densities. Assume 60 Hz.

   (a) A two-pole single-phase oscillating flux with peaks in the vertical direction and nulls in the horizontal direction. Assume that the flux is maximum at 0.5 T at $t = 0$. Give a time-domain expression.

   (b) A two-pole rotating flux that moves ccw and has its positive peak of 0.5 T at $\theta_m = 180°$ at $t = 0$. Give a frequency-domain expression.

P14.19. Write the equation for the magnetic-flux density in the air gap of a cylindrical magnetic structure having the following properties:

   • The machine has 18 coils, excited by three-phase ac currents to produce a rotating flux.

   • The synchronous speed of the flux pattern is 1500 rpm.

   • The maximum flux density is 0.8 T.

   • The stator coils are tapered to produce a sinusoidal spatial flux pattern.

   • The flux density maximum is horizontal at $t = 0$.

   Answer. $B(t, \theta_m) = 0.8 \cos (942t - 6\theta_m)$ T

P14.20. A three-phase cylindrical magnetic structure has $P = 4$ poles.

   (a) The windings are excited with balanced three-phase 60-Hz currents. Describe in words the character and behavior of the flux in the machine. State assumptions.

   (b) Give a mathematical expression for the flux in the time domain as a function of mechanical angle, using the standard notation. Assume the maximum of the air-gap flux in at $\theta_m = 0°$ at $t = 0$. Let B_m = the maximum flux density in the air gap.

   (c) Give a phasor representing the flux density at $\theta_m = 30°$.

   (d) If each coil has 100 turns and carries a current of 2 A and the gap width is 2 mm, what is the maximum flux density for the machine?

P14.21. The phasor representing a ccw rotating wave at $\theta_m = 0°$ is $\underline{B}(0°) = 0.5\underline{/-45°}$ T. Assume 60 Hz.

   (a) What is $B(t, \theta_m)$ at $\theta_m = 0°$?

   (b) What is $B(t, \theta_m)$ generally, assuming a two-pole machine?

   (c) Find the phasor $\underline{B}(\theta_m)$ generally.

P14.22. The phasor representation of a rotating flux is

$$\underline{B}(\theta_m) = 1.2\underline{/-30°}e^{+j4\theta_m} \quad \text{T}$$

The machine is three phase, excited by 50-Hz currents.

   (a) How many magnetic poles does the machine have?

   (b) In which direction does the field rotate?

   (c) Find the first time after $t = 0$ when a flux maximum passes the angle $\theta_m = 0°$.

   Answers. (a) 8 poles; (b) clockwise; (c) $\frac{1}{600}$ s

P14.23. The circuit shown in Fig. P14.23 represents the start and run windings for a two-pole single-phase induction motor. During starting, the motor is like a two-phase machine, except that the currents in the windings are not 90° out of phase. The air gap width is 0.5 mm, and the number of turns for the two coils is given in Fig. P14.23. Determine the forward and reverse rotating fluxes.

Answers. $\underline{\mathbf{B}}_f = .508\underline{/24.3°}$ T and $\underline{\mathbf{B}}_r = .547\underline{/-26.3°}$ T

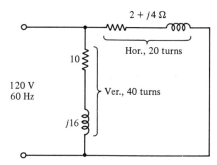

Figure P14.23

P14.24. For the phasor field,

$$\underline{\mathbf{B}}(\theta_m) = \underline{\mathbf{B}}_f e^{-j\theta_m} + \underline{\mathbf{B}}_r e^{+j\theta_m}$$

describe the situation which leads to the B's shown below. Your choices are (1) balanced three-phase currents, (2) single-phase current, (3) balanced two-phase currents with $B_h = B_v$, and (4) unbalanced two-phase currents. (More than one answer may be correct.)

(a) $\underline{\mathbf{B}}_f = 0$

(b) $\underline{\mathbf{B}}_r = 0$

(c) $\underline{\mathbf{B}}_f = \underline{\mathbf{B}}_r$

(d) $\underline{\mathbf{B}}_f = j\underline{\mathbf{B}}_r$

(e) $\underline{\mathbf{B}}_f = 2\underline{\mathbf{B}}_r$

15 Alternating-Current Motors

15.1 INDUCTION MOTORS

Introduction. The three-phase induction motor was patented by Nikola Tesla in 1888 and currently accounts for over 90% of the motors used in industry. In an induction motor, the stator poles rotate at synchronous speed and the rotor poles are induced by transformer action and also rotate at synchronous speed. The rotor rotates physically at a speed slightly slower than the synchronous speed, and slows down slightly as the output torque and power increase. In Chapter 14 we showed how the stator flux is produced, how the flux is rotated at synchronous speed, and how torque is developed between the stator flux and an assumed rotor flux. We turn now to the physical principles by which the rotor flux is induced. We then present an equivalent circuit based on Ampere's circuital law, Faraday's law, and conservation of energy. From this circuit, we determine overall motor characteristics such as efficiency, starting current, and torque as a function of speed.

15.1.1 Induction Principles

The rotor of the induction motor shown in Fig. 15.1a is an iron cylinder with large embedded conductors. The conductors, shown in Fig. 15.1b, run the width of the rotor and are shorted at both ends by large conducting rings. This is called a squirrel-cage rotor for obvious reasons. Figure 15.2a shows the rotor of a small induction motor with part of the metal removed to show the conductors. The

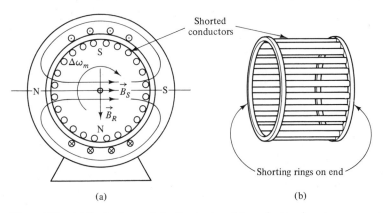

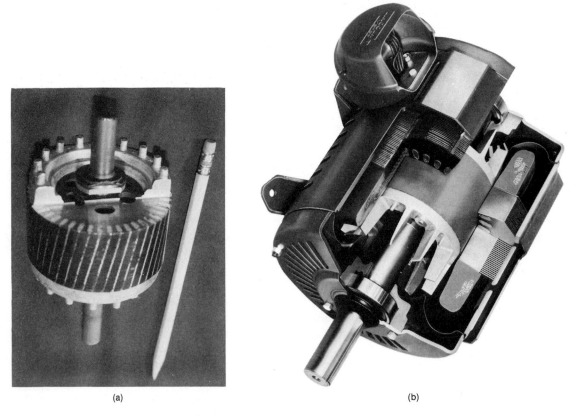

Figure 15.1 (a) Squirrel-cage induction motor; (b) conductors in rotor.

(a)

(b)

Figure 15.2 Cutaway views of (a) a squirrel-cage rotor and (b) a three-phase induction motor. (Photo courtesy of the Lincoln Electric Company.)

conductors are slanted slightly to eliminate resonance effects. The protrusions on the end rings enhance heat transfer. Figure 15.2b shows the construction of a small three-phase induction motor.

We will investigate the induction of the rotor poles through a mental experiment. We begin with nonrotating stator poles; thus we excite the stator windings with dc currents of appropriate magnitudes to create a field in the horizontal direction, as shown in Fig. 15.1a. We leave for a homework problem the determination of the appropriate magnitudes of field currents in the three-phase windings to produce such a field. We now (in our imagination) take hold of the rotor and slowly rotate it in the clockwise direction with an angular velocity $\Delta\omega_m$. This angular velocity, called the *slip speed*, is the angular velocity in the negative angular direction of the physical rotor relative to the stator flux, which is stationary in our mental experiment.

Induced currents. Figure 15.3 diagrams the effects of this rotation. The rotation induces a $\vec{u} \times \vec{B}$ electromotive force (emf) in the conductors, and currents flow in the shorted conductors. Applying the right-hand rule, you will determine that rotor currents flow out of the page on the right and into the page on the left, with the maximum current in the horizontal plane, under the stationary stator poles. If you consider a single conductor as it rotates under first one pole and then the other, you will realize that an ac current is induced in each conductor. Considering the conductors as electrical circuits, we would postulate an ac voltage induced in the conductor to drive these currents. For this reason we have shown in Fig. 15.3 that the steady motion produces a time-varying emf, $e(t)$.

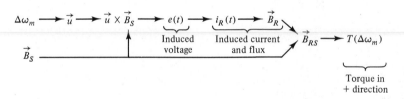

Figure 15.3 Cause and effect in an induction motor.

Induced flux. The induced rotor currents will produce a magnetic flux, $\vec{B}_R$. Applying the right-hand rule, you will determine that this magnetic flux is directed downward, as shown in Fig. 15.1a. If we associate magnetic poles with $\vec{B}_R$, we require a north pole at the bottom to act as source for the induced flux in the gap, and a south at top, as shown in Fig. 15.1a. The stator flux and induced rotor flux will combine to a rotor–stator flux in the gap, $\vec{B}_{RS}$, and a torque will be produced. Due to the location of the magnetic poles, the torque will be counterclockwise, in the positive θ_m direction, because the rotor magnetic poles tend to align with their opposites on the stator. From the viewpoint of Eq. (14.40), the power angle δ_{RS} will be $-90°$ and the magnetically generated torque will thus be in the positive θ_m direction. Consequently, the magnetically generated torque is a countertorque, opposing the torque we apply to cause the rotation in the negative direction. Here are some observations and consequences based on these effects:

- The induced currents and the resulting flux ($\vec{B}_R$) are caused by the stator flux ($\vec{B}_S$). Although the rotor turns physically, the pattern of currents and induced magnetic flux does not rotate. Thus the rotor–stator flux pattern $(\vec{B}_{RS})$ is constant in time and space for a given rotation speed $(\Delta\omega_m)$. The power angle δ_{RS} will be $-90°$ for small slip speed.

- The magnetically generated torque (T) will be proportional to the induced currents and the sine of the power angle and hence will vary with slip speed:

$$T(\Delta\omega_m) \propto -i_R(\Delta\omega_m)\sin\delta_{RS} = -i_R(\Delta\omega_m)\sin(\theta_R - 0°) \quad (15.1)$$

where $i_R(\Delta\omega_m)$ represents the rotor current, θ_R the position of the rotor north pole, and the position of the stator pole, $\theta_S = 0°$.

- The mechanical input power that we must supply to turn the rotor against torque is

$$P_{\text{in}} = P_R = \Delta\omega_m T(\Delta\omega_m) \quad (15.2)$$

This power will be converted into rotor-copper losses because resistive loss in the rotor conductors is the only energy conversion mechanism at work if we ignore mechanical losses.

- In a given rotor conductor, the induced currents will be sinusoidal at an electrical radian frequency of $\omega_e = \Delta\omega_m$. If there were more than $P = 2$ poles, the rotor electrical frequency would be $\omega_e = (P/2)\Delta\omega_m$.

Developed torque, $T(\Delta\omega_m)$. The developed torque, which counters our applied torque causing the rotation, will depend on the slip speed, $\Delta\omega_m$, through three factors:

1. The induced voltage will increase in proportion to the slip speed because the speed of the conducting bars relative to the magnetic flux will increase. This increase in voltage will cause the induced current to increase as slip speed increases.

2. The impedance of the rotor circuit will increase with slip speed because of the increase in the frequency of the ac induced voltage. At very small slip speeds the impedance is largely resistive, but at larger slip speeds the inductance of the rotor will dominate. Thus the current tends to reach a maximum value, where the increase in ac voltage is canceled by the corresponding increase in impedance.

3. The rotor–stator power angle between rotor and stator poles (fluxes) increases beyond the optimum value of $-90°$ because of the phase delay of the rotor currents associated with the inductance. This increase in power angle causes a lessening of the torque at high slip speeds, as shown by Eq. (15.1).

Combined effect. If we examine the dependence of torque given in Eq. (15.1) and then consider the dependence of induced current and power angle on slip speed, we predict the torque versus mechanical speed $(\omega_m = -\Delta\omega_m)$ characteristic shown in Fig. 15.4.† The developed torque, being a countertorque, has

---

† The *slip speed* is numerically positive for cw rotation. The *mechanical speed* is positive for ccw rotation. For the stationary field, the mechanical speed is the negative of the slip speed.

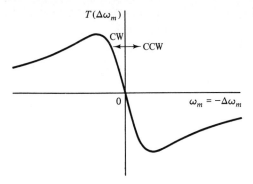

Figure 15.4 Torque versus speed with a stationary stator flux.

a sign opposite to that of the mechanical speed. For small slip speeds, the torque magnitude increases with slip speed. This reflects the linear increase of emf and the resulting increase in current. However, the curve levels off and eventually decreases as slip speed increases because the induced current levels off and because the power angle increases beyond $-90°$. Hence the torque reaches a maximum and then decreases due to the increasing power angle.

Summary. This completes our mental experiment. We have examined the physical principles at work in the induction motor. We looked at the emf and the resulting current and magnetic flux of the rotor. We examined the frequency of the emf and currents in the rotor. We showed how torque is developed. We developed a formula for the rotor-copper losses, Eq. (15.2). Finally, we anticipated how the developed torque depends on slip speed. In the next section we rotate the stator magnetic flux to create motor action.

15.1.2 Check Your Understanding

1. For maximum torque, rotor and stator poles should be aligned (0°) or spatially orthogonal (90°). (Which?)
2. In the mental experiment, all input power goes into stator-Cu loss, rotor-Cu loss, or the dc source supplying the stator flux. (Which?)
3. The slip speed is defined as positive in the negative angular direction. (T/F?)
4. The effect of rotor inductance is to increase or decrease developed torque as slip speed increases. (Which?)

Answers. 1. 90°; 2. rotor-Cu loss; 3. true; 4. decrease.

15.1.3 Three-phase Induction Motor Characteristics

Effect of rotation. We now apply three-phase currents to the stator field windings. As shown in Chapter 14, the stator flux will rotate at synchronous speed. In comparison with our mental experiment, the stator currents change from dc to ac, and the synchronous speed changes from $\omega_s = 0$ to $\omega_s = \omega$, where ω is the electrical angular frequency.† We assume now that the rotor is turning in the

---

† For more than $P = 2$ poles, the synchronous speed is $\omega/(P/2)$.

positive direction slower than synchronous speed, such that it is slipping backward by $\Delta\omega_m$ relative to the synchronous speed; thus the rotor speed changes from $\omega_m = -\Delta\omega_m$ to

$$\omega_m = \omega_s - \Delta\omega_m = \omega_s - s\omega_s = (1 - s)\omega_s \tag{15.3}$$

where s is the normalized slip speed, or more simply the *slip*:

$$s = \frac{\Delta\omega_m}{\omega_s} = \frac{\omega_s - \omega_m}{\omega_s} \tag{15.4}$$

Although the rotor is now turning at a high speed, the electrical voltage and currents in the rotor are still generated by the *relative* motion between the rotor conductors and the stator flux, that is, by the slip; hence the rotor voltage and current still have an electrical frequency of

$$\omega_e = \Delta\omega_m \times \frac{P}{2} = s\omega \tag{15.5}$$

where ω is the stator electrical frequency, normally $2\pi \times 60$ rad/s. Even though the rotor is not turning at synchronous speed, the rotor flux is still locked to the stator flux, as in our mental experiment, and hence the rotor *flux* rotates at synchronous speed.

From the viewpoint of the rotor, there is electrically no difference between stationary stator flux and rotating stator flux, provided the slip speed is the same. For the same slip speed, the emf, currents, and rotor magnetic flux are the same in both cases, and hence the developed torque will also be the same. The torque characteristic of the motor, therefore, is identical to Fig. 15.4, except that it now is shifted to the right by the synchronous speed, as shown in Fig. 15.5. Note in Fig. 15.5 that we have given the speed in units of mechanical speed (ω_m) and also in slip (s). Zero slip corresponds to rotation at synchronous speed and a slip of unity corresponds to zero mechanical speed.

Figure 15.5 Developed torque versus rotational speed and slip with rotating stator flux.

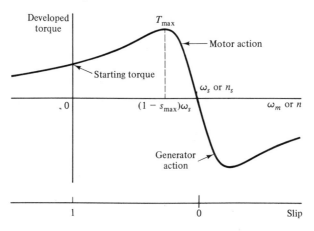

Alternating-Current Motors   Chap. 15

From Fig. 15.5, we note the following motor characteristics:

- Mechanical speeds below synchronous ($\omega_m < \omega_s$) produce motor action because the developed torque is in the same direction as the rotation.
- Mechanical speeds exceeding synchronous ($\omega_m > \omega_s$) produce generator action because the developed torque is opposite the direction of rotation. External mechanical torque would be required to drive the rotor against this countertorque. The electrical power generated would have the same frequency as the currents in the stator field windings. This generator would produce power when connected to a three-phase power source, but would have poor characteristics when connected only to a load. Such induction generators find application in wind-power generators and small power plants.
- When the mechanical speed is equal to the synchronous speed ($\omega_m = \omega_s$), neither motor nor generator action results; the system is merely idling. In this condition an external mechanical drive would be required to supply the mechanical losses.
- For $\omega_m \approx \omega_s$, the torque is proportional to slip. This is the region where the motor is normally operated.
- A point of maximum torque (T_{max}) exists at a speed of $(1 - s_{max})\omega_s$. This speed is slightly below the speed for maximum output power.
- The starting condition corresponds to $s = 1$. The starting torque is positive and hence this motor is self-starting.

Power relationships. We can derive several useful power relationships from basic principles and from the results of our mental experiment, which gave an expression for the rotor-copper losses in Eq. (15.2). At this point it is helpful to distinguish between the *developed* torque and power and the *output* torque and power. The developed torque is the magnetically generated torque and is the torque we have been discussing hitherto. The output torque would be smaller due to the torque required to overcome bearing and windage losses. Similarly, the developed power is $P_{dev} = \omega_m T_{dev}$, and represents the power converted from electrical to mechanical form. The output power would be $P_{out} = \omega_m T_{out}$ and is smaller than the developed power because of the mechanical losses.

The developed power is

$$P_{dev} = T_{dev} \times \omega_m = (1 - s)\omega_s T_{dev} \tag{15.6}$$

where T_{dev} is the developed torque and we have introduced the synchronous speed through the relationship $\omega_m = (1 - s)\omega_s$. As shown in Eq. (15.2), the power lost in the rotor-copper losses is

$$P_R = T_{dev} \times \Delta\omega_m = s\omega_s T_{dev} \tag{15.7}$$

where we have replaced the T used in Eq. (15.2) by T_{dev}. The power crossing the air gap from the stator to the rotor is called the *air-gap power*, P_{ag}:

$$P_{ag} = P_{dev} + P_R = (1 - s)\omega_s T_{dev} + s\omega_s T_{dev} = \omega_s T_{dev} \tag{15.8}$$

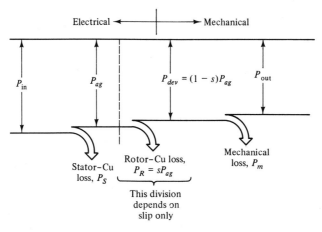

Figure 15.6 Power flow in the induction motor.

Equation (15.8) shows that air-gap power divides between the rotor-copper losses and the developed power. The meaning of the air-gap power is illustrated by Fig. 15.6, which accounts for the real power in the motor. A portion of the input power becomes copper loss in the stator windings. We have yet to focus on this loss, but it will be represented in the equivalent circuit we will develop in the next section. The remainder of the input power crosses the air gap into the rotor, similar to power moving from primary to secondary in a transformer. The air-gap power splits into two terms: the rotor-copper loss, which heats the rotor, and the developed power, which turns the rotor and load. From Eqs. (15.6) through (15.8), it follows that the air-gap power divides between rotor-copper loss and developed power in a ratio that depends only on the slip speed.

$$P_{ag} = \underbrace{(1 - s)P_{ag}}_{P_{dev}} + \underbrace{sP_{ag}}_{P_R} \Rightarrow \frac{P_{dev}}{P_R} = \frac{1 - s}{s} \qquad (15.9)$$

Equation (15.9) will be important when we develop an equivalent circuit for the motor.

Example. A three-phase, four-pole 60-Hz induction motor has a full-load output power of 5 hp at 1740 rpm. The motor efficiency is 87.5% at full load. The mechanical losses account for 20% of the total losses. Determine all the power quantities in Fig. 15.6.

The input power is $5 \times 746/0.875 = 4263$ W, and hence the combined losses are $4263 - 5 \times 746 = 533$ W. Of these losses, 20% are mechanical, so the mechanical losses are 106.6 W and electrical losses are 426.3 W. The developed power is

$$P_{dev} = P_{out} + P_m = 5 \times 746 + 106.6 = 3837 \quad \text{W} \qquad (15.10)$$

where P_m is the mechanical loss. To determine the air-gap power from the developed power, we need the slip. The synchronous speed, from Table 14.1, is 1800 rpm; hence the slip is

$$s = \frac{n_s - n}{n_s} = \frac{1800 - 1740}{1800} = 0.0333 \qquad (15.11)$$

From Eq. (15.9), the air-gap power is

$$P_{ag} = \frac{P_{dev}}{1 - s} = \frac{3837}{1 - 0.0333} = 3969 \ \ \text{W} \qquad (15.12)$$

Again from Eq. (15.9), the rotor-Cu loss is

$$P_R = sP_{ag} = 0.0333 \times 3969 = 132 \ \ \text{W} \qquad (15.13)$$

and hence the stator-Cu loss is 294 W. Notice the central role of the air-gap power in these calculations.

Summary. In Table 15.1, we compare the mental experiment with motor action. All torques are developed torques.

TABLE 15.1 COMPARISON OF MENTAL EXPERIMENT WITH MOTOR ACTION

Aspect	Mental Experiment, DC Stator Currents	Motor Action, AC Stator Currents
Synchronous speed	0	$\omega_s = \omega/(P/2)$
Slip speed	$\Delta\omega_m$	$\Delta\omega_m = s\omega_s$
Mechanical speed	$-\Delta\omega_m$	$\omega_s - \Delta\omega_m = (1 - s)\omega_s$
Developed torque	$T(\Delta\omega_m)$	$T(\Delta\omega_m) = T(s) = \text{same}$
Rotor-Cu loss	$\Delta\omega_m T(\Delta\omega_m)$	$\Delta\omega_m T(\Delta\omega_m) = s\omega_s T(s) = \text{same}$
Air-gap power	0	$\omega_s T(s)$
Developed power	$-\Delta\omega_m T(\Delta\omega_m)$	$\omega_m T(\Delta\omega_m) = (1 - s)\omega_s T(s)$

15.1.4 Check Your Understanding

1. What is the slip if the rotor is turning at synchronous speed for a six-pole machine?
2. Determine the output speed in rpm of a four-pole, three-phase induction motor operating at 60 Hz and having a slip of 3%.
3. The more poles an induction motor has, the faster it turns. (T/F?)
4. In a two-pole, three-phase induction motor, the frequency is 60 Hz and the slip is 5%. Find:
   (a) Rotational speed of the rotor (rpm)
   (b) Rotational speed of the rotor flux (rpm)
   (c) Frequency of the stator currents (Hz)
   (d) Frequency of the rotor currents (Hz)
5. If the developed power in a three-phase induction motor is 15 times the rotor-copper loss, what is the slip?

Answers. 1. Zero; 2. 1746 rpm; 3. false; 4. (a) 3420 rpm; (b) 3600 rpm; (c) 60 Hz; (d) 3 Hz; 5. 6.25%.

15.1.5 Equivalent Circuit for Three-phase Induction Motor

Energy processes. Our equivalent circuit should account for every major energy process in the motor. We need resistors to represent rotor- and stator-copper losses, and we need inductors to represent magnetic energy stored in the stray rotor and stator fields and in the flux coupling rotor and stator across the air gap. We must also have a resistor to account for the energy leaving the electrical circuit as developed mechanical power.

Stator circuit. The stator circuit in Fig. 15.7 accounts for stator-copper losses (R_S) stray magnetic fields (X_S), and the electromotive force ($\mathbf{E}_S$) induced in the stator coils by the rotating air-gap flux. The electrical frequency of stator voltage and currents is that of the ac power connecting to the stator, normally 60 Hz. Three times the complex power $\mathbf{E}_S\mathbf{I}_S^*$ into the stator emf represents the power and stored energy leaving the stator and passing into the air gap. The real part of this power must account for rotor-copper losses and developed mechanical power. The reactive power must account for stored energy in the air gap and stray fields in the rotor.

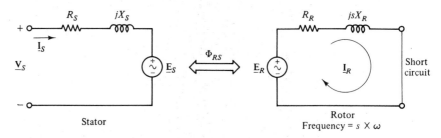

Figure 15.7 Per-phase equivalent circuit for rotor and stator. The region between the circuits represents the air gap, which stores magnetic energy and provides coupling between the stator and rotor.

Rotor circuit. The rotor circuit shown in Fig. 15.7 shows an emf ($\mathbf{E}_R$), a resistor (R_R), an inductor (sX_R), and a short circuit. The electrical frequency of the rotor emf and current is s times the stator frequency, and hence the reactance of the rotor is shown to be proportional to slip. The reactance X_R is the *blocked rotor reactance* because the slip would be unity with the rotor stationary (blocked). The rotor-copper loss would be

$$P_R = 3\,|\,\mathbf{I}_R\,|^2\,R_R \tag{15.14}$$

where the factor of three accounts for the three phases.

Mechanical output. The equivalent circuit shown in Fig. 15.7 is at present unable to account for the stored energy in the air gap or for the developed mechanical power. Application of Ampere's circuital law introduces a term to account for stored magnetic energy in the air gap. Application of Faraday's law *plus* conservation of energy introduces a resistor into the stator circuit to account for the developed mechanical power.

Ampere's circuital law. Ampere's circuital law may be applied to the induction motor in the same way it was applied to electrical transformers [Eq. (13.17)]. We introduce n_S and n_R as the equivalent turns for the stator and rotor, respectively, and integrate the magnetic field around a suitable path. The results can be put into the form

$$\underline{I}_S = \frac{n_R}{n_S} \underline{I}_R + \frac{\underline{E}_S}{jX_i} \tag{15.15}$$

where X_i is a reactance accounting for the stored energy in the air gap. Thus, *for the currents only*, the coupling between rotor and stator are represented by the ideal transformer shown in Fig. 15.8.

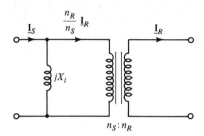

Figure 15.8 Equivalent circuit showing currents in stator and rotor. The reactance represents stored energy in the air gap.

Faraday's law. Again in analogy with the transformer, the emfs in stator and rotor are related by Faraday's law. However, we must reduce the emf in the rotor by the slip because the electrical frequency in the rotor is $s\omega$. Hence

$$\underline{E}_S = n_S \omega \underline{\Phi}_{RS} \tag{15.16}$$

and

$$\underline{E}_R = n_R s \omega \underline{\Phi}_{RS} \tag{15.17}$$

where Φ_{RS} is the air-gap flux. We may eliminate the flux to obtain

$$\frac{\underline{E}_S}{n_S} = \frac{\underline{E}_R}{s n_R} \tag{15.18}$$

which also suggests an ideal transformer *except* that the equivalent turns ratio is lower than that shown in Fig. 15.8 by the factor s.

Rotor impedance. The rotor voltage and current are related through

$$\underline{I}_R = \frac{\underline{E}_R}{R_R + jsX_R} \tag{15.19}$$

where R_R is the equivalent rotor resistance per-phase and X_R is the per-phase reactance of the blocked rotor.

Conservation of energy. We now have satisfied all circuit requirements except conservation of energy. Equations (15.15) and (15.18) require an ideal transformer with a voltage turns ratio $n_S : s n_R$ and a current turns ratio of $n_S : n_R$. Such a transformer would not obey conservation of energy, and of course it *should* not because electrical energy is not conserved in this device due to the mechanical output.[†]

We can solve our difficulty by scaling up the rotor voltage to "impose" conservation of *electrical* energy on the circuit. This is accomplished by dividing the numerator and denominator of the right side of Eq. (15.19) by s

$$\underline{I}_R = \frac{\underline{E}_R/s}{(R_R/s) + jX_R} \tag{15.20}$$

where $\underline{E}_R/s$ is the scaled-up rotor voltage. The transformation in Eq. (15.20) was first proposed by Carl Steinmetz (1865–1923). We may now use a normal ideal transformer to express voltage and current relationships between stator and "rotor" circuits. We place "rotor" in quotation marks because the secondary circuit in Fig. 15.9 now accounts for all the power passing from stator to rotor, including the mechanical power. We make the following observations:

- In transforming Eq. (15.19) into Eq. (15.20), rotor voltages and impedance values were scaled upward by $1/s$ but rotor current was unchanged. Thus the rotor current in Fig. 15.9 is the true rotor current.
- The rotor reactance (X_R) is the per-phase reactance of the blocked rotor, as if the frequency in the rotor circuit were the same as the frequency in the stator circuit. Of course, the frequencies differ in stator and rotor circuits, but the impedance scaling compensates for the different frequencies.
- The rotor part of Fig. 15.9 now accounts for all the power crossing the air gap into the rotor. This power divides between rotor-copper loss and developed mechanical power. The scaled-up resistor R_R/s must therefore account for both losses. Since the rotor current appears at its true value, we may calculate and subtract the rotor-copper loss from the air-gap power to establish the developed mechanical power:

$$P_{dev} = 3\left[|\underline{I}_R|^2\,\frac{R_R}{s} - |\underline{I}_R|^2\,R_R\right] = 3\,|\underline{I}_R|^2\,R_R \times \frac{1-s}{s} \tag{15.21}$$

Hence, $[(1-s)/s] \times R_R$ is an equivalent resistance accounting for the developed mechanical power per phase. We have thus modeled the developed mechanical power in our equivalent electrical circuit, as identified in

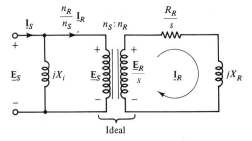

Figure 15.9 Equivalent circuit with rotor voltage and impedance scaled by a factor of $1/s$. This circuit accounts for electrical energy and energy converted to mechanical form.

† Of course, conservation of energy applies to the overall system. What we are saying here is that *electrical* energy is not conserved. We must include the mechanical system to account for all energy.

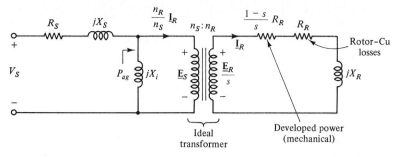

Figure 15.10 Equivalent circuit with the R_R/s term divided into two resistors, one accounting for rotor electrical loss and the other accounting for developed mechanical power.

Fig. 15.10. The first resistor in the rotor circuit represents power leaving the electrical circuit and entering the mechanical world as ordered mechanical energy. The second resistor represents power leaving the electrical circuit and entering the mechanical world as disordered mechanical energy, that is, as heat. In Fig. 15.10 we also inserted into the stator circuit a resistor and a reactance to account for stator copper loss and magnetic energy storage due to leakage flux.

- Note that the rotor circuit in Fig. 15.10 divides the air-gap power into developed mechanical power and rotor-copper losses in the ratio $(1 - s):s$, as we established earlier [Eq. (15.9)] from our mental experiment and conservation of energy. This correspondence supports the validity of the equivalent circuit in Fig. 15.10. We may simplify Fig. 15.10 through the impedance-transforming properties of an ideal transformer. Specifically, we multiply secondary (rotor) impedances by $(n_S/n_R)^2$ to obtain the primed values in Fig. 15.11, which represent rotor impedances referred to stator. Because energy-related quantities are unchanged by this transformation, we can with the equivalent circuit in Fig. 15.11 determine many quantities of interest, such as motor efficiency and output power.

- The development of the per-phase equivalent circuit in Fig. 15.11 has been "loose" in a number of ways. We have been loose in defining rotor and stator turns and in specifying means for determining the rotor resistance and reactance. With a more refined analysis, these factors could be defined and calculated, but we do not require such detail. We are interested in the motor's performance and overall characteristics. Our purposes are served by knowing the nature of the equivalent circuit and identifying the critical parameters. These machine parameters, such as the rotor resistance transformed into

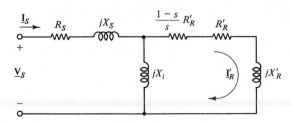

Figure 15.11 In this equivalent circuit, rotor impedance elements are moved to the primary.

the stator, can be measured or deduced from the external characteristics of the machine. From the equivalent circuit, we will be able to predict how motor characteristics depend on the various circuit model parameters, and thus we can gain an understanding of some of the design decisions that must be made in developing an induction motor.

Example. We will continue to investigate the equivalent circuit model through an example. Consider a 230-V, three-phase, 60-Hz motor operating at 1760 rpm. Clearly, this is a four-pole motor since the speed is slightly below 1800 rpm. In this example, we are given the per-phase circuit parameters: stator resistance = 0.40 Ω; stator reactance = 0.33 Ω; magnetizing reactance = 15 Ω; rotor resistance referred to stator = 0.12 Ω; blocked-rotor reactance referred to stator = 0.33 Ω; mechanical losses = 350 W. We will determine the output power and torque, efficiency, power factor, and input current at 1760 rpm. We also calculate starting torque and current for this motor.

Figure 15.12 shows the equivalent circuit. The normalized slip is (1800 − 1760)/1800 = 0.0222; therefore, the scaled-up rotor resistance is 0.12/0.0222 = 5.40 Ω. Note in Fig. 15.12 that we have divided the 5.40 Ω between the actual rotor resistance (0.12 Ω), which represents the rotor-copper losses, and the remainder (5.28 Ω), which represents the developed mechanical power. Notice that for small slip the rotor impedance is dominated by the resistor representing the mechanical output. Motor properties derive from a straightforward circuit analysis of the equivalent circuit model. We begin with $\mathbf{Z}_{ag}$, defined in Fig. 15.12.

$$\mathbf{Z}_{ag} = j15 \parallel (5.40 + j0.33) = 4.60 + j1.94 \quad \Omega \tag{15.22}$$

Thus the input impedance in steady state is

$$\mathbf{Z}_{in} = 0.40 + j0.33 + \mathbf{Z}_{ag} = 5.00 + j2.27 \quad \Omega \tag{15.23}$$

and the input current is

$$\mathbf{I}_s = \frac{\mathbf{V}_s}{\mathbf{Z}_{in}} = \frac{230/\sqrt{3}\underline{/0°}}{5.00 + j2.27} = 22.0 - j10.0 = 24.2\underline{/-24.4°} \quad \text{A} \tag{15.24}$$

The input power is

$$P_{in} = 3\frac{230}{\sqrt{3}} \times 24.2 \cos(24.4°) = 8769 \quad \text{W} \tag{15.25}$$

and the air-gap power will be the input power minus the loss in the stator windings.

$$P_{ag} = 8769 - 3(24.2)^2(0.40) = 8067 \quad \text{W} \tag{15.26}$$

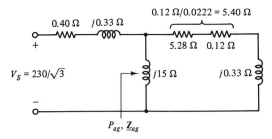

Figure 15.12 Per-phase equivalent circuit for the motor at s = 2.22%.

An alternative way to determine the air-gap power and check the results in Eq. (15.26) calculates the power into the real part of the air-gap impedance:

$$P_{ag} = 3 \mid \underline{I}_s \mid^2 R_{ag} = 3(24.2)^2(4.60) = 8067 \quad \text{W} \tag{15.27}$$

The air-gap power divides between rotor-copper loss and developed power in the ratio of s to $1 - s$; hence the rotor-copper losses are $8067(0.0222) = 179.3$ W, and the developed power is $8067(1 - 0.0222) = 7888$ W. The output power is the developed power minus the mechanical losses: $P_{out} = 7888 - 350 = 7538$ W, slightly more than 10 hp. The efficiency is the output power divided by the input power, or $\eta = 7538/8769 = 86.0\%$. The power factor is the cosine of the angle of the input impedance: PF $= \cos(24.4°) = 0.910$. Output torque is output power divided by angular speed in radians/second.

$$T_{out} = \frac{P_{out}}{\omega_m} = \frac{7538}{1760 \times (2\pi/60)} = 40.9 \text{ N-m} \tag{15.28}$$

This completes our analysis of the steady-state operation of the motor at 1760 rpm. Given the machine parameters and the equivalent circuit, motor analysis reduces to an exercise in circuit theory, followed by the proper interpretation of the calculated quantities.

Starting current and torque. To calculate the starting torque, we set $s = 1$ and calculate the developed torque. A stationary rotor has no mechanical losses; hence, the developed torque is the starting torque. We can determine the developed torque from the air-gap power and the synchronous speed. In general,

$$T_{dev} = \frac{P_{dev}}{\omega_m} = \frac{(1 - s)P_{ag}}{(1 - s)\omega_s} = \frac{P_{ag}}{\omega_s} \tag{15.29}$$

and specifically for the starting torque

$$T_{st} = \frac{P_{ag}(s = 1)}{\omega_s} \tag{15.30}$$

since $s = 1$ corresponds to stand-still. We show the equivalent circuit for $s = 1$ in Fig. 15.13a. The analysis proceeds along the same lines as above, with the following results: the starting current is $159.6\underline{/-51.8°}$, and the input power is 39,328 W. After we subtract stator-copper losses, we have an air-gap power of 8775 W; hence the torque is $8775/60\pi = 46.5$ N-m.

Effect of rotor resistance. The motor torque as a function of speed (or slip) can be derived from the equivalent circuit by repeating the analysis we have just illustrated. We show in Fig. 15.13b the effect of the rotor resistance presented to the stator. Near the synchronous speed, the torque curve is linear; that is, $T(s) = K \times s$, with K dependent on the rotor resistance. Because the motor normally operates in this linear region, a small resistance in the rotor leads to operation near the synchronous speed with essentially constant speed and high efficiency. Unfortunately, small rotor resistance produces relatively low starting torque. We note further that the maximum torque is independent of the rotor resistance, but the speed for maximum torque decreases as the rotor resistance increases.

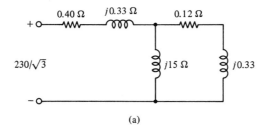

(a)

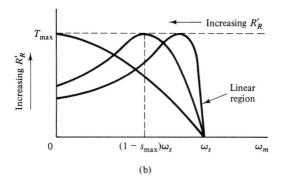

Increasing R'_R

T_{max}

Linear region

0 $(1 - s_{max})\omega_s$ ω_s ω_m

(b)

Figure 15.13 (a) Per-phase equivalent circuit for $s = 1$; (b) torque characteristics for several values of rotor resistance.

Small-slip region. The linear region and the maximum torque may be investigated using Thévenin's equivalent circuit. Hence we replace everything but the rotor circuit in Fig. 15.11 by a Thévenin equivalent circuit, as shown in Fig. 15.14a. The equations for the per-phase Thévenin voltage, $\underline{V}_T$, and the equivalent impedance presented to the rotor are given in Fig. 15.14b. Although the Thévenin voltage is a phasor, we will make it our phase reference and treat it as a scalar.

For small slip, the series impedance of the circuit is dominated by R'_R/s; hence most of the Thévenin voltage goes to R'_R/s, and the air-gap power in the small-slip region is approximately

$$P_{ag} \approx 3 \times \frac{(V_T)^2}{R'_R/s} = 3s \times \frac{V_T^2}{R'_R} \qquad (15.31)$$

Figure 15.14 (a) The per-phase Thévenin equivalent circuit to the rotor; (b) formulas for deriving the Thévenin circuit.

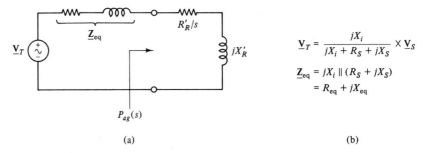

$\underline{Z}_{eq}$

R'_R/s

$\underline{V}_T$

jX'_R

$P_{ag}(s)$

(a)

$$\underline{V}_T = \frac{jX_i}{jX_i + R_S + jX_S} \times \underline{V}_S$$

$$\underline{Z}_{eq} = jX_i \parallel (R_S + jX_S)$$

$$= R_{eq} + jX_{eq}$$

(b)

The developed torque is the air-gap power divided by the synchronous speed, Eq. (15.29). Consequently, we may approximate the developed torque in the linear region near synchronous speed as

$$T_{dev}(s) = 3s \times \frac{V_T^2}{\omega_s R_R'} \tag{15.32}$$

Thus the torque with small-slip region is proportional to slip and inverse to rotor resistance, as shown in Fig. 15.13b. The developed mechanical power in the small-slip region would therefore be

$$P_{dev}(s) = \omega_m T_{dev}(s) = (1 - s)\omega_s T_{dev}(s) = 3(1 - s)s \frac{V_T^2}{R_R'} \tag{15.33}$$

The dependence of developed torque on slip allows the prediction of output power and speed throughout the small-slip region.

Continuing the previous example. Let us determine the no-load speed of the motor in the previous example. The mechanical losses would not vary greatly in the small-slip region because the speed is nearly constant. Therefore, with no load the developed power would be nearly 350 W. We may determine the no-load slip (s_{NL}) from Eq. (15.33) by proportion.

$$\frac{350}{7888} = \frac{(1 - s_{NL})s_{NL}}{(1 - 0.0222)(0.0222)} \Rightarrow s_{NL} = 9.65 \times 10^{-4} \tag{15.34}$$

Thus the no-load speed is $(1 - s_{NL})1800 = 1798$ rpm. In a similar manner we may determine the slip at any prescribed output power, provided the motor is operating in the small-slip region.

Summary. This completes our analysis of a three-phase induction motor. We have derived an equivalent circuit from which we can calculate steady-state operation and also starting characteristics. We have determined an approximate relationship relating torque and slip in the normal operating region. We turn now to discuss some considerations that influence the design of induction motors.

Effect of rotor resistance. High efficiency and relatively constant operating speed require small values of the rotor resistance, but high starting torque and moderate starting currents require relatively high values of rotor resistance. This conflict can be resolved in two ways. The first requires a different type of rotor from the squirrel-cage type. For a *wound rotor*, slots are milled in the rotor and three-phase coils are wound in the slots. These coils are connected to external resistors through a brush-slip ring assembly. With a wound rotor, we can maximize starting torque with external resistors and then remove all resistance for running. We can also achieve limited control of motor speed with a wound-rotor induction motor.

Shaped rotor conductors. The alternative to the wound rotor is the squirrel-cage rotor with shaped conductors. Figure 15.15 shows two extremes. With

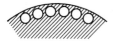

Big R, small L

(a)

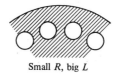

Small R, big L

(b)

Figure 15.15 (a) Small, shallow rotor conductors give good starting but poor running characteristics; (b) large, deep rotor conductors give poor starting but good running characteristics.

relatively small conductors near the surface of the rotor, we have relatively large resistance and small leakage inductance because the nearby air gap limits the leakage flux that encircles the currents in the rotor conductors. Thus the conductors in Fig. 15.15a would give good starting characteristics but poor run characteristics. On the other hand, the larger, deeper conductors in Fig. 15.15b have relatively small resistance and large leakage inductance. Such a rotor would give excellent run characteristics but would produce a low starting torque.

However, we can have both types at once; or rather we can shape the conductors to have a high-resistance, low-inductance portion near the surface and a low-resistance, high-inductance portion deeper in the rotor. At starting, the rotor currents have a high frequency, and the large inductance therefore shields the deeper conductors. The starting currents flow mostly in the surface conductors, which have relatively large resistance. As the motor comes up to speed, however, the rotor frequency decreases, and most of the rotor current shifts to the deeper, low-resistance conductors. Thus for starting we have relatively high resistance and for running relatively low resistance in the rotor.

Figure 15.16a shows some of the common classes of shaped conductors, with typical torque characteristics resulting from each shown in Fig. 15.16b. Clearly, the designer can, within limits, control the characteristics of the squirrel-cage motor by shaping the rotor conductors. The design classes in Fig. 15.16 refer to standard designations† of squirrel-cage three-phase induction motors to meet typical run and starting requirements:

- Designs A (not shown in Fig. 15.16b) and B have normal starting torque, but design B has lower starting current. Both have low slip (less than 5%) at rated output power. Typical applications are fans, blowers, rotary pumps, unloaded compressors, some conveyors, metal-cutting machine tools, and miscellaneous machinery.
- Design C has high starting torque and relatively low starting current, and runs with low slip at rated output power. Typical applications are starting of high-inertia loads such as large centrifugal blowers, flywheels, and crusher drums. Loaded starting of piston pumps, compressors and conveyors also require this design class.
- Design D has high starting torque and low starting current, but runs with relatively high slip (up to 11%) at rated output power. This design is required for very high inertia and loaded starts and also for loads having considerable variation in load speed throughout a load cycle. Typical applications are

† The National Electrical Manufacturers Association (NEMA) defines industry standards for electrical motors.

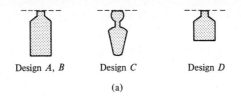

Design A, B   Design C   Design D

(a)

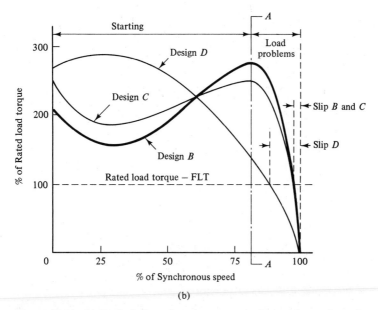

(b)

Figure 15.16 (a) Typical slots shaped to compromise starting and running characteristics; (b) typical torque characteristics corresponding to the shaped slots in part (a) (Figure courtesy of The Lincoln Electric Company).

punch presses, shears and forming machine tools, cranes, hoists, elevators, and oil-well pumping jacks.

15.1.6 Check Your Understanding

1. In the per-phase equivalent circuit shown in Fig. 15.17, identify what element(s) account for the specified physical effects. If nothing on the circuit is a suitable answer, indicate "none."
   (a) Developed mechanical power
   (b) Leakage flux in the stator
   (c) Rotor-copper loss
   (d) Mechanical loss
   (e) Stored energy in the air-gap flux
   (f) Iron loss
   (g) Stator-copper loss
   (h) Input three-phase voltage

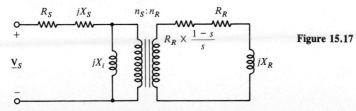

Figure 15.17

599

2. To operate with small slip, an induction motor should have rotor resistance that is large or small, or does not matter. (Which?)

3. For high starting torque, an induction motor should have rotor resistance that is large, small, or does not matter. (Which?)

4. As the load torque requirements vary, the induction motor has approximately constant speed, constant torque, constant power, or constant rotor losses. (Give one or more answers.)

Answers. 1. (a) $R_R(1 - s)/s$; (b) X_S; (c) R_R; (d) none; (e) X_i; (f) none; (g) R_S; (h) none directly, although V_S is the input voltage divided by $\sqrt{3}$; 2. small; 3. large; 4. constant speed.

15.1.7 Single-phase Induction Motors

Introduction. The three-phase induction motor is rugged, reliable, long-lived, self-starting, smooth-running, and relatively cheap. But three-phase power is not available everywhere, so if possible we need single-phase motors with the same characteristics. In fact, single-phase induction motors have excellent characteristics and outnumber the three-phase variety. Most of the small electric motors used in home, farm, or office are single-phase induction motors of one type or another.

In this section, we show first that the single-phase induction motor will run if started. We look then into the ways that single-phase motors are started, and finally we survey miscellaneous types of single-phase induction motors.

Forward and reverse slip. As we showed in Chapter 14, a single-phase flux can be resolved into two equal counterrotating flux waves. If the rotor is stationary, it will have a slip of unity relative to both waves and experience equal but opposite torque from each. Hence no starting torque will be produced.

Let us assume that the rotor is rotating in the forward direction with an angular velocity ω_m. The slip relative to the forward wave would be

$$s_f = \frac{\omega_s - \omega_m}{\omega_s} = 1 - \frac{\omega_m}{\omega_s} \tag{15.35}$$

where s_f is the slip relative to the forward flux wave. Relative to the forward direction, the synchronous speed of the reverse flux wave would be $-\omega_s$, and hence the slip of the rotor relative to the reverse flux wave would be

$$s_r = \frac{-\omega_s - \omega_m}{-\omega_s} = 1 + \frac{\omega_m}{\omega_s} = 2 - s_f \tag{15.36}$$

where s_r is the slip relative to the reverse flux wave. However, since both slips can be expressed in terms of the forward slip, we drop the subscripts and use s for the forward slip and $2 - s$ for the reverse slip.

Torque characteristic. We may estimate the torque characteristic of the single-phase induction motor from our results for the three-phase motor. Recall from Eq. (14.32) that an oscillating flux can be divided into two counterrotating fluxes of half-strength. Therefore, if $T_{3\phi}$ is the torque characteristic of the corresponding three-phase motor, then the torque as a single-phase motor for the same maximum flux, $T_{1\phi}$, would be

$$T_{1\phi}(s) = \tfrac{1}{2}[T_{3\phi}(s) - T_{3\phi}(2 - s)] \tag{15.37}$$

where the $\tfrac{1}{2}$ comes from the equal division of the flux, and the two terms represent coupling to forward and reverse fluxes. The torque characteristic of the single-phase motor would therefore be as shown in Fig. 15.18. We note that the motor has no starting torque but, once started, produces a torque in the direction in which it is rotating. Thus, it will run if started.

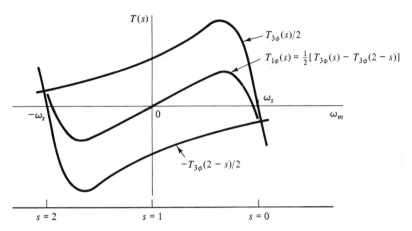

Figure 15.18 The single-phase run characteristic can be derived from the three-phase characteristic. These curves assume that the motor is excited at constant current.

Example. Consider a 4-pole single-phase motor running at 1725 rpm. The slip would be $(1800 - 1725)/1800 = 0.0417$ relative to the forward flux and $2 - 0.0417 = 1.9583$ relative to the reverse flux. The rotor frequencies would be $s \times 60 = 2.5$ Hz and $(2 - s) \times 60 = 117.5$ Hz. Due to the effect of rotor inductance, the 2.5-Hz currents will be much stronger than the 117.5-Hz currents in the rotor and hence coupling will be stronger to the forward flux wave. One problem with the single-phase motor is vibration at 120 Hz that results from an unavoidable coupling between the counterrotating fluxes and currents.

Constant-current supply. The argument supporting the use of Eq. (14.32) in explaining single-phase torque depends on having rotating fluxes that are not influenced by the currents in the rotor. This would require supplying power to the motor from a constant current source. The characteristic shown in Fig. 15.18 is unrealistic for normal ac sources.

Constant-voltage supply. In practice, a single-phase induction motor would be driven from a constant-voltage supply. As for the transformer, the applied voltage determines the flux in the stator winding, but in this case the flux is composed of two counterrotating fluxes, which are not necessarily equal.

Sec. 15.1   Induction Motors

601

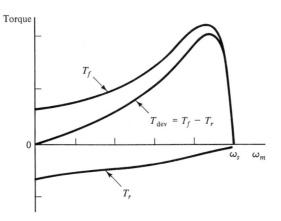

Figure 15.19 Torque characteristic for constant-voltage supply.

An analysis of the single-phase motor driven on a constant-voltage supply is beyond the scope of this text. Figure 15.19 shows the torque characteristic of a representative single-phase motor calculated for a constant-voltage supply. We show coupling to forward and reverse waves separately. Clearly, the reverse wave has a slight effect in the small-slip region; the single-phase motor runs well. Our problem is to get it started.

Starting methods for single-phase motors. The single-phase induction motor requires an auxiliary (or starting) winding. Such windings are physically separated by $90°/(P/2)$ in space from the main winding. Starting torque is created by the out-of-phase currents in the main and auxiliary windings because they produce rotating flux waves that are unequal. In effect, the motor is started as a two-phase motor. In all cases, the motor can be started in either direction by reversing the polarity of one of the windings. The auxiliary winding may or may not be designed for continuous operation. We distinguish between several types of single-phase induction motors on the basis of how the phase shift is created between the currents in the main and auxiliary windings, and whether the auxiliary windings are used continuously or only for starting.

Split-phase motors. The split-phase motor, whose circuit is shown in Fig. 15.20a, has an auxiliary winding that would burn out if used continuously. Normally, the starting winding is disconnected with a centrifugal switch that opens at about 75% of the rated speed. Occasionally, the auxiliary winding is switched with a relay that is activated by the current in the main winding. Figure 15.20b shows the current in the main winding and the region where the relay would engage the auxiliary winding. Such arrangements are used in compressor motors, in which the motor is enclosed in a refrigerant system.

The phase shift between main and auxiliary windings in a split-phase motor is created by their differing ratios of resistance to inductance. The auxiliary winding would use small wire and hence would have a higher-resistance than the main winding. Thus the auxiliary current would be smaller and more in phase with the line voltage, as shown in Fig. 15.21. This method yields a relatively small phase shift between main and auxiliary currents, and hence gives low starting torque.

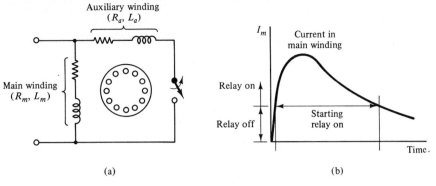

(a)   (b)

Figure 15.20 (a) For the split-phase motor, the auxiliary or starting winding is disconnected after the motor starts; (b) a current-sensing relay can be used to connect and disconnect the auxiliary winding.

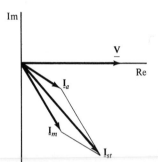

Figure 15.21 The split-phase motor has a small phase difference.

Split-phase motors serve applications requiring low starting torque, such as fans and bench grinders. Typical power ratings are $\frac{1}{20}$ to $\frac{1}{3}$ hp.

Capacitor-start/induction-run. In the circuit of the capacitor-start motor, shown in Fig. 15.22a, a capacitor in series with the auxiliary winding produces a leading current. Ideally, the auxiliary current would lead the main current by 90°

Figure 15.22 (a) A capacitor may be used to increase the phase shift between the currents in the main and auxiliary windings; (b) phasor diagram showing the 90° shift due to the capacitor.

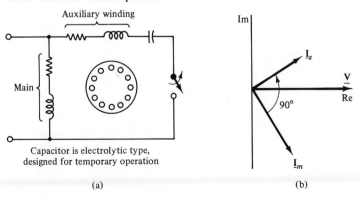

Capacitor is electrolytic type,
designed for temporary operation

(a)   (b)

of phase for maximum starting torque, as shown in Fig. 15.22b. The capacitor would be of the electrolytic type, designed for short-time operation, and the auxiliary winding would be disconnected after the motor is started. Such motors have good starting torque and are used for large appliances, some power tools, and large fans. Typical power ratings are $\frac{1}{4}$ to 10 hp.

Capacitor-start/capacitor-run. The capacitor-start/capacitor-run motor shown in Fig. 15.23 has two capacitors. One capacitor is rated for short-time operation, for starting the motor; the other is rated for long-term operation, for improving the run characteristics of the motor. By giving a more pure rotating wave, the run capacitor improves the torque characteristics of the motor and reduces vibration. Capacitor-start/capacitor-run motors are available in sizes from $\frac{1}{4}$ to 10 hp and are used for conveyors, air compressors, and other devices where heavy loads must be started.

Permanent-split capacitor (PSC) motors. In the PSC motor shown in Fig. 15.24, the auxiliary winding is designed for continuous operation. This motor has a continuous-rated capacitor for start and run, and no switch. The auxiliary winding gives some help in starting torque and some help in run characteristics. The main virtues of the PSC motor are high efficiency and smoother operation. Power ratings of $\frac{1}{8}$ to $\frac{3}{4}$ hp are typical of PSC motors, which are often used for fans and direct-drive blowers.

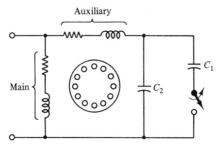

Figure 15.23 Smoother operation results if the auxiliary winding is used continuously.

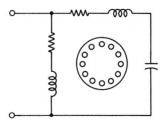

Figure 15.24 The PSC motor has no switch.

Shaded-pole motors. Figure 15.25 shows two geometries for shaded-pole motors. This motor is characterized by having part of the stator poles encircled by a conducting ring. As flux increases, current will flow in the "shading ring" to delay its buildup within the shaded portion of the pole. Later, as flux decreases, the current induced in the ring will delay the decrease in the shaded portion of the pole. Hence the flux maximum moves $1 \rightarrow 2 \rightarrow 3 \rightarrow 4 \rightarrow 1$, and so on. The motors shown in Fig. 15.25 turn in one direction only, but shaded-pole motors can be made reversible if both sides of the pole are shaded and the shading coils brought out for external switching. As shown, the motors are fixed speed, but multipole motors can be made to run at different speeds by exciting poles in

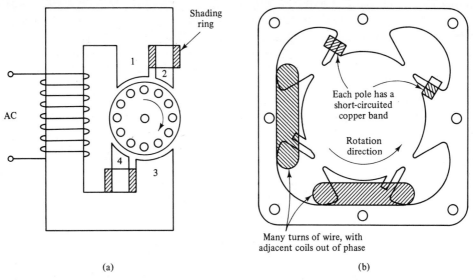

Figure 15.25 (a) A simple two-pole shaded-pole motor; (b) the magnetic structure for a larger, four-pole shaded-pole motor.

various patterns. Shaded-pole motors are widely used in small fans and in household appliances such as can openers.

15.1.8 Check Your Understanding

1. The single-phase induction motor will run in either direction, depending on which way it is started. (T/F?)
2. In a single-phase induction motor, the slip of the rotor relative to one rotating wave of stator flux is 6%. What is its slip relative to the other wave of stator flux?
3. A 60-Hz single-phase induction motor turns 1740 rpm.
   (a) How many poles does the motor have?
   (b) What is (are) the frequency (frequencies) of the stator current(s)?
   (c) What is (are) the frequency (frequencies) of the rotor current(s)?
4. The unloaded ideal (no mechanical loss) single-phase induction motor will run slower than synchronous speed. (T/F?)
5. Why does a capacitor-start, single-phase induction motor have greater starting torque than a split-phase motor?
6. A permanent split-capacitor (PSC) motor should have less vibration at 120 Hz than a capacitor-start/induction-run motor. (T/F?)
7. Why does the standard shaded-pole motor turn only one way?

Answers. 1. True; 2. 194%; 3. (a) four poles, (b) 60 Hz, (c) 2 Hz and 118 Hz; 4. true, because of the reverse torque; 5. the current in the start winding is more out of phase with the current in the main winding; 6. true, because the rotating flux has a smaller counterrotating flux; 7. because the mechanism to make the flux rotate is built into the iron of the poles.

Importance of the synchronous machine. The synchronous machine can be used as a generator or motor. As a motor, it has highly specialized properties and serves a narrow range of applications. The synchronous generator, by contrast, is the workhorse of the electric power industry for generating ac electric power. The "alternator" in an automobile is a small three-phase synchronous generator with six or eight diodes to rectify the ac to dc for charging the battery and supplying the electrical system.

Contents of this section. We are interested primarily in the synchronous motor. We begin by describing its construction with either cylindrical or salient-pole rotor. We then give without derivation the equivalent circuit for the cylindrical-rotor machine and explore its characteristics as a motor. We present the characteristics of the salient-pole motor, and finish with a brief discussion of the synchronous generator.

15.2.1 Synchronous Motor Construction and Equivalent Circuit

Stator construction. The stator of the synchronous is identical in principle to that of a three-phase induction motor. The three-phase currents produce a magnetic flux that rotates at a synchronous speed fixed by the electrical frequency and the number of poles. Electrically, the stator is the armature in the sense that it has a few large conductors that carry large currents.

Rotor construction. The rotor is a dc electromagnet. The dc current may be supplied to the rotor through a brush-slip ring assembly or may be supplied by rectification of an ac voltage that is induced in a separate winding on the rotor. Figure 15.26 shows both cylindrical and salient-pole rotor construction. The magnetic structure of the cylindrical rotor is symmetric, and a sinusoidally tapered rotor flux is produced by unequal distribution of the rotor mmf. The magnetic structure of the salient-pole rotor is unsymmetrical: the dc coil mmf is concentrated by coils wound around the pole† pieces, and a sinusoidal flux distribution is accomplished by tapering the width of the air gap. The same number of poles are used on stator and rotor. Electrically, the rotor carries the field circuit, in the sense that it holds many turns of small wire carrying relatively small current.

Operating principle. To develop motor action, the rotor must rotate at the synchronous speed. The stator currents set up rotating poles, and the rotor poles lock into synchronism and thus rotate at the same speed. The torque characteristic is that shown in Fig. 14.6b, the angle of displacement between stator and rotor poles being the rotor–stator power angle. Equation (14.40) gives the developed torque. At no load, the rotor–stator power angle is zero because no torque is required. As the mechanical load is applied, the rotor poles lag behind

---

† As for the dc motor stator (Fig. 14.10), the rotor "poles" refer both to the magnetic poles and also the iron protrusions around which coils are wound.

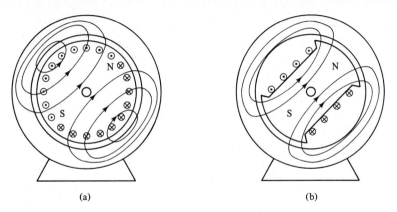

(a)   (b)

Figure 15.26   (a) Cylindrical rotor; (b) salient-pole rotor.

the stator poles, and the rotor–stator power angle becomes negative. If torque is applied by a mechanical drive, the rotor poles move ahead of the stator poles, mechanical power is supplied, and the machine acts as a generator. Thus in a motor the electrical system pulls on the mechanical system, and in a generator the mechanical system pulls on the electrical system.

Per-phase equivalent circuit.   Figure 15.27a shows a per-phase equivalent circuit for the synchronous motor that ignores loss in the stator circuit, which thus acts as a large inductance. The effect of the flux due to the field is represented by the *excitation voltage*, $\underline{E}_f$, as shown in Fig. 15.27b. However, we should not consider the excitation voltage as physically present in the machine because the flux due to the field on the rotor occupies the same space as the flux due to the stator currents represented by the *synchronous reactance*, X_s. The flux that exists physically in the machine is the rotor–stator flux, which is represented by the combined effects of the excitation voltage and the voltage across the synchronous reactance. By Faraday's law, the per-phase voltage induced in the stator coils by the motion of the total flux around the structure must match the applied per-phase voltage, $\underline{V}$.

Figure 15.27   (a) Per-phase equivalent circuit for synchronous motor; (b) the field determines the magnitude of the excitation voltage, $\underline{E}_f$.

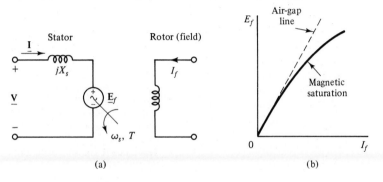

(a)   (b)

Summary. The stator equivalent circuit represents the rotating rotor flux by the excitation voltage and the rotating stator flux by an inductive reactance. The sum of the rotor and stator fluxes, which is the only physical flux in the machine, is represented by the sum of the excitation voltage and the voltage drop across the synchronous reactance and is equal to the applied voltage (Faraday's law). Losses in the stator circuit have been ignored. The magnitude of the excitation voltage is controlled by the field current that is supplied to the rotor through a brush–slip-ring assembly. The equivalent circuit in Fig. 15.27a applies only to cylindrical-rotor machine because it ignores the asymmetry in the magnetic structure of a salient-pole rotor.

15.2.2 Characteristics of the Cylindrical-rotor Motor

Analysis of the equivalent circuit. Kirchhoff's voltage law for the stator circuit is

$$\underline{V} = \underline{E}_f + j\underline{I}X_s \Rightarrow \underline{E}_f = \underline{V} - j\underline{I}X_s \tag{15.38}$$

The phasor diagram reflecting Eq. (15.38) is shown in Fig. 15.28, where ϕ is the phase angle between per-phase voltage and current and is shown negative in Fig. 15.28 (lagging current). The angle δ_R is the rotor power angle defined in Fig. 14.21 and is the physical angle (multiplied by $P/2$ for machines with more than two poles) between the rotor–stator flux and the rotor flux. This power angle appears between the per-phase voltage and the excitation voltage because the former represents the total flux and the latter the rotor flux.

The input electrical power to the machine is

$$P = 3VI \cos \phi \tag{15.39}$$

where V and I are the magnitudes of the per-phase voltage and current, respectively. From the geometry of Fig. 15.28, it is evident that

$$IX_s \cos \phi = -E_f \sin \delta_R \tag{15.40}$$

where E_f is the magnitude of the excitation voltage, and the minus sign is required because the rotor power angle is negative for a motor. Substitution of Eq. (15.40) into Eq. (15.39) yields

$$P = -\frac{3VE_f}{X_s} \sin \delta_R \tag{15.41}$$

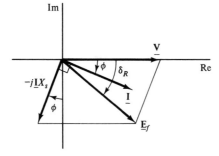

Figure 15.28 Phasor diagram for the synchronous motor. The applied per-phase voltage is the phase reference, and the current is shown lagging by a phase angle ϕ. The voltage across the synchronous reactance is shown lagging the current by 90°.

Our analysis ignores losses; hence Eq. (15.41) also gives the mechanical power developed by the machine. Motor speed is constant, so the developed power is proportional to the developed torque. Equation (15.41) is therefore another form of Eq. (14.42): V is proportional to B_{mRS}, E_f is proportional to B_{mR}, and $\sin \delta_R$ appears in both equations.

Using similar geometric reasoning, we may determine the reactive power into the machine to be

$$Q = \frac{3V^2}{X_s} - \frac{3VE_f}{X_s} \cos \delta_R \qquad (15.42)$$

The excitation voltage may be eliminated between Eqs. (15.41) and (15.42) to give the rotor power angle in terms of the real and reactive power into the machine.

$$\tan \delta_R = \frac{P}{Q - (3V^2/X_s)} = \frac{P}{Q - (V_L^2/X_s)} \qquad (15.43)$$

where V_L is the line voltage $(= \sqrt{3}V)$.

Example. A 480-V three-phase synchronous motor has 50-hp output power at unity power factor with a field current of 5 A dc. The field current is increased to 6 A dc. Find the new power factor and the new input current. Ignore losses, assume a synchronous reactance of 3.2 Ω, and assume that the excitation voltage is proportional to field current.

Our approach will be to find the excitation voltage for the first case and then for the second case, and from that change determine the new power factor. For 50 hp out, no losses, and unity power factor, the current would be

$$I = \frac{P}{3\,V} = \frac{50 \times 746}{3(480/\sqrt{3})} = 44.9 \quad \text{A} \qquad (15.44)$$

and would be in phase with the input voltage. By KVL in Eq. (15.38), the phasor excitation voltage would be

$$\mathbf{E}_{fu} = \frac{480}{\sqrt{3}} - j3.2 \times 44.9 = 312.1\underline{/-27.4°} \qquad (15.45)$$

where $\mathbf{E}_{fu}$ is the excitation voltage for unity power factor. Hence, for unity power factor, the magnitude of the per-phase excitation voltage is 312.1 V and the rotor power angle is $-27.4°$. When we increase the dc field current from 5 to 6 A dc, the magnitude of the excitation voltage will increase proportionally:

$$E_f' = \frac{6}{5} \times 312.1 = 374.5 \quad \text{V} \qquad (15.46)$$

where the prime refers to conditions after the field current is changed. Changing the field current will not change the real power output of the motor because that is established by the speed, which is constant, and the torque requirement of the load at that speed. Thus, we may determine the new rotor power angle from Eq. (15.41):

$$312.1 \sin(-27.4°) = 374.5 \sin(\delta_R') \Rightarrow \delta_R' = -22.5° \qquad (15.47)$$

And the new reactive power can be determined from Eq. (15.42):

$$Q' = \frac{3(480/\sqrt{3})^2}{3.2} - \frac{3(480/\sqrt{3})(374.5)\cos(-22.5°)}{3.2} = -17,873 \quad \text{VAR} \quad (15.48)$$

The reactive power is negative, meaning that the new current will lead the per-phase voltage. We will explore the reason for this presently. The new power factor may be determined from the real and reactive powers:

$$\text{PF} = \frac{P}{\sqrt{P^2 + Q^2}} = \frac{50 \times 746}{\sqrt{(50 \times 746)^2 + (-17,873)^2}} = 0.902 \quad (15.49)$$

Hence the new current magnitude is $44.9/0.902 = 49.7$ A, and the new phase angle is $\cos^{-1}(0.902) = 25.6°$ with current leading voltage.

Effect of varying the field current with constant output power. Figure 15.29 shows a phasor diagram for a synchronous motor that is said to be *overexcited* because the excitation voltage is greater than that required to give unity power factor. Let us consider the effect of varying the excitation voltage (varying field current) while the output power and per-phase voltage are kept constant. First consider the locus of the phasor excitation voltage. By Eq. (15.41), $E_f \sin \delta_R$ must remain constant, so the tip of $\mathbf{E}_f$ must move along a horizontal line. Likewise, for constant power, the in-phase component of the current, $I \cos \phi$ must remain constant; hence the tip of the current phasor moves along a vertical line. At point u, the tip of $\mathbf{E}_f (=\mathbf{E}_{fu})$ lies below $\mathbf{V}$ and the current will be in phase with the per-phase voltage. If the excitation voltage is increased beyond E_{fu}, the current phase must swing ahead of the per-phase voltage and a leading power factor is produced. If the excitation voltage is less than E_{fu}, the current phase must lag the per-phase voltage and a lagging power factor is produced.

The out-of-phase component of the current of the synchronous motor can

Figure 15.29 Phasor diagram for the synchronous motor. The motor draws leading or lagging current depending on the magnitude of the excitation voltage.

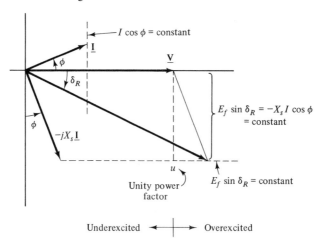

thus be controlled by the field current. This feature allows the synchronous motor to improve the overall power factor of a collection of loads that tend to draw lagging current. For example, in a factory with many induction motors, employment of overexcited synchronous motors is desirable to improve the power factor of the overall load.

Starting a synchronous motor. A synchronous motor has no starting torque. To develop a steady torque, the rotor must be rotating at the synchronous speed. This would appear to be a major defect of the synchronous motor, but the difficulty can be remedied by placing some shorted turns on the rotor. These shorted turns will produce torque by induction-motor action and accelerate the rotor to a speed just below the synchronous speed, whence the rotor field is energized and initiates synchronism. After the rotor field is energized, the shorted turns have no effect on the steady-state operation of the motor. Additionally, the shorted turns improve the dynamic performance of the motor by damping out oscillations in power angle caused by variations in mechanical load.

Continuing the example. For the previous example, the field current remains at 6 A dc and the load is now increased to 65 hp. Find the new input current. The increase in real power will increase the power angle. From Eq. (15.41), we calculate the new power angle to be

$$65 \times 746 = -\frac{3(480/\sqrt{3})(374.5)}{3.2} \sin \delta_R'' \Rightarrow \delta_R'' = -29.9° \qquad (15.50)$$

where double-primed quantities refer to conditions under the second change. Thus the new excitation voltage would be $\mathbf{E}_f'' = 374.5\underline{/-29.9°}$. Kirchhoff's voltage law, Eq. (15.38), now requires the current to be

$$\mathbf{I} = \frac{\mathbf{V} - \mathbf{E}_f''}{jX_s} = \frac{(480/\sqrt{3})\underline{/0°} - 374.5\underline{/-29.9°}}{j3.2} = 58.3 + j14.9$$

$$= 60.2\underline{/14.3°} \quad \mathrm{A} \qquad (15.51)$$

The principal effect is an increase in the in-phase component of the current.

15.2.3 Salient-pole Machines

Salient-pole rotors are used in slow-turning machines, such as hydroelectric generators and high-torque motors. In such cases where many poles are required, engineering and manufacturing considerations favor the salient-pole geometry. The asymmetry of the magnetic structure in this machine introduces two reactances in place of the synchronous reactance of the cylindrical rotor machine.

$$X_s \rightarrow X_d \text{ and } X_q \qquad (15.52)$$

where X_d is the direct-axis reactance and X_q is the quadrature-axis reactance. The direct-axis reactance applies to the axis aligned with the rotor poles; the quadrature-axis reactance applies to the axis between the rotor poles and is smaller because the equivalent air gap is larger between the poles.

Analysis of the salient-pole machine. An analysis of the salient-pole machine is beyond the scope of this book, but equations describing the performance of the machine will be given. No equivalent circuit exists, but the equations of the cylindrical-motor machine may be modified to describe the characteristics of the salient-pole machine. Equation (15.41) becomes

$$P = -\frac{3VE_f}{X_d} \sin \delta_R - \frac{3}{2} V^2 \left(\frac{1}{X_q} - \frac{1}{X_d}\right) \sin 2\delta_R \qquad (15.53)$$

and Eq. (15.42) becomes

$$Q = -\frac{3VE_f \cos \delta_R}{X_d} + 3V^2 \left(\frac{\cos^2 \delta_R}{X_d} + \frac{\sin^2 \delta_R}{X_q}\right) \qquad (15.54)$$

Equations (15.53) and (15.54) may be combined to eliminate the excitation voltage to yield

$$\tan \delta_R = \frac{P}{Q - (V_L^2/X_q)} \qquad (15.55)$$

where V_L is the line voltage ($= \sqrt{3}\, V$).

Example. A 2300-V, 60-Hz, 250-hp salient-pole synchronous motor has direct- and quadrature-axis reactances of 15 and 8 Ω, respectively. The excitation voltage is adjusted for operation at unity power factor with full load. Find the rotor power angle under this condition. The load is then removed, but the excitation voltage is unchanged. Find the reactive power into the motor under this new condition. Ignore losses.

For unity power factor, the reactive power is zero, and we may determine the rotor power angle directly from Eq. (15.55):

$$\tan \delta_R = \frac{250 \times 746}{0 - (2300)^2/8} = -0.282 \Rightarrow \delta_R = -15.8° \qquad (15.56)$$

Under the unloaded condition and with no losses, the power angle and real power will be zero, but the machine will still interact with the electrical system through reactive power flow, as indicated by Eq. (15.54) with $\delta_R = 0°$. The excitation voltage is unchanged from the loaded condition, so we may determine $\mathbf{E}_f$ from Eq. (15.53) with $\delta_R = -15.8°$.

$$250 \times 746 = -\frac{3(2300/\sqrt{3})E_f \sin(-15.8°)}{15} - \frac{3}{2}\left(\frac{2300}{\sqrt{3}}\right)^2\left(\frac{1}{8} - \frac{1}{15}\right)\sin(-2 \times 15.8°)$$

$$E_f = 1469 \quad V$$

$$(15.57)$$

Equation (15.55) is indeterminant with both P and δ_R equal to zero, but we may determine the reactive power from Eq. (15.54):

$$Q = -\frac{3(2300/\sqrt{3})(1469)}{15} + 3\left(\frac{2300}{\sqrt{3}}\right)^2\left(\frac{1}{15} + 0\right) = -37{,}384 \quad VAR \qquad (15.58)$$

The machine is overexcited and draws negative reactive power like a large capacitor.

Reluctance torque. The form of Eq. (15.53) reveals the presence of *reluctance torque* due to the salient poles. Since the speed is constant, the output torque of the machine is proportional to the output power, and hence from Eq. (15.53) the form of the machine torque is

$$T(\delta_R) = T_C \sin \delta_R + T_R \sin 2\delta_R \qquad (15.59)$$

where T_C is the maximum torque due to the rotor field and T_R is the maximum reluctance torque due to the asymmetry of the rotor magnetic structure. The reluctance torque arises because the rotor magnetic structure tries to come into alignment with the stator poles to minimize the stored magnetic energy in the system. In another sense, the reluctance torque is due to induced magnetic poles on the rotor protrusions and would be present if the rotor had no coils carrying dc currents.†

Consider the motor in the previous example. We assume 12 poles and 60 Hz, so the synchronous speed is 20π rad/s. Substitution of the excitation voltage into the first term in Eq. (15.53) yields a maximum torque of 6208 N-m due to the rotor dc current. The second term yields 2456 N-m for the maximum reluctance torque. For the rotor power angle calculated, the output torques due to the two components are 1685 N-m and 1283 N-m, respectively, and hence about 43% of the output torque is due to reluctance torque. The reluctance torque improves the characteristics of the motor because it increases the maximum possible torque and causes the motor to run with a smaller rotor power angle.

Synchronous generators. The vast majority of electric power plants employ two-pole (occasionally four-pole) cylindrical-rotor synchronous generators driven by steam turbines. Generators are normally operated as part of a large power system where many loads and many generators are interconnected over a geographic region. Such interconnections are used to enhance reliability, to permit maintenance on individual generators, and to permit exchange of electric power between participating power companies. In large power grids, all generators are synchronized and interconnected by long-distance transmission lines. From the viewpoint of an individual generator, the power system is the "load" into which it delivers real and reactive power.

The power system is considered an *infinite bus* because it maintains constant frequency and voltage (amplitude and phase), independent of the operation of the generator under consideration. The frequency of the system may be considered constant, being established by the controlled rotation of many generators. Thus, if we open the throttle of the drive to an individual generator, we do not increase the frequency of the system; rather, we increase the rotor power angle and contribute more real power to the system. Likewise, if we increase the dc field current to an individual generator, we do not change the real power but rather change

---

† Note that the excitation voltage does not appear in the second term in Eq. (15.53).

the reactive power contributed to the system. The effect of the dc field current on the reactive power output of the generator is similar to that of the motor. The generator is normally overexcited to produce lagging current, which is required for most power systems. The equations describing the synchronous generator are identical to those describing the synchronous motor except for some changes in sign to cover the power and current *out of* the machine. Since our emphasis in this text is on motors, we will not analyze the synchronous generator.

15.2.4 Check Your Understanding

1. The input current to a loaded synchronous motor will be minimum with the excitation voltage smaller than, equal to, or greater than the per-phase output voltage. (Which?)
2. A permanent magnet could be used as the rotor of a synchronous motor, provided a method of starting were provided. (T/F?)
3. In a cylindrical-rotor synchronous motor, increasing the dc field current increases the magnitude of the current into the motor. Is the current leading or lagging?
4. For a cylindrical-rotor synchronous motor, the torque is 500 N-m with a rotor power angle of $-30°$. What is the maximum torque available from this machine without changing the excitation voltage?
5. The effect of the reluctance torque is to degrade the performance of a salient-pole synchronous motor. (T/F?)
6. It is impossible for a cylindrical-rotor synchronous motor to operate with a rotor power angle beyond $-90°$. (T/F?)

Answers. 1. Greater than; 2. true; 3. leading; 4. 1000 N-m; 5. false; 6. true.

PROBLEMS

Section 15.1.1: Induction Principles

P15.1. What relative values of dc current in coils *a*, *b*, and *c* in Fig. 14.14 give a horizontal flux, as required for the mental experiment?

Section 15.1.3: Three-phase Induction Motor Characteristics

P15.2. A three-phase 60-Hz induction motor has a nameplate speed of 1160 rpm. Find:
   (a) Slip
   (b) Electrical frequency of the stator currents
   (c) Electrical frequency of the rotor currents
   (d) Mechanical rotational speed of the stator fields
   (e) Mechanical rotational speed of the rotor fields
   (f) Ratio of the developed power to the rotor-copper losses

P15.3. In a three-phase 60-Hz induction motor turning 1720 rpm, the following losses are known: stator-Cu loss = 50 W; rotor-Cu loss = 65 W; and mechanical losses = 40 W. Find:

(a) Number of poles
(b) Slip
(c) Air-gap power
(d) Output power
(e) Input power
(f) Efficiency
(g) Output torque
(h) Apparent power if the power factor is 0.85

Answers. (a) 4 poles; (b) 4.44%; (c) 1463 W; (d) 1358 W; (e) 1513 W;
(f) 89.8%; (g) 7.537 N-m; (h) 1779.4 VA

P15.4. A three-phase, 60-Hz induction motor has the speed–torque characteristic shown
in Fig. P15.4. The load characteristic is also shown. Find:
(a) Number of motor poles
(b) Operating speed
(c) The slip at maximum torque
(d) Motor starting torque
(e) Air-gap power at a slip of 1.0
(f) Output power at the operating speed (hp)
(g) Approximate rotor-copper losses at the operating speed

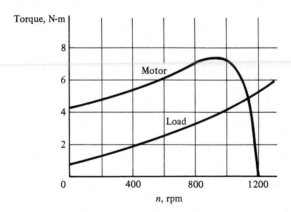

Figure P15.4 Motor and load characteristics.

P15.5. A motor catalog gives the following information about a motor: induction type,
three-phase, 60-Hz, ¼ hp, 1725 rpm, 220/440 V, 1.1/0.55 A, 66.8% efficiency,
NEMA frame type 56, 1.35 service factor, insulation class A, stock no. 2N101,
list price $194.00, wholesale $96.03, shipping weight 15 pounds, double-shielded
prelubricated ball bearings, automatic reset thermal protection, reversible shaft
rotation. Determine:
(a) Output torque at rated power
(b) Power factor at rated power
(c) Maximum power of which the motor is capable without overheating under
normal operating conditions

Answers. (a) 1.03 N-m; (b) 0.666; (c) 252 W

P15.6. A motor catalog gives the following information about a motor: induction type, single-phase, 60 Hz, 10 hp, 1740 rpm, 230 V, 44.0 A, prelubricated double-shielded ball bearings, NEMA type 215T, service factor 1.15, insulation class B, stock no. 6K100, list $1100.00, wholesale $733.70, shipping weight 144.0 pounds, reversible shaft rotation, gray finish. Determine:
(a) Output torque at nameplate conditions
(b) An equation relating the efficiency and the power factor at nameplate values
(c) Maximum power that the motor can produce without overheating under normal conditions

P15.7. A three-phase, 60-Hz induction motor has the following specifications: 5 hp, 1740 rpm, 230/460 V, 12.8/6.4 A, efficiency 87.5%, service factor 1.15.
(a) Find the slip of the motor at nameplate speed.
(b) Find the power factor under nameplate conditions.
(c) What is the air-gap power under nameplate conditions, assuming that 60% of the losses are electrical and 40% mechanical?
(d) If the motor were excited by 220 V, would the current be less than, equal to, or greater than 12.8 A with the same output power?

Answers. (a) 3.33%; (b) 0.836; (c) 4079 W; (d) greater than 12.8 A

P15.8. A four-pole, 60-Hz, three-phase induction motor operates from 460 V, draws 2.85 A at 0.76 PF (lagging). Of the input electrical power, 95% crosses into the rotor; of that power, 96% is converted to mechanical form; and of that 97% is output power. Find:
(a) Motor efficiency
(b) Motor speed in rpm
(c) Output power
(d) Output torque

P15.9. A three-phase, 60-Hz induction motor has the following nameplate information: 2 hp, 1725 rpm; 230/460 V, 5.7/2.85 A, 87% efficiency; service factor 1.15. If the machine is operating under nameplate conditions, determine the stator-copper loss and the mechanical losses, assuming they are equal.

Section 15.1.5: Equivalent Circuit for Three-phase Induction Motor

P15.10. A three-phase, four-pole 60-Hz induction motor is represented at one speed by the per-phase circuit shown in Fig. P15.10. The 8-Ω resistor represents the power being converted to mechanical form in the motor.
(a) What is the motor speed in rpm?
(b) Find the developed mechanical torque.
(c) What is the starting torque?

Answers. (a) 1756 rpm; (b) 33.6 N-m; (c) 74.5 N-m

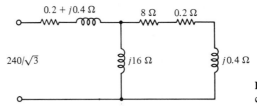

Figure P15.10 Per-phase equivalent circuit.

P15.11. A three-phase, 230-V, 60-Hz induction motor has the following nameplate characteristics: 10 hp full load, 90% efficiency, 0.82 power factor (lagging), 1740 rpm at full load.
   (a) Determine the apparent power rating of the machine.
   (b) What is the output torque at full load?
   (c) Find the reactive power required by the motor at full load.
   (d) The no-load speed is 1798 rpm. Estimate the stator-Cu loss at full load.

P15.12. The per-phase equivalent circuit of a three-phase 60-Hz induction motor is shown as a function of slip in Fig. P15.12. Ignore mechancial losses.
   (a) For an output power of 1 hp, find the speed of the two-pole motor. *Hint:* Use a Thévenin equivalent circuit.
   (b) For an output power of 1 hp, determine the input current and the efficiency of the motor.
   (c) Calculate the starting torque for this motor.

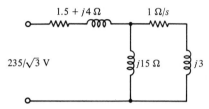

Figure P15.12   Per-phase equivalent circuit.

P15.13. A 60-Hz three-phase induction motor runs 3595 rpm at no load and 3500 rpm with 5 hp out. Assume operation in the small-slip region.
   (a) Find the slip at no load and at 5 hp out.
   (b) Find the mechanical losses at 5 hp out, assumed independent of speed. Assume that the developed torque is proportional to slip.
   (c) Find the air-gap power at 3500 rpm.

   Answers.   (a) 0.139%, 2.78%;   (b) 201.9 W;   (c) 4044 W

P15.14. The per-phase equivalent circuit for a 60-Hz three-phase, four-pole induction motor is shown in Fig. P15.14. At a slip of 4%, the input current is 47.4 A and the power factor is 0.9345.
   (a) What is the input power to the motor?
   (b) Find the air-gap power.
   (c) What is the developed torque for the motor?
   (d) Assuming half the losses of the motor to be electrical and half mechanical, find the efficiency at 4% slip.
   (e) If the input voltage dropped by 20%, would the current to the motor decrease, increase, or remain the same?
   (f) *Estimate* the input current for starting the motor. (Ignore $j10\ \Omega$)
   (g) *Estimate* the starting torque for the motor.

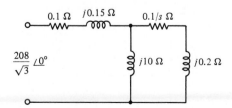

Figure P15.14

P15.15. The equivalent per-phase circuit shown in Fig. P15.15 represents a 60-Hz three-phase induction motor with six poles.

(a) *Estimate* the starting current. (Ignore $j20\ \Omega$)

(b) *Estimate* the slip for 25-hp developed power.

Answers. (a) 228 A; (b) 3.77% to 4.78%

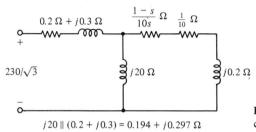

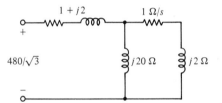

$j20 \parallel (0.2 + j0.3) = 0.194 + j0.297\ \Omega$

Figure P15.15 Per-phase equivalent circuit.

P15.16. A 6-pole, 60-Hz, three-phase induction motor has the per-phase equivalent circuit shown in Fig. P15.16.

(a) Determine the developed power at 1140 rpm.

(b) Find the efficiency at 1140 rpm, assuming 300 W of mechanical loss.

(c) At what speed does the motor produce 4 hp? Assume the same mechanical loss as above. Assume small-slip region.

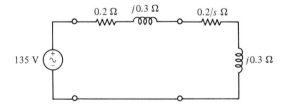

Figure P15.16

P15.17. Figure P15.17 shows the per-phase equivalent circuit for a six-pole, 60-Hz, three-phase induction motor. The stator circuit has been converted to a Thévenin equivalent circuit.

(a) Find the speed (rpm) for a developed power of 10 hp. Assume small-slip region.

(b) The full-load speed is 1135 rpm. Find the full-load output power, assuming 3% of the developed power is mechanical loss.

(c) From the previous information, estimate the no-load speed, assuming constant mechanical loss.

Figure P15.17

P15.18. A three-phase, 60-Hz induction motor has the following nameplate information: 5 hp, 1740 rpm, 230/460 V, 12.8/6.4 A, efficiency 87.5%, service factor 1.15, blocked-rotor current 92 A at 230 V, blocked-rotor torque 185% of rated torque. "Blocked-rotor" means the same as "starting" since the rotor is not turning in either case.

(a) Find the power factor under rated conditions.

(b) Find the air-gap power under blocked-rotor conditions.

(c) What happens to the air-gap power under blocked-rotor conditions?

(d) Assuming that the rotor-copper losses are proportional to the square of the input current, find the mechanical losses at 1740 rpm.

Section 15.1.7: Single-phase Induction Motors

P15.19. A single-phase 120-V inductor motor takes a current of 6 A. The electrical loss is 5% of the input power and the mechanical loss is 5% of the output power. If the total losses are 52.5 W, find the power factor of the motor.

P15.20. A 115-V, 60-Hz single-phase induction motor produces 2 hp at 3500 rpm with an input current of 18.8 A. Assume losses at full load to be 20% of the output power, equally divided between mechanical and electrical losses. Find the power factor, the efficiency, and estimate the no-load speed.

P15.21. A 60-Hz single-phase induction motor has the following specifications: 115/230 V, 8.2/4.1 A, 1725 rpm, $\frac{3}{4}$ hp. Assume a power factor of 0.82, lagging.

(a) Determine the efficiency of the motor at full load.

(b) What is the output torque at full load?

(c) What are the electrical frequencies in the rotor at full load?

(d) Explain how the motor can be connected for either 115- or 230-V operation.

(e) Would you expect the motor to have better speed regulation when connected for 115- or 230-V operation? Explain.

P15.22. A single-phase 60-Hz induction motor has the following specifications: $\frac{3}{4}$ hp, 1725 rpm, 115/230 V, 11.6/5.8 A, ball bearings, reversible.

(a) Assuming a power factor of 0.77, determine the efficiency of the motor.

(b) Find the output torque under nameplate conditions.

(c) Explain how to reverse the motor.

(d) If the motor was excited by 113 V, would the current be less than, the same as, or greater than 11.6 A for the same output power?

P15.23. A high-efficiency single-phase 60-Hz induction motor has the following nameplate information: $\frac{1}{2}$ hp, 1725 rpm, 115 V, 5.5 A. This is a capacitor-start, capacitor-run motor.

(a) How many poles does this motor have?

(b) Assuming an efficiency of 75%, determine the power factor.

(c) What electrical frequencies exist in the rotor?

(d) What is the output torque?

(e) Draw the motor circuit that exists during starting.

P15.24. A 60-Hz, 120-V, single-phase induction motor has a linear region described by the equation $P_{dev} = a + bs$. The motor runs at 1795 rpm no load and 1740 rpm with 1-hp output power. Find the output power at 1785 rpm, assuming the same mechanical losses at all speeds in the problem.

P15.25. A capacitor-start, induction-run, 60-Hz, single-phase induction motor has the following nameplate information: 2 hp, 1725 rpm, 60 Hz, 230 V, 10.3 A, and efficiency = 84.0%. Find:

(a) Number of poles

(b) Slip at rated load

(c) Power factor

(d) Rated torque

(e) Yearly operating cost based on 6 cents/kW-hour, 40 h/week, 50 weeks/year.

Answers. (a) 4; (b) 4.17%; (c) 0.7498; (d) 8.259 N-m; (e) $213.14

P15.26. The circuit diagram for a 120-V, 60-Hz, single-phase capacitor-start, induction-run motor is shown in Fig. P15.26. The circuit for the main winding applies only with the rotor at rest. After the motor comes to full speed, the current in the main winding is one-third of its value at the instant of starting. The auxiliary winding is connected only during starting. Find the ratio of the maximum starting current to the run current.

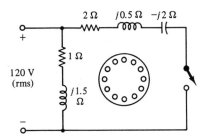

Figure P15.26

P15.27. A single-phase 60-Hz capacitor-start, induction-run motor has the equivalent circuit (for starting) shown in Fig. P15.27.

(a) Find C for 90° phase shift between starting currents in run and auxiliary windings at starting.

(b) Determine the starting current for this value of C.

(c) Assuming the nameplate run current is 3.9 A, the speed is 1740 rpm, and power is $\frac{1}{4}$ hp, estimate the power factor and output torque of the motor under nameplate conditions. Ignore mechanical losses.

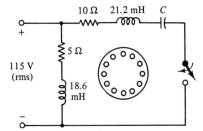

Figure P15.27

P15.28. A motor catalog lists a 60-Hz high-efficiency (capacitor-start, capacitor-run) single-phase motor with the following nameplate information: 1 hp, 1725 rpm, 115/230 V, 9.2/4.6 A. It also says this motor saves 100 W compared to the normal capacitor-start, induction-run motor, which has the following nameplate information: 1 hp, 1725 rpm, 115/230 V, 14.8/7.4 A. Assume that electrical losses are

$K(I_{in})^2$, where K is the same for both motors. Also assume that mechanical losses are 10% of the output power for both motors. Determine the power factors of the two motors.

P15.29. A $\frac{1}{4}$-hp, 120-V, 60-Hz single-phase induction motor runs at 3570 rpm no-load speed and 3450 rpm at rated output power. The mechanical losses are 25 W. *Hint:* Assume $P_{dev} = a + bs$.
(a) Find the two electrical frequencies in the rotor at $\frac{1}{4}$ hp out.
(b) *Estimate* the slip for $\frac{1}{8}$ hp out.
(c) Find the mechanical power required to drive the motor at 3600 rpm.

Answers. (a) 2.5, 117.5 Hz; (b) 2.5%; (c) 46.6 W

P15.30. A capacitor-start, induction-run 115-V, 60-Hz single-phase induction motor has stator field parameters as shown in Fig. P15.30. What is the capacitor required to produce a 90° phase shift between main and auxiliary currents for starting?

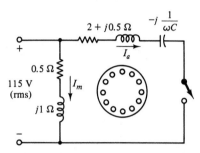

Figure P15.30 A capacitor-start motor.

P15.31. Figure P15.31 shows a shaded-pole induction motor. The conducting bands "shade" a portion of the poles.
(a) Estimate the flux in the iron.
(b) Estimate the input current to the motor if its output power is 25 W.
(c) Mark the direction that the motor will turn.

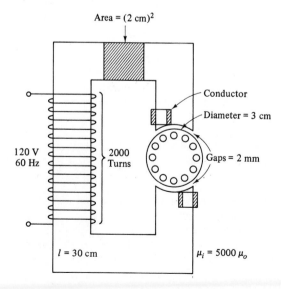

Figure P15.31

Section 15.2: Synchronous Motors

P15.32. Figure P15.32 shows a phasor diagram for a three-phase synchronous machine.
  (a) Is the machine operating as a motor or a generator?
  (b) What is the voltage and apparent power into/out of the machine?
  (c) Determine the synchronous reactance of the machine.
  (d) For the same real power, what magnitude of excitation voltage yields unity power factor?

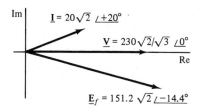

$\underline{E}_f = 151.2 \sqrt{2} \angle -14.4°$ **Figure P15.32**

P15.33. A 50-hp, 480-V, 8-pole, 60-Hz, three-phase synchronous motor has a power factor of 0.8 leading at full-load output. If the mechanical load is removed, the current into the motor is 40 A. Neglect losses. Find the synchronous reactance.

Answer. 2.28 Ω

P15.34. A cylindrical-rotor, 60-Hz, three-phase, 12-pole synchronous motor operates from 2300 V and produces 500 hp. The motor operates with unity power factor with an excitation voltage of $E_f = 1620$ V per phase. Determine the current, the synchronous reactance, the torque, and the rotor power angle. Neglect losses.

P15.35. A 100-hp, 480-V, 8-pole, 60-Hz, three-phase synchronous motor has rated power at unity power factor with a field current of 5 A dc. What field current corresponds to 80 hp out and 0.9 PF leading? Neglect losses. The synchronous reactance is 1 Ω. Assume excitation voltage proportional to field current.

P15.36. The phasor diagram for a three-phase synchronous motor is shown in Fig. P15.36. Note that all sides and two angles of the triangle are shown. The current is 21 A (rms) and the motor has four poles. The diagram shows per-phase vaues.
  (a) Is the motor overexcited or underexcited?
  (b) What is the rotor power angle?
  (c) What is the power factor and is it leading or lagging?
  (d) Determine the synchronous reactance per phase.
  (e) Determine the output power and torque, neglecting mechanical losses.

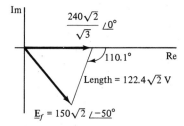

Figure P15.36 Phasor diagram for a synchronous motor.

P15.37. A 10-kVA, 600-V, three-phase synchronous motor operates at rated kVA with 10-hp output power. The field current is 5 A and the synchronous reactance is 30 Ω/phase. Find the field current for unity power factor, assuming that 5 A over-excites the machine. Ignore losses and assume excitation voltage proportional to field current.

P15.38. A 460-V, 50-kVA, 50-hp three-phase, 60-Hz synchronous motor has $X_s = 6 \Omega$. At full load out and unity power factor, the efficiency is 94% with losses divided equally between field, armature, and mechanical.

(a) What is the excitation voltage for 50 hp out and unity power factor?

(b) What is the excitation voltage for rated kVA in, neglecting any variation in the losses? Assume overexcited operation.

(c) What is the excitation voltage for rated kVA in, considering variation in losses? *Hints:* (1) The field losses play no role in the power passing through the machine from three-phase ac power to mechanical power. (2) The equivalent circuit in Fig. 15.27 must be modified by adding a resistance in series with the synchronous reactance. This per-phase resistance may be determined from the armature losses and ac current.

Answers. (a) 392 V/phase; (b) 579 V/phase; (c) 577 V/phase

P15.39. A test is performed on a three-phase, 460-V synchronous motor. The load is kept constant at rated load, and input current is measured while the field current is varied. The results are shown in the Fig. P15.39. The rated current is marked on the graph at 24 A. Ignore losses, and assume that the excitation voltage is proportional to field current.

(a) What is the apparent power rating of the motor?

(b) What is the rated output power in hp?

(c) Mark the range of field currents corresponding to overexcited operation.

(d) Determine the synchronous reactance of the motor.

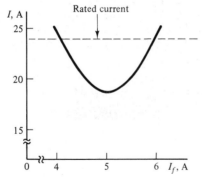

Figure P15.39

16 Direct-Current Motors

16.1 PRINCIPLES OF DC MACHINES

Importance. The dc machine can be used either as a motor or a generator. However, since semiconductor rectifiers can generate dc voltage from ac with electronic power supplies, dc generators are unneeded except for remote operations. Even in the automobile, the dc generator has been replaced by the *alternator* (a synchronous generator plus diodes for rectification). On the other hand, generator operation must still be considered because motors operate in the generator mode in braking and reversing.

Portable devices operating from battery power require dc motors, as for auto starters, window-lifts, and portable tape players. The wide range of speed and torque of the dc motor leads to many applications. More important, the dc machine is readily controlled in speed and torque and hence is useful for control systems. Examples are robots, elevators, machine tools, rolling mills, and large power shovels.

In this chapter we will examine the fundamental principles of dc motors and explore the wide range of characteristics that can be achieved through the various motor configurations. DC generators will be mentioned briefly, primarily because of the role of the back-generated voltage in the operation of dc motors.

16.1.1 How DC Motor Action Is Achieved

Figure 16.1 shows a cylindrical magnetic structure with stator and rotor poles. The relevant equations describing motor effects are shown in Eq. (16.1):

624

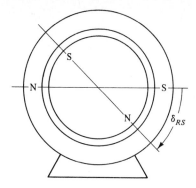

Figure 16.1 Cylindrical magnetic structure.

$$T_{dev} = -2 \ K\Phi_R\Phi_S \sin \delta_{RS} \quad \text{and} \quad P_{dev} = \omega_m T_{dev} \qquad (16.1)$$

In dc motors, the stator magnetic poles remain fixed in space; normally this is accomplished with salient poles, as shown in Fig. 16.2a. The mechanical rotor rotates relative to the rotor magnetic poles, which remain stationary in space. The rotor conductors carry "ac" currents (actually, chopped dc currents) under the control of the brush–commutator system. Thus, for the dc machine, the field is located on the stator and the armature on the rotor.

Brush–commutator system. The dc machine requires a brush-commutator system, indicated in Fig. 16.2a. Figure 16.2b shows a cutaway automobile generator. The commutator has a cylindrical surface of wedge-shaped segments connected to the rotor conductors. The commutator is part of the rotor and participates in its rotation. The brushes are stationary and rub against the commutator as the rotor rotates. The brush–commutator system provides, therefore, two related functions:

1. Electrical connection is made with the moving rotor.
2. Switching of the rotor currents is accomplished mechanically in a way that automatically synchronizes switching with rotor motion.

Due to the brush–commutator system, the *pattern* in space of the rotor currents is always the same, independent of the physical rotation of the rotor.

Many of the problems with dc machines arise out of commutation. Not only must the brush–commutator system carry potentially large currents across a sliding contact, but the switching of currents in the individual coils causes an inductive effect that limits performance. These problems have been solved in measure, with the result that we have available dc machines with excellent characteristics, although they require regular maintenance.

16.1.2 Circuit Model

Field circuit. The field circuit of the dc motor is wound on the stator and is modeled in Fig. 16.3a as a series connection of resistance and inductance, with a field-voltage supply, V_f. The motor is said to be separately excited because the

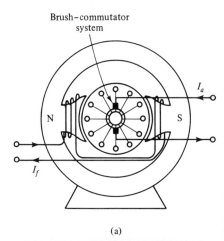

Brush–commutator
system

N   S

I_a

I_f

(a)

Figure 16.2 (a) A dc machine with
salient stator poles and a brush–
commutator system; (b) photograph
of an automobile generator.

(b)

field circuit is independent of the armature circuit. In steady-state operation, the
field current is established from Ohm's law:

$$I_f = \frac{V_f}{R_f} \tag{16.2}$$

where I_f is the field current and R_f the resistance of the field windings. The field
current produces a flux through the magnetic structure comprised of stator, rotor,
and air gaps. Neglecting magnetic saturation and residual magnetism, we may
relate the flux to the field current through the reluctance of the magnetic structure,
$\mathcal{R}$:

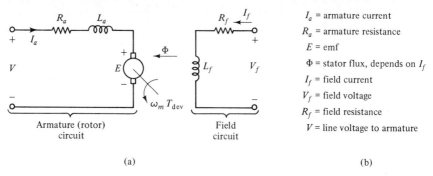

Figure 16.3 (a) Circuit model for a separately excited dc machine; (b) definition of motor variables.

$$\Phi = \frac{nI_f}{\mathcal{R}} \qquad (16.3)$$

where Φ is the flux and n is the equivalent number of turns. Normally, magnetic saturation and residual magnetism are significant, however, and the relationship between field current and flux is nonlinear. As we will see, this relationship is given as a *magnetization curve* that gives emf versus field current with the armature rotating at constant speed.

The number of poles runs from 2 or 4 for a small motor to as large as 24 to 30 for a large motor. As we will see, the number of poles does not determine the speed of the motor.

Armature circuit.   The armature circuit is located on the rotor. Our circuit model for the rotor consists of a resistance and inductance in series with an emf, E. Figure 16.3a shows the armature resistance and inductance outside the brush–commutator system, which introduces the emf (E) into the armature circuit. The armature resistance (R_a) and inductance (L_a) are located physically between the brushes but customarily are shown outside. In either event it does not matter for a series connection. We indicate the mechanical output with symbols indicating rotation and developed torque. Figure 16.3b defines the circuit quantities relevant to steady-state operation. As a consequence of its rotation in a magnetic field, the rotor will also have iron losses; however, these customarily are not represented in the electrical equivalent circuit but are combined with the mechanical losses.

Electromotive force and torque relationships.   When the rotor is rotated in the flux produced by the field, an ac voltage is produced in each rotor conductor. These voltages are rectified and summed by the brush–commutator system to produce a dc emf. This emf is proportional to flux and rotation speed, and hence may be expressed by the relationship

$$E = K_E \Phi \omega_m \qquad (16.4)$$

where K_E is a constant that depends on the number of rotor turns, details of how these are interconnected, and armature geometry.

If a dc armature current flows through the brush–commutator system, this current is applied to the rotor conductors, and a torque is developed. This developed torque is proportional to flux and armature current and hence may be expressed as

$$T_{dev} = K_T \Phi I_a \tag{16.5}$$

where K_T is a constant that also depends on the number of rotor turns, details of how these are interconnected, and armature geometry. Indeed, as we will soon demonstrate, conservation of energy requires that the two constants in Eqs. (16.4) and (16.5) be the same constant.

16.1.3 Power Flow in DC Machines

Motor or generator?   In steady state, KVL applied to the armature circuit yields

$$V = R_a I_a + E \Rightarrow I_a = \frac{V - E}{R_a} \tag{16.6}$$

where V is the armature voltage, R_a is the armature resistance, and E is the armature emf. If the armature voltage exceeds the emf, the armature current is positive relative to the reference direction shown in Fig. 16.3a, electrical power is delivered to the armature, and the machine acts as a motor. In this case the developed torque has the same direction as the direction of rotation, and mechanical power is delivered to the mechanical load. If the emf exceeds the armature voltage, the current is negative relative to the reference direction in Fig. 16.3a and flows out of the + polarity mark on the armature. The machine then acts as a generator, and the developed torque will be opposite to the direction of rotation. In many applications, the machine will alternatively act as motor or generator, depending on changes in the mechanical load or armature voltage.

Conservation of energy.   We may convert Eq. (16.6) into a power equation by multiplying by the armature current:

$$\underbrace{V I_a}_{P_{in}} = \underbrace{R_a I_a^2}_{\substack{\text{Armature} \\ \text{Cu} \\ \text{loss}}} + \underbrace{E I_a}_{\substack{\text{developed} \\ \text{power}}} \tag{16.7}$$

Equation (16.7) shows that the input power divides between armature-Cu losses and EI_a, which represents power leaving the electrical circuit and converting to mechanical power. Thus the developed power is

$$P_{dev} = EI_a = \omega_m T_{dev} \tag{16.8}$$

The output power would be less than the developed power by the mechanical losses:

$$P_{out} = P_{dev} - P_m \tag{16.9}$$

where P_{out} is the output power and P_m is the mechanical losses.

If we substitute Eqs. (16.4) and (16.5) for E and T_{dev}, respectively, Eq. (16.8) takes the form

$$K_E \Phi \omega_m I_a = K_T \Phi I_a \omega_m \qquad (16.10)$$

and thus $K_E = K_T = K$, the machine constant,† as asserted earlier.

Magnetization curve. Equation (16.4) allows the relationship between stator flux and field current to be represented as a *magnetization curve* of emf versus field current for a constant speed, as shown in Fig. 16.4. From the magnetization curve, we can determine the machine constant and the voltage at other speeds, because the generated voltage is strictly proportional to rotational speed. Furthermore, we can also derive torque information since Eqs. (16.4) and (16.5) can be combined to the form

$$T_{dev} = \frac{E}{\omega_m} \times I_a \qquad (16.11)$$

where E/ω_m depends on field current and can be determined from Fig. 16.4.

Two additional features of Fig. 16.4 deserve mention. We have indicated a small voltage at zero field current. This results from a residual (permanent) magnetism in the stator magnetic structure and can play an important role in the buildup of voltage if the machine is operated as a generator. Also, we note the effect of saturation in the iron of the magnetic structure. The air-gap line is important in the following sections because, to derive approximate motor characteristics with the various means of excitation, we often assume that the motor is operating with stator flux proportional to field current. The results of such an analysis suggest the general features of the motor characteristics, but any resulting calculations are approximate.

Example. A 120-V dc motor has an armature resistance of 0.70 Ω. It requires 1.1 A armature current and runs at 1000 rpm with no load. Find the output power and torque at 952 rpm output speed. Assume constant flux.

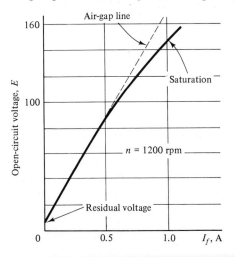

Figure 16.4 The magnetization curve gives emf versus field current at fixed speed.

† Often $K\Phi$ is called the machine constant. The context should make it clear which is meant.

From the no-load condition, we can calculate the machine constant and the mechanical losses. The input power at no load is 120 V × 1.1 A = 132 W, and the armature loss is $0.70(1.1)^2 = 0.85$ W; hence the mechanical losses at 1000 rpm are 131.2 W. The emf at this speed can be calculated from Eq. (16.6) as 120 − 0.70(1.1) = 119.2 V. Hence the product of the machine constant and the stator flux is $K\Phi = 119.2/1000$.†

At 952 rpm, the emf is reduced to $E' = K\Phi \times 952 = 119.2 \times 952/1000 = 113.5$ V. This reduced voltage implies an input current of $I_a' = (120 − 113.5/0.70 = 9.28$ A; and hence the developed power is $P_{dev} = E'I_a' = 113.5 \times 9.28 = 1052.9$ W. Mechanical losses at 952 rpm will be approximately the same as at 1000 rpm, and therefore the output power would be 1052.9 − 131.2 = 921.7 W (1.23 hp). The output torque would be

$$T_{\text{out}} = \frac{P_{\text{out}}}{\omega_m} = \frac{921.7}{952(2\pi/60)} = 9.25 \text{ N-m} \tag{16.12}$$

By repeated analysis, we could derive a torque versus speed curve for the motor, although this would require an assumption about the variation of mechanical losses with speed.

16.1.4 Check Your Understanding

1. On a dc machine, the magnitude of the rotor-stator power angle is about 90°. (T/F?)
2. DC motors are still used because they (a) can be portable, (b) have a good power factor, (c) can be controlled for speed, (d) can be controlled for torque, and (e) have low maintenance. Which? (May be more than one.)
3. DC motor torque is proportional to (a) armature current, (b) field current, (c) speed, or (d) developed power. (Which? May be more than one.)
4. DC motor emf is proportional to (a) armature current, (b) field current, (c) speed, or (d) developed power. (Which? May be more than one.)
5. To increase the speed of a separately excited dc motor, we would increase the (a) input voltage, (b) field current, (c) output torque, or (d) number of poles. (Which?)
6. The armature circuit of a separately excited dc motor is shown in Fig. 16.5. Determine the input power and developed power of the motor.

Answers. 1. True; 2. a, c, and d; 3. a and b; 4. b and c; 5. a; 6. 600 and 580 W, respectively.

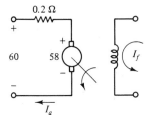

Figure 16.5

† In this expression, the units for K include the conversion between rpm and radians/second. Because we will work basically from scaling, such conversion factors cancel.

16.2.1 Shunt-connected Field

Circuit. Figure 16.6 shows a shunt connection of the field circuit. Here the field current is controlled by a field rheostat, R_F, and the field current is excited by the input voltage, in parallel or *shunt* with the armature circuit. Normally, the field has many turns of small wire, so the field current is small compared with the armature current. A variation on the shunt-connected motor is the separately excited motor. In this case, the field current is independent of the armature voltage, as in Fig. 16.3a.

Analysis. We will derive the torque as a function of speed with fixed input voltage (V) and fixed field current (I_f). In this case the nonlinear behavior of the magnetic structure is not a factor because the field current is constant. We begin with KVL in the armature circuit, Eq. (16.6), eliminate I_a through Eq. (16.5) and eliminate E through Eq. (16.4). The results are

$$V = R_a I_a + E = R_a \frac{T_{dev}}{K\Phi} + K\Phi\omega_m \qquad (16.13)$$

Solving for developed torque, we obtain

$$T_{dev} = \frac{K\Phi}{R_a}[V - K\Phi\omega_m] \qquad (16.14)$$

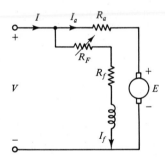

Figure 16.6 Circuit model for a shunt-connected motor.

Since $K\Phi$ is constant, Eq. (16.14) has the form of a straight line:

$$T_{dev}(\omega_m) = C_1 - C_2\omega_m \qquad (16.15)$$

where C_1 and C_2 are constants.

Example. Consider the machine whose characteristic is shown in Fig. 16.4. Assume the line voltage to be 120 V, the armature resistance to be 0.5 Ω, and the total field resistance to be 120 Ω, such that the field current is 1.0 A. The mechanical losses are 25 W, assumed constant. First we will find the speed for no load; then we will determine the torque and speed for 1 hp out. From these calculations, we can determine the torque as a function of speed.

With no external load, the power into the armature must supply armature resistive loss and mechanical losses. For the moment, we ignore the electrical loss; hence, the input power to the armature will be 25 W, and the required current is 25 W/120 V = 0.208 A. From KVL, we calculate the emf to be 120 − 0.208 × (0.5 Ω) = 119.9 V. From Fig. 16.4, we have, for a field current of 1.0 A, a generated voltage of 146 V at 1200 rpm. Hence, we scale the speed proportional to the generated voltage and estimate the no-load speed as

$$n_{NL} = \frac{119.9}{146} \times 1200 = 985.4 \text{ rpm} \tag{16.16}$$

This is one point on our torque–speed characteristic: 985.4 rpm at no output torque.

Assume now an output power of 1 hp. The developed power in the armature must now be 746 + 25 = 771 W. We can determine the armature current through conservation of energy. We use Eq. 16.7:

$$120I_a = (0.5)I_a^2 + EI_a \tag{16.17}$$

The last term, EI_a, is 771 W; hence Eq. (16.17) is a quadratic equation in I_a.

$$(0.5)I_a^2 - 120I_a + 771 = 0 \tag{16.18}$$

The quadratic has two solutions: I_a = 233.4 A or 6.61 A. Both solutions are theoretically possible, but the larger current value is unrealistic because the motor would burn up long before this value were reached. Hence we choose the smaller value as the realistic solution. The emf for this armature current would be 120 − (0.5)(6.61) = 116.7 V, and it follows from the line of reasoning that we used earlier that the speed must drop to

$$n = \frac{116.7}{146} \times 1200 = 959.1 \text{ rpm} \tag{16.19}$$

The output torque for 1 hp and 959.1 rpm would be

$$T_{\text{out}} = \frac{746}{959.1(2\pi/60)} = 7.43 \text{ N-m} \tag{16.20}$$

The torque–speed characteristic is shown in Fig. 16.7. We note that the speed is nearly constant. The motor torque is limited by the ability of the armature to dissipate the armature-Cu loss without damage, or perhaps of the brush–com-

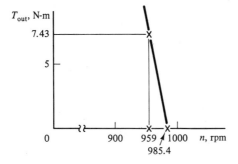

Figure 16.7 The torque–speed characteristic for the example. The shunt-connected motor maintains nearly constant speed.

mutator system to handle the required current. Incidentally, the *input* current for the motor with 1-hp output power would include the field current in addition to the armature current and hence would be 7.61 A. The total input power would be 120(7.61) = 913 W, and the efficiency would be 746/913 = 81.7%. Also, note that we could have used the analysis based on conservation of energy for the no-load condition. The exact analysis of the no-load condition, left as a homework problem, fully justifies our ignoring the electrical losses in the no-load case.

Another example. A separately excited dc motor (Φ = constant) has the following specifications: 50 hp, 200 V, 1200 rpm, armature resistance of 0.05 Ω, and a full-load current of 200 A. We will determine the speed and output power if the voltage is lowered to 150 V and the current remains at 200 A. Assume that the mechanical losses are proportional to speed. First we will analyze the nameplate information. At the nameplate values of voltage and current, the emf is

$$E = V - I_a R_a = 200 - 200(0.05) = 190 \text{ V} \tag{16.21}$$

and the machine constant in volts/rpm would be

$$K\phi = \frac{190 \text{ V}}{1200 \text{ rpm}} = 0.158 \text{ V/rpm} \tag{16.22}$$

Thus the developed power and mechanical losses would be

$$P_{dev} = 190 \times 200 = 38,000 \text{ W} \Rightarrow P_m = 38,000 - 50 \times 746 = 700 \text{ W} \tag{16.23}$$

Now we change the input voltage to 150 V. The armature current is unchanged, so the new emf is

$$E' = 150 - 200(0.05) = 140 \text{ V} \tag{16.24}$$

We may determine the new speed by scaling:

$$n' = n \times \frac{E'}{E} = 1200 \times \frac{140}{190} = 884 \text{ rpm} \tag{16.25}$$

This same result could be obtained from the machine constant in Eq. (16.22). The new developed power would be

$$P'_{dev} = 28,000 \text{ W} \tag{16.26}$$

We have assumed the losses proportional to speed, so the new mechanical losses are 516 W. Hence the new output power is 27,484 W (36.8 hp). This would be the rated output power of the machine at 150 V.

Summary. The characteristics of the shunt-connected dc motor are as follows:

- For fixed field current and input voltage the speed is nearly constant. This is true because the emf (E) is approximately equal to the input voltage (V).
- For fixed field current, the speed is approximately proportional to armature voltage. Since the emf is proportional to the product of the rotational speed

and the stator magnetic flux, for fixed armature voltage the speed is inversly affected by field current. To increase motor speed, we must decrease field current.

- Reversing the input voltage to the shunt-connected motor will not reverse the direction of rotation because both stator flux and armature current reverse. To reverse the motor directions, we must reverse the polarity of either field or armature.

Speed control.   As indicated above, we can control the speed of the dc motor by varying the armature voltage while keeping the field current constant, or by varying the field current while keeping the armature voltage constant. Both these methods are effective, but the control of the armature voltage offers more desirable dynamic properties for control applications. The armature voltage may be controlled by several methods, as shown in Chapter 17.

Permanent-magnet (PM) DC motors.   Large numbers of dc machines are manufactured with fields provided by permanent magnets. Applications include fan and window-lift motors on automobiles, small appliances such as electric toothbrushes, tape recorders, and hair dryers, instruments like tachometers and novelty devices such as toy trains. Some large machines, up to 200 hp, are designed with permanent magnetic fields to meet special requirements for size, weight, or efficiency. Figure 16.8 shows a permanent-magnet fan motor for an automotive heater–air conditioning system.

Characteristics.   Machines with permanent magnet fields have characteristics similar to separately-excited machines.

$$V \approx E_a = \quad K\Phi n \tag{16.27}$$

Figure 16.8 Permanent-magnet fan motor for automotive heater–air conditioning system.

To a good approximation, the machine constant $K\Phi$ would be constant; hence the machine would operate with speed proportional to input voltage.

16.2.2 Series-Connected Field

Circuit. Figure 16.9 shows a series-connected dc motor. Here the armature and field carry the same current, and hence the field normally consists of a few turns of large wire.

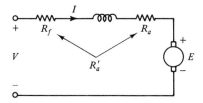

Figure 16.9 Circuit model for series-connected motor.

Analysis. We now derive the torque–speed characteristic of the series-connected motor. We ignore magnetic saturation by assuming that the stator flux is proportional to field current; thus Eq. (16.5) becomes

$$T_{dev} = K\Phi I = K_1 K_2 I^2 \tag{16.28}$$

where I is the current, K_1 is the machine constant, and K_2 is related to the slope of the air-gap line, as in Fig. 16.4. We let $R_a' = R_a + R_f$ represent the combined field and armature resistance; hence KVL in the armature circuit becomes

$$V = R_a' I + K_1 \Phi \omega_m = (R_a' + K_1 K_2 \omega_m)I \tag{16.29}$$

We may eliminate the current through Eq. (16.28), with the result

$$V = (R_a' + K_1 K_2 \omega_m) \sqrt{\frac{T_{dev}}{K_1 K_2}} \tag{16.30}$$

To obtain the torque–speed characteristic, we solve Eq. (16.30) for the developed torque:

$$T_{dev} = K_1 K_2 \left[\frac{V}{R_a' + K_1 K_2 \omega_m} \right]^2 \tag{16.31}$$

The speed–torque characteristic in Eq. (16.31), modified to include mechanical losses, is shown in Fig. 16.10. We have shown a finite no-load speed at zero output torque because we will have a nonzero developed torque due to the mechanical losses. On the other end of the characteristic in Fig. 16.10, we have a limit on the torque due to armature and field resistance, and also because of saturation effects in the magnetic structure (although saturation is not included in our model).

Example. Consider a 50-V series-connected motor with $R_a' = 0.05\ \Omega$. The motor has an output power of 1 hp at 500 rpm. We are to find the speed and motor current for an output power of 3 hp, ignoring mechanical losses.

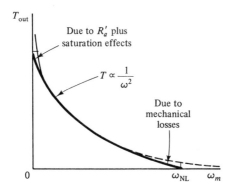

Figure 16.10 Output torque–speed characteristic for a series-excited motor.

An approximate solution is easy. If we assume we are in that part of the characteristic where torque is inversely proportional to the square of the speed, it follows that the power, $P = \omega_m T$, is inversely proportional to speed. Thus

$$P = \omega_m T \approx \frac{C}{n} \tag{16.32}$$

where C is a constant and n is motor speed in rpm. From the given information, we know that $C = 500$ when power is expressed in hp; hence for 3 hp the speed (n') must be

$$3 = \frac{C}{n'} \Rightarrow n' = \frac{500 \times 1}{3} = 167 \text{ rpm} \tag{16.33}$$

The current (I') can be determined from conservation of energy:

$$50I' = 0.05(I')^2 + 3 \times 746 \tag{16.34}$$

The quadratic has two solutions, and we pick the smaller root as the realistic current: $I' = 47.0$ A. We might note that, although the speed was derived from an approximate analysis, the current was derived from conservation of energy and thus is exact.

We begin a more accurate analysis by calculating the current (I) for 1 hp out.

$$50I = 0.05I^2 + 746 \tag{16.35}$$

Hence $I = 15.1$ A. Since the torque is

$$T = \frac{746}{500(2\pi/60)} = 14.2 \text{ N-m} \tag{16.36}$$

Eq. (16.28) gives $K_1 K_2$ as

$$K_1 K_2 = \frac{T}{I^2} = \frac{14.2}{(15.1)^2} = 0.06208 \text{ N-m/A}^2 \tag{16.37}$$

This value will be used in Eq. (16.31), which we convert to a power equation by multiplying by ω_m:

$$P = \omega_m T = K_1 K_2 \left[\frac{V}{R' + K_1 K_2 \omega_m} \right]^2 \times \omega_m \qquad (16.38)$$

With the known values of power in watts (3×746), input voltage (50 V), total resistance (0.05 Ω), and $K_1 K_2$ (0.06208), Eq. (16.38) is quadratic in ω_m. The two roots are $\omega_m = 0.0397$ and 16.3 rad/s. In this case, the larger root (156 rpm) is the realistic answer. The current for 3 hp was calculated earlier to be 47.0 A.

Features of the series-connected motor. Reversing the polarity of the input voltage will not reverse the direction of rotation because both field and armature currents are reversed. To reverse the motor, we must reverse field or armature polarity.

The sloping torque–speed characteristic of the series-connected motor offers benefits for many applications. The motor gives good torque for starting without excessive current. For this reason, series-connected motors are used as starter motors for automobiles.

16.2.3 Universal (AC/DC) Motors

Principles of operation. The series-connected dc motor will operate with alternating current. With the field in series with the armature, as shown in Fig. 16.11a, the flux is proportional to the current, and a time-average torque is produced by ac current, as shown in Fig. 16.11b. Thus we may time average Eq. (16.28)

$$\langle T(t) \rangle = K_1 K_2 \langle i^2(t) \rangle = K' I_{rms}^2 \qquad (16.39)$$

where $K' = K_1 K_2$ in Eq. (16.28). The torque–speed characteristic of the universal motor is similar to that for the series-connected dc motor in Fig. 16.10.

The features of the universal motor are:

- Although the machine will operate on ac or dc, most universal motors are designed to operate only on ac.

Figure 16.11 (a) Circuit model for a universal motor; (b) the ac current produces a time-average torque.

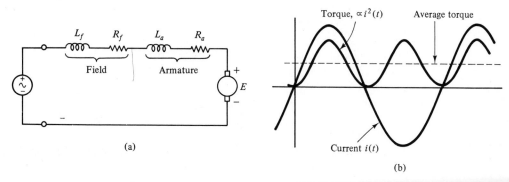

- Unlike other ac motors, whose speed is limited by the line frequency and number of poles, the universal motor has no such limitation. Universal motors routinely operate at high speeds, up to about 25,000 rpm.
- Because of the high speed, the universal motor produces more horsepower per pound of weight than other ac motors. For this reason it is favored for hand-held tools.
- The stator iron is laminated to reduce eddy current and hysteresis losses.
- The inductances of the field and armature are a factor in establishing the current in the machine. Because of the series connection, however, the flux and armature current are always in phase.

Applications. When you see an ac motor with a brush–commutator system, you know that you are dealing with a universal motor. When an application requires speeds higher than 3600 rpm, a universal motor is required. Also important is the ease with which the speed of a universal motor can be controlled with a "dimmer" circuit. For these reasons, universal motors are used for tools such as drills and routers and for household appliances such as mixers, blenders, and vacuum cleaners.

Also, the torque–speed characteristic of the series motor fits many applications. For example, in a hand drill we want a high speed for a small drill bit, but with the heavier load of a large bit we would like a slower speed. The overall characteristic of the universal motor is to slow down and increase torque as the mechanical load increases and to allow stall at a moderate torque. The motor is therefore tolerant of a wide variety of load conditions. Finally, the universal motor gives more horsepower per pound than other motor types because of its high speed, and thus universal motors have the edge in hand-held tools such as drills, sanders, and saws of various types.

16.2.4 Check Your Understanding

1. A shunt-connected dc motor will reverse if the polarity of the input voltage is reversed. (T/F?)
2. A shunt-connected dc motor runs at 800 rpm. The field current is doubled but the armature voltage remains constant. What is the approximate speed?
3. A shunt-connected dc motor draws 10 A at 24 V. The motor torque drops to half its former value. What is the new input power?
4. A series-connected dc motor with constant input voltage will run with essentially constant (a) speed, (b) current, (c) power, (d) torque, or (e) none of these. (Which?)
5. A series-connected dc motor will reverse if the polarity of the input voltage is reversed. (T/F?)
6. Name the motor most likely to be used in an electric chain saw.
7. The starter motor in an automobile would be (a) shunt-connected, (b) series-connected, or (c) have a permanent magnet field. (Which?)

Answers. 1. False; 2. 400 rpm; 3. 120 W; 4. e; 5. false; 6. universal motor; 7. series-connected dc motor.

Section 16.1: Principles of DC Machines

P16.1. A dc motor is shown in Fig. P16.1. The stator coil has 300 turns per pole and carries 5 A dc in the direction shown. The rotor is turning ccw as shown and the rotor currents come out on the right and go in on the left, as shown. The width of the air gap is 1.5 mm on each side and the iron has $\mu_i = \infty$.
   (a) Determine the direction of the stator flux and mark the poles N and S accordingly on the stator poles.
   (b) Find the flux density in the air gap due to the field current.
   (c) What is the direction of the magnetically generated torque on the rotor?
   (d) Is the machine acting as a generator or a motor?

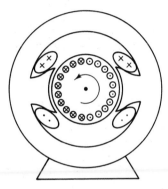

Figure P16.1

P16.2. A dc motor is separately excited, that is, has independent control over armature and field. The ratings are P watts, n rpm, I_a amperes, V volts on armature, T newton-meters, and I_f amperes of field current. Assuming a linear magnetic structure and ignoring losses, fill in Table P16.2.

TABLE P16.2

Field Current	Voltage	Current	Speed	Torque	Power
I_f	V	I_a	n	T	P
I_f	V			$\frac{1}{2}T$	
$\frac{1}{2}I_f$	V	I_a			
$2I_f$	$\frac{1}{2}V$	$\frac{1}{2}I_a$			
I_f	$2V$				P

P16.3. For the motor model in Fig. 16.3, the input voltage and current are 24 V and 10 A, respectively, the armature resistance is 0.10 Ω, the field voltage is 24 V, the field current is 1.0 A, the mechanical losses are 10 W, and the motor speed is 1200 rpm. Find:

(a) Input power, including that required for the field current
(b) emf
(c) Developed power
(d) Output power
(e) Output torque
(f) Efficiency
(g) What is the machine constant in volts/(radian per second)?

Answers. (a) 264 W; (b) 23 V; (c) 230 W; (d) 220 W; (e) 1.75 N-m; (f) 83.3%; (g) 0.183 V/(rad/s).

P16.4. An 80-V, 25-A dc motor is separately excited (from a constant voltage, thus has constant field current). The armature resistance is 0.100 Ω. The motor is used to hoist an elevator weighing 500 lb (Fig. P16.4). Neglect mechanical losses throughout this problem.

(a) What is the lifting rate in feet per second with the motor drawing rated current?
(b) The motor voltage is then reversed in polarity to lower the load. The voltage is adjusted to lower at the same rate as in part (a), again with rated current. What is the required voltage?
(c) What fraction of the energy used to raise the load is then returned to the electrical supply in lowering the load? Neglect mechanical loss and the effects of acceleration and deceleration.

Answers. (a) 2.86 ft/s; (b) −75 V; (c) 93.8%

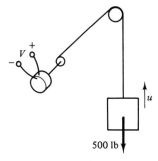

500 lb

Figure P16.4

P16.5. A separately-excited dc motor has the following nameplate information: 180 V, 9.5-A armature current, 1150 rpm, 1.03-Ω armature resistance, and 2 hp. Determine the no-load speed of the motor for the same field current, assuming the mechanical losses are proportional to the square of the motor speed.

P16.6. An 80-V dc motor has a constant field flux (separately excited) and a full-load speed of 1150 rpm with 710-W output power. The full-load armature current is 10 A and the no-load current is 0.5 A. Assume constant mechanical loss.

(a) Find the mechanical loss.
(b) Determine the armature resistance.
(c) What is the no-load speed in rpm?
(d) Find the machine constant in volts/(rad/s).
(e) Find the efficiency at full load, including field-circuit losses of 30 W.

Section 16.2: Characteristics of DC Motors

P16.7. A separately-excited dc motor runs 1000 rpm for a line voltage of 120 V dc, a field current of 2 A, and an output power of 2 hp. Give the effect on motor speed (speed up or slowdown, and whether the change is large or small) of the following changes, made one at a time and then restored before the next change is made. Assume negligible armature resistance and field flux proportional to field current.
   (a) Voltage changes to 60 V
   (b) Output power doubles
   (c) Output torque doubles
   (d) Field current doubles

P16.8. For the motor in Problem P16.7, the armature resistance is 0.8 Ω and the mechanical loss is 30 W.
   (a) Find the armature current for 2 hp out.
   (b) Find the no-load speed.

P16.9. A shunt-connected dc motor operates from 24 V and has an armature resistance of 0.30 Ω. The mechanical losses are 5% of the output power. The armature current is 10 A and the speed is 1200 rpm.
   (a) Find the input power. Ignore field losses.
   (b) Find the output power in horsepower.
   (c) Find the machine constant $K\Phi$.
   (d) Find the approximate no-load speed in rpm.

P16.10. A shunt-connected, 1-hp, 180-V dc motor puts out full-load power at 1800 rpm but runs 1850 rpm no load. The mechanical loss is a constant 50 W and field resistance is 810 Ω.
   (a) Draw a circuit diagram of the motor.
   (b) What is the field current?
   (c) Find the input current to the armature at full load.
   (d) Determine the armature resistance.
   (e) Find the efficiency, including field losses.
   (f) Find the torque out at full load.
   (g) What is the machine constant in N-m/ampere?

P16.11. The circuit model for a shunt-excited dc motor is shown in Fig. P16.11. The machine constant is 0.2 V/(rad/s). Find the speed in rpm for a developed power of 500 W.

   Answer.   4013 rpm

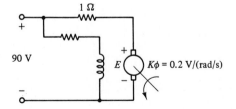

Figure P16.11

P16.12. A dc motor has the following nameplate information: 1.5 hp, 1750 rpm, 180 V on armature, 7.3 A in armature, 1.05-Ω armature resistance, 180 V for field, and 0.55 A for field. The motor is shunt connected. Assume constant mechanical losses in this problem.
(a) Find the mechanical losses at 1750 rpm.
(b) Find the developed torque at 1750 rpm.
(c) Determine the no-load speed.

P16.13. A 180-V, shunt-excited dc motor runs 1150 rpm at full load with 2 hp out and 1190 rpm at no load. Estimate the armature resistance. State assumptions.

Answer. 0.7 Ω, approximate

P16.14. For the motor with the magnetization curve shown in Fig. P16.14, find the voltage required for 5 hp out at 1000 rpm in shunt-connected operation with $I_f = 1.5$ A. The mechanical losses at 1000 rpm are 100 W, and $R_a = 0.8$ Ω.

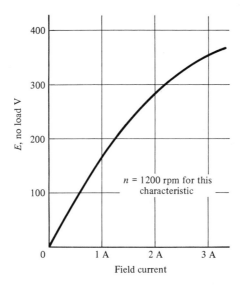

n = 1200 rpm for this characteristic

Field current

Figure P16.14

P16.15. For the motor with the magnetization curve shown in Fig. P16.14, determine the developed torque at the following currents: (a) $I_f = 2$ A, $I_a = 15$ A; (b) $I_f = 2$ A, $I_a = 10$ A; (c) $I_f = 1$ A, $I_a = 10$ A.

P16.16. A 5-hp shunt-connected 180-V dc motor has 0.25-Ω armature resistance and 200-W mechanical loss at the rated speed of 600 rpm. The field current is 0.5 A.
(a) What is the developed torque at the rated output power of 5 hp?
(b) What is the efficiency at 5 hp out, including field losses?
(c) What is the no-load speed? Assume same mechanical losses.

P16.17. A shunt-connected dc motor has an armature resistance of 0.2 Ω, a line voltage of 300 V, and a field resistance of 100 Ω. The magnetization curve is shown in Fig. P16.14. Ignore mechanical losses.
(a) At what speed would the motor run with no load?
(b) A load is applied and the speed drops 5%. Find the input current and developed power in horsepower.
(c) With the same torque requirement as in part (b), the input voltage is dropped to 250 V. Find the new speed.

P16.18. A dc motor with a permanent-magnet field structure has the following characteristics:

Condition	Armature V	Armature I (A)	Speed (rpm)
No load	12	0.40	1210
Full load	12	6.00	627

Find the armature resistance and the mechanical losses at no load. Assume mechanical losses are proportional to speed.

P16.19. A permanent-magnet-field, window-lift motor for an automobile runs at 2000 rpm and draws 10 A from a 12.6-V battery source. The stall current of the motor is 25 A.

(a) Find the developed power at 2000 rpm.

(b) What resistance must be placed in series with the motor to give 1000 rpm if the power required by the load, including mechanical losses, is 61% that for 2000 rpm?

P16.20. An engineer purchased a dc motor with a permanent-magnet field. The nameplate gives 90 V and 9.2 A but does not give the rated power. The engineer applied 90 V to the motor and measured the input current (0.5 A) and speed (1843 rpm). The engineer then loaded the motor mechanically until the current reached 9.2 A and measured the speed (1750 rpm). The engineer assumed that mechanical losses were constant, and determined from these data the output power at full load. What was the result?

Answer. 739 W

P16.21. A series-connected dc motor runs at 1200 rpm with an input voltage of 180 V and an output power of 1 hp. Ignore all losses in this problem. Fill in Table P16.21.

TABLE P16.21

Voltage (V)	Power (hp)	Speed (rpm)
180	1	1200
90	$\frac{1}{2}$	
180		1800
	1	900

P16.22. A series-connected dc motor has a combined field and armature resistance of 0.5 Ω, an input voltage of 24 V dc, and runs at 5000 rpm with $\frac{1}{4}$-hp output power. Ignore mechanical losses in this problem.

(a) Find the output torque.

(b) Find the input current.

(c) At twice the input current for part (b), what would be the speed? Assume magnetic flux is proportional to current.

P16.23. A dc series-connected motor drives a gear train whose no-load input torque requirement is proportional to speed. The combined armature and field resistance is 3 Ω. With an input voltage of 100 V dc, the motor input power is 50 W and the output speed of the unloaded gear train is 100 rpm. Derive the equation for the no-load speed at other input voltages. Assume that the flux of the field is proportional to the current. Ignore mechanical losses.

Answer. Check your equation with the point 62.7 rpm at 50 V.

P16.24. Does the output power of a series-excited dc motor increase, decrease, or remain the same as load torque increases? Explain.

P16.25. A 12-V series-connected dc motor has mechanical losses that are proportional to speed. At no-load the motor runs at 10,000 rpm and draws 0.83 A. Find the current and speed for an output power of 50 W. Assume no armature or field resistance.

P16.26. A 90-V, series-connected dc motor has a no-load speed of 20,000 rpm. Its rated output power of 1.5 hp occurs at a speed of 10,000 rpm. Assume mechanical loss is proportional to speed, but assume electrical losses are negligible. Find the no-load current.

Answer. 8.29 A

P16.27. A hand-held drill powered by a universal motor runs 1200 rpm with no load. With a ⅜-in. drill bit in operation, the drill slows down to 900 rpm. Ignore electrical loss and also the effects of inductance. Mechanical loss is assumed to be proportional to speed. Determine at the slower speed the following ratio: the output power to the bit divided by the mechanical loss.

P16.28. A 24-V series-connected dc motor has a combined armature and field resistance of 2 Ω. The motor runs at 5000 rpm with a developed power of 22 W. Find the maximum power that the motor can develop and the associated speed. Assume flux proportional to current.

P16.29. A 250-V dc motor is equipped for speed control with both a field rheostat in the shunt field (0 to 150 Ω) and a series armature rheostat (0 to 5 Ω). The resistance of the shunt field is 100 Ω and the armature resistance is 0.5 Ω. Figure P16.29 shows the motor circuit. The mechanical losses are 300 W at 5,000 rpm and are proportional to speed. Assume the field is proportional to field current.

   (a) If with both rheostats set to zero the motor draws 18.5 A total input current at 5,000 rpm, what is the output power?

   (b) Find the minimum no-load speed and the combination of field and armature rheostat resistances for minimum speed.

   (c) Find the maximum no-load speed and the combination of field and armature rheostat resistances for maximum speed.

Answers. (a) 4.79 hp; (b) 5024 rpm at minimum field resistance and maximum armature resistance; (c) 12,833 rpm at maximum field resistance and minimum armature resistance.

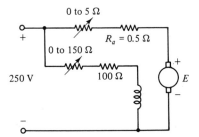

0 to 5 Ω

$R_a = 0.5$ Ω

0 to 150 Ω

250 V

100 Ω

E

Figure P16.29

17 Power Electronic Systems

Introduction. Power electronics applies electronic techniques to the control of electrical power. The field is not new because vacuum tubes have long been used for this purpose. But recently power electronics has grown rapidly because a class of solid-state devices has been developed to control electrical power. Important devices in this class are power transistors, silicon-controlled rectifiers (SCRs), and gate-turnoff thyristors (GTOs).

Ordinary electronics deals with the control of electrical power in small quantities, normally for its information content. Power is the product of voltage and current; hence ordinary electronic devices, such as transistors and diodes, must handle small voltages or small currents, or both. Power electronic devices must handle large voltages *and* large currents. How large depends on the context. The variable-speed hand-held drill requires voltages around 200 V (peak) and currents of a few amperes. By contrast 4500-V, 2500-A SCRs can be operated in tandem to handle conversion of 200 MW of ac power to dc power for a high-voltage direct-current (HVDC) transmission line. The principles would be the same for both applications.

This chapter introduces basic concepts of power electronics. We begin by discussing semiconductor switches. We then analyze a simple power controller such as might be used in a light dimmer or hand-held variable-speed drill. After giving some general principles of motor controllers, we discuss the electronic control of both dc motors and ac motors. Our overall purpose is to illustrate the

645

principles of power electronic systems through the analysis of representative applications.

17.1.1 Semiconductor Switches

Introduction. As discussed on page 241, a switch is a device that has two stable states, ON and OFF. When ON, the switch has an impedance much smaller than its load, and when OFF it has an impedance much larger than its load. Semiconductor switches have recently been developed to handle large amounts of power. In this section we describe principles of operation, terminal characteristics, and limitations of thyristors, gate-turn off thyristors (GTOs), and power transistors.

Four-level diode. Figure 17.1b shows the *pnpn* diode structure. The device consists of four alternating layers of *p*- and *n*-type semiconductor, forming three *pn* junctions. Two of the *pn* junctions face the same direction, and the middle one faces the opposite direction. Thus you might anticipate that the device will operate as three diodes in series, with the middle one turned around. Hence no current ought to flow in either direction, because at least one of the diodes will always be reverse biased. The characteristics displayed in Fig. 17.1c reveal that no current flows for negative voltages, where two *pn* junctions are reverse biased; neither does current flow for positive voltages, where the middle *pn* junction is reverse biased, until a threshold voltage, V_{th}, is reached. After this threshold voltage is exceeded, the four-level diode begins to conduct freely: it "fires," acting as if the middle *pn* junction disappears. The threshold phenomenon occurs because the doping levels in the four layers differ greatly. The outside *p* and *n* materials are doped heavily and hence have many carrier holes and electrons, respectively, available to diffuse into the middle *n* and *p* regions, which are lightly doped. Once the breakdown occurs in the middle junction at V_{th}, the holes from above and the electrons from below flood into the depletion region of the middle junction, where the internal battery-capacitor effect, described on page 228, reinforces their movement across the junction. Thus this middle depletion region effectively dis-

Figure 17.1 Four-level diode: (a) symbol, (b) structure, and (c) current–voltage characteristic.

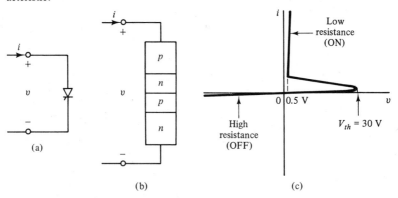

appears due to the carriers from the forward-biased junctions, and we are left with two forward-biased junctions in series. The small voltage for the *pnpn* diode in the ON state results from the contributions from each ON junction. Less than 0.7 V is required because the excess carriers from each junction help each other.

Thus the *pnpn* has two states. It is OFF for negative voltage, and remains OFF for positive voltage until the threshold voltage is reached, after which it turns ON. It will remain ON until the current is reduced to a small value, after which it will turn OFF again. The value of threshold voltage, V_{th}, can be controlled over a modest range by the semiconductor designer. Typical *pnpn* diodes have threshold voltages from 6 to 32 V; we have shown 30 V. Thus this device, like the *pn*-junction diode, is a voltage-controlled switch, except that the four-level diode requires much more than 0.7 V to turn ON.

Thyristor (SCR). As Fig. 17.2b reveals, the thyristor, or silicon-controlled rectifier (SCR), is a *pnpn* structure with an external connection to one of the internal layers. The extra input is called the *gate*, and its function is to turn the device ON. With no gate current, the SCR characteristic is that of a four-level diode, as shown. The important difference, however, is that the breakdown threshold occurs at a much higher voltage, indeed, high enough that the SCR should never conduct because the input voltage exceeds its threshold. On the contrary, the SCR should fire only when a pulse of current is delivered to the gate. This is shown by the $i_G > 0$ characteristics in Fig. 17.2c; in effect, the threshold is reduced to a very small value when the gate conducts. Physically, the gate injects holes into the lightly doped *p* region and floods the depletion region with carriers, thus initiating breakdown.

From the instant when voltage is applied to the gate, there is a small delay before the SCR turns ON, but this is not a problem at power frequencies. More problematic is the rate of rise of SCR current, which can cause device failure due to localized heating of the junction. If not limited by load inductance, an external inductor must be added in series to limit the di/dt of the SCR.

Also critical is the rate of voltage rise of the SCR in its OFF state. In certain applications, rapid voltage increase can initiate conduction, independent of the

Figure 17.2 Silicon-controlled rectifier (SCR): (a) symbol, (b) structure, and (c) current–voltage characteristic.

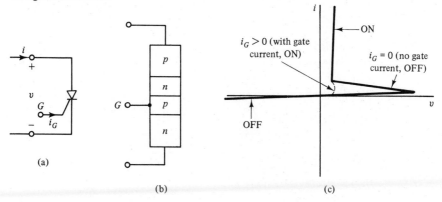

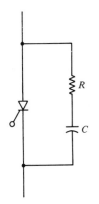

Figure 17.3 The *RC* snubber circuit limits the rate of voltage rise of the SCR in its OFF state.

gate signal, and cause device malfunction. The *RC* snubber shown in Fig. 17.3 limits the *dv/dt* across the SCR. With the SCR OFF, the small (10-Ω to 100-Ω) resistor in series with *C* and the inductance in the load circuit limit the *dv/dt* across the SCR, and the capacitor blocks the dc current. When the SCR fires, the energy stored in the capacitor is dissipated in *R* and the SCR.

To turn OFF, the SCR must be reverse biased for a sufficient period of time for carriers to recombine and reestablish the blocking junction. This time (t_{off}) can vary from 20 to 200 μs, depending on SCR size and type.

Line and forced commutation. *Commutation* refers to the switching of a conducting SCR from the ON state to the OFF state. As we have seen, an SCR may be commutated by reducing its current below the value required to sustain conduction. Normally, this is accomplished by reverse biasing the device for a period of time.

When in an ac cycle the voltage across an SCR changes from forward bias to reverse bias, the SCR ceases to pass current and is *line commutated*. When an auxiliary circuit is used to commutate the SCR independent of the line voltage, the SCR has a *forced commutation*.

Summary. In the previous section we have discussed use of SCRs as power switches. We have given the conditions for turning SCRs ON and OFF. The need for auxiliary circuits to force the commutation of the SCR is a serious drawback, and for this and other reasons alternative electronic switches have been developed.

Gate-turnoff thyristors (GTOs). Figure 17.4a shows the circuit symbol for a GTO, which is like an SCR symbol except for a mark on the gate, and Fig. 17.4b shows GTO construction. The GTO is turned ON by a pulse of positive current to its gate, but unlike the SCR it is turned OFF by a pulse of negative

Figure 17.4 (a) Circuit symbol for a gate-turnoff thyristor (GTO); (b) construction of a GTO.

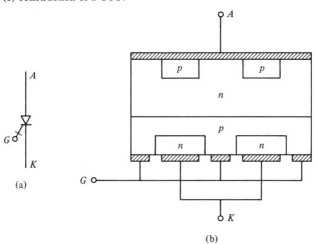

gate current. The negative current removes carriers from the cathode region, and the inner *pn* junction blocks the forward current.

The switching characteristics of the GTO are excellent, and this device is appearing in a variety of power controllers. Development in GTOs and power field-effect transistors is very active, and the circuit designer has an increasing number of semiconductor devices to consider in power controllers.

Power bipolar-junction transistors. Recent advances in bipolar power transistors (BJTs) have allowed their use in power electronic circuits. Although SCRs are still dominant in applications requiring high voltage and high power, transistors are replacing SCRs in ac motor controllers in the voltage range of 460 V and below, and power levels up to about 400 hp. The main advantage of transistor switches is that turn-on and turn-off are controlled by base current, and no commutation circuit is required. Transistors also offer fast switching time, which is important in many ac motor controller schemes. The disadvantages of transistors are that they require medium power base drive circuits and that they have higher switching and operating losses than SCRs. In this section, we consider transistor switches with resistive and inductive loads.

Transistor switching losses with resistive loads. Figure 17.5a shows a transistor being used as a switch for a resistive load, and Fig. 17.5b shows voltage and current during turn-on. Base current is applied at $t = 0$. After a delay time, t_d, the collector current begins to rise and the collector–emitter voltage begins to drop. The *rise time, t_r,* is defined as the time required for the collector current to rise from 10% to 90% of its final value, I_{Cp}. When the transistor is OFF, it has no collector current, and hence no power dissipation. After switching, the transistor is saturated and has a low collector–emitter voltage, and hence has low dissipation. During switching, however, the transistor has appreciable voltage *and* current, and hence the power level can become high. We will calculate the power during switching and the total energy dissipated by the transistor during switching. We assume linear voltage and current transitions and assume an origin (t') at the instant the current begins to rise. The voltage and current are therefore

Figure 17.5 (a) Transistor switch with resistive load; (b) transistor voltage and load during switching.

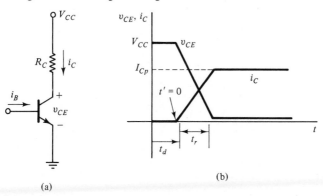

(a)

(b)

$$i_c(t') = 0.8t' \times \frac{I_{Cp}}{t_r} \quad \text{and} \quad v_{CE}(t') = V_{CC}\left(1 - \frac{0.8t'}{t_r}\right) \qquad (17.1)$$

where $t_r/0.8$ is the total time of transition. The expressions in Eq. (17.1) are valid only during the time $0 < t' < t_r/0.8$. The power into the transistor would be the product of voltage and current

$$p_C(t') = V_{CC}I_{Cp}\frac{0.8t'}{t_r}\left(1 - \frac{0.8t'}{t_r}\right) \qquad (17.2)$$

which has a maximum value of $V_{CC}I_{Cp}/4$ at $t' = 0.625t_r$. For a 300-V, 100-A device, the peak power into the transistor during switching is 7500 W. The energy given to the transistor in switching is the integral of the power.

$$W = \int_0^{t_r/0.8} v_{CE}(t')i_C(t')\,dt' = V_{CC}I_{Cp}\int_0^{t_r/0.8}\frac{0.8t'}{t_r}\left(1 - \frac{0.8t'}{t_r}\right)dt' = \frac{V_{CC}I_{Cp}t_r}{4.8}$$

$$(17.3)$$

which gives an energy of 18.8 mJ for each switching transition, assuming a 2-μs rise time. This appears small, but if the transistor is switching at a 2-kHz rate, this gives a required dissipation of 75 W from switching alone.

During saturation, the collector–emitter voltage drop would be about 1.2 V at high current levels, so the collector dissipation would be 120 W while the transistor is ON. In the switching mode, the transistor would be ON half of the time, so the average power would be 60 W. Therefore, the switching power would be an important contribution to the transistor dissipation.

Transistor switching losses with inductive loads. Figure 17.6a shows a transistor with an inductive load. The free-wheeling diode is required to protect the transistor from large inductive voltages. Figure 17.6b compares the switching trajectory in the $v_{CE}i_C$ plane for the resistive and inductive loads. To explain the trajectory for an inductive load, we must investigate the role of the transistor in

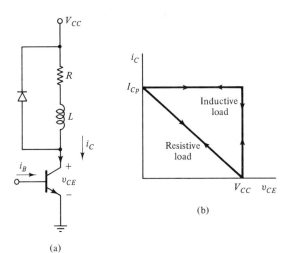

(a)

(b)

Figure 17.6 (a) Transistor switch with an inductive load; (b) switching trajectories for resistive and inductive loads.

Power Electronic Systems   Chap. 17

controlling the current through the inductive load, and also investigate the importance of the time-average voltage in controlling the current through the inductive load.

Inductive loads.   For the resistive load, the transistor turns the current on and off, and hence controls the average voltage and power to the load. For an inductive load, by contrast, the transistor is turned ON and OFF to control the average voltage across the load, and this average voltage determines the average current through the load. The current through the load does not cease because the inductance keeps it more or less constant, the transistor switching period being much smaller than the time constant of the $R–L$ load.

Average voltages.   We now explain why the average current through an $R–L$ load is controlled by the time-average voltage across the load. This is true for all periodic voltages and currents because the time-average voltage across the inductance must be zero. To see that this is true, consider the current through the $R–L$ load as a discrete spectrum of sinusoidal harmonics, plus a dc component. The voltage across the inductance consists therefore of sinusoidal components, and the time average of a sinusoid is zero. Therefore, the time-average voltage across the inductance is zero. It follows also that the time-average voltage across the $R–L$ load is equal to the dc component of the current times the resistance of the load, because the time average of the resistor voltage due to the harmonics of the current must also be zero. Thus the transistor controls the current in the $R–L$ load by controlling the time-average voltage across the load.

Returning to the Transistor.   The transistor in Fig. 17.6a switches ON and OFF, but the current through the inductor continues more or less constant. Therefore, while the transistor is OFF and any time the transistor is not carrying the full load current, some current will pass through the free-wheeling diode, and the collector voltage of the transistor will be equal to the supply voltage. It follows that the switching trajectory for an inductive load is that shown in Fig. 17.6b with constant voltage at V_{CC} while the collector current is moving from zero to I_{Cp} and back during the switching cycle.

An analysis of the power and energy, left to a homework problem, shows that the peak power into the transistor is four times that with a resistive load, and the energy per cycle is three times that required for switching a resistive load. The strain on the transistor is greatly increased, and auxiliary circuits are required to reduce the power requirement on the transistor.

Summary.   The section has introduced SCRs, GTOs, and power transistors as semiconductor switches. We have considered how to turn these devices ON and OFF and discussed their power limitations. In the next section, we analyze a light-dimmer circuit to show a common application of power electronic techniques.

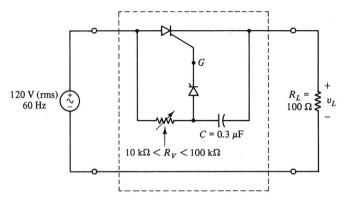

Figure 17.7 Typical SCR circuit.

17.1.2 Common Application of Power Electronics

Power electronic circuit.   In this section we examine the operation of the circuit shown in Fig. 17.7. The ac source represents a standard 120-V, 60-Hz supply. The circuit in the box is a power controller: the resistor with the arrow represents a variable resistor, and R_L represents the load receiving the power. The circuit uses a four-level diode and an SCR as semiconductor switches.

Common applications for this circuit.   Circuits of this type are used in light dimmers. In this application, the variable resistor is adjusted by a rotary mechanism, and an on–off switch is normally built into the same mechanism. The load would be a light of some sort. A variable-speed drill is controlled by this or a similar circuit. Here the variable resistor is adjusted by the squeeze trigger on the drill handle and the load is a universal motor, discussed in Section 16.2.3.

Four-layer diode function.   To explain the operation of the circuit in Fig. 17.7, we will start with the simpler circuit shown in Fig. 17.8. We begin with a discharged capacitor. Closing the switch at $t = 0$ will allow current to flow and the capacitor voltage will increase. The *pnpn* diode will remain OFF and no current will flow through the small resistor until the voltage across the capacitor reaches the threshold voltage of 30 V. Thus until the *pnpn* diode fires, we have a simple *RC* transient. The time constant is $R_V C$, the initial value of the voltage across

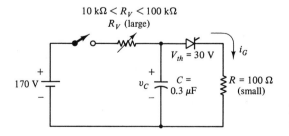

Figure 17.8   The four-level diode will fire when $v_C = 30$ V.

the capacitor is 0 V, and the final value would be 170 V if the transient reached completion. Using the techniques from Chapter 3, we determine the voltage across the capacitor to be

$$v_C(t) = 170(1 - e^{-t/R_V C}) \tag{17.4}$$

The capacitor voltage will increase according to Eq. (17.4) until the threshold voltage of the *pnpn* diode is reached. At this point the *pnpn* diode turns ON and allows current to flow through the small resistance. Current from the dc source is limited by the large value of R_V, but the current from the capacitor discharge can be quite large. The time at which the discharge will occur will be

$$30 = 170(1 - e^{-t_f/R_V C}) \Rightarrow t_f = -R_V C \times \ln\left(1 - \frac{30}{170}\right)$$

$$= +0.194 R_V C \tag{17.5}$$

The capacitor value is 0.3 µF and the value of the variable resistor lies between 10 and 100 kΩ. Thus the minimum firing time will be about 0.6 ms. The firing of the *pnpn* diode results in a current pulse through the small resistor. The peak magnitude of this current pulse is about $(30 - 0.5)/100 = 0.295$ A, and its duration would be approximately the time constant of the *RC* circuit containing the capacitor and the smaller resistor, about 30 µs. After the discharge, the *pnpn* diode will turn OFF, there being insufficient current flowing to keep it ON, and the cycle will begin again. Hence the capacitor voltage and the current through the small resistor will be as shown in Fig. 17.9, where we have shown a 5-ms cycle time.

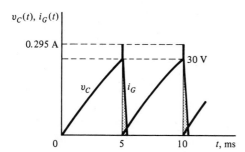

Figure 17.9 Capacitor voltage and load current waveforms.

With an AC source. We will now consider what would happen if the dc source were replaced by an ac source, as shown in Fig. 17.10. We have drawn the input voltage as a sine function and shown the capacitor voltage, v_C, as it would be if the *pnpn* diode did not conduct. If the capacitor were initially uncharged, and if the four-level diode never fired, there would be a start-up transient, and then the voltage of the capacitor would lag the input voltage with a phase shift between 0° and 90°, depending on the value of $R_V C$. However, the four-level diode will fire at the time labeled t_α, when the capacitor voltage reaches the threshold of 30 V.

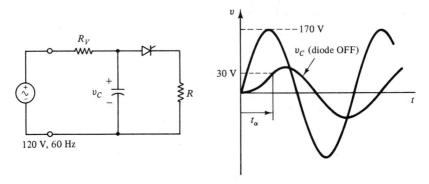

Figure 17.10 Circuit response with a sinusoidal source.

Analysis. The calculation of t_α combines transient and ac steady-state techniques. The steady-state voltage found by ac circuit techniques would be

$$v_C(t) = \frac{V_p}{\sqrt{1 + (\omega R_V C)^2}} \sin(\omega t - \phi) \tag{17.6}$$

where V_p is the peak ac voltage, $\phi = \tan^{-1} \omega R_V C$, and $t = 0$ when the input voltage is zero. The transient portion is $Ae^{-t/R_V C}$, where A is a constant to be determined from the initial conditions. The capacitor is initially uncharged so

$$0 = \frac{V_p}{\sqrt{1 + (\omega R_V C)^2}} \sin(0 - \phi) + Ae^{-0/R_V C} \Rightarrow A = \frac{V_p \sin \phi}{\sqrt{1 + (\omega R_V C)^2}} \tag{17.7}$$

The *firing time* (t_α) for the circuit occurs when the capacitor voltage equals the threshold voltage of the *pnpn* diode:

$$V_{th} = \frac{V_p}{\sqrt{1 + (\omega R_V C)^2}} [\sin(\omega t_\alpha - \phi) + e^{-t_\alpha/\tau} \sin \phi] \tag{17.8}$$

The *delay angle, $\alpha = \omega t_\alpha$,*† describes the position in the cycle when the *pnpn* diode fires. For the values of R_V and C that we show, the delay angle lies between 30° and 180°. Hence we can control the firing time, or delay angle, with R_V as we did with the dc source. When the four-level diode fires, the capacitor will discharge through the small resistor with a short time constant, as before. After the capacitor discharges, the four-level diode turns OFF, there being insufficient current passing through R_V to keep it conducting. The capacitor voltage again begins building up and may fire the four-level diode again before the input voltage goes negative, but once the input becomes negative the four-level diode will not fire again until the input again goes positive.

Summary. We can create a pulse of current with the four-level diode. The timing of this pulse of current is controlled by the $R_V C$ time constant and can be varied over a range of delay angles.

† The delay angle is often given in degrees.

Back to the original circuit. We now can investigate the operation of the circuit in Fig. 17.7, repeated in Fig. 17.11. Figure 17.11 shows the SCR gate to be the "load" to which the four-level diode delivers its current pulses. The load of the SCR, R_L, is presumably a low resistance and does not affect greatly the charging of the capacitor. Hence, when the input voltage goes positive, the capacitor begins charging through R_V and the load. When the capacitor voltage builds up to the threshold voltage, the four-level diode fires and the current pulse passes through the gate, turning ON the SCR. The voltage to the load would thus be as shown in Fig. 17.12. By adjusting the delay angle with R_V, we can control the conduction angle of the SCR and hence the power to the load. The current pulse may be too great for the gate of a small SCR, which typically requires only about 10 mA. A resistor in series with the four-level diode would limit the gate current. With the SCR gate for a load, repeated firing of the four-level diode would not matter. Once the SCR is turned ON, it will remain ON until the input voltage goes negative.

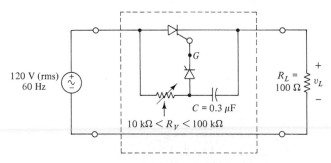

Figure 17.11 Typical SCR circuit (identical to Fig. 17.7).

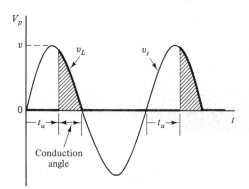

Figure 17.12 Load voltage for the circuit in Fig. 17.11.

Diacs and Triacs. Figure 17.13 shows a similar power controller. The device replacing the four-level diode is called a *diac*. It is like two parallel four-level diodes facing in opposite directions and will fire in either direction. Similarly, the device replacing the SCR is called a *triac*, and it functions like two parallel SCRs facing opposite directions. This circuit operates like the circuit in Fig. 17.11, except that it performs no rectification. Hence the load voltage would be as shown

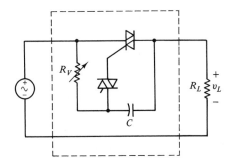

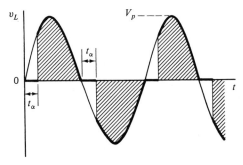

Figure 17.13 Diacs and triacs fire in both directions.

Figure 17.14 Load voltage for the triac circuit.

in Fig. 17.14. This is the preferred circuit for light dimmers and variable-speed drills, which do not require dc voltage.

17.1.3 Introduction to Motor Controllers

Power electronic converters. A power electronic converter controls the power exchange between an electrical supply and a load, as shown in Fig. 17.15. The information that controls the power electronics may be a simple manual control or may be part of a feedback control loop. The electrical supply may be ac or dc. The load may be a passive load, such as a resistance or an induction furnace, or may be an electromechanical system, such as a dc or ac motor. The load may even be another electrical supply. Normally, the power flows from the electrical supply to the load, but may for certain loads flow in the other direction. We introduce briefly the various types of systems:

- *DC–DC converters.* A converter that takes fixed-voltage dc voltage and produces variable-voltage dc is normally called a *chopper*. We could, for example, control a dc motor armature circuit and hence control motor speed with a dc–dc converter.
- *AC–DC converters.* An ac-to-dc converter is a controlled rectifier. In such a converter, the rectifier diodes are replaced by SCRs or other switches, and the dc voltage is controlled by the switches. We will deal with both single- and three-phase converters of this type.
- *DC–AC and AC–DC–AC converters.* A dc–ac converter is called an inverter. Combined with an ac–dc converter, we can develop an ac–dc–ac converter to make a variable-frequency ac source. For example, if we wished to control the speed of an induction motor, one approach would be to rectify

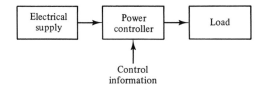

Figure 17.15 Block diagram for power controller.

the available 60-Hz ac to dc power and then invert the dc to ac at a controlled frequency to drive the motor. Such systems can also link together unsynchronized ac power systems for exchange of ac power between different power grids, sometime over a high-voltage dc (HVDC) transmission line. On a smaller scale, an ac–dc–ac converter can make a noninterruptable power supply if the dc portion maintains a battery bank to supply temporary dc power in case of a power failure.

- *AC–AC controllers.* In addition to the ac–dc–ac converters described above, there exist ac–ac converters that perform direct conversion from fixed-frequency, fixed-voltage ac power to variable-frequency, variable-voltage ac power.

Quadrants of operation. Our principal concern will be motor controllers. Figure 17.16 shows the four quadrants of motor operation. Because power is the product of torque and speed, the power out of the motor is positive in the first and third quadrants, which correspond to forward and reverse driving of the motor, respectively. In the fourth quadrant, the motor rotates in the forward direction, but the torque is supplied in the reverse direction, tending to slow down the motor. In this forward braking region, the power output of the motor is negative, and the motor acts as a source of electrical power. This power may be delivered back to the electrical supply or may be dissipated in the motor or the converter. Similarly, in the second quadrant the motor is braked from reverse rotation. Operation in the second and fourth quadrants is sometimes called "plugging," or *regenerative braking*. As we will see, motor controllers are classified according to the quadrants in which they operate.

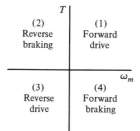

Figure 17.16 The four quadrants of motor operation.

17.1.4 Check Your Understanding

1. If the *pnpn* diode in Fig. 17.8 fired at 20 V, what would be the resistor value required for it to fire in 3 ms?
2. Turning ON an SCR requires (a) forward bias, (b) a pulse of current to the gate, or (c) both of the above. (Which?)
3. Turning OFF an SCR requires (a) reverse-biasing the SCR, (b) keeping the SCR reverse biased for a prescribed period of time, and/or (c) removing the gate signal. (Which? May be more than one.)
4. To operate successfully, the SCR must be protected from (a) high *dv/dt* in the OFF condition, (b) high *di/dt* in the ON condition, and/or (c) excessive reverse voltage in the OFF condition. (Which? May be more than one.)

5. Does the SCR in the light-dimmer circuit in Fig. 17.11 turn OFF by line or forced commutation?

6. The power dissipation limits of a power transistor affect the rate at which it can be cycled as a switch. (T/F?)

   Answers.   1. 79.9 kΩ;   2. c;   3. b and c;   4. a, b, and c;   5. line commutation;   6. true.

17.2 DC MOTOR CONTROLLERS

We now discuss power electronic controllers for dc motors. We begin with a discussion of the model we will use for the dc motor.

17.2.1 DC Motor Model

Time-average torque and power.   Figure 17.17a shows the model we will use for a dc load. This is the armature circuit of a separately excited dc motor. The armature emf is

$$E = K\Phi\omega_m \tag{17.9}$$

where $K\Phi$ is the machine constant. Although the applied voltage is periodic, the emf will be constant because of the inertia of the armature and mechanical load. The armature current is also periodic, but the time-average torque is

$$T = \langle K\Phi i_a(t) \rangle = K\Phi\langle i_a(t) \rangle = K\Phi I_a \tag{17.10}$$

where I_a is the dc component of the armature current. Because the emf is constant, the developed power also depends on the dc component of the armature current:

$$P_{dev} = \langle Ei_a(t) \rangle = E\langle i_a(t) \rangle = EI_a \tag{17.11}$$

Input voltage.   Although the armature voltage, $v_a(t)$, is periodic, the dc component of the armature voltage is simply related to the emf and the dc current. Kirchhoff's voltage law is

$$v_a(t) = Ri_a(t) + v_L(t) + E \tag{17.12}$$

Figure 17.17   (a) DC-motor model; (b) quadrants of operation for electrical drive.

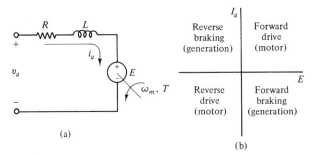

(a)

(b)

where $v_L(t)$ is the voltage across the inductor. As explained earlier on page 651, the time-average voltage across the inductance must be zero for steady-state operation, so the time average of Eq. (17.12) is

$$V_a = RI_a + E \qquad (17.13)$$

where $V_a = \langle v_a(t) \rangle$, the time-average voltage applied to the armature circuit.

Quadrants of operation. Because torque is proportional to current and emf is proportional to speed, the four quadrants of operation in Fig. 17.16 correspond to identical quadrants in the E–I_a plane shown in Fig. 17.17b. For example, in the first quadrant, the emf and dc current are positive, which corresponds to positive torque and speed, hence to the forward-drive condition. The input voltage to the armature circuit is not identical to the emf, but in steady state the emf follows the input voltage closely. Thus we may assume that motor operation in the first quadrant corresponds to positive input voltage and current, and so on.

Notation for switches. In the remainder of this chapter, we will symbolize a semiconductor switch as a "switch-in-a-box," as shown in Fig. 17.18. Since the SCR, GTO, and BJT switches allow current flow in one direction only and require forward bias in that direction to conduct, we have shown the allowed current direction by the orientation of the open end of the switch-in-a-box, as shown. By this symbol we imply the existence of circuits to switch the device ON and OFF (for forced commutation), including necessary protective circuits.

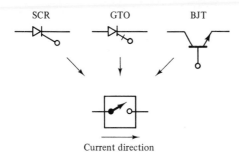

SCR   GTO   BJT

Current direction

Figure 17.18 The "switch-in-a-box" represents a semiconductor switch. The allowed current direction is shown. All switches require forward bias in this direction to conduct.

17.2.2 Single-phase Rectifier Circuits

Introduction. In this section we analyze single-phase controlled rectifiers that drive a dc motor from an ac source. The single-phase circuits are treated in detail because the three-phase circuits operate on similar principles; indeed, most single-phase equations can be modified for the analysis of three-phase rectifiers.

Single-phase controlled rectifier operation. Figure 17.19 shows a controlled full-wave rectifier circuit loaded by a dc motor. The two voltage sources represent the center-tapped secondary of a transformer. Figure 17.20 shows the ac voltage and the gating signals. Note that the time axis is expressed in electrical angle. The gating signals consist of a sequence of pulses beginning at $\omega t = \alpha$ for

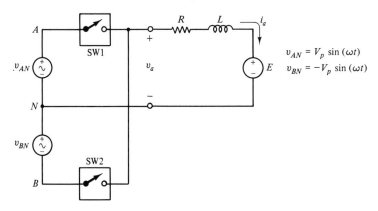

Figure 17.19 A switched full-wave rectifier with a dc motor for a load.

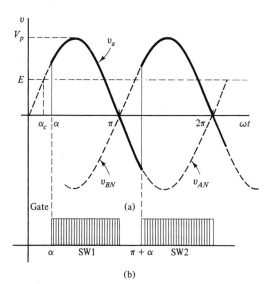

Figure 17.20 (a) Applied voltage and
(b) gating signal as a function of ωt.

SW1 and $\omega t = \pi + \alpha$ for SW2 and repeated with the ac period. This type of gating signal is required because the switches may not conduct when first gated because the emf may have them reverse biased. If SW1† were gated before α_c, it would fire at α_c, when the input voltage first exceeds the emf.

$$\alpha_c = \sin^{-1} \frac{E}{V_p} \qquad (17.14)$$

Equation (17.14) gives the critical delay angle, α_c, where V_p is the peak amplitude of the ac voltage. Our analysis must consider only values of $\alpha \geq \alpha_c$.

Figure 17.21 shows the waveform applied to the load for a specific case considered below. Firing the switch begins a buildup of current in the load. The load current lags due to the inductance, and thus the current will remain positive

---

† Any statement about SW1 is true for SW2 due to the symmetry of the circuit and the gating signals. Henceforth we will discuss principally the behavior of SW1.

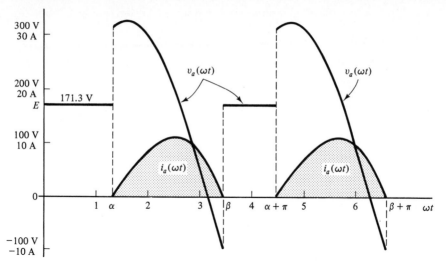

Figure 17.21 Voltage and current waveforms for an inductive load.

after the source voltage falls below the emf. When the load current goes to zero and tries to reverse, the switch will turn OFF by line commutation. With no load current, the load voltage is equal to the emf.

Analysis. The analysis of the circuit response involves a transient solution with a steady-state condition consisting of a sinusoidal and a dc response.

Load current. Once SW1 is fired, the circuit is that shown in Fig. 17.22. The DE for the circuit is

$$L \frac{di_a}{dt} + Ri_a = V_p \sin \omega t - E \tag{17.15}$$

and the solution will consist of a transient and a steady-state response.

$$i_a(t) = Ae^{-t/\tau} + \frac{V_p}{Z} \sin(\omega t - \phi) - \frac{E}{R} \tag{17.16}$$

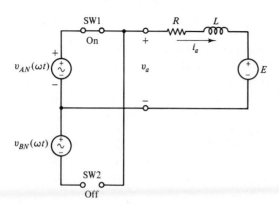

Figure 17.22 Circuit in Fig. 17.19 with SW1 ON and SW2 OFF.

where A is an unknown constant to be determined from the initial condition, $\tau = L/R$ is the time constant of the transient, $Z = \sqrt{R^2 + (\omega L)^2}$ is the magnitude of the impedance of the resistor and inductor in series, and $\phi = \tan^{-1}(\omega L/R)$ is the phase shift from the inductor. The sinusoidal portion of the current is determined from routine ac analysis, using the *sine* function for the ac source. The dc portion of the forced response involves only the emf and the resistor because the inductor and the ac source are short circuits at dc.

We assume that $\alpha > \alpha_c$ such that SW1 fires at $\omega t = \alpha$. Because of the inductor, the initial current must be zero, so the initial condition is

$$i(\alpha) = Ae^{-\alpha/\tan \phi} + \frac{V_p}{Z} \sin(\alpha - \phi) - \frac{E}{R} = 0 \qquad (17.17)$$

where in the exponential we have replaced t/τ by $\alpha/\omega\tau$, and then replaced $\omega\tau$ by $\omega L/R = \tan \phi$. Solving for A, we find

$$A = \left[\frac{E}{R} - \frac{V_p}{Z} \sin(\alpha - \phi)\right] e^{\alpha/\tan \phi} \qquad (17.18)$$

and hence the current is

$$i_a(\omega t) = \frac{V_p}{Z} [\sin(\omega t - \phi) - e^{(\alpha - \omega t)/\tan \phi} \sin(\alpha - \phi)] - \frac{E}{R} [1 - e^{(\alpha - \omega t)/\tan \phi}]$$
$$(17.19)$$

Equation (17.19) is valid as long as $i_a > 0$. When the current tries to reverse, the switch will turn OFF through line commutation, and the current will cease. Equation (17.19) crosses zero going negative at $\omega t = \beta$, where β satisfies Eq. (17.20).

$$i_a(\beta) = 0 = \frac{V_p}{Z} [\sin(\beta - \phi) - e^{(\alpha - \beta)/\tan \phi} \sin(\alpha - \phi)] - \frac{E}{R} [1 - e^{(\alpha - \beta)/\tan \phi}]$$
$$(17.20)$$

Figure 17.21 shows β. For a specific circuit and value of α, Eq. (17.20) may be solved for β with numerical techniques.

Discontinuous or continuous current? If β is less than $\pi + \alpha$, the current indeed does go to zero and is discontinuous. With discontinuous current, each pulse of current is identical to the first, and steady state is established immediately. If β is greater than $\pi + \alpha$, then the current does not go to zero and the next pulse of current will begin from a nonzero value. In this case the current will be continuous and will reach steady state after a transient period has passed. Our analysis must therefore consider both possibilities and also furnish a means for determining whether the current is continuous or discontinuous in a given application.

Discontinuous current. Figure 17.21 shows load voltage and current for $V_p = 230\sqrt{2}$ V, $E = 171.3$ V dc, $\alpha = 75°$, $R = 1.78$ Ω, $L = 30$ mH, and $\omega = 377$ rad/s. These parameters are appropriate for a 180-V, 1-hp, 1750-rpm dc motor at full load. The corresponding full-load armature current is 4.9 A. The armature impedance is $Z = 11.45$ Ω and the phase delay $\phi = 81.1°$. The calculated value for β, the *extinction angle*, is 197.7°, so the *conduction angle* is $\gamma = \beta - \alpha = 197.7° - 75° = 122.7°$.

Load voltage for discontinuous current. While the switch is ON, the load voltage is equal to the input sinusoidal voltage, and the load voltage is E for the remainder of the cycle. The average load voltage is therefore

$$V_a = \frac{1}{\pi} \int_\alpha^{\pi+\alpha} v_a(\omega t)\, d(\omega t) = \frac{1}{\pi} \int_\alpha^\beta V_p \sin(\omega t)\, d(\omega t) + \frac{1}{\pi} \int_\beta^{\pi+\alpha} E\, d(\omega t)$$

$$= \frac{V_p}{\pi}(\cos\alpha - \cos\beta) + E\left(1 - \frac{\gamma}{\pi}\right)$$

(17.21)

where $\gamma = \beta - \alpha$ is the conduction angle in radians. Notice that we may average from α to $\pi + \alpha$ because of the half-wave symmetry of the waveform. For the numerical values we are using, the average load voltage is 180.0 V. As we showed in Eq. (17.13), the dc current can be determined from the dc load voltage, the emf, and the resistance:

$$I_a = \frac{V_a - E}{R} = \frac{180.0 - 171.3}{1.78} = 4.90$$

(17.22)

The total power given to the emf would be given by Eq. (17.11):

$$P_{\text{dev}} = EI_a = 171.3 \times 4.90 = 837.3 \text{ W}$$

(17.23)

and the total power to the armature circuit would be slightly higher because of loss in the armature resistance. The developed power minus the mechanical losses gives the full-load output power of 746 W for this motor.

Load voltage for continuous current. For continuous current, the load voltage is equal to the source voltage for $\alpha < \omega t < \pi + \alpha$ as shown in Fig. 17.23; hence the average load voltage is

$$V_a = \frac{1}{\pi} \int_\alpha^{\pi+\alpha} V_p \sin(\omega t)\, d(\omega t) = \frac{2}{\pi} V_p \cos\alpha$$

(17.24)

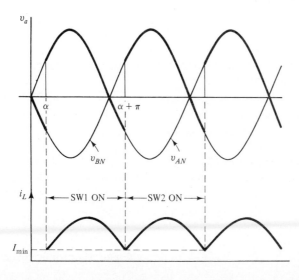

Figure 17.23 Voltages and load current for continuous current.

663

which is the limit of Eq. (17.21) as the conduction angle becomes 180°. The dc load current and developed power can be determined as above.

Continuous or discontinuous current? We may determine the condition for continuous current by considering the marginal situation where $\beta = \pi + \alpha$. From Eq. (17.20), this requires

$$0 = \frac{V_p}{Z} [\sin(\alpha + \pi - \phi) - e^{-\pi/\tan\phi} \sin(\alpha - \phi)] - \frac{E}{R} [1 - e^{-\pi/\tan\phi}] \quad (17.25)$$

which reduces to

$$E_{\text{mar}} = V_p \cos\phi \sin(\phi - \alpha) \left[\frac{1 + e^{-\pi/\tan\phi}}{1 - e^{-\pi/\tan\phi}} \right] \quad (17.26)$$

where E_{mar} is the marginal emf for continuous conduction. Any value of emf below E_{mar} will cause continuous conduction. A complication arises if E_{mar}, as determined by Eq. (17.26), is larger than $V_p \sin\alpha$, for then the effective value for α is α_c, which depends on E_{mar}. In that case we must go one step further and solve Eq. (17.26) for the critical value of α. We show how this is done in the example below.

Example. For the motor and ac voltage given above, we will determine the motor speed at which the current becomes continuous. According to Eq. (17.26), the emf for marginal continuous current is

$$E_{\text{mar}} = 230\sqrt{2} \cos(81.1°) \sin(81.1° - 75°) \left[\frac{1 + e^{-\pi/\tan 81.1°}}{1 - e^{-\pi/\tan 81.1°}} \right] = 22.0 \text{ V} \quad (17.27)$$

and thus the speed is $(22.0/171.3) \times 1750 = 225$ rpm. The current is marginally continuous, so the dc load voltage can be determined from Eq. (17.24):

$$V_a = \frac{2}{\pi} 230\sqrt{2} \cos 75° = 53.6 \text{ V} \quad (17.28)$$

The dc current is thus

$$I_a = \frac{V_a - E}{R} = \frac{53.6 - 22.0}{1.78} = 17.75 \text{ A} \quad (17.29)$$

which is well beyond the full-load current of 4.9 A. Thus the motor would normally operate with discontinuous current.

Extending the example. We now ask, for the same motor, what will be the output power and speed for continuous current if the delay angle is reduced to zero, that is, at "full throttle"? In this condition the delay angle will be the critical angle defined in Fig. 17.20b and Eq. (17.14). This requires that Eq. (17.26) satisfy the condition

$$\sin \alpha_c = \cos 81.1° \sin(81.1° - \alpha_c) \left[\frac{1 + e^{-\pi/\tan 81.1°}}{1 - e^{-\pi/\tan 81.1°}} \right] \quad (17.30)$$

where $\sin \alpha_c = E_{\text{mar}}/V_p$. If we solve this equation for α_c, we obtain

$$\tan \alpha_c = \frac{\sin \phi \cos \phi}{\left[\dfrac{1 - e^{-\pi/\tan \phi}}{1 + e^{-\pi/\tan \phi}}\right] + \cos^2 \phi} = 0.5763 \Rightarrow \alpha_c = 30.0° \quad (17.31)$$

and thus the maximum emf for continuous conduction at "full throttle" is $230\sqrt{2} \sin 30.0° = 162.4$ V. From Eq. (17.24) the dc armature voltage is 179.4 V, and hence the dc current is 9.55 A, over twice the limit for the machine. The developed power is 1550 W. We conclude that the machine should never operate in continuous mode with single-phase, 230-V power.

Continuous current. With continuous current, the machine analysis is relatively simple compared with the analysis for discontinuous current. The difficulty for discontinuous current is that the extinction angle must be determined from Eq. (17.20) by numerical techniques before the load voltage, current, and power can be determined. Even more difficult is the inverse problem, that of determining the required value of delay angle for a prescribed voltage. For these reasons, we usually resort to graphs relating the various machine parameters and operating levels or develop computer routines to handle the specifics. Our purposes do not require further investigation of these details.

Uncontrolled rectified drives. A motor that does not require speed control can be driven by an uncontrolled rectifier. The gated semiconductor switches would in that case be replaced by ordinary *pn*-junction diodes, sized to accommodate the power level required. The diodes would be self-gating at the critical angle, and the analysis of the "full throttle" condition would apply.

17.2.3 Single-phase Rectifier Drive with Free-wheeling Diode

Free-wheeling diode. The current to the load can be increased by placing a free-wheeling diode across the load, as shown in Fig. 17.24. If SW1 is ON and SW2 OFF, the diode is in parallel with the source and will be reverse biased while

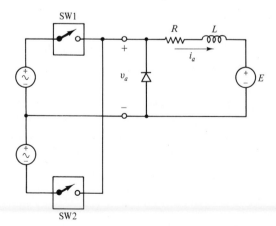

Figure 17.24 The free-wheeling diode commutates the switches and prevents negative load voltage.

the source voltage is positive. When the source voltage goes negative, the diode will turn ON, and SW1 will turn OFF by line commutation. The load current will continue through the diode. Here again the current can be continuous or discontinuous.

Analysis for discontinuous current. The current for the period $\alpha < \omega t < \pi$ is given by Eq. (17.19) if we assume zero current when SW1 is fired. When at $\omega t = \pi$ SW1 turns OFF, the current is

$$i_a(\pi) = \frac{V_p}{Z}[\sin(\pi - \phi) - e^{(\alpha - \pi)/\tan \phi}\sin(\alpha - \phi)] - \frac{E}{R}[1 - e^{(\alpha - \pi)/\tan \phi}] \quad (17.32)$$

At $\omega t = \pi$, the free-wheeling diode conducts, SW1 is line-commutated OFF, and the current begins to decay in a simple dc transient. The initial value is $i_a(\pi)$, and the final value would be $-E/R$; hence the current for $\pi < \omega t < \beta$ is

$$i_a(\omega t') = -\frac{E}{R} + \left[i_a(\pi) + \frac{E}{R}\right]e^{-t'/\tau} \quad (17.33)$$

where $t' = 0$ at $\omega t = \pi$. When this current is zero at $\omega t = \beta$, the free-wheeling diode turns OFF, and the load current remains at zero until a switch is again gated ON in the next cycle. From Eq. (17.33) we may find the time when the current reaches zero:

$$0 = -\frac{E}{R} + \left[i_a(\pi) + \frac{E}{R}\right]e^{-t'_\beta/\tau} \Rightarrow t'_\beta = \tau \ln\left[1 + \frac{i_a(\pi)R}{E}\right] \quad (17.34)$$

where t'_β is the time the current reaches zero. The extinction angle is related to t'_β through

$$\beta = \pi + \omega t'_\beta \quad (17.35)$$

For discontinuous current, β must be less than $\pi + \alpha$, and in that case the following pulses of current will be identical to the first. This is the situation pictured in Fig. 17.25.

Figure 17.25 The load voltage and current for discontinuous load current. The free-wheeling diode conducts between $\pi < \omega t < \beta$.

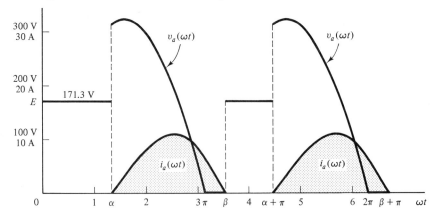

For example, with $E = 171.3$ V, $\alpha = 75°$, and the circuit parameters given above, we calculate from Eq. (17.32) $i_a(\pi) = 6.20$ A, and from Eq. (17.34) an extinction time of $t'_\beta = 1.053$ ms. Equation (17.35) gives an extinction angle of 202.7°, and thus the conduction angle, γ, is 127.7°. These results correspond to Fig. 17.25.

DC load voltage for discontinuous current. The load voltage will be equal to the source voltage while the switch is ON, will be zero while the diode is ON, and will be equal to E while both switch and diode are OFF, as shown in Fig. 17.25. The average load voltage is

$$V_a = \frac{1}{\pi} \left[\int_\alpha^\pi V_p \sin(\omega t)\, d(\omega t) + \int_\beta^{\pi+\alpha} E\, d(\omega t) \right] = \frac{V_p}{\pi}(1 + \cos \alpha) + E\left(1 - \frac{\gamma}{\pi}\right) \tag{17.36}$$

which yields a value of 180.0 V for $E = 171.3$ V, $\alpha = 75°$, and the circuit parameters given above. Equation (17.13) gives the average load current to be 4.94 A. Equation (17.11) gives the power delivered to the emf as 845.4 W. The load current and developed power are slightly higher than before because of the freewheeling diode.

Average voltage for continuous current. If the current is continuous, a transient period passes before the current reaches steady state. In steady state the diode remains on while the switches are OFF, and the average voltage is

$$V_a = \frac{1}{\pi} \int_\alpha^\pi V_p \sin(\omega t)\, d(\omega t) = \frac{V_p}{\pi}(1 + \cos \alpha) \tag{17.37}$$

which is the same as Eq. (17.36) for a conduction angle of 180°. The average current and developed power can be determined with Eqs. (17.13) and (17.11).

Continuous or discontinuous? We now have means to analyze the circuit, provided we know if the current is continuous or discontinuous. One way to determine which case applies is to calculate the extinction angle from Eqs. (17.34) and (17.35). If β is greater than $\pi + \alpha$, the current is continuous. Another way to determine if the current is continuous is to analyze the marginal case between continuous and discontinuous current. This requires that $\beta = \pi + \alpha$. From Eq. (17.32) we have

$$i_a(\pi) = \frac{V_p}{Z}[\sin(\pi - \phi) - e^{(\alpha - \pi)/\tan \phi} \sin(\alpha - \phi)] - \frac{E}{R}[1 - e^{(\alpha - \pi)/\tan \phi}] \tag{17.38}$$

and from Eq. (17.34) we have

$$0 = -\frac{E}{R} + \left[i_a(\pi) + \frac{E}{R}\right] e^{-(\alpha)/\tan \phi} \Rightarrow i_a(\pi) = \frac{E}{R}[e^{(\alpha)/\tan \phi} - 1] \tag{17.39}$$

Here, as before, E_{mar} is the marginal value of emf between continuous and discontinuous current. When we substitute $E = E_{mar}$ in both equations, Eqs. (17.38) and (17.39) yield

$$E_{mar} = V_p \frac{\cos \phi [\sin \phi e^{-\alpha/\tan \phi} - \sin(\alpha - \phi)e^{-\pi/\tan \phi}]}{1 - e^{-\pi/\tan \phi}} \qquad (17.40)$$

Equation (17.40) gives the maximum emf for continuous conduction for a specified circuit and input voltage. As before, a complication arises if the marginal emf from Eq. (17.40) corresponds to a critical delay angle (α_c) larger than the specified value of α. In that case, we proceed as described in connection with Eq. (17.26). In this case, the equation can be solved only by numerical techniques.

Example. We will determine the emf required to draw continuous current for $\alpha = 75°$ and the circuit parameters used throughout the previous section. Substitution of $\alpha = 1.309$ rad (75°), $\phi = 81.16°$, and $V_p = 230\sqrt{2}$ V into Eq. (17.40) yields $E_{mar} = 112.6$ V. Equation (17.37) gives an average load voltage of 130.3 V, so the dc current is 9.98 A, about twice the rated value. Again we conclude that the motor should run with discontinuous current for this value of α.

Summary. The controlled full-wave rectifier delivers dc voltage and current to the $R-L$ load with an emf. The current may be continuous or discontinuous. A free-wheeling diode across the load increases the load voltage and tends to make the current continuous.

Harmonics in the AC line. The current that flows in the ac power system is clearly nonsinusoidal, containing in general all the even harmonics of 60 Hz. The harmonics constitute noise for other users of the ac power circuit and can disturb sensitive electronic circuits supplied from the line. Three-phase controlled rectifiers lessen problems with harmonics, as we show in the next section.

Operation in the second quadrant. Equation (17.24) indicates that the circuit without a free-wheeling diode can put out negative load voltage for $\alpha >$ 90°. As α is advanced more and more, the switches are gated ON and continue conducting when the input voltage is negative, making the load voltage negative. However, the load current must be kept positive in the indicated direction because the switches in the controlled rectifier in Fig. 17.19 permit only a positive load current. This can occur when $V_a > E$. If E is positive, V_a has to be more positive. This can also occur if E is negative and V_a is less negative. This circuit controls motor torque, therefore, in the first and second quadrants, as shown in Fig. 17.26.

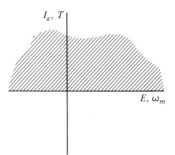

Figure 17.26 The controlled rectifier in Fig. 17.19 operates in the first and second quadrants.

The controller is incapable of drive in the reverse direction, but could operate as a brake if the motor direction were reversed by the mechanical system.

Inverter operation. In the circumstance where the emf is negative, but the current is positive, the emf acts as a generator and delivers power to the ac source. It thus can function as an inverter, converting dc power to ac power. The same may be accomplished in the four-quadrant converter described below, where the current may reverse.

Four-quadrant controller. Figure 17.27 shows a controlled rectifier that operates in all four quadrants. The original switches (SW1 and SW2) are gated for positive load current, and SW3 and SW4 are gated for negative load current. We can envision the rectifier operating with continuous positive current for small α and going into discontinuous positive current as α is increased. Somewhere near $\alpha = 90°$ the current will reverse and the motor will slow and reverse. Further increases of α will drive the motor in the reverse direction. At some point the current may become continuous in the negative direction.

The circuit in Fig. 17.27 is completely symmetrical and may be analyzed by the methods given above; indeed the various formulas may be adapted for this type of operation. However, the control of the switches is complicated by the possibility of reversal of the load current, for the proper switches to gate depend on which way the current flows.

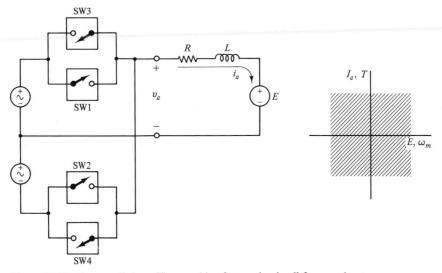

Figure 17.27 A controlled rectifier capable of operation in all four quadrants.

17.2.4 Three-phase Rectifier Circuits

A three-phase power system may act as a power source for a controlled rectifier to drive a dc motor. In this section we show that three-phase systems behave similar to single-phase systems. We begin with a three-phase uncontrolled rectifier with a resistive load.

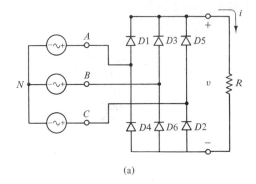

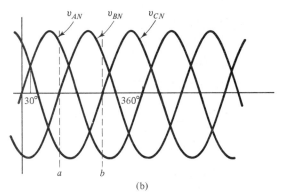

(a)

(b)

Figure 17.28 (a) Three-phase uncontrolled rectifier with a resistive load; (b) three-phase line-to-neutral voltages.

Three-phase uncontrolled rectifier with resistive load. Figure 17.28a shows a three-phase bridge rectifier with a resistive load. The diodes are arranged such that the line with the most positive voltage is connected to the positive load terminal and the most negative voltage is connected to the negative load terminal. For example, at the point marked a in Fig. 17.28b, the most positive voltage is v_{AN} and the most negative voltage is v_{CN}, hence diodes $D1$ and $D2$ are ON and the other four diodes are reverse biased. Consequently, at point a the voltage across the load is $v_{AN} - v_{CN} = v_{AC}$. Similarly, at the point marked b, we can determine that diodes $D3$ and $D4$ are ON, and the voltage across the load is v_{BA}, which is positive at that time. The diodes are numbered in the order that they come ON, as careful study of Fig. 17.28b will confirm.

Load voltage and current. The line voltages are shown in Fig. 17.29 with $V_{AB} = V_p \sin(\omega t)$. We begin our description at $\omega t = 60°$, when $D1$ comes ON. At this point $D1$ comes ON and $D6$ remains ON, and the load voltage is v_{AB}. At $\omega t = 120°$, $D6$ goes OFF and $D2$ comes ON, and the load voltage is v_{AC}. The load voltage therefore follows the peaks of the line voltages, the darker line in Fig. 17.29. The period of the load current is one-sixth that of the ac line, and hence the frequencies in the load current are the zeroth, sixth, twelfth, and so on, harmonics of the line frequency. The time-average load voltage, V_{dc}, may be determined by averaging over one-sixth of the period:

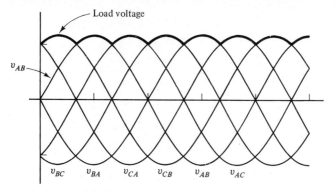

Figure 17.29 Three-phase line voltages.

$$V_{dc} = \frac{6}{2\pi} \int_{\pi/3}^{2\pi/3} V_p \sin(\omega t) \, d(\omega t) = \frac{3}{\pi} V_p \tag{17.41}$$

With a resistive load, the instantaneous load current is proportional to the instantaneous load voltage. Figure 17.30a shows the load current with the diodes that are conducting for the various pulses of current. Figure 17.30b shows the current in line A. Note that while D1 is ON, i_A is positive, and while D4 is ON, i_A is negative. The line current contains no dc component, has a strong fundamental, and has many higher harmonics. However, many of the harmonics cancel if a wye-connected transformer couples the rectifier to the three-phase ac source.

Three-phase controlled rectifier with DC motor load. Figure 17.31a shows a three-phase controlled rectifier with a dc motor for a load. The rectifier circuit is the same as Fig. 17.28a with the diodes replaced by controlled switches. The delay angle for a three-phase converter is customarily defined as zero at the

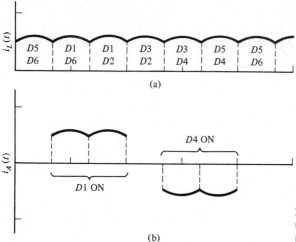

Figure 17.30 (a) Load current, with the diodes that are ON; (b) current in line A.

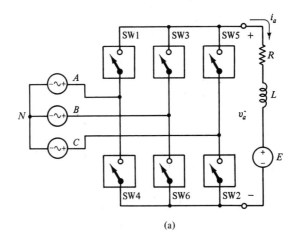

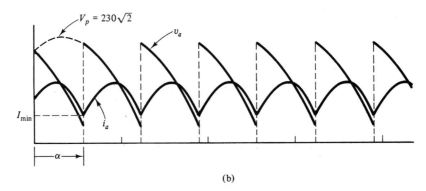

(b)

Figure 17.31 (a) Three-phase controlled rectifier with dc motor load. The switches are numbered in the order they are fired. (b) Load voltage and current for $\alpha = 42.2°$.

point where the switch would become forward biased for a resistive load. We use v_{AB} as a phase reference and hence have $\alpha = 0$ at $\omega t = \pi/3$ (60°). Here as before the current can be continuous or discontinuous, but the current tends to be continuous due to the rapid gating of the switches. The voltage waveform shown in Fig. 17.31b is valid for continuous current, where two of the switches are always conducting. For continuous current, the time-average load voltage is

$$V_a = \frac{6}{2\pi} \int_{\pi/3+\alpha}^{2\pi/3+\alpha} V_p \sin(\omega t) \, d(\omega t) = \frac{3}{\pi} V_p \cos \alpha \qquad (17.42)$$

Figure 17.31b shows load voltage and current for $\alpha = 42.2°$. The time-average current may be determined from Eq. (17.13). We give an example after discussing the analysis of three-phase circuits.

Analysis. The analysis of the three-phase circuit is adapted from that of a single-phase circuit.

Power Electronic Systems   Chap. 17

- The delay angle for the three-phase circuit is measured from a different point in the ac cycle, as stated above. Thus the α's differ by 60° between single-phase and three-phase converters.
- Equation (17.19) gives the current, assuming zero initial current. For discontinuous conduction, the extinction angle, β, and conduction angle, γ, are determined the same as for a single-phase circuit. For the three-phase case, however, any conduction angle greater than 60° indicates continuous current, which is normal.
- The three-phase controlled rectifier shown in Fig. 17.31a can operate in the first and second quadrants. For negative emf and positive current, the dc motor delivers power to the ac source.

Example. We will consider a separately excited 230-V, 3-hp dc motor with the following parameters: $R = 1.43\ \Omega$, $L = 10.4$ mH, $I_a = 11.0$ A, 1150 rpm. The motor will operate on a 230-V three-phase controlled rectifier. We will determine the value of α for full load operation and confirm that the armature current is continuous.

The full-load conditions require 230 V dc and 11.0 A. The emf is therefore $230 - 11.0 \times 1.43 = 214.3$ V. Assuming continuous conduction, we determine from Eq. (17.42) the firing angle for 230-V dc load voltage to be

$$230 = \frac{3}{\pi} \times 230\sqrt{2} \cos \alpha \Rightarrow \alpha = 42.2° \tag{17.43}$$

The critical value of delay angle would be

$$\alpha_c = \sin^{-1} \frac{E}{V_p} - 60° = \sin^{-1} \left(\frac{214.3}{230\sqrt{2}} \right) - 60° = -18.8° \tag{17.44}$$

and thus the value of α calculated in Eq. (17.43) will control the switches.

Conditions for continuous conduction. We must now confirm that the current is continuous. One approach would be to examine the first cycle, assuming zero initial current. In this case Eq. (17.19) gives the current, provided we use $60° + 42.2°$ for the delay angle. We therefore substitute the motor parameters into Eq. (17.19) and calculate the resulting current at $120° + 42.2°$, when the next switch fires. Routine calculations give $Z = 4.17\ \Omega$ and $\phi = 69.96°$; hence the current when SW2 fires is

$$i_a(120° + 42.2°) = \frac{230\sqrt{2}}{4.17} [\sin(92.3°) - e^{-\pi/3\tan\phi} \sin(32.3°)]$$

$$- \frac{214.3}{1.43} [1 - e^{-\pi/3\tan\phi}] = 1.91\ \text{A} \tag{17.45}$$

and hence the current is continuous. If the calculated current at $120° + 42.2°$ had been negative, we would have concluded that the current is discontinuous because negative current is impossible.

Minimum current. Another way to determine the mode of conduction is to assume steady state and calculate the minimum current. Here the requirement is that each pulse of current begin and end with the same value, $I_{\min}$. Returning to Eq. (17.17), we require as an initial condition

$$I_{\min} = \frac{V_p}{Z} \sin\left(\frac{\pi}{3} + \alpha - \phi\right) - \frac{E}{R} + Ae^{-(\pi/3 + \alpha)/\tan\phi} \qquad (17.46)$$

and thus the constant A is

$$A = e^{(\pi/3 + \alpha)/\tan\phi}\left[\frac{E}{R} + I_{\min} - \frac{V_p}{Z} \sin\left(\frac{\pi}{3} + \alpha - \phi\right)\right] \qquad (17.47)$$

The instantaneous current must therefore be

$$i_a(\omega t) = \frac{V_p}{Z}\left[\sin(\omega t - \phi) - e^{(\pi/3 + \alpha - \omega t)/\tan\phi} \sin\left(\frac{\pi}{3} + \alpha - \phi\right)\right]$$

$$(17.48)$$

$$- \frac{E}{R}\left[1 - \left(1 + \frac{I_{\min}R}{E}\right) e^{(\pi/3 + \alpha - \omega t)/\tan\phi}\right]$$

which is the same as we had before except for an additional term for $I_{\min}$ and the change in origin for α. For steady-state operation with this value of $I_{\min}$, we require the same current at the end of the cycle:

$$i_a\left(\frac{2\pi}{3} + \alpha\right) = I_{\min} \qquad (17.49)$$

Solving Eqs. (17.48) and (17.49) for $I_{\min}$, we obtain

$$I_{\min} = \frac{\dfrac{V_p}{Z}\left[\sin\left(\dfrac{2\pi}{3} + \alpha - \phi\right) - e^{-\pi/3\tan\phi} \sin\left(\dfrac{\pi}{3} + \alpha - \phi\right)\right]}{(1 - e^{-\pi/3\tan\phi})} - \frac{E}{R}$$

$$(17.50)$$

Substitution of the parameters for our conditions yields $I_{\min} = +6.02$ A. Thus on this basis we again conclude that the current is continuous.

Completing the example. We have derived the value of α for $V_a = 230$ V and confirmed that the current is continuous for $E = 214.3$ V. The dc current would be $(230 - 214.3)/1.43 = 11.0$ A. The developed power is $11.0 \times 214.27 = 2,357.0$ W. These are rated conditions, so the output power of the 3-hp motor is 3×746 W.

Equation (17.42) indicates that the load voltage is negative with values of α exceeding 90°; thus the rectifier in Fig. 17.31a can operate in the first and second quadrants. In the second quadrant, the motor acts as a generator and the "controlled rectifier" becomes an inverter.

Inverter operation. In the circumstance where the emf is negative, but the current is positive, the emf acts as a generator and delivers power to the ac source. The same may be accomplished in the four-quadrant converter circuit, which would have parallel, reversed switches in the manner of Fig. 17.27. The four-quadrant converter is completely symmetrical and may be analyzed by the methods given above; indeed, the various formulas are still valid. However, the control of the switches is complicated by the possibility of reversal of the load current, because the gate pulses must be supplied to the switches that are carrying current.

Summary. We have in this section analyzed three-phase controlled rectifiers. Such circuits can supply power with variable voltage to a dc load. The current may be continuous or discontinuous, but is usually continuous. The same circuit can invert the dc voltage to deliver ac power to the three-phase system.

Check Your Understanding

1. A free-wheeling diode increases or decreases the armature current to a dc motor. (Which?)
2. If a dc motor draws continuous current from a source at a given speed, it will draw continuous current for higher speeds, everything else being held the same. (T/F?)
3. The peak of the current in Fig. 17.21 corresponds roughly to what condition between input voltage and emf?
4. For a uncontrolled rectifier drive of a dc motor, the value of α is always the critical value. (T/F?)
5. For a controlled single-phase rectifier, the extinction angle cannot be greater than the firing angle by 180°. (T/F?)

Answers. 1. Increases; 2. false; 3. the current maximum occurs roughly when the input voltage and emf are equal; 4. true; 5. true.

7.3 AC MOTOR CONTROLLERS

Introduction. Many types of ac motor controllers are currently in use, and this is an active field of research and development, with continual innovation in response to improvements in semiconductor devices and control techniques. Controllers are tailored to specific types of motors, such as induction motors, synchronous motors, reluctance-type motors, and others. We will limit our discussion to the control of induction motors, and examine the most common types of power controllers.

Speed control of induction motors. Of the various means for controlling the speed of a three-phase induction motor, we will consider the two most common: variation of applied voltage and variation of frequency of the applied voltage.

Figure 17.32 shows the effect of lowering the voltage applied to an induction motor. The principal effect is to lower the torque but with certain loads this leads to a measure of speed control.

Figure 17-33 shows the torque characteristics of a three-phase induction

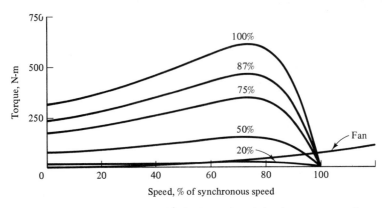

Figure 17.32 Torque characteristic of a three-phase induction motor at various percentages of rated voltage. The torque scales with the square of the applied voltage. As the voltage is decreased, the intersection with load requirement moves to lower speeds.

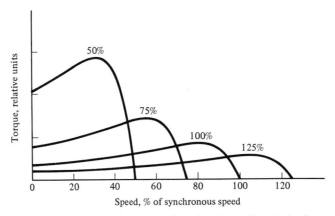

Figure 17.33 Torque characteristics of a three-phase induction motor at various percentages of rated frequency. The applied voltage is kept constant.

motor as the frequency of the applied voltage is varied. The voltage amplitude is kept constant for these characteristics. At low frequencies the impedance of the motor is low, current is increased, and torque is high. However, high currents cause magnetic saturation and increased losses. For this reason, motor controllers reduce applied voltage at frequencies below rated frequency to keep the current roughly constant. This is known as *constant volts/hertz drive*.

17.3.1 Voltage Control of Induction Motors

Figure 17.34 shows a representation of a fixed-frequency variable-voltage controller for an induction motor. Such a controller is useful for loads whose torque requirements increase strongly with speed, such as fans and pumps. The controller consists of six back-to-back switches that permit the gating of the ac waveform in the manner of Fig. 17.31a, except that both halves of the ac waveform are

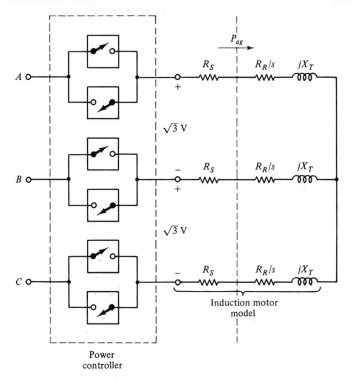

Figure 17.34 Circuit model of a fixed-frequency variable-voltage motor controller and model of three-phase induction motor. The slip is s.

gated because rectification is unwanted. The output of the converter is an ac waveform that can be varied in amplitude, plus harmonics. The model for the induction motor is like that in Fig. 15.11, except that we have for simplicity omitted the magnetizing inductance and combined the rotor and stator leakage inductance in each phase.

Torque characteristic. Considering only the fundamental in the voltage applied to the induction motor, we have for the air-gap power

$$P_{ag} = \frac{3V^2(R_R/s)}{[R_S + (R_R/s)]^2 + X_T^2} \tag{17.51}$$

where V is the per-phase voltage, R_S is the stator per-phase resistance, R_R is the rotor per-phase resistance, X_T is the combined leakage reactance of rotor and stator, and s is the slip. Thus, from Eq. 15.29, the developed torque is

$$T_{dev} = \frac{P_{ag}}{\omega_s} = \frac{3V^2}{\omega_s} \times \frac{R_R/s}{[R_S + (R_R/s)]^2 + X_T^2} \tag{17.52}$$

We note that the torque is proportional to the voltage squared, and thus the motor speed can be controlled if the load torque varies strongly with speed. Figure 17.32 shows the result of variations in voltage when the load (fan) torque is proportional to the square of the speed.

Sec. 17.3   AC Motor Controllers

Example. Consider a six-pole, 230-V, 60-Hz, three-phase induction motor with $R_S = 0.2\ \Omega$, $R_R = 0.2\ \Omega$, and $X_T = 1.0\ \Omega$. The load torque varies as the square of speed. We assume a slip of 4% at rated voltage. The developed torque at 4% slip is therefore

$$T(s = 0.04) = \frac{3(230/\sqrt{3})^2}{40\pi} \times \frac{0.2/0.04}{[0.2 + (0.2/0.04)]^2 + 1^2} = 75.1\ \text{N-m} \quad (17.53)$$

and hence the load torque is generally

$$T_L(s) = 75.1 \times \frac{(1 - s)^2}{(0.96)^2} \quad (17.54)$$

Combining Eq. (17.52) and (17.54), we have for the system

$$75.1 \frac{(1 - s)^2}{(0.96)^2} = \frac{3V^2}{40\pi} \frac{0.2/s}{[0.2 + (0.2/s)]^2 + 1} \quad (17.55)$$

which may be solved for slip or voltage if either is known. Figure 17.35 shows speed versus voltage from Eq. (17.55), and also the speed characteristics of a high-slip motor ($R_R = 1.0\ \Omega$). We include the latter to demonstrate its improved speed-control characteristics.

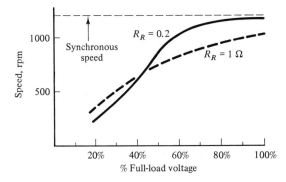

Speed, rpm

1000 — Synchronous speed — $R_R = 0.2$

$R_R = 1\ \Omega$

500

20% 40% 60% 80% 100%
% Full-load voltage

Figure 17.35 Speed versus applied voltage for a normal and a high-slip induction motor. The larger rotor resistance of the high-slip rotor improved controllability but increases motor losses.

Discussion of fixed-frequency voltage control. This method is limited to smaller loads with favorable torque characteristics. The efficiency is low, and harmonics are generated in the power system. The controller will operate in the first and second quadrants, although the latter does not occur with typical loads.

17.3.2 DC-link Variable-frequency Controllers

Basic configuration. Figure 17.36 represents a generic dc-link variable-frequency motor controller. The input ac power is converted to dc, filtered, and then converted to variable-frequency ac by an inverter. The switches are numbered in the order that they are turned ON, with the complementary switch turned OFF simultaneously, that is, when switch 1 is ON, switch 4 is OFF, and so on. We will examine several means by which the switches may convert the dc to three-phase ac power to drive an induction motor. The bypass diodes are required for reactive power flow and to limit the voltage to that of the dc supply. The filter

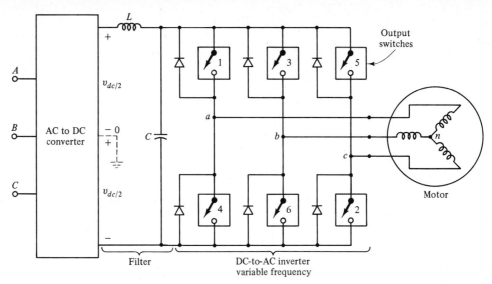

Figure 17.36 DC-link motor controller.

we have shown supplies to the inverter a dc voltage that is largely independent of load current. If the capacitor is omitted, the dc voltage may vary, but the inductor tends to keep the current constant, which offers some advantages. The ac-to-dc converter output may be fixed or variable voltage (or current), depending on the type of inverter scheme and filter used. We have shown the neutral as a dashed ground because no actual ground is required.

Square-wave inverter. In the square-wave inverter, each input is connected alternatively to the positive and negative power-supply outputs to give a square-wave approximation to an ac waveform at a frequency of ω_e determined by the gating of the switches. The voltage in each output line is phase shifted by 120° to synthesize a three-phase source. Voltage waveforms for the square-wave inverter are shown in Fig. 17.37a. The circuit in Fig. 17.37b shows the configuration when a and c are connected to the positive dc voltage and b is connected to the negative dc voltage. The two phases in parallel give half the impedance of an individual phase and thus get one-third the total dc voltage, whereas the unpaired phase gets two-thirds the dc voltage. Thus, for the first 30°, $v_{an} = \frac{1}{3}V_{dc}$, and for the next 30°, $v_{an} = \frac{2}{3}V_{dc}$, and so on. In this manner, the stair-step voltage for each motor phase is produced. The fundamental component in each waveform is about $0.637 \times V_{dc}$.

Need for constant voltage/frequency. At frequencies below the rated frequency of the motor, the applied voltage must be reduced. Otherwise, the current to the motor will be excessive and cause magnetic saturation. This follows from Faraday's law:

$$\underline{V} = j\omega_e n\underline{\Phi} \tag{17.56}$$

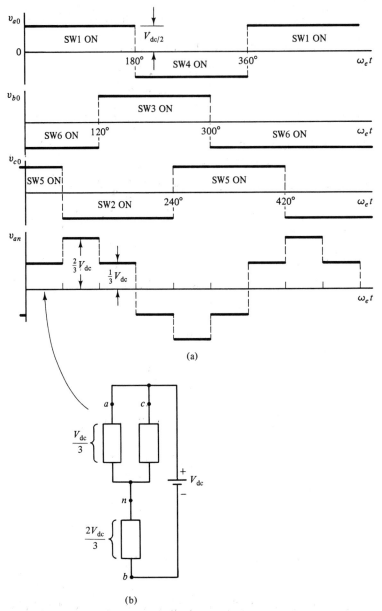

(a)

(b)

Figure 17.37 (a) The synthesis of the line-to-neutral voltage from a square-wave inverter. The stair-step waveform results from the changing neutral voltage on the motor. (b) Circuit configuration when a and c are connected to the positive and b is connected to the negative dc voltage.

where ω_e is the electrical frequency, which requires decreasing the voltage level if the frequency is reduced and the peak flux is to be kept constant. Thus the square-wave inverter requires decreasing dc voltage as motor speed is reduced below rated speed. This can be accommodated by a controlled rectifier, but this leads to problems with harmonics in the power system supplying the controller.

Harmonics. The square-wave inverter has two types of problems with harmonics. At the input, the controlled rectifier creates harmonics that constitute noise in the power system. These can be filtered, but the added complexity degrades efficiency and power factor, which is already somewhat poor for a controlled rectifier.

The output waveforms also have serious problems with harmonics. The stair-step output waveforms in Fig. 17.37a contain only odd harmonics. The third, ninth, and so on, cause no problems because they are in phase and thus self-cancel at the input to the wye-connected motor. But all the rest, principly the fifth and seventh, cause currents that increase losses in the motor and produce no torque. Although these harmonics are filtered somewhat by the inductance of the motor, the combination of these problems has caused designers to look for better methods. The pulse-width modulation (PWM) system discussed below conquers these problem with a more complicated switching sequence.

Pulse-width modulation (PWM). Figure 17.38b shows 180 electrical degrees of a waveform that has been generated through pulse-width modulation. The basic idea in PWM is to chop pieces out of the wave to control the fundamental in the output, while in the same operation shifting the harmonics to high frequencies that are easily filtered by the inductance of the motor.

Figure 17.38 (a) The switching sequence is generated by comparing a triangular wave with a sinusoidal wave. The modulation index is m. (b) The resulting waveform has a sinusoidal component proportional to m at the frequency of the sinusoid in part (a).

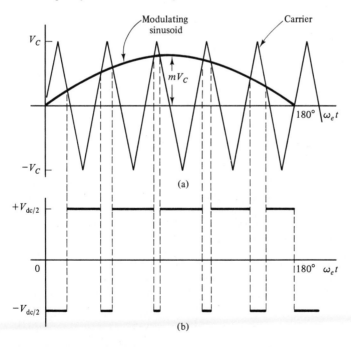

Sine/triangle modulation.   Figure 17.38a shows one means for controlling the switches. The triangle wave is the carrier, and its frequency is much higher than the electrical frequency applied to the motor. The sinusoidal waveform, of which only half is shown, is the modulating waveform that controls the amplitude and frequency of the fundamental applied to the motor. Specifically, the electrical frequency of the ac applied to the motor is equal to that of the modulating sinusoid, and the amplitude of the ac is proportional to the modulation index, m, defined in Fig. 17.38a.

The relationship between the carrier, modulating sinusoid, and switched waveform is evident: whenever the modulating sinusoid exceeds the carrier, the switches connect the motor to the positive dc voltage, and whenever the modulating sinusoid is below the carrier, the switches connect the motor to the negative dc voltage. Note that, if m were small, the output would have roughly equal amounts of positive and negative voltage for this half-cycle and thus produce a small component of the fundamental. But as m increases, the positive voltage dominates. The opposite would be true for the second half of the cycle. In this way the modulating sinusoid controls the amplitude and frequency of the output.

Harmonics in PWM.   The harmonics are shifted to frequencies comparable to the carrier frequency and thus are much higher than with the square wave inverter. The higher harmonics are filtered by the inductance of the motor, and the resulting current is sinusoidal with a small ripple on it.

We have shown the modulated waveform for only one phase of the three-phase inverter. For the other two phases, the modulating sinusoids would be phase shifted by 120° and 240° in the usual fashion. In this way the inverter produces three-phase voltage of controlled frequency and amplitude.

Hysteresis current control.   Figure 17.39 illustrates yet another method for deriving the PWM switching sequence. The output current is compared with a reference signal of the desired amplitude and frequency. When the output current falls below the reference signal by a specified error, the output is connected to the positive dc bus, and when the output current rises above the reference signal by the specified error, the output is connected to the negative dc bus. This method

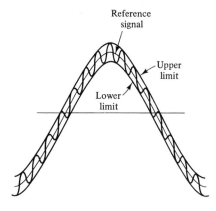

Figure 17.39 Hysteresis current control keeps the current within a prescribed range of a reference waveform.

controls current, and thus torque, directly and is not greatly affected by the variation of the motor impedance with drive frequency.

Summary. Pulse-width modulation controls motor voltage or current, depending on the modulation scheme, by connecting the motor to the positive or negative buses of a fixed-voltage dc supply. Such a supply has good efficiency and power factor and does not generate excessive harmonics in the power system. The PWM systems shift the harmonics to high frequencies that are filtered by the inductance of the motor.

17.3.3 Check Your Understanding

1. The speed of an induction motor can be controlled by varying the drive voltage or frequency, but not both. (T/F?)
2. Figure 17.38 shows a sine/triangle modulation scheme. Sketch the voltage waveform if the modulation index, m, equals 1.5.
3. The speed of a synchronous motor cannot be controlled by a power electronics system because the slip is always zero. (T/F?)

Answers.   1. False;   2. almost a square wave;   3. false: the rotor position, and hence its speed, can be controlled precisely.

PROBLEMS

Section 17.1: Introduction to Power Electronics

P17.1. A snubber circuit, shown in Fig. P17.1, is placed in parallel with an SCR to limit the rate of voltage rise across the OFF device. If $120\sqrt{2}$ V were suddenly applied to the SCR and snubber circuit with the load shown, what would be the required value of R to keep the dv/dt below 100 V/μs? Treat the capacitor as a short circuit.

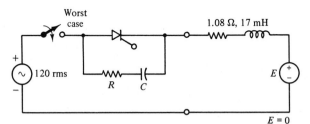

E = 0   **Figure P17.1**

P17.2. Verify the result of the integration in Eq. 17.3.

P17.3. Calculate the power in the transistor for the inductive switching trajectory shown in Fig. 17.6b. In both transitions the collector voltage is at the power supply voltage while the current makes its transition in $t_r/0.8$. What is the energy per switching cycle (two transitions) for the transistor?

Problems   **683**

P17.4. The SCR in Fig. P17.4 has basically infinite resistance when OFF and a voltage of 0.3 V when ON. A pulse of current occurs at the gate at $t = 0$. Find:
  (a) Current in the circuit before the gate pulse
  (b) Maximum current that can flow in the circuit
  (c) Time constant

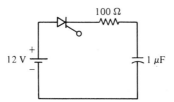

Figure P17.4

P17.5. Commercial airplanes carry flashlights that have a light-emitting diode (LED) that flashes occasionally to show that the batteries are still good (and also presumably to allow one to find the flashlights in the dark). A circuit that accomplishes this is shown in Fig. P17.5. Assume that the *pnpn* diode fires at 2.2 V and has 0.7 V when ON. The light-emitting diode (LED) requires at least 10 mA of current at 0.7 V to glow. Find R_1 and R_2 so that the LED glows for 4 ms every 10 s.

Answers. $R_1 = 36.1$ kΩ; $19.7 < R_2 < 100.9$ Ω

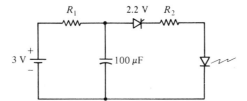

Figure P17.5

P17.6. If the gate of the SCR in Fig. 17.11 is limited to 30 mA of current, what value of resistance should be placed in series with the *pnpn* diode to limit the gate current to this value. Assume that the *pnpn* diode fires at 30 V and that the equivalent resistance of the gate is 15 Ω.

P17.7. For the circuit in Fig. 17.8, determine the exact minimum and maximum firing time if the *pnpn* diode has a threshold voltage of 26 V.

P17.8. For the circuit in Fig. 17.11, what would be the maximum $R_V C$ time constant for the *pnpn* diode to reach 30 V and fire the *pnpn* diode? *Hint:* The exact solution is difficult due to the nonlinear character of the equations. A good numerical approach would be to ignore the exponential term in Eq. (17.8) for a first approximation and then include it for subsequent approximations.

Answer. 25.8 ms

P17.9. For $V_{th} = 30$ V and $R_V = 30$ kΩ, determine the firing time, t_α, from Eq. 17.8. This requires graphical or numerical solution unless you have a calculator that solves nonlinear equations. Assume 60 Hz.

P17.10. Normally, the firing time in SCR operation, t_α, shown in Fig. 17.12, is described in terms of a "conduction angle," where

$$\gamma = 180° - 360° \times \frac{t_\alpha}{T}$$

where T is the period of the frequency, usually $\frac{1}{60}$ s. Clearly, γ is the amount of phase angle during which the SCR fires each cycle.

(a) For a conduction angle of 90°, $V = 120$ V (rms), and $R_L = 100$ Ω, what is the dc current and the total power in the load? Note that the total power would be a measure of the total heating effect on the load and would be more than the dc power.

(b) Repeat part (a) for a conduction angle of 120°. Part (a) can be solved without evaluating any integrals, but part (b) requires two integrations.

Answers. (a) 0.270 A, 36.0 W: (b) 0.405 A, 57.9 W

Section 17.2: DC Motor Controllers

For P17.11–P17.22, assume a separately-excited 180-V dc motor with armature resistance of 1.05 Ω and inductance of 22 mH. Nameplate ratings are: 1750 rpm, 1.5-hp, and 7.3 A armature current. Assume no free-wheeling diode unless stated otherwise.

P17.11. The 180-V dc motor is driven by an uncontrolled full-wave rectifier. The emf of the motor is 175 V. The current is discontinuous and flows for 90% of the time. What is the extinction angle β in radians?

P17.12. The dc motor requires 180 V to run full speed at rated power and is driven by an uncontrolled full-wave rectifier. The input voltage to the rectifier is 230 V (rms). Assuming that the current to the motor flows continuously, what is the value of the emf to give 180 V to the motor? Is the current below its rated value?

Answers. 160.8 V; no

P17.13. The dc motor is separately excited for rated conditions. The motor is supplied by a single-phase, full-wave rectifier and has no free-wheeling diode. Determine the minimum input voltage (rms) such that the armature current flows continuously at rated speed. Find the current under these conditions.

P17.14. The dc motor is driven by a controlled single-phase rectifier. The ac source is 240 V rms.

(a) If $\alpha = 0°$, what is the no-load speed of the motor? Ignore losses.

(b) With no load, what range of α has no effect on the speed?

(c) With $\alpha = 90°$, what range of emfs gives continuous conduction?

(d) With $\alpha = 90°$ and $E = 160$ V, the extinction angle, β, is 123.2°.

  (1) What is the conduction angle, γ, in radians?

  (2) What is the dc armature voltage?

  (3) Find the developed torque in the motor.

Answers. (a) 3447 rpm; (b) $0 < \alpha < 90°$; (c) < -27.3 V: (d) 0.579, 189.6 V, 26.6 N-m

P17.15. The dc motor is driven by a single-phase controlled rectifier with $V_p = 120\sqrt{2}$ V.

(a) If $\alpha = 0°$, what is the no-load speed of the motor? Ignore losses.

(b) With no load, what range of α has no effect on the speed?

(c) With $\alpha = 0°$, what range of emfs gives continuous conduction?

(d) With $\alpha = 0°$ and $E = 150$ V, the extinction angle, β, is 137.4.

  (1) What is the conduction angle, γ, in radians?

  (2) What is the dc load voltage?

  (3) Find the developed torque in the motor.

P17.16. Consider a single-phase, controlled rectifier drive. The input voltage is 220 V rms for each half of the transformer supplying the rectifier. Consider that the motor is running at full-load speed with rated power.

(a) Determine the critical α for the circuit, α_c.

(b) Assuming $\alpha \geq \alpha_c$, write the equation for the current in the first cycle of current.

(c) Determine if the current is continuous or discontinuous for $\alpha = \alpha_c$. *Hint:* Calculate E_{mar} for this case and compare with the actual emf.

Answers. (a) 33.6°; (c) discontinuous since $E_{mar} = 151$ V

P17.17. For the motor described above operating at rated speed:

(a) Determine the critical angle for marginal continuous current from Eq. 17.31.

(b) For this delay angle, what peak input voltage gives the marginal emf?

(c) If the motor is running at rated speed, what is the motor current? From this calculation, what conclusion can be made?

Answers. (a) 30.4°; (b) 340.2 V; (c) 13.7 A, too much

P17.18. For the dc motor described above, what is the maximum speed that can be obtained with 120 V (rms) input voltage to a single-phase controlled rectifier if the motor output power is 1 hp. Neglect the mechanical loss. *Hint:* Set $\alpha = 0°$, which means that the effective $\alpha = \alpha_c$ and thus depends on E. Assume continuous current.

P17.19. Equation (17.26) gives the maximum emf for continuous conduction for the full-wave controlled rectifier without a free-wheeling diode. As stated in the paragraph following the equation, the minimum value of α is the critical value, which depends on the emf. If follows that, for a given value of ϕ, there is a maximum value of emf for which the current is continuous, even if the SCRs are gated at $t = 0$. Confirm that the critical value of α obeys Eq. (17.31). The maximum marginal value of the emf is $E_{mar} = V_p \sin \alpha_c$.

P17.20. Repeat Problem P17.17 if the motor has a free-wheeling diode. Part (a) requires the numerical solution of Eq. (17.40) with $E_{mar}/V_p = \sin \alpha_c$ and $\alpha = \alpha_c$ on the right-hand side. The solution lies between the 30° and 35°, so evaluate the equation for both extremes and interpolate.

P17.21. Consider a three-phase controlled rectifier, 230 V, 60 Hz, driving the motor at rated voltage and rated load.

(a) Determine the required value of α, assuming continuous conduction, and confirm that this is greater than α_c.

(b) Confirm that the current is continuous by calculating the minimum current from Eq. (17.50).

(c) Determine the average current and the developed power in the motor.

(d) What would be the no-load speed of the motor? Ignore losses.

P17.22. From Eq. (17.50), determine the marginal value of the emf for continuous conduction by setting I_{min} to zero. For the rectifier circuit in Problem P17.21, at what speed does this occur if the rectifier is putting out 180 V dc, and what is the dc current to the motor?

Answers. 1797 rpm, 2.85 A

Section 17.3: AC Motor Controllers

P17.23. Determine the speed versus percent full-load voltage curve for the motor described in the example in Section 17.3.1 if the full-load slip were 11% and the rotor resistance were 0.5 Ω. The stator resistance and the total reactance are unchanged. *Hint:* Assume values of s and then calculate the voltage required.

P17.24. Figure P17.24 shows an approximate per-phase circuit for a three-phase, two-pole, 60-Hz induction motor at rated frequency. At a slip of 3%, the developed torque is 8.058 N-m for rated voltage.

(a) Find the torque at 80% of rated voltage if the slip is constant.

(b) Find the torque at 80% of rated voltage and 80% of rated frequency if the slip is changed to 5%.

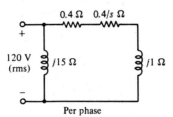

Per phase   **Figure P17.24**

P17.25. The load for a three-phase induction motor requires constant power. The motor is driven by a power electronic circuit to control speed. The motor operates with a slip of 3% at rated voltage and frequency. Estimate the slip at 80% of rated frequency and 80% of rated voltage. Ignore mechanical losses in this problem. *Hint:* Assume the motor is operating in the small-slip region.

P17.26. In Fig. 17.33, the various curves show increasing slope in the small-slip (linear) region. Explain why this is true, and how it relates to power electronic control of induction motors.

APPENDIX

Useful Constants

1. Velocity of light in vacuum, $c = 2.998 \times 10^8$ meters/second
2. Charge on the electron, $e = -1.602 \times 10^{-19}$ coulombs
3. Mass of the electron, $m_e = 9.110 \times 10^{-31}$ kilogram
4. Electron charge/mass ratio, $e/m_e = 1.759 \times 10^{+11}$ coulomb/kilogram
5. Mass of the proton, $m_p = 1.673 \times 10^{-27}$ kilogram
6. Permeability of vacuum, $\mu_o = 4\pi \times 10^{-7}$ henry/meter
7. Permittivity of vacuum, $\epsilon_o = 8.854 \times 10^{-12}$ farad/meter
8. Boltzmann's constant, $k = 1.381 \times 10^{-23}$ joules/kelvin
9. Voltage equivalent of temperature, $kT/e = 25.9$ millivolts at 300 kelvin
10. Typical mass density of magnetic steel, $\rho = 7.65$ grams/cubic centimeter

Index

N

O

Overdamped behavior, 471
Overexcited motor, 610

P

Parallel connection
  resistors, 31
Particular integral, 89
Partition noise, 398
Per-phase equivalent circuit, 531, 607
Per-phase representation, 530
Percent regulation
  transformer, 526
Periodic signals, 318, 337
Permanent magnet, cylindrical systems,
    562
Permanent-magnet dc motor, 634
Permanent-split capacitor motor, 604
Permeability, 498
Phase, 112
Phase rotation, three phase systems, 186
Phasor, 122, 128, 319, 521
Phasor, rotating waves, 567
Pinchoff voltage, 247
Plugging, 657
pn-junction
  application, 370
  diode, 230
  resistance, 382
  transistor, 233
pnpn diode, 646
Polar form, 118, 121
Polarity
  transformer, 522
Polarity convention, 11
Polarization, 446, 452
Pole
  salient, 560, 606, 626
Poles, 466
Position transducers, 388
Potential, 10
  electrical, 10
  voltage, 47
Potentiometer, 78
  loading, 420
  transducer, 388
Power, 15
  frequency domain, 323
  mechanical analogy, 17

reference direction, 16
resistor, 25
three phase, 187
time average, 331
Power angle, 572
Power distribution system, 170, 211
Power electronics, 645
Power factor, 162
Power factor, correction, 170, 172
Power spectrum, 322, 332
Power supply, 214–15
Power triangles, 169
Precision, ADC, 405
Pressure transducers, 389
Programming, 306
PROM, 306
Proton, 4
PSC motor, 604
Pulse-Width Modulation, 681
Pulses, 99
PWM, 681

Q

Quadrature-axis reactance, 611

R

Radar, 446
Radar cross section, 453
Radar equation, 454
Radiation efficiency, 447
Radio receiver, 432
RAM, 306
Ramp ADC, 406
Random Access Memory, 306
Random signal, 329, 431
Range, ADC, 405
R-C snubber, 648
Reactance, 136
Reactive, 136
Reactive power
  definition, 165
  importance, 171
  sign, 167
  synchronous generator, 613
  synchronous motor, 609